AF767208

Protein Surface Recognition

Protein Surface Recognition

Approaches for Drug Discovery

Edited by

ERNEST GIRALT

Department of Organic Chemistry, University of Barcelona and Institute for Research in Biomedicine, Barcelona, Spain

MARK W. PECZUH

Department of Chemistry, University of Connecticut, USA

XAVIER SALVATELLA

ICREA and Institute for Research in Biomedicine, Barcelona, Spain

WILEY

A John Wiley and Sons, Ltd., Publication

This edition first published 2011
© 2011 John Wiley & Sons, Ltd

Registered office

John Wiley & Sons Ltd, The Atrium, Southern Gate, Chichester, West Sussex, PO19 8SQ, United Kingdom

For details of our global editorial offices, for customer services and for information about how to apply for permission to reuse the copyright material in this book please see our website at www.wiley.com.

The right of the author to be identified as the author of this work has been asserted in accordance with the Copyright, Designs and Patents Act 1988.

All rights reserved. No part of this publication may be reproduced, stored in a retrieval system, or transmitted, in any form or by any means, electronic, mechanical, photocopying, recording or otherwise, except as permitted by the UK Copyright, Designs and Patents Act 1988, without the prior permission of the publisher.

Wiley also publishes its books in a variety of electronic formats. Some content that appears in print may not be available in electronic books.

Designations used by companies to distinguish their products are often claimed as trademarks. All brand names and product names used in this book are trade names, service marks, trademarks or registered trademarks of their respective owners. The publisher is not associated with any product or vendor mentioned in this book. This publication is designed to provide accurate and authoritative information in regard to the subject matter covered. It is sold on the understanding that the publisher is not engaged in rendering professional services. If professional advice or other expert assistance is required, the services of a competent professional should be sought.

The publisher and the author make no representations or warranties with respect to the accuracy or completeness of the contents of this work and specifically disclaim all warranties, including without limitation any implied warranties of fitness for a particular purpose. This work is sold with the understanding that the publisher is not engaged in rendering professional services. The advice and strategies contained herein may not be suitable for every situation. In view of ongoing research, equipment modifications, changes in governmental regulations, and the constant flow of information relating to the use of experimental reagents, equipment, and devices, the reader is urged to review and evaluate the information provided in the package insert or instructions for each chemical, piece of equipment, reagent, or device for, among other things, any changes in the instructions or indication of usage and for added warnings and precautions. The fact that an organization or Website is referred to in this work as a citation and/or a potential source of further information does not mean that the author or the publisher endorses the information the organization or Website may provide or recommendations it may make. Further, readers should be aware that Internet Websites listed in this work may have changed or disappeared between when this work was written and when it is read. No warranty may be created or extended by any promotional statements for this work. Neither the publisher nor the author shall be liable for any damages arising herefrom.

Library of Congress Cataloging-in-Publication Data

Protein surface recognition : approaches for drug discovery / editors, Ernest Giralt, Mark Peczuh, Xavier Salvatella.
p. ; cm.
Includes bibliographical references and index.
ISBN 978-0-470-05905-0 (cloth)
1. Drugs–Design. 2. Protein-protein interactions 3. Proteins–Inhibitors. 4. High throughput screening (Drug development) I. Giralt, Ernest. II. Peczuh, Mark. III. Salvatella, Xavier.
[DNLM: 1. Proteins–metabolism. 2. Drug Discovery–methods. 3. Enzyme Inhibitors–pharmacology. 4. Protein Binding. 5. Surface Properties. QU 55 P9689 2010]
RS420.P76 2010
615'.19–dc22

2010018781

A catalogue record for this book is available from the British Library.

ISBN 978-0-470-05905-0

Set in 10/12 pt Times by Thomson Digital, Noida, India
Printed in Singapore by Fabulous Printers Pte Ltd

Contents

Preface

By and large, current drugs fall into two broad categories: small molecules and protein therapeutics (biologics). While specific notions of 'small molecule' may vary, they can generally be characterized by their low (<1 kD) molecular weight, high functional group density, and often the presence of heterocycles as part of the core structure. As such, small molecules may be derived from, or be inspired by, natural products or they may be the product of organic synthesis. Such 'synthetic drugs' are at the origin of the pharmaceutical industry itself.

From a financial perspective, small molecules are presently the *bread and butter* of the industry with worldwide annual sales in the hundreds of billions (USD). Biologics, however, are themselves a multibillion dollar annual market and are seen by some as having a high potential for growth. The success of biologics has been mainly the result of advances in biotechnology that have facilitated the identification and subsequent expression of the appropriately tailored proteins. To be active, both small molecules and protein therapeutics must bind to a target biomolecule. It is *how* each of these types of molecules binds its partner that further differentiates them. Small molecules usually bind at an interior active site whereas proteins are involved in protein-protein interaction (PPI) that involve the exterior surfaces of proteins.

An example of each is illustrative of this point. Atorvastatin (Lipitor), a second generation statin derived from the related fungal metabolite pravastatin (Prevachol) is the number one small molecule therapeutic in the US as determined by 2009 retail sales. It is a competitive inhibitor of HMG-CoA reductase, the rate-limiting enzyme in cholesterol biosynthesis. This reductase focuses functional groups of the peptide backbone and side chains in a convergent manner inward toward the drug. In contrast, binding to the protein therapeutic Trastuzumab (Herceptin *ca* 1.3b USD/yr) by the HER2 cellular receptor covers a sizeable exterior surface area (ca. 1600 $Å^2$). Functional group presentation by the two binding partners in the Herceptin-Her2 interaction is more divergent in nature, especially when compared to the interaction between Lipitor and HMG-CoA reductase.

A growing body of evidence suggests that a middle ground – using small molecules to bind the exterior protein surface and inhibit PPIs – can be a powerful strategy for the development of new tools for chemical biology and medicinal chemistry. A small molecule with these properties, a binder of Bcl-X_L and other anti-apoptotic Bcl proteins, is already in Phase 2a clinical trials (ABT-263); notably, the molecule is a result of a fragment-based drug discovery effort. It binds the Bcl proteins in a groove usually occupied by an α-helix of the native protein binding partners. Molecules such as ABT-263 can combine the mode of action of protein therapeutics with the synthetic accessibility, ease of administration, bioavailability, and robustness of traditional small molecule drugs. The development of protein surface binders by structure-based drug design or similar approaches should be contrasted with the discovery of small molecules that inhibit a given PPI by an allosteric mechanism, which have largely arisen via serendipity and are testament to our still crude understanding of the physico-chemical principles that govern the interactions between proteins.

In between the contrasting worlds of small molecules and therapeutic proteins, peptides are likely to play an important role as therapeutic agents that modulate protein-protein interactions. Compared to protein therapeutics, the most important advantages offered by small molecules are their relatively straightforward synthetic accessibility and bioavailability. Protein therapeutics, by contrast, have specificity as a key asset, which arises from their ability to establish a large number of noncovalent interactions with the surface of the target. Nowadays, peptides combine the advantages of therapeutic proteins with those of small molecules due, among other developments, to recent progress in their modification to improve their bioavailability profile.

This book provides both a context and a guidepost for the development of molecules that alter protein function by inhibiting protein-protein interactions (PPIs) as opposed to conventional active site inhibition. The subject material has been broken into four broad sections: principles, approaches, techniques, and case studies. The principles section provides a general description of the biophysical properties of PPIs with an emphasis on those that are relevant to drug design; in Chapter 1, 'The Discovery and Characterization of Protein–Protein Interactions', we provide an overview of the methods used for identifying and characterizing PPIs, a survey of the main structural and dynamical properties of protein-protein complexes and a discussion of the challenges and opportunities inherent to inhibiting their formations whereas in Chapter 2, 'Biophysics of Protein–Protein Interactions', we provide, instead, a detailed account of the noncovalent interactions that provide the driving force for complex formation and of the thermodynamics and kinetics of the process.

Following this overview, the approaches section reviews established strategies for the inhibition of PPIs in terms of the small molecule inhibitor. Chapter 3, 'On the Logic of Natural Product Binding in Protein–Protein Interactivity' presents a rationale on how natural products bind protein surfaces and the functional consequences of the interactions. Chapters 4 and 5, 'Interface Peptides', and 'Inhibition of Protein–Protein Interactions by Peptide Mimics', detail the progression of a strategy whereby peptide sequences from the protein-protein interface are used as inhibitors and then subsequently serve as models for the development of peptide mimics with the same activity. Secondary structures such as turns and α-helices are common in the collection of interface peptides. As such, mimicry of these elements has received significant attention. Chapter 6, 'Discovery of Inhibitors of Protein–Protein Interactions by Screening Chemical Libraries', collects examples of small molecule inhibitors of protein–protein interactions that have come about via screening efforts.

A review of technologies that enable the evaluation of protein surface binding constitutes the next section of the book. This details aspects that range from organic synthesis to screening methods. Chapter 7 'High-throughput Methods of Chemical Synthesis Applied to the Preparation of Inhibitors of Protein-Protein Interactions', describes methods for the preparation of small molecule inhibitors of PPIs and the strategies behind their synthesis. Chapters 8, 'In Silico Screening', and 9.1 '*In Vitro* Screening: Screening by Nuclear Magnetic Resonance', provide accounts of how computational tools and Nuclear Magnetic Resonance can provide key information, while aspects of high throughput screening in terms of *in vitro* and cell-based assays are outlined in Chapter 9.2, '*In Vitro* Screening: Methods of High-throughput Screening'.

Finally, the integration of the previous concepts is illustrated through two case studies in the final section of the book. These case studies include 'Inhibitors of the MDM2-p53 Protein–Protein Interaction' (Chapter 10) and 'The Discovery of Potent LFA-1 Antagonists' (Chapter 11).

We trust that the readers of the book will find it a source of valuable information in addressing the challenges and potential rewards associated with the inhibition of protein-protein interactions and we wish to thank all our co-workers and co-authors for their enthusiastic contributions in making the book possible. We specifically would like to thank Brendan Orner and David Bolstad for very valuable discussions and Paul Deards for initiating the project.

Ernest Giralt, Mark W. Peczuh and Xavier Salvatella

List of Contributors

Xavier Barril, ICREA and University of Barcelona, Barcelona, Spain

Jorge Becerril, Department of Chemistry, Yale University, New Haven, CT, USA

Denzil Bernard, Comprehensive Cancer Center and Departments of Internal Medicine, University of Michigan, Ann Arbor, MI, USA

C. W. Bertoncini, Department of Chemistry, University of Cambridge, UK

Richard T. Desmond, Department of Chemistry, University of Connecticut, Storrs, CT, USA

Annaliese K. Franz, Department of Chemistry, University of California at Davis, Davis, CA, USA

Tom Gadek, SARcode Corporation, San Francisco, CA, USA

Carlos García-Echeverría, Novartis Institutes for Biomedical Research, Basel, Switzerland

Ernest Giralt, Department of Organic Chemistry, University of Barcelona and Institute for Research in Biomedicine, Barcelona, Spain

Andrew D. Hamilton, Department of Chemistry, Yale University, New Haven, CT, USA

A. Higueruelo, Department of Biochemistry, University of Cambridge, UK

James J. La Clair, Xenobe Research Institute, San Diego, CA, USA

Qing Lin, Department of Chemistry, State University of New York at Buffalo, Buffalo, NY, USA

Francisco Javier Luque, Department of Physical Chemistry, Faculty of Pharmacy, University of Barcelona, Barcelona, Spain

Irene Luque, Department of Physical Chemistry and Institute of Biotechnology, Faculty of Sciences, University of Granada, Granada, Spain

Mark W. Peczuh, Department of Chemistry, University of Connecticut, Storrs, CT, USA

Johanna M. Rodriguez, Department of Chemistry, Yale University, New Haven, CT, USA

Xavier Salvatella, ICREA and Institute for Research in Biomedicine, Barcelona, Spain

Sanjeev Shangary, Comprehensive Cancer Center and Departments of Internal Medicine, University of Michigan, Ann Arbor, MI, USA

Jared T. Shaw, Department of Chemistry, University of California at Davis, Davis, CA, USA

Wenjiao Song, Department of Chemistry, State University of New York at Buffalo, Buffalo, NY, USA

Yuchen Tang, Department of Chemistry, University of California at Davis, Davis, CA, USA

Shaomeng Wang, Comprehensive Cancer Center and Departments of Internal Medicine, University of Michigan, Ann Arbor, MI, USA

Pauline N. Wyrembak, Department of Chemistry, Yale University, New Haven, CT, USA

Part I
Principles

1

The Discovery and Characterization of Protein–Protein Interactions

C. W. Bertoncini[1], *A. Higueruelo*[2] *and X. Salvatella*[3]
[1]*Department of Chemistry, University of Cambridge, UK*
[2]*Department of Biochemistry, University of Cambridge, UK*
[3]*ICREA and Institute for Research in Biomedicine, Barcelona, Spain*

1.1 Introduction

The regulation of protein–protein interactions (PPIs) is fundamental for cellular function because PPIs are involved in virtually all biological processes. A complete and detailed description of the interaction map for proteins, known as interactome, is therefore one of the most important challenges in molecular biology, one that will provide great opportunities for therapeutic intervention in the complex diseases that challenge the biomedical community and the pharmaceutical industry. In this chapter we provide an overview of the different techniques that are currently available for the discovery and structural and thermodynamic analysis of PPIs as well as a survey of the general structural and dynamical properties of proteins and protein complexes that affect drug design. Rather than a comprehensive survey of the technical literature on methods to screen and characterize PPIs we present here a general discussion of these tools and refer the reader to the reviews and examples of application that we cite to identify the primary literature.

Protein Surface Recognition: Approaches for Drug Discovery Ernest Giralt, Mark W. Peczuh and Xavier Salvatella. © 2011 John Wiley & Sons, Ltd

1.2 Techniques to Identify Protein–Protein Interactions

Many methods have been developed for the isolation and characterization of protein complexes, both *in vitro* and *in vivo*. Among them five methodologies are particularly suitable for high-throughput, and account for the majority of proteome-wide studies.

1.2.1 The Yeast Two Hybrid Assay (Y2H)

This system exploits the formation of a stable complex between interacting proteins to bring together two modules of a cis-acting transcriptional promoter, stimulating the expression of a reporter gene. It requires the construction of two hybrid genes, one encoding the DNA-binding domain (BD) of the transcription factor fused to a target protein (the bait) and a second encoding its transcription-activation domain (AD) fused to a different protein (the prey). If the prey and bait proteins interact through a PPI the two modules of the transcription factor (BD and AD) are brought together to reconstitute the transcription activity. Provided that the interaction between the prey and bait proteins is sufficiently strong, the now functional transcription factor will bind to the promoter sequence in the proximity of the reporter gene, via its DNA binding domain (BD), and recruit the transcriptional machinery, via its transcription-activation domain (AD, Figure 1.1A). The most commonly employed DNA-binding domains are derived from the yeast Gal4 and LexA transcription factors, while activating domains come also from Gal4 or from the viral activator VP16. Expression of the reporter gene gives the yeast a unique characteristic which allows identification of a successful PPI interaction between the bait and prey proteins. Reporter genes commonly employed are *lacZ*, that codifies for the enzyme β-galactosidase, that metabolizes X-gal (5-bromo-4-chloro-3-indolyl-β-D-galactoside) to give a distinctive blue color, or auxotrophic genes such as *HIS3*, *LEU2* or *URA3*, which confer positive colonies the ability to grow in media lacking specific nutrients [1].

One key advantage of the Y2H assay is its *in vivo* nature, that allows the investigation of PPIs under physiological conditions. Additional advantages of this method are its high sensitivity – it can detect very weak and therefore transient interactions, with Kd as low as 10^{-7} M – its scalibility and its easy automation. The Y2H assay can also be used in a quantitative fashion to determine the strength of the interaction between the bait and prey proteins by monitoring the amount of reporter protein produced, for example by measuring, when using the lacZ reporter gene, the β-galactosidase activity. The main disadvantage of the Y2H approach to the identification of PPIs is the number of control experiments that it requires, that are mainly aimed at determining whether the bait and prey proteins have affinity for DNA and are indeed capable of self-activating the transcription of the reporter. An additonal concern when using this approach, one that is directly linked to the its *in vivo* nature, is the possibility that a third protein mediates, in the assay, the interaction between the pray and bait proteins; it is therefore important to validate all PPIs derived from this assay by other methods, including those discussed in this chapter. Other limitations of the Y2H assay are due to its use of yeast, as some post-translational modifications are different in yeast to those in other eukaryots, and to the localization of of interactions in the nucleus, where some target PPIs may experience an incorrect cellular environment. Membrane proteins are obviously not suitable for this assay, but interactions between the cytoplasmic domains of extracellular receptors can be screened, using this approach, to study signal transduction pathways. PPIs

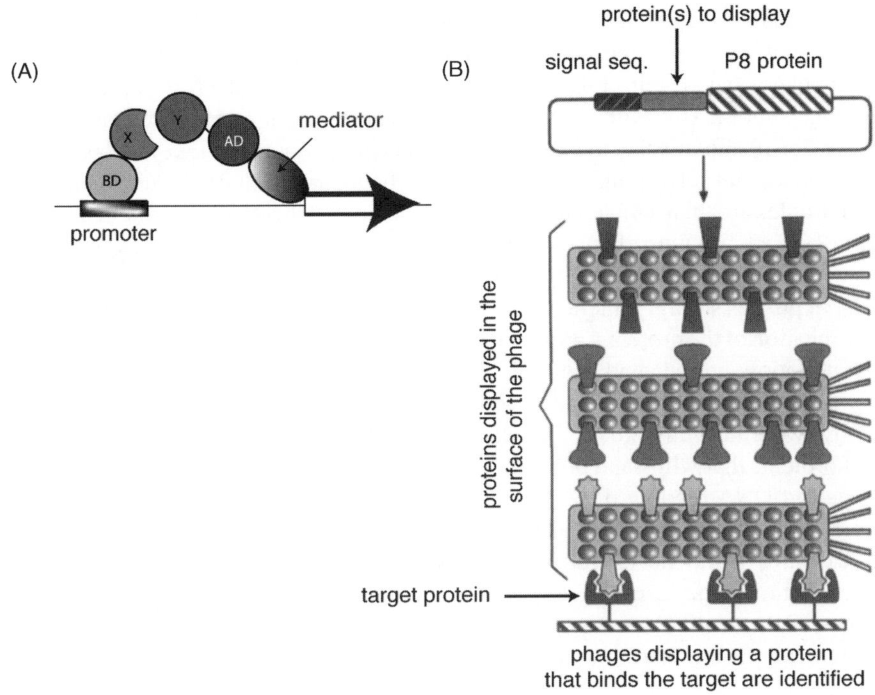

Figure 1.1 *Methods to study protein-protein interactions I. (A) Yeast two hybrid. The formation of a complex between the bait (X) and prey (Y) proteins brings together the binding domain (BD) and the activation domain (AD) of the transcription factor, which stimulates the expression of a reporter gene (adapted from Shoemaker and Panchenko [4]) (B) Phage display. The genome of a bacteria-specific virus, a phage, is engineered to carry the DNA of an exogenous protein. This protein is also displayed on the outer envelope of the phage, making the phage particle a unique carrier of both the genetic information and functional polypeptide for a given gene. Positive interaction partners are isolated by incubating the phages with the target protein immobilized on a solid support (adapted from [5] with permission from Elsevier)*

identified from genomic-scale Y2H analysis are expected to have a success rate of 50%, and bioinformatic analysis tools to refine the results with co-expression and co-localization analysis can very significantly increase the accuracy of the results [2, 3].

1.2.2 Phage Display

This method was one of the earliest tools developed for screening PPIs, before the recent spread of mass spectrometry-assisted protein identification. Phages are bacteria-specific viruses which carry the viral DNA enclosed in an envelope of viral proteins. Phage particles are therefore unique in that they contain both DNA and protein copies of a given gene in a single entity [6]. This singularity provided molecular biologist with a unique tool to isolate simultaneously both the protein displayed in the exterior of a phage particle and its DNA sequence.

The phage display technique involves the construction of a DNA library where the sequences that code for the proteins to be screened are fused to the sequence of a bacteriophage coat protein (P8 or P3) in a plasmid containing the rest of the components of the phage genome (6.5 Kbp for the filamentous bacteriophage M13). Upon infecting an *E. coli* host the phages display the chimeric proteins in their outer surface and bear inside the DNA sequences that correspond to such proteins. Phages are produced in *E. coli* individually, and the particles are then assayed for binding to the immobilized target protein in an ELISA (Enzyme-Linked Immunsorbent Assay) fashion. In order to reduce background bound phages are collected and re-amplified in *E. coli* and, after two to three rounds of binding, the DNA of phages strongly interacting with the target is isolated and sequenced, leading to the identification of the proteins interacting with the target protein [7]. A concise recollection of several random peptide and genomic libraries constructed in different phage vectors, as well as different kind of proteins successfully displayed in filamentous phages was published in 1997 by Smith and Petrenko [6]. Despite being slightly outdated, this survey highlights the range of proteins that withstand phage display that includes enzymes, hormones, receptors, cytokines and DNA binding proteins and account for more than 50 publications.

The range of target molecules that have been subjected to phage display-based screening is very wide, and is by no means restricted to polypeptides. Smith and Petrenko also collected published data on the range of target proteins that have been screened, that includes antibodies, Calmodulin, the tumor suppressor P53, Hsc70, integrins and hormones [6]. Three applications are worth mentioning in the context of PPIs:

(i) the identification of epitopes to monoclonal antibodies, by constructing naive phage peptide libraries of 10 to 40 amino acids [8];
(ii) the analysis of interfaces at the residue level by alanine scanning mutagenesis [9];
(iii) the high-throughput determination of surfaces and free energies of binding [10, 11];
(iv) the identification and construction of new scaffolds for PPIs, like single domain β-sandwich proteins (FN3 and V_HH), ankyrin repeats, WW or SH3 domains, and four helix boundels [7, 12].

Phage display is not, however, without disadvantages. An important one is, for certain applications, the need for the constructed library to be a representative sample of the whole genome, that may be challenging for some laboratories [5]; the libraries can, however, now be obtained commercially, and once produced, can be replicated by passage through a bacterial host. An additional major limitation arises when the displayed proteins fold rapidly because the chimeric fusion protein needs to be remain unfolded for efficient secretion to the bacterial periplasm, prior to display; this has however been recently overcome by the use of alternative translocation pathways and by signal recognition particles [13]. A third major potential problem concerns the immobilization of the target protein, which may hinder the interaction surface: GST or His tags are therefore desirable to aid in immobilization.

1.2.3 Protein Microarrays

A recently proposed method to analyse PPIs on a genomic scale uses functional protein microarrays [14, 15], where thousands of recombinantly expressed and purified proteins are individually spotted on a surface, by chemical derivatization, to constitute the panel of proteins to be screened (Figure 1.1B). A single fluorescently labelled protein (or ligand), is then put in contact with the array in buffered aqueous solution, and subsequently washed with incrementing stringency. Following this the microarray slide is read by a scanner with laser excitation and fluorescence detection capabilities to identify fluorescent spots indicative of the occurrence of a PPI between the labelled ligand and a protein of the microarray. An important advantage of this method is that variable solution conditions can be easily assayed; this makes it possible, for example, to characterize binding at different concentrations in a high-throughput fashion to report on the thermodynamic stability of the PPIs detected. This approach has been successfully employed to identify and characterize proteins interacting with the Erb receptor family, where affinities using the protein microarray where comparable to those determined by surface plasmon resonance [16].

The main advantage of this technique lies on its ability to screen thousand of interactions simultaneously on a single chip [17]. However care must be taken when interpreting some interactions, in particular low affinity ones, as the chemical derivatization process may affect the properties of immobilized proteins. In addition, checks for correct expression and adequate immobilization have to be carried out; for this purpose and it is common for proteins to carry an extra peptide tag which allows identification in western blots and in the microarray slide.

1.2.4 Affinity-based Methods

A number of methods have been developed to specifically isolate protein complexes formed *in vivo* and further analyse them by mass spectrometry [18]. The main idea is to fuse the protein of interest (bait) to a peptide tag which confers affinity to a ligand immobilized on a solid support. Proteins that establish a PPI with the tagged protein can in this way be co-isolated upon incubation with the ligand matrix, and complexes can then be eluted by incubation with free ligand. Modern MS methodologies are key in this approach as they are used to analyse the bound proteins.

The general procedure involves the construction of the gene for the chimera that fuses the coding sequence of the bait to the desired tag. The plasmid is then transfected into a eukaryotic cellular host, where it is expressed, producing large amounts of the protein. Cells are then lysed, and the lysates are subjected to affinity chromatography, where protein complexes involving the tagged bait are specifically isolated. Proteins composing the complexes are resolved by polyacrilamide gel electrophoresis (SDS-PAGE), bands are excised and then subjected to tryptic digestion to produce peptides suitable to MS analysis. Such standard methodologies include the use of Matrix Assisted Laser Desorption Ionization (MALDI) MS, or liquid chromatography coupled to Electro Spray Ionization (ESI) MS [19].

This is a simple methodology that is recommended for most laboratories, as it is relatively inexpensive, requires neither complex equipment nor commercial services and can be carried out with the help of commercial kits. Affinity-based methods normally identify high affinity interactions i.e. with slow kinetics of dissociation, and one of their great advantages is that they allow the isolation of multiprotein complexes, that is not possible when using the Y2H or phage display assays or protein microarrays. It is however important to acknowledge that the use of a peptide tag can promote or impair certain PPIs, affect the normal localization of the bait protein as well as impair the isolation of the protein–protein complexes if the tag becomes buried as they form. All these problems are, however, easily overcome by the use of a second unrelated tag in further similar experiments aimed at confirming the PPI. Depending on the nature of the tag, it is useful to classify affinity methods in three groups:

1.2.4.1 Single Tag Affinity Purification

This method involves the use of a unique peptide motif at the N- or C-terminus of the bait protein to detect protein–protein interactions that occurr *in vivo* by co-sedimentation of the interacting partners. One of the most extended methodologies involves the use of the gluthatione-S-transferase (GST) protein as a fusion of one of the assayed proteins [1, 21, 22]. The GST tag confers the bait protein high affinity to gluthatione, which is immobilized on agarose beads to *pull down* interacting proteins from cellular extracts. The disadvantages of this technique lie in the considerable size of GST (27 KDa) that can perturb the structure of the fused protein, and in the co-isolation of proteins interacting with GST itself rather than with the bait. Similar approaches employ a poly-Histidine tagged protein with high affinity to metal-chelated beads; this tag only slightly perturbs the structure of proteins, but usually results in the isolation of His-rich proteins that are false positives. Other motifs widely used as tags include maltose binding protein (MBP), immunoglobulin binding domains (protein A or G), and the Strep-tag, which is based on the high affinity biotin/ streptavidin interaction [23].

1.2.4.2 Tandem Affinity Purification (TAP)

TAP is a modified version of the single tag affinity method, and involves two different peptide motifs in tandem, separated by a protease cleavage site (Figure 1.2B) [20]. The improvement in TAP in respect to the single tag method lies on the usage of two affinity purification steps which reduces the presence of spurious interacting proteins that can lead to false positives. The initial combination of tags featured a tandem of protein A and a Calmodulin binding peptide (CaMBP), separated by a Tobacco Etch Virus (TEV) protease cleavage site [24]. Protein complexes involving the tagged-bait protein are first isolated with immunoglobulin-agarose beads that have high affinity for protein A. Digestion with TEV protease releases the complex and exposes the Calmodulin Binding Peptide (CaMBP). Incubation with CaM-coated beads followed by and elution with EGTA or free CaMBP allows isolation of the purified complex. A new generation of tags involves high efficiency cloning vectors, the use of inducible promoters of expression, tetracysteine motifs suitable for in cell fluorescence imaging, and streptavidin tags [18, 25]. Proteome wide scale studies in yeast by the TAP method have recently identified more than 500 protein complexes of physiological relevance, demonstrating the high-throughput capabilities of the technique [26, 27].

1.2.4.3 Co-immunoprecipitation (Co-IP)

A protein complex stabilized by PPIs present in a cellular or tissue homogenate can be isolated by means of an appropriate antigen-antibody pair followed by affinity chromatography with protein A or G-coated beads [28]. Antibodies suitable for Co-IP studies can be raised against the protein of interest, or against a small peptide tag fused to the protein of interest; the second option is preferable since it ensures the absence of cross reactions and allows the use of already characterized commercial antibodies. Commonly employed tags for Co-IP studies are HA ($_{1}$YPYDVPDY$^{A_{9}}$), c-Myc ($_{1}$EQKLISEED$^{L_{10}}$), FLAG ($_{1}$DYKDDDD$^{K_{8}}$), all of which have plasmids and antibodies commercially available.

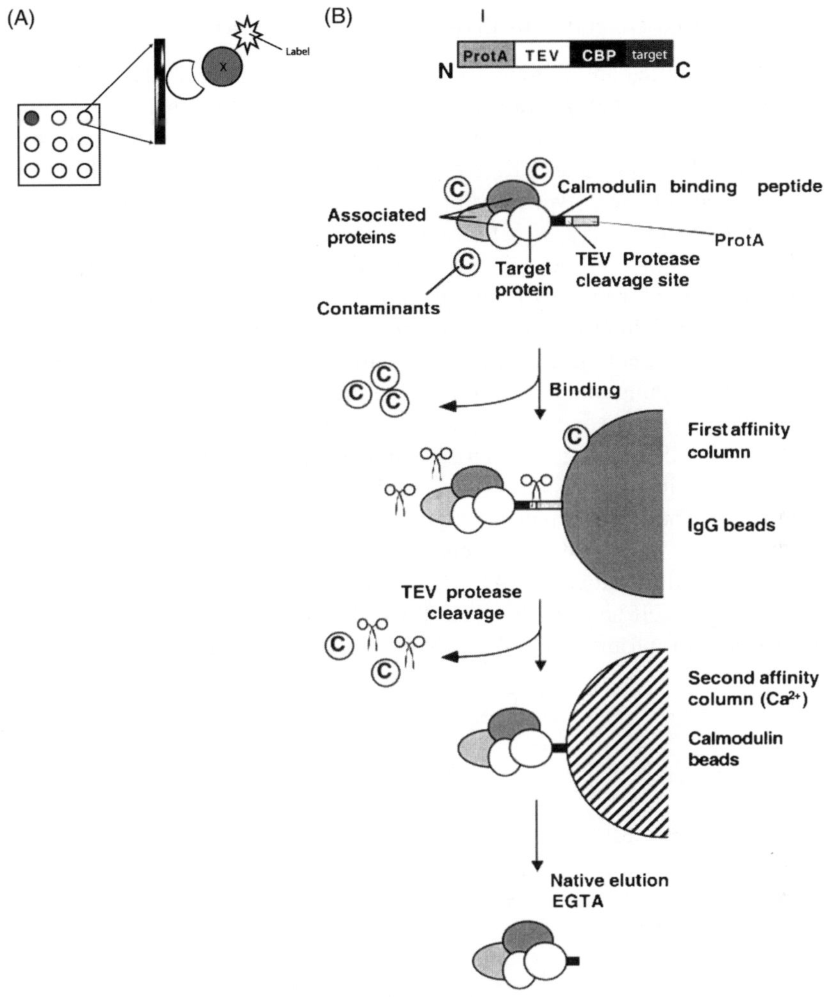

Figure 1.2 *Methods to study protein-protein interactions II (A) Protein microarrays. Proteins are immobilized on a slide by chemical derivatization, and the array of proteins is screened by a fluorescently labelled protein X (or ligand). Fluorescent spots, indicative of a protein complex, are identified by laser-based scanning (adapted from Shoemaker and Panchenko [4]) (B) Tandem affinity purification (TAP). In order to isolate interaction partners for a desired protein, the TAP method genetically fuses the protein of interest to a tandem of peptide tags. The standard TAP tag possesses a protein A (ProtA) and a Calmodulin binding peptide (CBP), linked by a Tobacho Etch Virus (TEV) protease cleavage site. Complexes with the tagged protein are purified in first instance with immunoglobulin-coated beads (affinity to protein A) followed by digestion with TEV protease. The released protein complex is subjected to a second purification step with CaM-coated beads (affinity to CBP) and eluted by addition of EGTA or free peptide (adapted from [20] with permission from Elsevier). Proteins are resolved by gel electrophoresis and identified by MALDI-MS*

1.2.5 FRET-based Detection of Protein–Protein Interactions

When two fluorophores are sufficiently close in space, a nonradiative transfer of energy termed Föster resonance energy transfer (FRET) can occur between them. The efficiency of FRET varies strictly with the sixth power of the distance between the two fluorophores (Figure 1.3A) and this provides FRET-based methodologies with the ability to efficiently assess the distance between two appropriately labelled molecules. For most biologically useful fluorophores, FRET occurs in the range of distances between 10 and 80 nm, that is the same order of magnitude as the size of macromolecules, thus certifying that FRET can be used for detecting the formation of protein complexes through PPIs. FRET has indeed long been exploited to study interactions between proteins by exploiting the intrinsic fluorescence of amino acids or by labeling reactive groups with extrinsic fluorophores [29]. However, it was not until the cloning and expression of the green fluorescent protein (GFP) from the jellyfish Aequorea victoria and its variants [30, 31] that the possibility of utilizing FRET to study protein–protein interactions *in vivo* became a reality [32, 33]. The use of GFPs is now well established for imaging protein complexes but the size of this protein (27 KDa) makes it desirable to design new tools, such as genetic tags with small dyes [34, 35], to fluorescently label proteins in the cellular environment.

The screening and identification of PPIs using fluorescent technologies has promising use in high-throughput cell analysis but FRET still has limitations, in particular concerning acquisition time and automation [36]. However recent technological advances have allowed

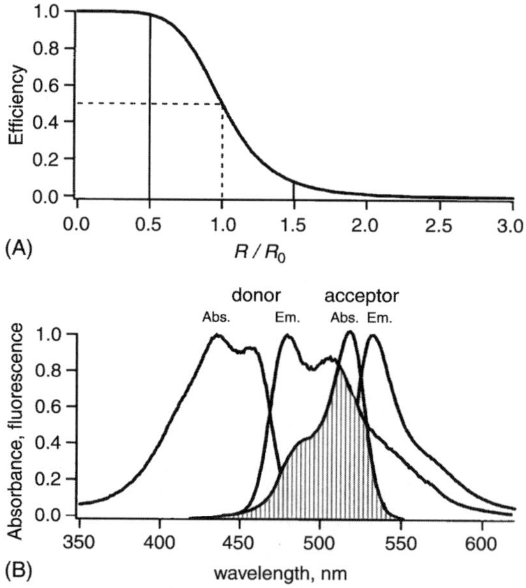

Figure 1.3 *FRET as a tool to identify protein-protein interactions (A) Forster theory shows that FRET efficiency (**E**) varies inversely with the sixth power of the distance between donor and acceptor molecules (**R**), where $\boldsymbol{R_0}$ is the characteristic distance at which transfer efficiency is 50%. $\boldsymbol{R_0}$ (nm) depends on the relative orientation between the transition dipoles of the donor and acceptor, on the spectrum overlap integral between the region of emission of the donor and the region of excitation of the acceptor (B) as well on the refractive index of the medium and the quantum yield of the donor (adapted from [39] with kind permission from Springer Science + Business Media)*

the combination of FRET with high resolution optical microscopy and flow cytometry, providing the capability of observing and screening PPIs in *E. Coli* [37]. In the near future this will be certainly applied to screening protein complexes in yeast, mammalian cells in culture and possibly in organisms such as *C. Elegans*, in particular with the use of fluorescence lifetime-based methods [38]. Published methodologies involve the creation of a library of plasmid DNA containing the protein to be studied fused to YFP and a variable gene fused to CFP. Upon transfection and protein expression, cells displaying high FRET efficiency, and hence reporting an interaction between the two labelled proteins, are sorted and isolated in a flow cytometer. Positive cells are cultured further and subjected to a second sorting step, to reduce background. Clones with high FRET are plated individually in a well plate and the plasmid DNA is isolated and sequenced to identify the identity of the interaction partners [37].

1.3 Techniques to Characterize Protein–Protein Interactions

Once a protein–protein complex is unequivocally identified, several biophysical methods can be used to characterize the macromolecular assembly, understand its properties and suggest, in a structure-based fashion, strategies to inhibit its formation. Low resolution techniques, that are key to determine the stoichiometry and overall topology of the complex, should ideally be combined with high resolution approaches that report at atomic resolution on the structure and dynamics of the binding interface as well as with methods to determine its thermodynamic stability. Since proteins often operate as multiprotein complexes it may be necessary to use multidisciplinary approaches to dissect how multiprotein complexes are formed; one recent example of such an approach has been the determination of the molecular architecture of the nuclear pore complex, a 50 MDa macromolecular machine consisting of 456 proteins. The authors of this study employed a combination of high-resolution structures, cryo-electron microscopy, mass spectrometry and analytical ultracentrifugation to determine the stoichiometry and position of each protein in the complex. This information was then used in a computational analysis to reveal the overall morphology of the nuclear pore complex, as determined by cryo-electron microscopy [40, 41].

It is not within the scope of this chapter to provide a thorough description of the biophysical methods applicable to the study of PPIs, as more specific chapters in this book will address them; we just intend to comparatively assess the information that each technique is capable to provide, consider their applications and limitations, and describe how they can be combined to provide a complete analysis of a given protein–protein complex.

1.3.1 X-ray Crystallography

It is the most widely employed method to determine structure of macromolecular complexes with atomic resolution. The method strongly depends on the quality and diffraction capabilities of the crystals obtained, but nowadays commercial kits are available for the screening of various crystallization conditions. In addition, modern crystallography experiments need to be performed at synchrotrons, where the flux and characteristics of the X-ray beams are optimal for biomolecules. When resolution permits, analysis of the crystals provides a full snapshot of the subunit-subunit contact surface, including side-chain to side-chain contacts. In some cases the stoichiometry of the complex can be affected by crystal contacts in the lattice, it is therefore necessary to use a complementary biophysical method to confirm the stoichiometry of the protein–protein complex in solution. There appears not to be a limit in the size or nature of proteins capable of being crystallized, as the recent study showing a 340 KDa multimeric transmembrane ion channel at atomic resolution permits infer [42]. However, poorly structured polypeptides, which represent almost 30% of the genome, or flexible regions in folded proteins, which are dynamically important, are not visualized by this method and need to be characterized by other techniques such as Nuclear Magnetic Resonance.

1.3.2 Nuclear Magnetic Resonance (NMR)

This spectroscopic method can be used to provide atomic resolution models of macromolecules in solution. For the study of folded proteins NMR methodologies have a narrower range of applicability than X-ray crystallography because the NMR signals of proteins containing more than 400 amino acids are too broad to be detected efficiently. NMR spectroscopy is however an extremely powerful tool for the study of macromolecular complexes that present flexible regions and for the dynamical characterization of macromolecules. A detailed description of this technique is provided in chapter 9.1, authored by E. Giralt.

1.3.3 Isotermal Titration Calorimetry (ITC)

This method is based on the measurement of the heat absorbed or generated upon the interaction of macromolecules in solution. As heat quantities involved in binding events are considerably small, the ITC equipment relies on the accurate quantization of such heat changes. The information generated by a single ITC experiment comprises the association constant Ka (or its inverse, the dissociation constant Kd) and the stoichiometry of binding (n). In addition, thermodynamics of the binding event are also characterized, as both changes in free energy (ΔG) and enthalpy ($\Delta;H$) are measured [43]. If titrations at varying temperatures are performed a third thermodynamic variable can be determined, the heat capacity upon binding (ΔCp), which can be related to changes in solvent accessible area upon complex formation [44]. This technique as well as its potential for the thermodynamic characterization PPIs are described in detail in chapter 2, authored by I. Luque.

1.3.4 Other Techniques

From a drug discovery perspective the most important aspects of protein–protein complexes that need to be characterized are their structure at high resolution (Section 1.4.4) and thermodynamic stability as a function of sequence (Sections 1.4.4.7 and 1.5.1.1 of this chapter). A number of complementary techniques are however having an important impact in the characterization of PPIs in that they allow to determine the structure of complexes that are challenging to crystallize, such as cryo-electron microscopy and mass spectrometry [45–47], and to determine very accurately the kinetics of protein–protein binding, such as surface plasmon resonance [48].

1.4 Structure and Dynamics of Protein Complexes

Structure-based strategies for the inhibition of PPIs for therapeutic purposes rely on an accurate understanding of the structure of the binding interface and require an awareness of the general properties of protein complexes. These are, as will become evident in this section, clearly distinct from complexes formed by proteins and small organic molecules such as enzyme inhibitors.

1.4.1 Functional Classification of Protein–Protein Complexes

Functional classifications of protein–protein complexes stabilized by PPIs typically divide them in four groups [49, 50]: antibody-antigen complexes, enzyme-inhibitor complexes, electron-transfer complexes and complexes involved in signal transduction and cell cycle regulation.

1.4.1.1 Antibody-Antigen Complexes

These complexes play a key role in the immune system. The structures of antibodies contain six complementary-determining regions (CDR) that identify the antigen with high specificity; although enriched in Ser and Thr these regions are highly variable. Interaction surfaces are of medium size (1200–2000 $Å^2$) [49] and, in most cases, binding occurs with minimal conformational change in the antigen, suggesting that structural adaptation operates on the surface of the antibody.

1.4.1.2 Enzyme-Inhibitor Complexes

These complexes can be further divided in two subsets depending on their interface size, small (1200–2000 $Å^2$) or large (>2000 $Å^2$) [51]. Usually small interfaces show a single recognition patch, whereas larger interfaces present more than one recognition site.

1.4.1.3 Electron-Transfer Complexes

These complexes are transient and of low stability. They are therefore difficult to obtain in crystal form. Most electron-transfer proteins characterized until now have interfaces of between 900 and 1200 $Å^2$ [49].

1.4.1.4 Complexes Involved in Signal Transduction and Cell Cycle Regulation

These include G-proteins and protein-receptor assemblies. These complexes exhibit exquisite sensitivity to environmental changes, usually forming transient interactions and presenting low to medium affinity range (low mM to high nM) [49].

1.4.2 Differentiation Between Crystallographic and Functional Complexes

Although X-ray crystallography is an extremely powerful tool to extract high-resolution structural formation about the protein–protein interface it is important to acknowledge that crystal contacts can, due to the high concentration of samples, lead to the formation of crystallographic complexes that are present, in solution, neither *in vitro* nor *in vivo*. Discerning whether a complex is crystallographic or functional solely from knowledge of the sequence of the binding partners and, possibly, of the structure of the complex is highly nontrivial. In an interesting study of the determinants of specificity in PPIs Janin and coworkers compared a set of specific interactions (without including short-lived assemblies or electron-transfer) against a set of nonspecific interactions; in their analysis crystal contacts were found to have a smaller interface than specific ones (with an average interface area of 570 Å^2), presented a lower number of hydrogen bonds per surface unit. and were less closely packed than interfaces from specific protein–protein contacts [52]. Predictive models based on the size and composition of the protein–protein interface that use this information can distinguish these two types of complexes with a certain degree of accuracy, especially when sequence conservation from related proteins is utilized.

1.4.3 Classification Based on the Nature of the Constituents and the Lifetime of the Complex

Nooren and Thornton defined three categories to classify PPIs according to these criteria [53]. The main division is that between homo-oligomeric (composed by identical chains) and hetero-oligomeric (nonidentical chains) complexes. Homo-oligomers can be further subdivided into those that are isologous, where binding interfaces are composed by the same region of each monomer, and those that are heterologous, where monomers interact through different regions. Heterologous homo-oligomers can form cyclic structures or aggregate into an endless repeated structure. Another important distinction Nooren and Thornton considered is one between complexes where the monomers can exist independently *in vivo* (nonobligate) and those where they cannot (obligate). The latter need to be denatured to dissociate, whereas the former can form from stable self-standing monomers. Examples of nonobligate complexes include antibody-antigen, enzyme-inhibitor and signal transduction complexes. Finally a third distinction can be made based on lifetime, to distinguish between permanent and transient complexes *in vivo*. Usually obligate interactions are permanent, e.g. those between monomers in homodimers, whereas transient interactions present a wide range of affinities and kinetics.

It is apparent from these definitions that interfaces of permanent complexes are more similar to the protein interior than the rest of the protein surfaces. Indeed permanent interfaces are dryer, more hydrophobic and larger than the interfaces of transient complexes [49, 54]. However, as mentioned before, the vast diversity in function, flexibility, affinity and specificity that protein assemblies present is difficult to capture in general rules.

1.4.4 Descriptors and Topology of Protein Complexes

We will here provide a brief review of the structural features typically studied in protein–protein interfaces. Classical computational characterization of interfaces include size, shape, packing, electrostatic interactions such as hydrogen bonds and salt bridges, amino acid composition, amino acid pairing preferences and hotspots.

1.4.4.1 Size

The size of the interfaces is commonly expressed as the change in the ASA (solvent-accessible surface area) between monomers and complex. For example, for a dimer, the interface size B, is $B = ASA_1 + ASA_2 - ASA_{12}$. Some authors prefer to report B/2 in spite of the fact is not exactly half B for both surfaces unless they are flat. Standard sizes for protein complexes interfaces lie between 1200 and 2000 Å^2 averaging 23 residues in each monomer [49].

1.4.4.2 Shape

Although interacting surfaces are most often flat, they can be concave or convex. In general, the partner of smaller size shows some convexity, binding to the concave cavity in the partner of larger size. An exception to this trend are antibody-antigen complexes where the antibody antigenic site is generally convex independently of the antigen size [49]. For interfaces larger than 2000 Å^2 it has been found that the binding site is closer to the centre of mass of the protein than the average location of the surface [55].

1.4.4.3 Packing

Another structural feature of the interacting protein surfaces that is often computed is the degree packing as it is useful order to estimate the degree of steric complementarity between protomers. The most reported packing indices are Shape Complementarity score (Sc) [56] and Gap Volume index (GV) between proteins [57]. It has been found that homodimers, enzyme-inhibitor and permanent hetero-complexes are more closely packed than antibody-antigen and transient hetero-complexes.

1.4.4.4 Electrostatic Interactions

It is clear that electrostatic complementarity between partners in protein–protein complexes confers specificity. On average, there is one hydrogen bond (HB) per 200 Å^2 of interface area (B), or 100 Å^2 if one references to a single component (B/2)[58, 59]. Permanent protein complexes have typically less intermolecular hydrogen bonds per buried ASA than non-obligate complexes, 0.9 HB per 100 Å^2 in homodimers where enzyme-inhibitor complexes show 1.4 HB and antibody-antigen 1.1 per 100 Å^2 [58]. Additionally, protein–protein interfaces also have water-mediated hydrogen bonds, which present the same average distribution as the direct protein–protein hydrogen bonds, that is 10 water molecules per 1000 Å^2 (B/2) [60]. Salt bridges or hydrogen bonds involving at least one charged residue do occur: Lo Conte *et al.* found in their data set that 30% of the hydrogen bonds accounted at the interfaces are salt bridges [61]. However, almost half of the homodimeric structures analysed do not present this type of interaction. Disulfide bonds can also be found between interacting proteins but they are quite rare [58].

1.4.4.5 Amino Acid Composition

In analyzing the amino acid composition of protein–protein interfaces and their pairing preferences different studies find different frequencies due to the sets they analyse and how the interfaces were defined. Ofran and Rost, for example, divided their data set in six different types of protein–protein interfaces and, while they found some generalities, their main finding was that each interface type had distinct residue propensities [62]. For example, while Lys was underrepresented in all types of interfaces, Arg was overrepresented (Arg is common in all protein surfaces, not only interfaces). Large hydrophobic amino acids such as His, Met and Tyr were favoured in all interfaces while Ser, Ala and Gly were underrepresented. The authors corroborated previous findings that hydrophobic residues were more frequent at homo-multimers than hetero complexes; however, when they further divided into transient and obligate interaction, this distinction no longer held.

1.4.4.6 Pairing Preferences

With respect to interactions at the interface, Ofran and Rost found that hydrophobic-hydrophilic contacts were preferred at intradomain, domain-domain and transient hetero-complexes interfaces and disulfide bonds between Cys residues occurred more often than expected [62]; they also found that salt bridges were less frequent at the interfaces of homocomplexes and that interactions between identical amino acid were favored at obligate homo-complexes. More recently, Headd and colleagues, studied 135 transient hetero-complexes and found that 32% of all contacts at the interfaces involve backbone atoms [63] and, focusing on side chains, he found Glu, Ser, Asp, Lys and Arg to be the most frequent interacting side chains at the interface, each having more than 7% presence, and Met, Cys, Trp and His to be the least frequent, with less than 3.5% representation. Concerning interchain pair contacts in this data set, the most frequent occurring pairs were salt bridges (Glu-Arg, Asp-Arg, Glu-Lys and Asp-Lys); this evidence highlights the importance of electrostatic complementarity between interacting surfaces. After the charge–charge interaction the next most frequent interactions were found to be Tyr with Arg, Asn, Lys and Glu, followed by Arg with Trp and Asn.

1.4.4.7 Hot Spots

The most striking feature of protein–protein interacting surfaces is the existence of hot spots. In 1995 Clackson and Wells, using a technique called Ala scanning mutagenesis, systematically mutated to Ala all the receptor residues at the interface between the human growth hormone and its receptor and measured the free energy of binding of the resulting complex mutants [64]. In this pioneering work, the authors found that certain residues were responsible for most of the interaction free energy of the complex. A number of other experimental studies have proved that this is a common characteristic for almost all interfaces of the protein complexes [65]. Moreover, a public accessible database (ASEdb – http://nic.ucsf.edu/asedb/) holds most of the current experimental data for Ala scanning mutagenesis [66]. The accepted criterion to define a residue as a hot spot is that upon its mutation to Ala the free energy of binding increases by at least 2 kcal.mol^{-1}. Bogan and Thorn analysed a data set from alanine scanning mutagenesis and found all the hot spots shared common characteristics, which led them to postulate the 'O-ring' arrangement for the hot spot residues in protein–protein binding interfaces. [67] Hot spot residues are usually clustered at the centre of the interface and are surrounded by energetically neutral residues; the role of these neutral residues is to maintain the hot spots shielded from the solvent by creating a micro environment around the hotspot with lower dielectric constant, enhancing electrostatic interactions and reducing the desolvation cost of binding. It is no surprise then that the most frequent hot spots residues (Trp, Tyr and Arg) are capable of both hydrophobic and electrostatic interactions. They also found that hot spots are self-complementary across the interfaces. Nussinov and coworkers found that structural conserved residues at the interfaces of protein complexes correlate with experimental Ala scanning data and studied the organization of these computational hotspots [68]. They found hot spots are not evenly distributed in the interface, but rather they cluster together in 'hot regions'. These regions are highly packed and, within a region, hot spots form networks of cooperative interactions, whereas between hot regions the contribution to the global energy of binding is additive. They suggested that the clustering of hot spots in dense hot regions makes it easier to remove water molecules and strengthens electrostatic interactions in agreement with the O-ring arrangement. Furthermore these regions are more rigid and therefore have a relatively low entropic penalty upon binding. In conclusion, PPIs are locally optimized in these hot regions, whereas the rest of the interface is more tolerant and lees specific, a fact that could explain the diversity in protein binding.

1.5 Protein–Protein Complexes as Therapeutic Targets

The ubiquitous nature and central role of PPIs make them very attractive targets for therapeutic intervention. However, PPIs have long been believed to be undruggable, supported by the logical assumption that a small molecule will be unable to substitute one of the partners in a multiprotein complex where the average standard interfaces are 2000 $Å^2$, with an average of 23 residues in each interacting polypeptide unit.

1.5.1 Challenging Undruggability [69, 70]

As recently summarized by Witty and Kumaravel [70] two main risks need to be assessed in the selection of therapeutic targets for a given indication. Biological risk tries to determine the probability that the modulation of the activity of the target will lead to the desired pharmacological effect whereas chemical risk tries to determine the probability that it will be possible to identify a bio-available small molecule that will effectively bind the surface of the therapeutic target and affect its function. From the biological risk point of view PPIs are very attractive targets for drug discovery, i.e. are of low biological risk, because their ever-presence in biological processes suggests that their modulation will be therapeutically relevant. In fact, in the case of extracellular targets, antibody drugs represent a key validation of this concept. In the case of PPIs the key question is therefore to assess the chemical risk for protein complexes, or in other words how likely is to find a small molecule capable of disrupting the interactions between proteins. Without considering the possibility of allosteric modulation, two experimental findings have lowered this chemical risk: the existence of energetic hot spots at the interfaces and site adaptability of surface patches.

1.5.1.1 Hot Spots

The existence of localized regions responsible for most of the binding free energy in PPIs (Section 1.4.4.7) suggests that small molecules that target the key regions of the interface can have the ability to inhibit PPIs and modulate the activity of protein complexes, i.e. decreases the chemical risk of using PPIs as therapeutic targets. An increasing number of studies report that small molecules can bind directly to protein interfaces [71, 72]; examples include the inhibition of the p53-MDM2 interaction [73], antagonists for the Bcl-2 anti-apoptotic family of proteins [74], inhibitors of ZipA-FtsZ interaction [75] and disruption of the interaction between IL-2 and its receptor IL-2Ra [76]. Alanine scanning is a relatively costly approach to the detection of hot spots in protein–protein interactions but knowledge of the structure of the complex from either X-ray crystallography or NMR spectroscopy can be exploited for this purpose by using programs that predict hot spots.

1.5.1.2 Site Dynamics

As previously described, PPIs surfaces tend to be quite flat. This represents an additional challenge for drug design because it limits the number of noncovalent interactions that can be established between potential drugs and protein surfaces. Recent structural evidence of flexible adaptability in these regions, like in the classical example of IL-2 [77, 78], opens the prospect for more druggable protein complexes because they suggest that dynamics in the surface of the free proteins will offer conformations that are less flat and therefore more druggable that the corresponding bound structure. Indeed surface flexibility is now included in assessments of the druggability of surface patches, as shown recently [79, 80]. The availability of methods to describe proteins as dynamics ensembles rather than as rigid structures [81–83] will undoubtedly play a key role in future developments in drug discovery, especially in the discovery of compounds that target surface patches, as they explicitly incorporate flexibility in the structure determination process.

1.6 Conclusions

The technologies for the systematic screening and structural characterization of PPIs are now very mature and are providing a large number of very attractive therapeutic targets for the drug discovery and pharmaceutical industries. PPIs continue however to be considered very challenging therapeutic targets, i.e. non druggable, because the interaction surfaces involved in their stabilization are large, relatively flat and difficult to address using small molecules. As we have seen, however, two key features of PPIs, namely the uneven distribution of free energy and the dynamics of the free state, suggest that it will be possible, in the near future, to identify synthetically accessible molecules to harness the potential of inhibiting PPIs.

References

1. Singh CR, Asano K (2007) Localization and characterization of protein–protein interaction sites. *Meth Enzymol* 429:139–161.
2. Kemmeren P. *et al.* (2002) Protein interaction verification and functional annotation by integrated analysis of genome-scale data. *Mol Cell* 9:1133–1143.
3. von Mering C. *et al.* (2002) Comparative assessment of large-scale data sets of protein–protein interactions. *Nature* 417:399–403.
4. Shoemaker BA, Panchenko AR (2007) Deciphering protein–protein interactions. Part I. Experimental techniques and databases. *PLoS Comput Biol* 3:337–344.
5. Sidhu SS, Lowman HB, Cunningham BC, Wells JA (2000) Phage display for selection of novel binding peptides. *Meth Enzymol* 328:333–363.
6. Smith GP, Petrenko VA (1997) Phage display. *Chem Rev* 97:391–410.
7. Sidhu SS, Koide S (2007) Phage display for engineering and analyzing protein interaction interfaces. *Curr Opin Struct Biol* 17:481–487.
8. Yip YL, Ward RL (1999) Epitope discovery using monoclonal antibodies and phage peptide libraries. *Comb Chem High Throughput Screen* 2:125–138.
9. Morrison KL, Weiss GA (2001) Combinatorial alanine-scanning. *Curr Opin Chem Biol* 5:302–307.
10. Kotz JD, Bond CJ, Cochran AG (2004) Phage-display as a tool for quantifying protein stability determinants. *Eur J Biochem* 271:1623–1629.
11. Pál G, Kouadio JL, Artis DR, Kossiakoff AA, Sidhu SS (2006) Comprehensive and quantitative mapping of energy landscapes for protein–protein interactions by rapid combinatorial scanning. *J Biol Chem* 281:22378–22385.
12. Magliery TJ, Regan L (2004) Combinatorial approaches to protein stability and structure. *Eur J Biochem* 271:1595–1608.

13. Paschke M, Höhne W (2005) A twin-arginine translocation (Tat)-mediated phage display system. *Gene* 350:79–88.
14. Zhu H. *et al.* (2001) Global analysis of protein activities using proteome chips. *Science* 293:2101–2105.
15. Zhu H, Snyder M (2001) Protein arrays and microarrays. *Curr Opin Chem Biol* 5:40–45.
16. Jones RB, Gordus A, Krall JA, MacBeath G (2006) A quantitative protein interaction network for the ErbB receptors using protein microarrays. *Nature* 439:168–174.
17. Phizicky E, Bastiaens PI, Zhu H, Snyder M, Fields S (2003) Protein analysis on a proteomic scale. *Nature* 422:208–215.
18. Berggård T, Linse S, James P (2007) Methods for the detection and analysis of protein–protein interactions. *Proteomics* 7:2833–2842.
19. Aebersold R, Mann M (2003) Mass spectrometry-based proteomics. *Nature* 422:198–207.
20. Puig O. *et al.* (2001) The tandem affinity purification (TAP) method: a general procedure of protein complex purification. *Methods* 24:218–229.
21. Kaelin WG, Pallas DC, DeCaprio JA, Kaye FJ, Livingston DM (1991) Identification of cellular proteins that can interact specifically with the T/E1A-binding region of the retinoblastoma gene product. *Cell* 64:521–532.
22. Brymora A, Valova VA, Robinson PJ (2004) Protein–protein interactions identified by pull-down experiments and mass spectrometry in *Current Protocols in Cell Biology* (Ed: Bonafino JS *et al.*) Wiley and Sons, Chapter 17:Unit 17.5.
23. Junttila MR, Saarinen S, Schmidt T, Kast J, Westermarck J (2005) Single-step Strep-tag purification for the isolation and identification of protein complexes from mammalian cells. *Proteomics* 5:1199–1203.
24. Rigaut G. *et al.* (1999) A generic protein purification method for protein complex characterization and proteome exploration. *Nat Biotechnol* 17:1030–1032.
25. Giannone R.J. *et al.* (2007) Dual-tagging system for the affinity purification of mammalian protein complexes. *BioTechniques* 43:296, 302.
26. Krogan N.J. *et al.* (2006) Global landscape of protein complexes in the yeast Saccharomyces cerevisiae. *Nature* 440:637–643.
27. Gavin A.C. *et al.* (2006) Proteome survey reveals modularity of the yeast cell machinery. *Nature* 440:631–636.
28. Masters SC (2004) Co-immunoprecipitation from transfected cells. *Methods Mol Biol* 261:337–350.
29. Wu P, Brand L (1994) Resonance energy transfer: methods and applications. *Anal Biochem* 218:1–13.
30. Chalfie M, Tu Y, Euskirchen G, Ward WW, Prasher DC (1994) Green fluorescent protein as a marker for gene expression. *Science* 263:802–805.
31. Heim R, Cubitt AB, Tsien RY (1995) Improved green fluorescence. *Nature* 373:663–664.
32. Mitra RD, Silva CM, Youvan DC (1996) Fluorescence resonance energy transfer between blue-emitting and red-shifted excitation derivatives of the green fluorescent protein. *Gene* 173:13–17.
33. Pollok BA, Heim R (1999) Using GFP in FRET-based applications. *Trends Cell Biol* 9:57–60.
34. Giepmans BN, Adams SR, Ellisman MH, Tsien RY (2006) The fluorescent toolbox for assessing protein location and function. *Science* 312:217–224.
35. Jares-Erijman EA, Jovin TM (2006) Imaging molecular interactions in living cells by FRET microscopy. *Curr Opin Chem Biol* 10:409–416.
36. Ramm P (2005) Image-based screening: a technology in transition. *Curr Opin Biotechnol* 16:41–48.
37. You X. *et al.* (2006) Intracellular protein interaction mapping with FRET hybrids. *Proc Natl Acad Sci USA* 103:18458–18463.
38. Esposito A, Dohm CP, Bähr M, Wouters FS (2007) Unsupervised fluorescence lifetime imaging microscopy for high content and high throughput screening. *Mol Cell Proteomics* 6:1446–1454.
39. Hink MA, Bisselin T, Visser AJ (2002) Imaging protein–protein interactions in living cells. *Plant Mol Biol* 50:871–883.

40. Alber F. *et al.* (2007) Determining the architectures of macromolecular assemblies. *Nature* 450:683–694.
41. Alber F. *et al.* (2007) The molecular architecture of the nuclear pore complex. *Nature* 450:695–701.
42. Jasti J, Furukawa H, Gonzales EB, Gouaux E (2007) Structure of acid-sensing ion channel 1 at 1.9 A resolution and low pH. *Nature* 449:316–323.
43. Leavitt S, Freire E (2001) Direct measurement of protein binding energetics by isothermal titration calorimetry. *Curr Opin Struct Biol* 11:560–566.
44. Wintrode PL, Privalov PL (1997) Energetics of target peptide recognition by calmodulin: a calorimetric study. *J Mol Biol* 266:1050–1062.
45. Taverner T. *et al.* (2008) Subunit architecture of intact protein complexes from mass spectrometry and homology modeling. *Acc Chem Res* 41:617–627.
46. Robinson CV, Sali A, Baumeister W (2007) The molecular sociology of the cell. *Nature* 450:973–982.
47. Beck M, Lucić V, Förster F, Baumeister W, Medalia O (2007) Snapshots of nuclear pore complexes in action captured by cryo-electron tomography. *Nature* 449:611–615.
48. Hartmann-Petersen R, Gordon C (2005) Quantifying protein–protein interactions in the ubiquitin pathway by surface plasmon resonance. *Meth Enzymol* 399:164–177.
49. Janin J, Rodier F, Chakrabarti P, Bahadur RP (2007) Macromolecular recognition in the Protein Data Bank. *Acta Crystallogr D Biol Crystallogr* 63:1–8.
50. Cho KI, Lee K, Lee KH, Kim D, Lee D (2006) Specificity of molecular interactions in transient protein–protein interaction interfaces. *Proteins* 65:593–606.
51. Chakrabarti P, Janin J (2002) Dissecting protein–protein recognition sites. *Proteins* 47:334–343.
52. Bahadur RP, Chakrabarti P, Rodier F, Janin J (2004) A dissection of specific and non-specific protein–protein interfaces. *J Mol Biol* 336:943–955.
53. Nooren IM, Thornton JM (2003) Diversity of protein–protein interactions. *EMBO J* 22:3486–3492.
54. De S, Krishnadev O, Srinivasan N, Rekha N (2005) Interaction preferences across protein–protein interfaces of obligatory and non-obligatory components are different. *BMC Struct Biol* 5:15.
55. Nicola G, Vakser IA (2007) A simple shape characteristic of protein–protein recognition. *Bioinformatics* 23:789–792.
56. Lawrence MC, Colman PM (1993) Shape complementarity at protein/protein interfaces. *J Mol Biol* 234:946–950.
57. Laskowski RA (1995) SURFNET: a program for visualizing molecular surfaces, cavities, and intermolecular interactions. *J Mol Graph* 13:323–30, 307-8.
58. Jones S, Thornton JM (2000) in *Protein-protein recognition*, (Ed: Kleanthous C) Oxford University Press, Oxford, pp 33–59.
59. Janin J (2000) in *Protein-protein recognition*, (Ed: Kleanthous C) Oxford University Press, Oxford, pp 1–32.
60. Rodier F, Bahadur RP, Chakrabarti P, Janin J (2005) Hydration of protein–protein interfaces. *Proteins* 60:36–45.
61. Lo Conte L, Chothia C, Janin J (1999) The atomic structure of protein–protein recognition sites. *J Mol Biol* 285:2177–2198.
62. Ofran Y, Rost B (2003) Analysing six types of protein–protein interfaces. *J Mol Biol* 325: 377–387.
63. Headd J.J. *et al.* (2007) Protein–protein interfaces: properties, preferences, and projections. *J Proteome Res* 6:2576–2586.
64. Clackson T, Wells JA (1995) A hot spot of binding energy in a hormone-receptor interface. *Science* 267:383–386.
65. Reichmann D, Rahat O, Cohen M, Neuvirth H, Schreiber G (2007) The molecular architecture of protein–protein binding sites. *Curr Opin Struct Biol* 17:67–76.
66. Thorn KS, Bogan AA (2001) ASEdb: a database of alanine mutations and their effects on the free energy of binding in protein interactions. *Bioinformatics* 17:284–285.

67. Bogan AA, Thorn KS (1998) Anatomy of hot spots in protein interfaces. *J Mol Biol* 280:1–9.
68. Keskin O, Ma B, Nussinov R (2005) Hot regions in protein–potein interactions: the organization and contribution of structurally conserved hot spot residues. *J Mol Biol* 345:1281–1294.
69. Hopkins AL, Groom CR (2002) The druggable genome. *Nat Rev Drug Discov* 1:727–730.
70. Whitty A, Kumaravel G (2006) Between a rock and a hard place? *Nat Chem Biol* 2:112–118.
71. Arkin MR, Wells JA (2004) Small-molecule inhibitors of protein–protein interactions: progressing towards the dream. *Nat Rev Drug Discov* 3:301–317.
72. Pagliaro L. *et al.* (2004) Emerging classes of protein–protein interaction inhibitors and new tools for their development. *Curr Opin Chem Biol* 8:442–449.
73. Murray JK, Gellman SH (2007) Targeting protein–protein interactions: lessons from p53/MDM2. *Biopolymers* 88:657–686.
74. Papadopoulos K (2006) Targeting the Bcl-2 family in cancer therapy. *Semin Oncol* 33:449–456.
75. Tsao D.H. *et al.* (2006) Discovery of novel inhibitors of the ZipA/FtsZ complex by NMR fragment screening coupled with structure-based design. *Bioorg Med Chem* 14:7953–7961.
76. Blundell T.L. *et al.* (2006) Structural biology and bioinformatics in drug design: opportunities and challenges for target identification and lead discovery. *Philos Trans R Soc Lond, B, Biol Sci* 361:413–423.
77. Thanos CD, DeLano WL, Wells JA (2006) Hot-spot mimicry of a cytokine receptor by a small molecule. *Proc Natl Acad Sci USA* 103:15422–15427.
78. Arkin M.R. *et al.* (2003) Binding of small molecules to an adaptive protein–protein interface. *Proc Natl Acad Sci USA* 100:1603–1608.
79. Hajduk PJ, Huth JR, Fesik SW (2005) Druggability indices for protein targets derived from NMR-based screening data. *J Med Chem* 48:2518–2525.
80. Brown SP, Hajduk PJ (2006) Effects of conformational dynamics on predicted protein druggability. *ChemMedChem* 1:70–72.
81. Lindorff-Larsen K, Best RB, Depristo MA, Dobson CM, Vendruscolo M (2005) Simultaneous determination of protein structure and dynamics. *Nature* 433:128–132.
82. Richter B, Gsponer J, Várnai P, Salvatella X, Vendruscolo M (2007) The MUMO (minimal under-restraining minimal over-restraining) method for the determination of native state ensembles of proteins. *J Biomol NMR* 37:117–135.
83. De Simone A, Richter B, Salvatella X, Vendruscolo M (2009) Toward an accurate determination of free energy landscapes in solution states of proteins. *J Am Chem Soc* 131:3810–3811.

2

Biophysics of Protein–Protein Interactions

Irene Luque
Department of Physical Chemistry and Institute of Biotechnology, Faculty of Sciences, University of Granada, Granada, Spain

2.1 Introduction

Most cellular processes, including cell proliferation, growth, differentiation, signal transduction or apoptosis, rely on the establishment of complex and tightly regulated networks of protein–protein interactions. The disruption or deregulation of these interactions is very frequently at the origin of disease and, as a consequence, the identification of molecules with the ability to inhibit or modulate protein–protein recognition has been the object of great interest. Nonetheless, because of the specific properties of protein–protein interfaces effectively interfering with protein–protein interactions remains today a considerable challenge.

Contrary to the binding of small molecules (substrates, inhibitors or effectors) to enzymes, which are generally buried in deep cavities with small interaction surfaces (300–1000 Å^2) [1, 2], most protein–protein interaction surfaces are large (1500–3000 Å^2), shallow and featureless lacking well-defined binding pockets [3, 4]. Moreover, with the exception of protein–protein interactions mediated by modular domains, most proteins recognize noncontiguous amino acids in their partners. Different strategies have been applied to the search for inhibitors of protein–protein interactions that include the use of miniature proteins or proteomimetics, although most efforts have been devoted to the identification of small nonpeptidic molecules with the required properties to become effective orally bioavailable drugs [5–8]. Nonetheless, obtaining validated leads for these

Protein Surface Recognition: Approaches for Drug Discovery Ernest Giralt, Mark W. Peczuh and Xavier Salvatella. © 2011 John Wiley & Sons, Ltd

systems is specially difficult due to the lack of small natural partners to take as a starting point for rational design and the low rate of success generally obtained with high-throughput techniques [5, 9, 10]. The situation is further complicated by the high level of plasticity and adaptability of protein–protein binding sites. The establishment of protein–protein complexes is frequently associated with significant conformational changes and, in many cases, the conformation induced by the natural protein partner is significantly different from that observed in the complexes with small-molecule inhibitors that induce the formation of new pockets at the interaction interface. Despite these difficulties very promising advances have been made in the last decade in the search for small drug-like inhibitors of protein–protein interactions that have proved the feasibility of this approach (see [6–8, 11–15] for recent reviews).

In this chapter we will focus on the most relevant biophysical aspects to protein recognition to be considered in the drug development process, including the nature of the forces governing protein interactions, the thermodynamic considerations that should guide lead optimization and some experimental and computational techniques of special interest for the thermodynamic characterization of protein interactions. Even though, in many cases, the guidelines that will be presented have been mostly applied to the binding of small molecules to enzymes, they are fully applicable to protein–protein interactions. Other aspects such as the influence of interfacial waters and the conformational equilibrium of the proteins in the determination of binding energetics and cooperativity will also be discussed.

2.2 Intermolecular Forces in Protein Recognition

In spite of the higher complexity inherent to biological macromolecules, the physical principles governing molecular recognition in proteins, either intramolecular as in protein folding or intermolecular as in protein-ligand or protein–protein interactions, are the same that drive the interactions between small molecules. In general, molecular recognition takes place as a result of a complex interplay of different noncovalent interactions established between the binding partners that, globally, need to overcome those established with the solvent in the unbound state. Nonetheless, the relative weight of each of the interacting forces might be different depending on the specific characteristics of the system (for example, electrostatic interactions seem to play a more important role in protein binding than in protein folding) [16, 17]. A detailed description of the nature of noncovalent interactions and the experimental procedures available to estimate their strength and range can be found in [18–20].

Noncovalent interactions have been traditionally classified into two main classes: electrostatic interactions (charge–charge, charge–dipole, dipole–dipole interactions, etc.) and the hydrophobic effect. Electrostatic forces in protein interactions span a wide range of distance and geometric dependencies and include: (i) van der Waals interactions, which arise from fluctuations in the electric dipole moments of the molecules that become correlated as they approximate. van der Waals forces can be attractive or repulsive, although they are generally attractive between molecules or materials of similar nature. In general, the strength of van der Waals interactions increases with the surface complementarity between the two binding partners [19, 21]. (ii) Hydrogen bonds, which are attractive interactions between a hydrogen donor (A-H) and an acceptor (B). It is generally accepted that these bonds arise from the electrostatic interaction between the positive end of the A–H dipole and the negative end of the B dipole, although its nature (percentage of covalent or ionic character) and definition remains under discussion [22–24]. Two main factors condition the strength of a hydrogen bond: the nature of the donor and acceptor groups and their spatial arrangement. In this way, strong hydrogen bonds are generally established between highly electronegative donors and acceptors while weaker interactions involve C-H groups as donors or π systems as acceptors [25, 26]. Also, small distances (1.2–1.5 Å) between the H atom and the acceptor B and $A\text{-}H \cdots B$ angles near 180° are typically associated to very strong interactions. Nonetheless, the strength of these interactions is also highly dependent on the dielectric properties of their microscopic environment and other factors, such as cooperative effects within hydrogen bond networks, have also been shown to be of relevance [27]. (iii) Coulombic interactions between charged residues, which depend mostly on the distance between the charges and the dielectric constant of the medium. Because of the high desolvation penalty entailed, controversy exists about the energetic benefits of buried salt bridges. Nonetheless, electrostatic complementarity is frequently exploited by biological macromolecules to enhance binding free energies and specificity as well as the rates of bimolecular association [28–33]. These forces are especially relevant in protein–protein interactions, which binding interfaces are highly enriched in charged residues [3, 4, 34]. (iv) Other less frequent interactions, such as π–π or cation–π interactions implicating the aromatic side chains of phenylalanine, tyrosine or tryptophan residues in proteins have been described to contribute significantly to the binding energy [35–37].

In addition to the establishment of specific noncovalent interactions between the partners, complex formation is generally associated to a reduction in the conformational and translational degrees of freedom of the interacting molecules [38–41] and to important changes in the properties of the water molecules solvating the free proteins and ligands that are released into the bulk solvent upon binding. This is the origin of the hydrophobic effect that, as is generally accepted, is associated to the rigidification induced by hydrophobic groups in the solvating water molecules that become organized into stiff 'cagelike' structures or clathrates, establishing strong water-water interactions. The increment in the degrees of freedom of these water molecules when the hydrophobic groups are buried from the solvent is one of the main driving forces for protein folding and for many protein–ligand interactions [42–44].

Identifying which are the specific forces stabilizing protein complexes and quantifying their contributions to the binding affinity has been the goal of numerous investigations for many years, since the answer to this question will provide important mechanistic information on protein recognition and a rationale for effective ligand design. In spite of this, directly quantifying or predicting the contribution of specific interactions to the binding energy remains today controversial. Nonetheless, during the last decade, substantial advances in instrumentation and analytical methodologies have generated an ideal scenario for the application of thermodynamic methods and calorimetric techniques to the study of molecular recognition. These approaches have demonstrated an enormous potential for drug discovery, rational design and high-throughput screening, since they can provide very valuable insights into the nature and magnitude of the forces governing protein interactions.

2.3 Basic Binding Thermodynamics

From a thermodynamic perspective, the association of two molecules M and L to form a noncovalent complex ML is governed by the association constant (or binding affinity) K_a, which reflects the ratio between the concentrations of the different species at equilibrium:

$$M+L \overset{K_a}{\leftrightarrow} ML \quad K_a=\frac{1}{K_d}=\frac{[ML]}{[M][L]} \tag{2.1}$$

At constant pressure, the binding affinity is dictated by the Gibbs energy (ΔG), so that $K_a = e^{-\Delta G/RT}$, where R is the gas constant and T the absolute temperature. The binding Gibbs energy is a key parameter since its value under a particular set of conditions and reactant concentrations determines the direction of the equilibrium (a spontaneous process is characterized by a negative ΔG value) and the extent of the reaction. ΔG is, in turn, determined by the enthalpy (ΔH) and entropy (ΔS) changes upon binding ($\Delta G = \Delta H - T\Delta S$), which vary with temperature according to the heat capacity change, ΔCp, defined as the temperature derivative of the enthalpy and entropy changes at constant pressure:

$$\Delta C_p=\left(\frac{\partial \Delta H}{\partial T}\right)_P=T\left(\frac{\partial \Delta S}{\partial T}\right)_P \tag{2.2}$$

The Gibbs energy is, thus, completely specified at any temperature T if the enthalpy (ΔH), entropy (ΔS) and heat capacity (ΔCp) changes are know at some reference temperature T_R:

$$\Delta G(T)=\Delta H(T)-T\cdot\Delta S(T) \tag{2.3}$$

$$\Delta H(T)=\Delta H(T_R)+\int_{T_R}^{T}\Delta C_p dT=\Delta H(T_R)+\Delta C_p(T-T_R) \tag{2.4}$$

$$\Delta S(T)=\Delta S(T_R)+\int_{T_R}^{T}\Delta C_p d\ln T=\Delta S(T_R)+\Delta C_p \ln\left(\frac{T}{T_R}\right) \tag{2.5}$$

Traditionally, the binding affinity has been the only discriminating parameter considered in most screening procedures and rational design strategies. This constitutes a very limited perspective of the binding process, since the Gibbs energy alone does not provide information about the nature and magnitude of the forces driving the association. This information is, nonetheless, encoded in the enthalpic and entropic contributions to the Gibbs energy of binding, which arise from different types of interactions and, thus, report on the relative weight of the different intermolecular forces in a particular binding event.

There are two major and opposing contributions to the binding enthalpy [45–47]: a favorable contribution arising from the establishment of hydrogen bonds, van der Waal's contacts and other noncovalent interactions between the binding partners and an unfavorable contribution arising mostly from the desolvation of polar groups at the interacting surfaces (the desolvation enthalpy of apolar groups is one order of magnitude smaller [48]). Thus, a favorable (negative) binding enthalpy indicates that strong interactions are established between the two molecules that compensate the unfavorable enthalpy associated with desolvation. On the other hand unfavorable binding enthalpies indicate that the predominant effect is the desolvation of polar groups that are not implicated in strong interactions in the complex.

In turn, the binding entropy results mostly from the balance between two main contributions: the change in solvation entropy and the change in conformational entropy [38–40]. The solvation entropy term arises from the increment in the configurational degrees of freedom of the solvating water molecules released to the bulk upon burial of the interacting surfaces in the complex. This contribution is often favorable (positive) and large and is the most important force directing the interactions between hydrophobic molecules. On the other hand, since binding usually involves a reduction in the conformational degrees of freedom of the interacting molecules, the conformational entropy contribution generally opposes binding. Additional terms, related to changes in the overall rotational degrees of freedom and the number of particles in solution, also contribute to the binding entropy although in a smaller extent and with very similar values for most binding events [41].

Lastly, the heat capacity is a measure of the enthalpic fluctuations of the system associated to hindered internal rotations, low frequency conformational fluctuations, high frequency bond stretching and bending as well as noncovalent interactions and hydration effects. Even though these last two terms contribute a small factor to the absolute heat capacity of the system, they account for most of the changes in heat capacity associated to binding events. In summary, the binding heat capacity, which is a weak function of temperature, originates mainly from changes in hydration although it also contains small contributions associated to the establishment of noncovalent interactions [49, 50].

Because the binding Gibbs energy (ΔG) can arise from many different combinations of enthalpic and entropic contributions, molecules with the same binding affinity can interact with their partners for very different reasons. This is clearly illustrated in Figure 2.1 for two structurally similar inhibitors that differ in the terminal substituent interacting in the S3/S4 pocket of the thrombin binding site (a cyclopentyl group in molecule 1 and a cyclohexyl group in molecule 2). Although both molecules bind with the same affinity, they present different enthalpic and entropic contributions, indicating that small structural modifications in this region can significantly alter the way in which the molecules interact with the protein. Molecule 1 establishes stronger interactions that result in a more favorable binding enthalpy while the binding of molecule 2 is dominated by entropic contributions. Inspection of the electron density maps for these complexes clearly shows that, in this case, the favorable entropic contributions for molecule 2 are associated to a higher conformational flexibility of the cyclohexyl moiety that does not fit properly into the binding pocket and maintains a significant residual mobility that compensates for the loss of interactions. This and other examples in the literature [51–53], clearly demonstrate that knowledge of the thermodynamic signature of an interaction (i.e. the proportion in which the binding enthalpy and entropy contribute to the binding affinity) provides an invaluable insight into the forces governing the binding process. Additionally, because different thermodynamic signatures reflect distinct chemical properties in the ligands this information can be of use for the identification of the most appropriate leads or the selection of the most adequate

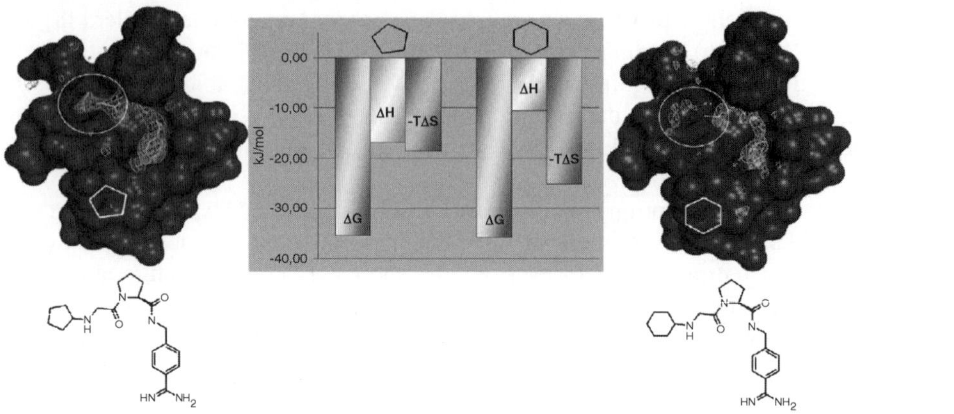

Figure 2.1 *The same binding affinity can arise from different combinations of enthalpic and entropic contributions. The central panel summarizes the binding energetics of two closely related thrombin inhibitors bearing a cyclopentyl or cyclohexyl moiety as terminal substituent that binds at the S3/S4 pocket in the thrombin binding site. Even though both molecules bind with the same binding affinity, significant differences are observed in the entropic and enthalpic contributions to the Gibbs energy. The corresponding crystal structures are shown in which the structure of the binding pocket is shown as a dark gray surface and the electron density maps for both ligands are shown as white chicken-wire contours. (See Plate 1.) (Reprinted from [52] with permission from Elsevier)*

optimization strategy in each situation. Today, it is well established that taking into account thermodynamic considerations can increase the efficiency, robustness and reliability of lead optimization and drug discovery procedures [53–59].

2.4 Thermodynamically Driven Drug Design

The structure-based optimization of a lead compound relies on the modification of its chemical structure with the ultimate goal of improving its binding affinity; that is, making ΔG more negative. Most commonly, initial leads are characterized by weak or moderate binding affinities, typically on the micromolar range ($\Delta G = -8.2\,\text{kcal}\cdot\text{mol}^{-1}$ for a $K_d = 1\,\mu\text{M}$). For a compound to become an effective drug the binding affinity frequently needs to be brought down to the subnanomolar level, which means improving ΔG in more than $6\,\text{kcal}\cdot\text{mol}^{-1}$. This can be achieved by making the binding enthalpy more negative or the binding entropy more positive. An enthalpic optimization will imply introducing polar groups in the molecule that establish strong interactions at the binding interface. An entropic optimization can be based in two different lines of action: (a) reducing the conformational entropic penalty upon binding by restricting the conformational flexibility of the molecule and (b) making the solvation entropy more favorable by increasing the hydrophobicity of the ligand. Because the interactions being optimized in each case are different, the choice of a particular optimization strategy can have important consequences in the properties of the designed drug and in its potential for further optimization. In this sense, the pioneering work carried out by the Freire laboratory during the last decade has provided useful thermodynamic guidelines for drug screening and optimization. [57, 58, 60–62].

2.4.1 Entropic Optimization of Lead Compounds

Optimizing the entropic contributions is a relatively straightforward process. Obtaining improvements in binding affinity by increasing the hydrophobicity of the ligand is easy since the burial of hydrophobic groups results in a significant increase in solvent entropy with very small opposing enthalpic effects, so the entropic benefits translate almost fully to changes in free energy (it has been estimated that the burial of one carbon atom contributes $25\ \mathrm{cal{\cdot}mol^{-1}\ \AA^{-2}}$ to the Gibbs energy of binding [63]). A similar situation is found for the conformational entropy (the immobilization of one rotatable bond in the free ligand will reduce the conformational entropy penalty in $0.5\ \mathrm{kcal{\cdot}mol^{-1}}$ [38]). Consequently, traditional drug design approaches have been mostly directed towards the generation of highly hydrophobic and conformationally constrained molecules. Nonetheless, as pointed out by Freire and coworkers [54, 57, 62], this optimization strategy has important drawbacks. Since the hydrophobic effect is nonspecific, the selectivity in this type of compounds, which bind to the target mostly because of repulsive interactions with the solvent, relies almost exclusively on shape complementarity. Even though this might be enough to confer specificity for some enzymes, it is problematic when trying to specifically target a single member in protein families with highly similar binding sites [64]. Additionally, rigid molecules designed to fit tightly into a specific binding site are very sensitive to small changes in binding site geometry and show a high susceptibility to target heterogeneity. As a consequence, resistance against this type of compounds is readily developed, especially when directed against highly mutable targets such as viral and bacterial proteins. This is the case for the entropically optimized first generation inhibitors of the HIV-1 protease, characterized by markedly favorable entropy changes and unfavorable enthalpic contributions, which are strongly affected by binding site mutations. These rigid inhibitors cannot adapt to the distorted binding site geometry and experience a significant loss in binding enthalpy [65–68]. Increasing conformational flexibility in those regions of the inhibitor matching highly variable sites in the target has been proposed as a feasible strategy for the design of adaptive inhibitors capable of maintaining an adequate level of binding affinity when facing target heterogeneity, either arising from naturally occurring polymorphisms or associated to drug-resistant mutations [62, 64, 69]. These concepts have been successfully applied to the development of high affinity inhibitors of the plasmepsins, a family of highly homologous antimalarial targets [70].

2.4.2 Guidelines for Enthalpic Optimization of Ligands

In any case, the introduction of flexibility in the ligand necessarily implies a loss in binding affinity that would need to be compensated by the engineering of enthalpically favorable interactions. The need for a combined enthalpic and entropic optimization is stressed by the fact that solubility and bioavailability requirements frequently set a limit to the tolerable degree of hydrophobicity of a drug (the maximal binding affinity achievable exclusively from entropic contributions while maintaining a minimal level of solubility has been estimated to be around 55 nM [58]. Consequently, focusing the design process exclusively on the optimization of entropic contributions can easily lead to a dead end. In the practice, to achieve subnanomolar binding, favorable enthalpic and entropic contributions to the Gibbs energy are required [61]. This is clearly illustrated by the evolution of the thermodynamic

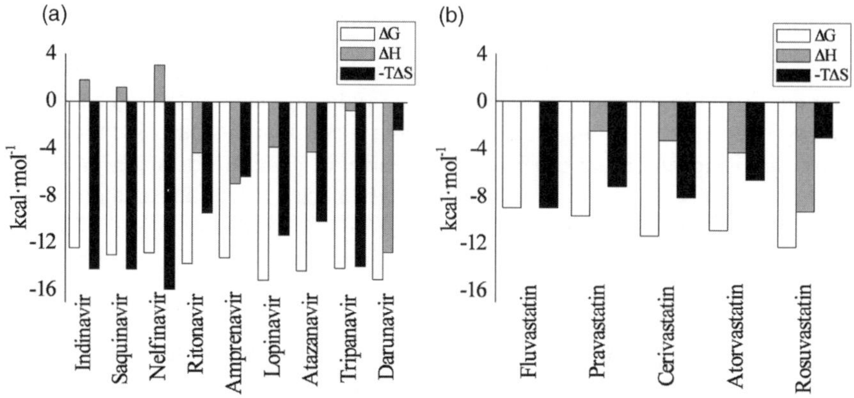

Figure 2.2 *Evolution of thermodynamic signature from first to second generation drugs. Panel A shows the thermodynamic signature for all HIV-1 protease inhibitors approved by the FDA from 1995 (Indinavir) until 2006 (Darunavir) (Data from [58, 69, 166]). This figure illustrates how second generation inhibitors, optimized in terms of binding affinity, selectivity and drug resistance profiles are characterized by favorable enthalpic contributions. A similar situation is found for HMG-CoA reductase inhibitors currently in use as cholesterol lowering drugs. Panel B shows how newer, more potent statins, such as Atorvastatin or Rosuvastatin, are enthalpically optimized with respect to older ones [58, 167]*

signature of HIV-1 protease and HMG-CoA reductase inhibitors presented in Figure 2.2. From this data, it is apparent that highly potent second generation inhibitors resulting from many rounds of optimization over more than ten years and, in many cases, selected by their superior selectivity and drug resistance profiles, show a clear tendency towards a well-balanced thermodynamic signature in which both the binding enthalpy and entropy contribute favorably to the binding affinity [58, 71]. This clearly indicates that those molecular features leading to favorable enthalpic contributions are desirable for the development or selection of the best-behaved ligands.

Nonetheless, contrary to the entropic optimization, rationally engineering new interactions that contribute favorably to the binding enthalpy is not trivial. The establishment of new interactions, such as hydrogen bonds, at the binding interface is associated to a significant enthalpic penalty arising from the desolvation of polar groups, to the point that weak hydrogen bonds can actually contribute unfavorably to the binding enthalpy. Engineering hydrogen bonds strong enough to achieve favorable enthalpic contributions would require a level of precision in the modeling of angles and distances between donors and acceptors generally unattainable for current structure-based design techniques. The situation is further complicated by the fact that even when optimal geometries are achieved, the enthalpic benefits are very frequently counterbalanced by compensatory entropic contributions arising from structuring effects induced by the new interactions on the ligand and/or protein. These effects are sometimes accompanied by suboptimal burial of nonpolar groups that result in further solvent entropy losses. It has been proposed that these enthalpy/entropy compensation effects can be minimized when new interactions are established between conformationally constrained hydrogen bond donors and acceptors in the ligand that target highly structured and stable regions in the protein [60].

The evolution of HIV-1 protease inhibitors towards the thermodynamically balanced second-generation drugs has taken more than 10 years and a huge investment, reflecting the difficulties inherent to the process of enthalpic optimization. Nonetheless, devising adequate screening and design strategies can significantly simplify and accelerate the process. In this sense, it has been proposed that the selection for enthalpically driven leads in high-throughput screening procedures at early stages of the process can facilitate the development of thermodynamically balanced high affinity drugs since the binding affinity of these enthalpic leads could be easily improved by increasing hydrophobicity and introducing conformational constrains [61]. During the last years, the combination of structure-activity relationships with thermodynamic information has emerged as a very potent tool for drug discovery. In this sense, monitoring the enthalpic and entropic impact of the introduction of functionalities at different points of the ligand can aid in the identification of the optimal locations for hydrogen bond donors and acceptors. This is illustrated in Figure 2.3 for a series of plasmepsin II inhibitors [58]. Recent improvements in experimental techniques for the measuring of binding energetics and the development of computational approaches that allow the prediction of the different thermodynamic parameters from structure have generated a very favorable scenario for the implementation of these thermodynamic guidelines into the drug discovery process.

2.5 Measurement of Binding Energetics

2.5.1 Calorimetric vs Noncalorimetric Techniques

The binding affinity, K_a, can be measured using different experimental techniques. Some, like equilibrium dialysis or ultracentrifugation, are based on the direct determination of the concentration of the different species in equilibrium. Others, including most spectroscopic techniques and calorimetry, rely on the measurement of an observable reporting on the degree of completion of the binding reaction, i.e. proportional to the degree of saturation. The latter are the most commonly used in biophysics because they are less time- and sample-consuming.

In addition to the measurement of the binding affinity, a full thermodynamic characterization of a binding reaction requires the determination of the enthalpy and heat capacity changes upon binding. This can be achieved using noncalorimetric techniques by carrying out binding experiments at different temperatures and making use of the van't Hoff relationship for the temperature dependency of the association constant:

$$\frac{\partial \ln K_a}{\partial (1/T)} = -\frac{\Delta H}{R}; \quad \ln\frac{K_{a1}}{K_{a2}} = \frac{\Delta H}{R}\left(\frac{1}{T_1} - \frac{1}{T_2}\right) \tag{2.6}$$

These equations assume ΔH to be temperature independent, which is indeed a rough approximation considering that for most reactions ΔH varies considerably with temperature. Thus, the use of more complicated expressions, including nonzero ΔC_p values, is frequently required.

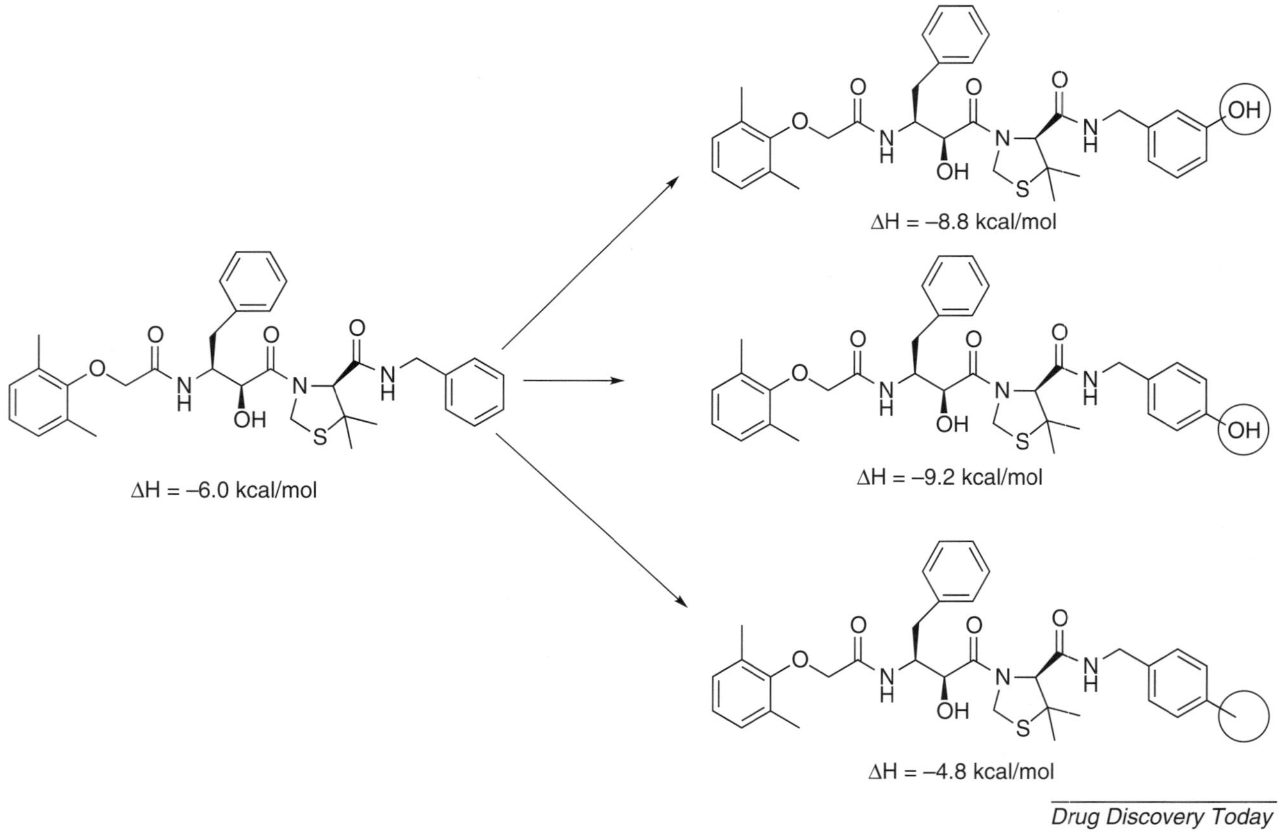

Figure 2.3 *Structure-activity-thermodynamic relationships as potent tools for drug discovery. Monitoring the impact on the binding energetics of the introduction of functionalities at different positions in the ligand can aid in the identification of the optimal locations for hydrogen bond donors and acceptors. In this example, the best location of a hydrogen donor (highlighted by circles) was identified using isothermal titration calorimetry (ITC) for a series of plasmepsin II inhibitors. (Reprinted from [58] with permission from Elsevier)*

$$\frac{\partial^2 \ln K_a}{\partial(1/T)^2} = -\frac{T^2 \Delta C_p}{R}; \quad \ln\frac{K_{a1}}{K_{a2}} = \frac{\Delta H - T_1 \Delta C_p}{R}\left(\frac{1}{T_1} - \frac{1}{T_2}\right) + \frac{\Delta C_p}{R}\ln\frac{T_2}{T_1} \tag{2.7}$$

In any case, the estimation of ΔH and ΔCp of binding from a van't Hoff analysis presents serious problems. Even though the enthalpy and entropy of binding are strongly temperature dependent, because of enthalpy/entropy compensation effects the binding affinity does not change considerably over the narrow temperature range accessible experimentally. This, together with the inherent experimental error associated to the measurement of binding affinities that propagates into large errors in ΔH, ΔS and ΔC_p and the fact that, frequently, the estimated values for ΔH and ΔC_p are strongly dependent upon each other, has led to serious discrepancies between the calorimetric and van't Hoff binding enthalpies [72–75].

In contrast, Isothermal Titration Calorimetry (ITC) can provide direct and accurate information about the binding affinity (ΔG), the binding enthalpy (ΔH) and the stoichiometry of the complex (n) from a single titration experiment. From this data the change in entropy upon binding (ΔS) can be easily derived. Also, the binding heat capacity, ΔC_p, can be accurately determined by performing titrations at different temperatures. Nonetheless, the application of calorimetric techniques to the study of molecular recognition and drug design has been very limited until recently due mainly to the lack of commercially available microcalorimeters that allowed a fast, reliable and accurate thermodynamic characterization of biological systems with small sample requirements and little technical expertise. The first commercial instruments became available a decade ago and today most modern isothermal titration calorimeters (e.g. VP-ITC from Microcal (http://www.microcalorimetry.com) and INC from Calorimetry Sciences (http://www.calscorp.com) require less than 500 µg of protein to carry out a complete calorimetric titration and are characterized by accuracies in the determination of the thermodynamic parameters better than 0.1 kcal·mol^{-1}. Additionally, attempts are being made to improve the throughput of calorimetric techniques by generating fully automated instruments that allow fast characterization of a high number of samples (e.g. auto-iTC$_{200}$ from Microcal) or by developing miniature devices with small sample requirements [76, 77].

2.5.2 Principles of Isothermal Titration Calorimetry

ITC measures directly the heat associated with a chemical reaction triggered by the mixing of two components. Detailed descriptions of the instrument design, principles of microcalorimetry and operational details have been described elsewhere [55, 59, 78–82]. Briefly, an ITC experiment is carried out by the stepwise addition at constant temperature of a small volume (typically a few micro-liters) of one binding partner into the reaction cell (aprox. 1 mL) containing the other. Figure 2.4 shows an example of a typical ITC experiment. After each injection, a certain amount of heat is either released or absorbed that is proportional to the amount of ligand bound to the protein. What is actually measured in most modern calorimeters is the power required to maintain a constant temperature difference between the reaction cell (generally containing the protein) and a twin reference cell filled with buffer. An exothermic reaction generates heat and, consequently, induces a decrease in power while endothermic reactions increase the feedback power. The total heat associated to each injection is derived from the area under each peak (upper panel of

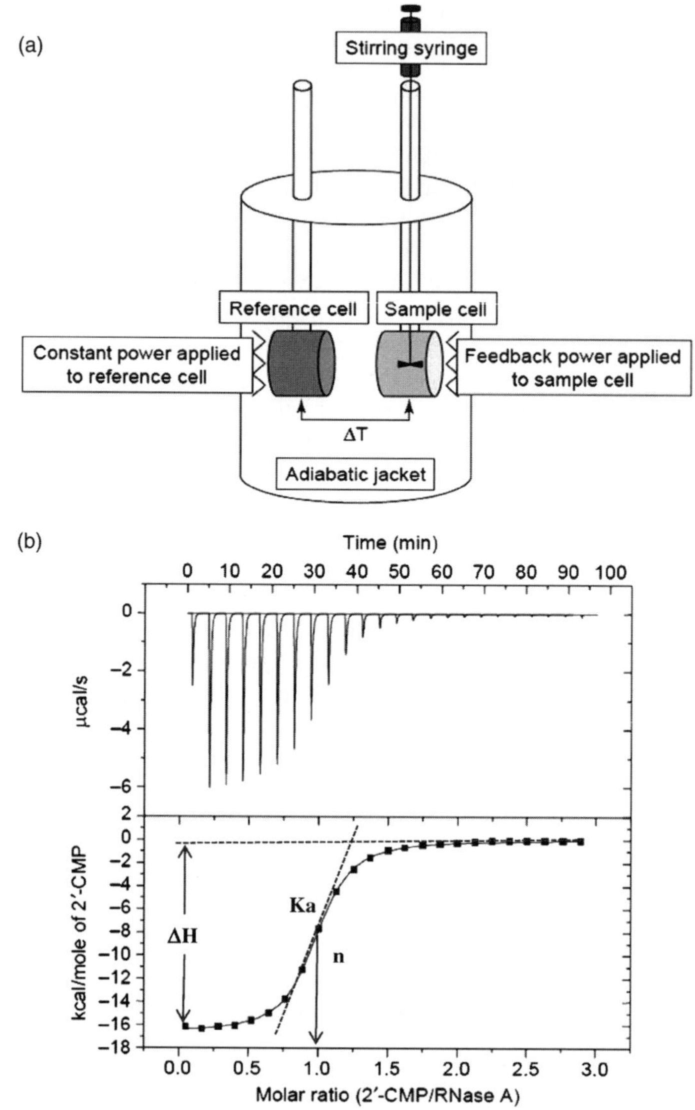

Figure 2.4 *ITC measurement. (a) Schematic representation of a power compensation Isothermal Titration Calorimeter. (b) Typical ITC experiment corresponding to the titration of 2′-CMP into RNAse A. The top panel shows the power applied by the instrument throughout the titration experiment. In the first injections most of the ligand is bound to the protein, as the titration progresses the protein becomes increasingly saturated so less binding occurs and the size of the peaks decreases. When the protein is saturated the heat of dilution is measured. The bottom panel shows the normalized heats obtained from the integrals of the peaks from the top panel together with the best fit corresponding to a model of one set of sites, from which the stoichoimetry (n), the binding affinity (Ka) and the binding enthalpy (ΔH) can be estimated. (Adapted from [59] with permission from Elsevier)*

Figure 2.4b). By integrating each of the peaks a binding isotherm is obtained, most frequently in terms of normalized heat per mole of injected ligand versus the molar ratio of the two binding partners (shown in the lower panel of Figure 2.4b). The shape of the binding isotherm is closely related to the binding affinity (K_a) while its scale is proportional to ΔH. Also, from the midpoint of the sigmoidal binding curve the binding stoichoimetry, n, can be obtained. The specific values for the different parameters are usually estimated by nonlinear regression analysis of the binding isotherm according to an adequate binding model. Some relevant aspects related to the fitting of ITC data can be found in [83–86].

2.5.3 The ITC Experiment

With currently available instrumentation, it is possible to estimate both K_a and ΔH with standard errors below 1%. Nonetheless, this requires a careful design of the titration experiment as well as extreme care in sample preparation and data analysis. Even though the nature of the system under study may dictate in good extent the error with which the different thermodynamic parameters can be determined, there are several aspects of the experimental setup that are under the direct control of the experimentalist. These include the concentration of the reactants, the injection volume and the number of injections. Careful choice of these parameters can significantly improve the accuracy in the determination of the thermodynamic parameters [81, 87–89].

To obtain reliable binding constants, it is important that the concentrations of the binding partners are such that both the free species and the complex are significantly populated during the experiment. This condition is met when the c-value (defined as the product between the binding site concentration and the association constant, $c = [M] \cdot K_a$ [79]) is between 1 and 1000. This ensures that the curvature of the titration isotherm is adequate for the accurate determination of K_a. In this way, for a protein concentration in the cell of 10 μM direct titration experiments allow the reliable measurements of binding constants ranging between 10^4 and $10^8\,M^{-1}$. In this optimal range of binding affinities, K_a, ΔH and n can be simultaneously and accurately determined from a single titration experiment. Nonetheless, outside this interval of binding affinities the situation is more complicated since the protein and ligand concentrations required to obtain reasonable c values are not easily accessible. Thus, for tight binding complexes with K_a values over $10^8\,M^{-1}$ the binding isotherm becomes too sharp for the determination of the binding affinity, although ΔH and n can still be determined very accurately. In the opposite situation, for c values below 1, the binding isotherm becomes too shallow, full saturation is difficult to approach and the ΔH and n values obtained from the analysis are strongly correlated, precluding the reliable determination of the binding parameters.

The use of properly designed profiles of variable injection volumes in the titration experiments have been shown to be useful to improve the shape of the binding isotherms in low affinity systems (c values below 1) allowing a better definition of the initial regions and a higher degree of saturation [89–91]. Also, systems characterized by very high or low binding affinity can, in principle, be brought into the optimal affinity region by changing experimental conditions, such as temperature or pH [92, 93]. Nonetheless, this approach is not always effective due to the weak temperature-dependence of the binding affinity and the sensitivity of biological macromolecules to changes in pH. Alternatively, the binding affinity can be modulated by the coupling to a second binding equilibrium, so that competition experiments can be performed in which a high-affinity ligand is titrated into protein saturated with a weaker one or vice-versa. Rigorous protocols for the analysis of this type of displacement titrations have been recently developed [94] and successfully implemented for several high and low affinity systems [54, 55, 82, 95].

From a different perspective, it is important to keep in mind that the heat measured in an ITC experiment is a global property of the whole system. As a consequence, the measured binding enthalpy ($\Delta H_{apparent}$) might contain, in addition to the heat associated to the binding reaction, other contributions arising from additional heat generating processes. These include contributions from nonspecific effects associated to the heats of dilution of the ligand and macromolecule solutions, mechanical heats derived from stirring as well as other effects arising from mismatches in temperature or composition between the solutions in the sample cell and the syringe. To minimize these effects it is important to extensively dialyze the samples prior to the experiment and to carry out control dilution experiments that can be used to correct the normalized heats in the binding isotherm. The most important effects are those associated with the dilution of the ligand that can be determined by performing an identical titration experiment into a sample cell containing only dialysis buffer.

Additionally, many binding reactions are coupled to the protonation or deprotonation of ionizable groups both in the protein or the ligand that are associated to significant enthalpic contributions that would be reflected in the measured apparent binding enthalpy. Once corrected for dilution effects, the calorimetric binding enthalpy, ΔH_{app}, can be decomposed in:

$$\Delta H_{app} = \Delta H_{bind} + n_H \cdot \Delta H_{ion}, \tag{2.8}$$

where ΔH_{bind} is the enthalpy associated with the reaction that is independent of the buffer used in the experiment although pH dependent, n_H is the number of protons absorbed (if positive) or released (if negative) upon formation of the complex, and ΔH_{ion} is the enthalpy of ionization of the buffer used in the titration experiment. The influence of coupled ionization equilibria can be evaluated experimentally by performing ITC titrations in several buffers with different ionization enthalpies (ΔH_{ion}), so that the number of protons exchanged upon binding, $n_{H,}$ and the intrinsic binding enthalpy ΔH_{bind} can be obtained, respectively, from the slope and the y-axis intercept in a representation of ΔH_{app} vs. ΔH_{ion} [96–98]. The values of ΔH_{ion} for most buffers have been tabulated [99] or can be measured by ITC [100]. Additionally, by carrying titrations at different pH values, a pH-independent value for ΔH_{bind}, corrected for the enthalpy of ionization of the groups in the protein and ligand, can be obtained. This type of analysis, together with structural information, can provide relevant information about the identity of the groups undergoing ionization events [98].

2.6 Structure-based Calculation of Protein Binding Energetics

In addition to the experimental determination of the binding energetics, the ability to accurately predict thermodynamic parameters from structure is important for virtual screening and structure-based lead optimization. Over the years, a number of potential functions have been developed for the calculation of binding Gibbs energy from structural considerations that include force-field methods, empirical scoring functions and knowledge-based potentials (see [20, 101] for recent reviews). Ideally, it would be desirable that these functions were able to predict binding affinities with an accuracy similar to experiments (about 1 kcal·mol^{-1}). However, this is rarely achieved. Important limitations arise from the fact that, frequently, the Gibbs energy of binding is calculated as the difference between two large and opposing contributions: one related to the interactions established between the ligand and the protein in vacuum or implicit solvent and the other containing apolar solvation effects. Because of the difficulties inherent to the evaluation of the energetic contributions of each individual interaction (for example, the strength of a particular hydrogen bond is strongly dependent on its environment) and the contributions from solvation effects, the errors affecting both terms are very high with respect to the magnitude of the Gibbs energy itself. Moreover, since ΔG is calculated directly, these approaches do not generally provide information about the enthalpic and entropic components to the binding affinity or its temperature dependency. As we have mentioned before, being able to accurately predict binding enthalpies from structure is crucial to tackle the enthalpic optimization of a lead compound by structure-based rational design.

In this sense, the development of accurate empirical parameterization based on structure/thermodynamic correlations constitutes an interesting alternative for the prediction of the different thermodynamic parameters from structure [102, 103]. These phenomenological approaches do not require a detailed knowledge of the different forces and interactions involved in the stabilization of proteins or protein complexes but rely on the statistical analysis of data sets containing high-quality structural and thermodynamic information. This is the case of the structural parameterization of the protein folding energetics developed by the Freire laboratory, which is based on the separate parameterization of the three components of the Gibbs energy: the enthalpy, entropy and heat capacity changes in terms of structural parameters such as changes in solvent accessible surface area or packing densities [39, 41, 45, 49, 104]. Conformational entropy contributions, which do not scale in terms of ASA, are evaluated explicitly from tabulated values derived from the computational analysis of model systems [39]. Even though these functions do not consider the contributions of specific elementary forces explicitly, they have been shown to effectively capture the magnitude of the different energetic terms. This approach has been successfully applied to protein stability and cooperativity [45, 105–108] and its predictive ability has been validated by several laboratories [105, 106, 109, 110]. This parameterization has also been successfully applied to the estimation of the binding energetics in some systems [65, 98, 111–114].

In the protein folding parameterization, the enthalpy change is calculated as a function of changes in solvent accessibility and atomic packing densities [45] according to the equation:

$$\Delta H(T) = a(T) \cdot \Delta ASA_{ap} + b(T) \cdot \Delta ASA_{pol}, \tag{2.9}$$

where ΔASA_{ap} and ΔASA_{pol} are the changes in solvent accessible surface area for nonpolar and polar atoms and a(T) and b(T) are empirically determined coefficients obtained from the analysis of the structural-thermodynamic databases. Analysis of 60 proteins yielded a value of a $(60\,°C) = -1.9 \pm 2.6\,\text{cal·mol}^{-1}\text{·Å}^{-2}$ and $b(60\,°C) = 20.6 \pm 4.1\,\text{cal·mol}^{-1}\text{·Å}^{-2}$ [106]. This simple parametric equation is one of the most accurate ways to predict the enthalpy changes associated to protein folding, with typical standard errors on the order of $4.8\,\text{kcal·mol}^{-1}$. Even though this level of accuracy is more than acceptable for protein folding (characterized by enthalpic changes on the order of $120\,\text{kcal·mol}^{-1}$), it is too high for the reliable estimation of binding enthalpies, of much smaller magnitude. Also, there are additional factors, not relevant for folding but very important for binding, which need to be taken in to account.

To address these issues, a binding-specific parameterization has been developed that considers the experimentally measured binding enthalpy as composed of at least three terms: (a) the intrinsic binding enthalpy that reflects the nature of the interactions between ligand, target and solvent; (b) the enthalpy associated with any possible conformational change in the protein or the ligand upon binding; and (c) the enthalpy associated with protonation/deprotonation events coupled to the association.

$$\Delta H_{app} = \Delta H_{binding} + \Delta H_{protonation} = \Delta H_{\text{int}\,rinsic} + \Delta H_{conformation} + \Delta H_{protonation} \tag{2.10}$$

$\Delta H_{protonation}$ needs to be independently determined by carrying out experiments at several buffers with different ionization enthalpies, as previously described. In the absence or after correction of ionization contributions, the binding enthalpy can be accounted for in terms of an intrinsic binding enthalpy and a conformational enthalpy change. $\Delta H_{intrinsic}$ is parameterized in terms of changes in accessible surface area and packing density in a similar way as the folding enthalpy, although, in this case, changes in accessible surface area need to be calculated taking into consideration buried water molecules at the binding interface. $\Delta H_{conformation}$ can be estimated from the global analysis of a set of complexes in which different ligands induce the same bound conformation in their target. This simple approach allows the calculation of the binding enthalpy of small ligands with a much higher accuracy (the binding enthalpy for a set of 25 small molecular weight peptide and nonpeptide ligands in seven different protein systems was calculated with a standard error of $0.3\,\text{kcal·mol}^{-1}$ [47]. Thus, its incorporation to rational design strategies will contribute to the development of more efficient, thermodynamically guided, ligand design strategies.

2.7 Interfacial Water Molecules in Protein Recognition

Even though proteins and other biological macromolecules carry out their functions in water, the role of solvent in the analysis of their stability, dynamics and functional properties have been traditionally ignored. Nonetheless, during the last decade, an increasing number of works have been reported about the structural details of water and their behavior at protein surfaces and cavities [115, 116] as well as about their impact in the stability and specificity of protein–protein and protein-ligand complexes [117–120] using a wide variety of experimental and computational techniques [121–125]. These reports have made patent the versatility and relevance of protein-solvent interactions, beyond the widely studied hydrophobic effect.

With respect to protein recognition, it was common practice until very recently to completely disregard the presence of interfacial water molecules at the binding interfaces of protein complexes. Nonetheless, as the number of high resolution crystal structures of protein complexes increased it became evident that buried water molecules are very frequently found at the binding interfaces (on average, one water molecule is observed per 100 Å^2 in protein–protein complexes [4]) mediating the interactions between the binding partners by the establishment of multiple hydrogen bonds. In fact, it has been shown that water-mediated hydrogen bonds are as frequent as those established directly between protein and ligand to the point that 40% of the residues at the binding interfaces interact through water-mediated interactions [126]. These results clearly indicate that water, far from being a mere inert matrix, is part of the recognition code and needs to be considered for a complete and precise description of the binding interfaces. This is clearly illustrated by the recognition of proline-rich sequences by SH3 domains that traditionally have been presented as highly hydrophobic interactions, although, contrary to what would be expected for hydrophobic binding, these complexes are invariantly characterized by highly favorable binding enthalpies. A detailed structural and thermodynamic analysis has revealed that interfacial waters are consistently found at SH3 complexes (Figure 2.5) and constitute a critical factor for the understanding of their binding energetics [91, 127, 128]. A database of protein complexes including water-mediated interactions in the definition of binding interfaces have been recently developed (www.scowlp.org) [126, 129]. It is interesting to note that water molecules at protein–protein interfaces are frequently organized in a ring around the center of the interface that remains dry [121].

The role of interfacial water molecules and their impact on the binding process has been widely studied from a thermodynamic and functional standpoint. Interfacial water

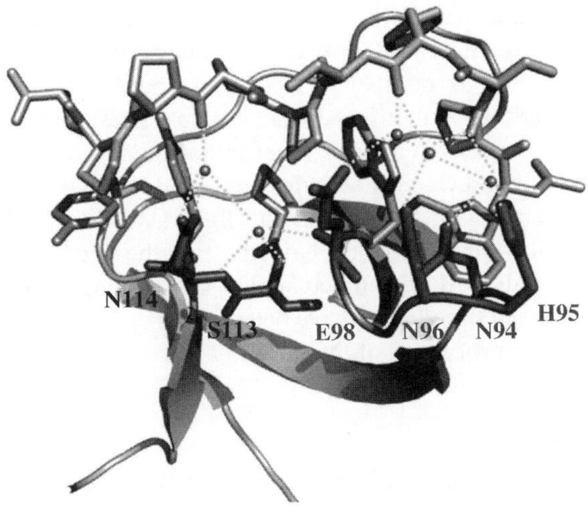

Figure 2.5 *Water molecules at the Abl-SH3/p41 binding interface. The structure of the Abl-SH3 domain is shown in a grey cartoon representation. Residues defining the canonical binding site for polyproline recognition are shown as grey sticks. The structure of the p41 peptide is shown as white sticks. Fully buried water molecules at the binding interface are shown as light grey spheres. Water-mediated hydrogen bonds are depicted as dotted light grey lines (See Plate 2.)*

molecules act as adaptors filling empty spaces and optimizing van der Waals interactions, participate in the establishment of hydrogen bonds between the binding partners and assist in the dissipation of charges. All these terms are expected to contribute favorably to the binding enthalpy. Conversely, fixing a water molecule at the binding interface entails a considerable entropic penalty that counterbalances these favorable enthalpic effects so that, generally, the impact of interfacial water molecules on the Gibbs energy of binding is modest [47, 130]. A wide spectrum of energetic contributions have been theoretically predicted [131–134] and reported experimentally (see [135] for a recent review) for water-mediated interactions that seem to be strongly dependent on the specific characteristics of the system making patent the need for a detailed structural, thermodynamic and dynamic analysis for each specific situation.

Recently, new docking and rational design algorithms have been developed that allow the consideration of water mediated interactions, showing a significant improvement in the predictions when these are included in the calculations [47, 136–142]. Interfacial water molecules can be used in two main ways in rational design: (a) by modifying the lead towards optimization of its interactions with existing water molecules at the binding interface and (b) by introducing new moieties on the ligand that displace interfacial waters from the binding site releasing them into the bulk solvent. In the first case, a significant energetic benefit will only be obtained if the enthalpic benefits associated to the establishment of new interactions overcome the entropic penalties arising from structuring effects. This could only be achieved when these new interactions are directed against highly immobilized water molecules coordinated by very stable regions of the protein [60, 127] that can be considered as an integral part of the protein. Conversely, these tightly bound water molecules are bad candidates for displacement. Recently, several algorithms, such as Consolv [143], Waterscore [141], HINT/RANK [144, 145], have been presented for the classification of interfacial water molecules and the selection of those water molecules to be maintained at the binding interfaces and those that could be readily eliminated from the binding site according to different indicators such as number of hydrogen bonds, crystallographic B factors, proximity to polar residues or geometrical considerations. These algorithms indicate that, in general, the free energy associated to each water molecule depends on the environment so that, frequently, tightly bound water molecules are found in polar cavities and establish at least three hydrogen bonds with the protein and ligand, while weakly bound waters are typically found in apolar environments. Nonetheless, the predictive capacity of these methodologies is still limited. More detailed analysis combining structural, thermodynamics and molecular dynamics simulations have been shown to provide a much more accurate description of the specific properties of the different interfacial water molecules and their role in a particular complex [127, 146].

2.8 The Linkage Between Binding and Conformational Equilibrium in Proteins

The ability of proteins to bind multiple targets and propagate binding information to distal regions in the molecule by cooperative interactions is crucial for the establishment and regulation of complex interactions networks and signal transduction. These functional properties of proteins (transmission of binding effects, cooperative interactions, molecular signalling and allosterism) are encrypted in the distribution of stabilizing interactions throughout the protein that determines, not only the overall structure and stability of the molecule, but also the propensity of the different regions in the molecule to undergo conformational changes [147]. As a consequence, understanding the mechanisms eliciting conformational changes or cooperativity between different sites in the proteins associated to a molecular recognition event is of relevance for the development of inhibitors or modulators of a specific signal transduction pathway.

2.8.1 The Native State Ensemble

Proteins are not rigid molecules that adopt a unique conformation. On the contrary, as revealed by an increasingly high number of NMR detected hydrogen/deuterium exchange studies [148, 149], under native conditions proteins must be described in terms of a statistical ensemble of conformational states resulting from the occurrence of local unfolding events scattered throughout the molecule. These local unfolding reactions, which can involve only a few amino acids, are characterized by a Gibbs energy change that determines their relative probability under equilibrium conditions, according to a Boltzmann distribution:

$$P_i = \frac{\exp(-\Delta G_i/RT)}{Q} = \frac{\exp(-\Delta G_i/RT)}{\sum_j \exp(-\Delta G_j/RT)}, \tag{2.11}$$

where $\exp(-\Delta G_i/RT)$ is the statistical weight for state i and Q is the partition function defined as the sum of statistical weights for all states in the conformational ensemble. This means that, under native conditions, proteins are not globally cooperative and, thus, the main equilibrium is not established exclusively between the native and the unfolded state but, on the contrary, it involves a large number of conformational states of different energetic and structural and functional properties [149–153]. A direct consequence of the statistical nature of the native state is that the Gibbs energy of stabilization of a given protein is not evenly distributed throughout the structure. Consequently, within the same protein there are regions of low stability that readily undergo conformational fluctuations and regions of high stability that are rarely involved in local unfolding events. In the context of this statistical distribution, it is possible to define a stability constant per residue as the ratio between the summed probabilities of all conformational states in which a particular residue j is folded (Pf_j) over the summed probabilities of those states in which this residue is not folded (Pnf_i) [108, 152, 153]:

$$k_{f,j} = \frac{\sum_i Pf_{j,i}}{\sum_i Pnf_{j,i}} \tag{2.12}$$

The analysis of the structural distribution of protein stability can be addressed experimentally by measuring hydrogen exchange protection factors by NMR that, under the appropriate conditions, are related to the individual stability constants [108]. Also, computational algorithms have been developed that allow the estimation of the individual stability constants from structure. This is the case of the COREX algorithm (http://www.bmb.

Figure 2.6 The native state ensemble for the SH3 domain of alpha-spectrin. Shown are the most probable conformations according to the COREX algorithm. Only states with Gibbs energies lower than 5 kcal·mol-1 have been depicted. Native regions are shown in gray while regions unfolded in each state are shown in white. States are ordered according to their Gibbs energies. (Reprinted with permission from [150]. Copyright 1999 American Chemical Society)

utmb.edu/hilser/corexbest.htm) that uses the high resolution structure of the protein as a template to generate a conformational ensemble assuming that folded regions in the different conformational states maintain a native-like conformation [108, 152, 153]. The probabilities of the different states within this conformational distribution are estimated using the structural parameterization of the folding energetics described above. This is illustrated in Figure 2.6 for the SH3 domains of alpha spectrin. In spite of the drastic assumptions implicit, this algorithm has been shown to quantitative account for hydrogen exchange data obtained for several proteins [108, 150, 152–155].

Considering that the high plasticity of protein–protein interaction surfaces is one of the main problems encountered in the development of small molecule inhibitors of protein–protein interactions, the ability to identify which residues are likely to maintain the same conformation in the free and bound states and which are more prone to undergo conformational changes is of obvious interest.

2.8.2 The Structural Stability of Binding Sites

From a biological perspective, it is also important to assess whether the lack of global cooperativity in proteins and the uneven distribution of structural stability respond to specific functional requirements. Recent works have clearly shown that this is indeed the case. For example, the application of the COREX algorithm to a set of structurally different protein complexes revealed that in all cases studied the binding sites have a dual character and contains regions of high stability and regions of low stability [156] (Figure 2.7). This setup

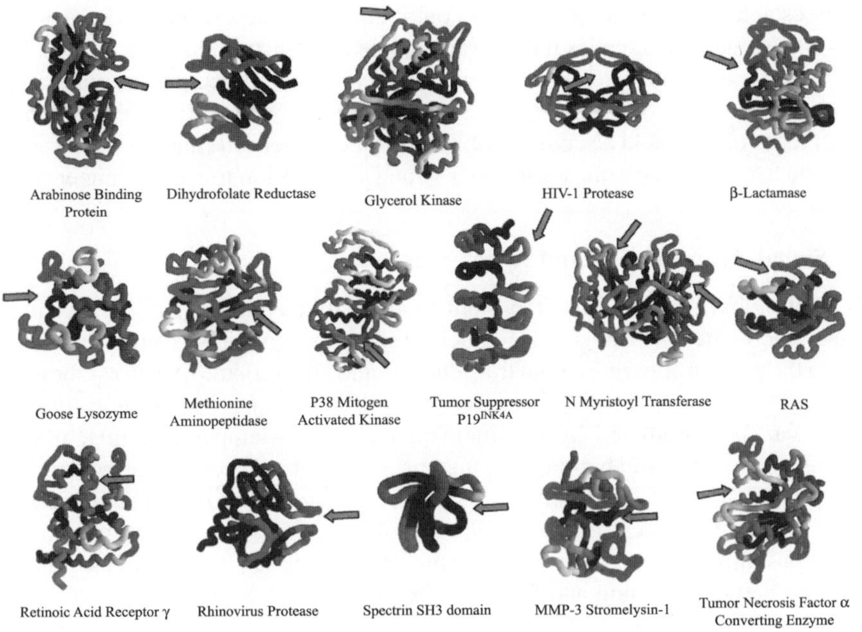

Figure 2.7 *The distribution of structural stability for 16 different proteins. The structures have been color-coded according to the magnitude of the individual stability constants per residue using a normalized grey scale ranging from 1 to 100 according to the overall Gibbs energy of stabilization for each protein. Dark grey regions to residues with stability constants greater than 75. Medium grey residures define the middle of the scale and correspond to residues with stability constants of 50. The arrows indicate the location of the binding sites. (See Plate 3.) (Reprinted with permission from [156]. Copyright 2000 John Wiley & Sons., Inc)*

provides important functional advantages. In many cases, low stability regions correspond to loops that become stabilized upon binding, shielding the ligand from the solvent and, thus, providing the means to obtain high affinity binding for small molecules. This dual character can, thus, be an adequate discrimination criterion for binding site identification. Moreover, from the drug development perspective, careful consideration should also be paid to the structural stability of the different binding site regions. As we have mentioned before, in the process of enthalpic optimization of lead compounds targeting regions of high structural stability for the establishment of new interactions, such as hydrogen bonds, will maximize the energetic benefits by reducing undesirable enthalpy/entropy compensation effects.

In most protein–protein interactions the binding energy is generally not evenly distributed throughout the binding interface, so some residues (hotspots) have been found to contribute very strongly to the Gibbs energy of binding while other make only a marginal contribution [157–159]. It has been reported that, on average, about 10% or the residues contribute more than $2\,kcal{\cdot}mol^{-1}$ to the binding affinity and can be, thus, classified as hot spots [160] Systematic analysis show that hot spots show nonrandom composition (aromatic residues and arginine being significantly predominant) and are highly shielded from the solvent in the complex by surrounding residues forming an o-ring structure ([159] and references therein). All this will favour the establishment of strong enthalpic contributions, but these would not have a significant impact on the binding energetics if they were coupled to structuring effects in highly flexible regions of the protein. Interestingly, hot spots are generally well structured and preferably found in preexisting pockets in the free protein that are not affected by ligand induced conformational changes [161–163]. It could be argued that, among other factors, the high structural stability of these residues is, thus, determinant for their strong contribution to the binding energy.

2.8.3 Signal Transduction and Allosterism

The lack of global cooperativity in proteins under native conditions raises also important questions about the mechanisms by which binding information is propagated to distal regions in the protein allowing signal transduction and allosterism. To address these issues it is important to consider that the energies of the different states within the conformational ensemble can be modulated by environmental changes (temperature, pH, presence of denaturing agents or ligand binding), so that the most probable conformational distribution under a particular set of conditions may not be the same under different conditions. The influence of these physical or chemical environmental factors on the properties of the statistical ensemble can be formulated by linkage equations [147, 149, 164] that, for the relationship between binding and folding have the form:

$$\Delta G_i = \Delta G_i^0 - RT\ln\frac{(1+K_{B,i}[L])}{(1+K_{B,0}[L])} \qquad P_i = \frac{\exp(-\Delta G_i^0/RT)\cdot\frac{(1+K_{B,i}[L])}{(1+K_{B,0}[L])}}{\sum_j \exp\left(-\Delta G_j^0/RT\right)\cdot\frac{(1+K_{B,j}[L])}{(1+K_{B,0}[L])}}, \tag{2.13}$$

where ΔG_i^0 is the Gibbs energy of state i in the absence of ligand, $K_{B,0}$ is the binding constant of the reference state (in our case the native fully folded conformation), $K_{B,i}$ is the binding constant for state i and [L] is the concentration of free ligand. According to this equation each state will be stabilized by the presence of the ligand depending on its binding affinity. Because those states with a fully competent binding site will be preferentially stabilized with respect to others showing partially unfolded or distorted binding sites, the binding of the ligand not only will induce a change in energy but also can result in a significant redistribution of the population of states reflected in changes in the average properties of the conformational ensemble.

This has important implications for signal transduction, cooperativity and allosteric regulation. Even though ligands interact with a small subset of residues at the binding site, binding effects propagate to distal regions in the molecule, not in direct contact with the ligand. It has been observed that this propagation takes place through a small subset of residues critical for the transmission of binding information, reflecting the lack of global cooperativity under native conditions and the uneven distribution of cooperative interactions within the protein. Thus, different sites in the protein (regulatory and catalytic sites in allosteric proteins, for example) are communicated by a small number of residues defining cooperative pathways for the transmission of binding information. The linkage equations have been implemented into the CORE_BIND algorithm that, as illustrated in Figure 2.8, allows the evaluation of the

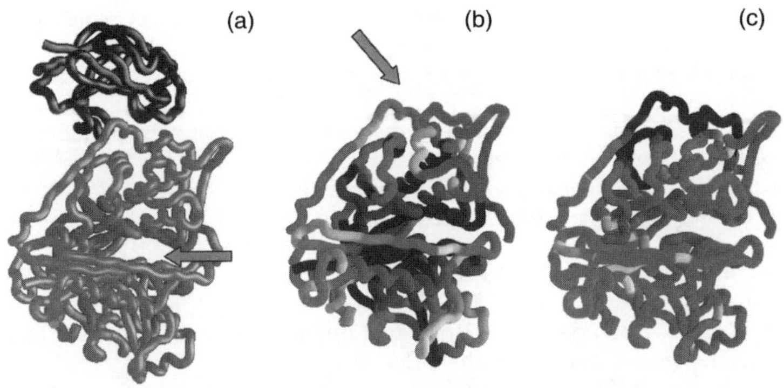

Figure 2.8 *(A) The complex between Glycerol Kinase (medium grey) and the allosteric regulator IIAGlc (dark grey). The arrow indicates the location of the catalytic site. (B) The structural distribution of the stability of unbound Glycerol Kinase. The residues at the binding site for IIAGlc are intrinsically unstable and do not interact strongly with the rest of the molecule in the absence of the regulator. (C) Effect on the structural stability of the binding of IIAGlc. The structure of Glycerol Kinase has been color coded in a grey scheme according to the changes in stability constants induced by the binding of IIAGlc (dark grey regions are the most affected and light grey regions are unafected). Binding of the regulator triggers the propagation of cooperative interactions trough a stretch of residues that connect the regulatory and catalytic domains. (Reprinted with permission from [156]. (See Plate 4.) Copyright 2000 John Wiley & Sons., Inc)*

propagation of binding induced cooperative effects and has been successfully applied to the identification of cooperative pathways within several proteins [149, 151, 156, 165].

Ligands induce conformational changes or transitions between different functional states by modulating the probability distribution of conformational states by selectively stabilizing certain conformations over others. This would only take place if the binding competent conformations are not significantly populated in the absence of the ligand. This is accomplished when some regions of the binding site show low structural stability or exist in a nonbinding conformation in the unliganded protein [164]. The same stands for allosteric regulation that requires the correlation between the structural stability of the active and regulatory sites. The switch between the two functional states implies the existence of two functionally different subensembles, one noncompetent for effector binding and predominantly populated in the free state and the other, binding competent, that becomes selectively stabilized in the presence of the effector. This clearly indicates that the dual character of the binding sites not only provides advantages in terms of binding affinity, as described before, but also in terms of functional cooperativity. In this sense, knowledge of the cooperative pathways within a protein and the structural stability properties of the different components of the interaction surface is relevant for ligand design.

References

1. Smith, R.D., *et al.*, Exploring protein-ligand recognition with Binding MOAD. *J Mol Graph Model*, 2006. **24**(6): 414–25.

2. Cheng, A.C., *et al.*, Structure-based maximal affinity model predicts small-molecule druggability. *Nat Biotechnol*, 2007. **25**(1): 71–5.
3. Jones, S. and J.M. Thornton, Principles of protein–protein interactions. *Proc Natl Acad Sci USA*, 1996. **93**(1): 13–20.
4. Lo Conte, L., C. Chothia, and J. Janin, The atomic structure of protein–protein recognition sites. *J Mol Biol*, 1999. **285**(5): 2177–98.
5. Cochran, A.G., Antagonists of protein–protein interactions. *Chem Biol*, 2000. **7**(4): R85–94.
6. Fletcher, S. and A.D. Hamilton, Targeting protein–protein interactions by rational design: mimicry of protein surfaces. *J R Soc Interface*, 2006. **3**(7): 215–33.
7. Wells, J.A. and C.L. McClendon, Reaching for high-hanging fruit in drug discovery at protein–protein interfaces. *Nature*, 2007. **450**(7172): 1001–9.
8. White, A.W., A.D. Westwell, and G. Brahemi, Protein–protein interactions as targets for small-molecule therapeutics in cancer. *Expert Rev Mol Med*, 2008. **10**:e8.
9. Spencer, R.W., High-throughput screening of historic collections: observations on file size, biological targets, and file diversity. *Biotechnol Bioeng*, 1998. **61**(1): 61–7.
10. Hopkins, A.L. and C.R. Groom, The druggable genome. *Nat Rev Drug Discov*, 2002. **1**(9): 727–30.
11. Arkin, M., Protein–protein interactions and cancer: small molecules going in for the kill. *Curr Opin Chem Biol*, 2005. **9**(3): 317–24.
12. Fry, D.C., Protein–protein interactions as targets for small molecule drug discovery. *Biopolymers*, 2006. **84**(6): 535–52.
13. Fry, D.C., Drug-like inhibitors of protein–protein interactions: a structural examination of effective protein mimicry. *Curr Protein Pept Sci*, 2008. **9**(3): 240–7.
14. Fletcher, S. and A.D. Hamilton, Protein–protein interaction inhibitors: small molecules from screening techniques. *Curr Top Med Chem*, 2007. **7**(10): 922–7.
15. Yin, H. and A.D. Hamilton, Strategies for targeting protein–protein interactions with synthetic agents. *Angew Chem Int Ed Engl*, 2005. **44**(27): 4130–63.
16. Xu, D., S.L. Lin, and R. Nussinov, Protein binding versus protein folding: the role of hydrophilic bridges in protein associations. *J Mol Biol*, 1997. **265**(1): 68–84.
17. Hu, Z., *et al.*, Conservation of polar residues as hot spots at protein interfaces. *Proteins*, 2000. **39**(4): 331–42.
18. Burley, S.K. and G.A. Petsko, Weakly polar interactions in proteins. *Adv Protein Chem*, 1988. **39**:125–89.
19. Leckband, D. and J. Israelachvili, Intermolecular forces in biology. *Q Rev Biophys*, 2001. **34**(2): 105–267.
20. Gohlke, H. and G. Klebe, Approaches to the description and prediction of the binding affinity of small-molecule ligands to macromolecular receptors. *Angew Chem Int Ed Engl*, 2002. **41**(15): 2644–76.
21. Israelachvili, J.N., ed *Intermolecular and Surface Forces*. 2nd Edition ed. 1991, Academic Press: London & New York.
22. Rozas, I., On the nature of hydrogen bonds: an overview on computational studies and a word about patterns. *Phys Chem Chem Phys*, 2007. **9**(22): 2782–90.
23. Fleming, P.J. and G.D. Rose, Do all backbone polar groups in proteins form hydrogen bonds? *Protein Sci*, 2005. **14**(7): 1911–17.
24. Baldwin, R.L., In search of the energetic role of peptide hydrogen bonds. *J Biol Chem*, 2003. **278**(20): 17581–8.
25. Perrin, C.L. and J.B. Nielson, 'Strong' hydrogen bonds in chemistry and biology. *Annu Rev Phys Chem*, 1997. **48**:511–44.
26. Desiraju, G.R., C-H···O and other weak hydrogen bonds. From crystal engineering to virtual screening. *Chem Commun (Camb)*, 2005 (24): 2995–3001.
27. Scheiner, S. and T. Kar, Effect of solvent upon CH···O hydrogen bonds with implications for protein folding. *J Phys Chem B*, 2005. **109**(8): 3681–9.
28. Honig, B. and A. Nicholls, Classical electrostatics in biology and chemistry. *Science*, 1995. **268**(5214): 1144–9.

29. Sheinerman, F.B. and B. Honig, On the role of electrostatic interactions in the design of protein–protein interfaces. *J Mol Biol*, 2002. **318**(1): 161–77.
30. Sheinerman, F.B., R. Norel, and B. Honig, Electrostatic aspects of protein–protein interactions. *Curr Opin Struct Biol*, 2000. **10**(2): 153–9.
31. Shaul, Y. and G. Schreiber, Exploring the charge space of protein–protein association: a proteomic study. *Proteins*, 2005. **60**(3): 341–52.
32. Schreiber, G., G. Haran, and H.X. Zhou, Fundamental Aspects of Protein–protein Association Kinetics. *Chem Rev*, 2009.
33. Schreiber, G., Y. Shaul, and K.E. Gottschalk, Electrostatic design of protein–protein association rates. *Methods Mol Biol*, 2006. **340**:235–49.
34. Tsai, C.J., *et al.*, Studies of protein–protein interfaces: a statistical analysis of the hydrophobic effect. *Protein Sci*, 1997. **6**(1): 53–64.
35. Burley, S.K. and G.A. Petsko, Aromatic-aromatic interaction: a mechanism of protein structure stabilization. *Science*, 1985. **229**(4708): 23–8.
36. Zacharias, N. and D.A. Dougherty, Cation-pi interactions in ligand recognition and catalysis. *Trends Pharmacol Sci*, 2002. **23**(6): 281–7.
37. Dougherty, D.A., Cation-pi interactions in chemistry and biology: a new view of benzene, Phe, Tyr, and Trp. *Science*, 1996. **271**(5246): 163–8.
38. D'Aquino, J.A., E. Freire, and L.M. Amzel, Binding of small organic molecules to macromolecular targets: evaluation of conformational entropy changes. *Proteins*, 2000. **Suppl 4**:93–107.
39. D'Aquino, J.A., *et al.*, The magnitude of the backbone conformational entropy change in protein folding. *Proteins*, 1996. **25**(2): 143–56.
40. Lee, K.H., *et al.*, Estimation of changes in side chain configurational entropy in binding and folding: general methods and application to helix formation. *Proteins*, 1994. **20**(1): 68–84.
41. Murphy, K.P., *et al.*, Entropy in biological binding processes: estimation of translational entropy loss. *Proteins*, 1994. **18**(1): 63–7.
42. Kauzmann, W., Some factors in the interpretation of protein denaturation. *Adv Protein Chem*, 1959. **14**:1–63.
43. Tanford, C., The hydrophobic effect and the organization of living matter. *Science*, 1978. **200**(4345): 1012–8.
44. Chandler, D., Interfaces and the driving force of hydrophobic assembly. *Nature*, 2005. **437**(7059): 640–7.
45. Hilser, V.J., J. Gomez, and E. Freire, The enthalpy change in protein folding and binding: refinement of parameters for structure-based calculations. *Proteins*, 1996. **26**(2): 123–33.
46. Lazaridis, T., G. Archontis, and M. Karplus, Enthalpic contribution to protein stability: insights from atom-based calculations and statistical mechanics. *Adv Protein Chem*, 1995. **47**:231–306.
47. Luque, I. and E. Freire, Structural parameterization of the binding enthalpy of small ligands. *Proteins*, 2002. **49**(2): 181–90.
48. Cabani, S., *et al.*, Group contributions to the thermodynamic properties of non-ionic organic solutes in dilute aqueous solution. *Journal of Solution Chemistry*, 1981. **10**(8): 563–595.
49. Gomez, J., *et al.*, The heat capacity of proteins. *Proteins*, 1995. **22**(4): 404–12.
50. Murphy, K.P. and E. Freire, Thermodynamics of structural stability and cooperative folding behavior in proteins. *Adv Protein Chem*, 1992. **43**:313–61.
51. Vega, S., *et al.*, A structural and thermodynamic escape mechanism from a drug resistant mutation of the HIV-1 protease. *Proteins*, 2004. **55**(3): 594–602.
52. Klebe, G., Virtual ligand screening: strategies, perspectives and limitations. *Drug Discov Today*, 2006. **11**(13–14): 580–94.
53. Weber, P.C. and F.R. Salemme, Applications of calorimetric methods to drug discovery and the study of protein interactions. *Curr Opin Struct Biol*, 2003. **13**(1): 115–21.
54. Velazquez-Campoy, A., I. Luque, and E. Freire, The application of thermodynamic methods in drug design. *Thermochimica Acta*, 2001. **380**:217–227.
55. Velazquez Campoy, A. and E. Freire, ITC in the post-genomic era. . .? Priceless. *Biophys Chem*, 2005. **115**(2–3): 115–24.

56. Chaires, J.B., Calorimetry and thermodynamics in drug design. *Annu Rev Biophys*, 2008. **37**:135–51.
57. Freire, E., Isothermal titration calorimetry: controlling binding forces in lead optimization. *Drug Discov Today: Technologies*, 2004. **1**(3): 295–9.
58. Freire, E., Do enthalpy and entropy distinguish first in class from best in class? *Drug Discov Today*, 2008. **13**(19–20): 869–74.
59. Holdgate, G.A. and W.H. Ward, Measurements of binding thermodynamics in drug discovery. *Drug Discov Today*, 2005. **10**(22): 1543–50.
60. Lafont, V., *et al.*, Compensating enthalpic and entropic changes hinder binding affinity optimization. *Chem Biol Drug Des*, 2007. **69**(6): 413–22.
61. Ruben, A.J., Y. Kiso, and E. Freire, Overcoming roadblocks in lead optimization: a thermodynamic perspective. *Chem Biol Drug Des*, 2006. **67**(1): 2–4.
62. Nezami, A. and E. Freire, The integration of genomic and structural information in the development of high affinity plasmepsin inhibitors. *Int J Parasitol*, 2002. **32**(13): 1669–76.
63. Sharp, K.A., *et al.*, Extracting hydrophobic free energies from experimental data: relationship to protein folding and theoretical models. *Biochemistry*, 1991. **30**(40): 9686–97.
64. Ohtaka, H., *et al.*, Thermodynamic rules for the design of high affinity HIV-1 protease inhibitors with adaptability to mutations and high selectivity towards unwanted targets. *Int J Biochem Cell Biol*, 2004. **36**(9): 1787–99.
65. Luque, I., *et al.*, Molecular basis of resistance to HIV-1 protease inhibition: a plausible hypothesis. *Biochemistry*, 1998. **37**(17): 5791–7.
66. Todd, M.J., *et al.*, Thermodynamic basis of resistance to HIV-1 protease inhibition: calorimetric analysis of the V82F/I84V active site resistant mutant. *Biochemistry*, 2000. **39**(39): 11876–83.
67. Velazquez-Campoy, A., Y. Kiso, and E. Freire, The binding energetics of first- and second-generation HIV-1 protease inhibitors: implications for drug design. *Arch Biochem Biophys*, 2001. **390**(2): 169–75.
68. Muzammil, S., P. Ross, and E. Freire, A major role for a set of non-active site mutations in the development of HIV-1 protease drug resistance. *Biochemistry*, 2003. **42**(3): 631–8.
69. Ohtaka, H. and E. Freire, Adaptive inhibitors of the HIV-1 protease. *Prog Biophys Mol Biol*, 2005. **88**(2): 193–208.
70. Nezami, A., *et al.*, High-affinity inhibition of a family of Plasmodium falciparum proteases by a designed adaptive inhibitor. *Biochemistry*, 2003. **42**(28): 8459–64.
71. Freire, E., Overcoming HIV-1 resistance to protease inhibitors. *Drug Discov Today. Disease Mechanisms*, 2006. **3**(2): 281–6.
72. Naghibi, H., A. Tamura, and J.M. Sturtevant, Significant discrepancies between van't Hoff and calorimetric enthalpies. *Proc Natl Acad Sci USA*, 1995. **92**(12): 5597–9.
73. Liu, Y. and J.M. Sturtevant, Significant discrepancies between van't Hoff and calorimetric enthalpies. II. *Protein Sci*, 1995. **4**(12): 2559–61.
74. Liu, Y. and J.M. Sturtevant, Significant discrepancies between van't Hoff and calorimetric enthalpies. III. *Biophys Chem*, 1997. **64**(1–3): 121–6.
75. Horn, J.R., *et al.*, Van't Hoff and calorimetric enthalpies from isothermal titration calorimetry: are there significant discrepancies? *Biochemistry*, 2001. **40**(6): 1774–8.
76. Recht, M.I., *et al.*, Enthalpy array analysis of enzymatic and binding reactions. *Anal Biochem*, 2008. **377**(1): 33–9.
77. Torres, F.E., *et al.*, Enthalpy arrays. *Proc Natl Acad Sci USA*, 2004. **101**(26): 9517–22.
78. McKinnon, I.R., *et al.*, A twin titration microcalorimeter for the study of biochemical reactions. *Anal Biochem*, 1984. **139**(1): 134–9.
79. Wiseman, T., *et al.*, Rapid measurement of binding constants and heats of binding using a new titration calorimeter. *Anal Biochem*, 1989. **179**(1): 131–7.
80. Freire, E., O.L. Mayorga, and M. Straume, Isothermal titration calorimetry. *Anal. Chem.*, 1990. **62**: 950A–959A.
81. Jelesarov, I. and H.R. Bosshard, Isothermal titration calorimetry and differential scanning calorimetry as complementary tools to investigate the energetics of biomolecular recognition. *J Mol Recognit*, 1999. **12**(1): *p.* 3–18.

82. Leavitt, S. and E. Freire, Direct measurement of protein binding energetics by isothermal titration calorimetry. *Curr Opin Struct Biol*, 2001. **11**(5): 560–6.
83. Fisher, H.F. and N. Singh, Calorimetric methods for interpreting protein-ligand interactions. *Methods Enzymol*, 1995. **259**:194–221.
84. Indyk, L. and H.F. Fisher, Theoretical aspects of isothermal titration calorimetry. *Methods Enzymol*, 1998. **295**:350–64.
85. Tellinghuisen, J., Statistical error in isothermal titration calorimetry. *Methods Enzymol*, 2004. **383**:245–82.
86. Tellinghuisen, J., Statistical error in isothermal titration calorimetry: variance function estimation from generalized least squares. *Anal Biochem*, 2005. **343**(1): 106–15.
87. Pierce, M.M., C.S. Raman, and B.T. Nall, Isothermal titration calorimetry of protein–protein interactions. *Methods*, 1999. **19**(2): 213–21.
88. Tellinghuisen, J., Optimizing experimental parameters in isothermal titration calorimetry. *J Phys Chem B*, 2005. **109**(42): 20027–35.
89. Tellinghuisen, J., Optimizing experimental parameters in isothermal titration calorimetry: variable volume procedures. *J Phys Chem B*, 2007. **111**(39): 11531–7.
90. Tellinghuisen, J., Isothermal titration calorimetry at very low c. *Anal Biochem*, 2008. **373**(2): 395–7.
91. Palencia, A., *et al.*, Thermodynamic dissection of the binding energetics of proline-rich peptides to the Abl-SH3 domain: implications for rational ligand design. *J Mol Biol*, 2004. **336**(2): 527–37.
92. Doyle, M.L. and P. Hensley, Tight ligand binding affinities determined from thermodynamic linkage to temperature by titration calorimetry. *Methods Enzymol*, 1998. **295**:88–99.
93. Doyle, M.L., *et al.*, Tight binding affinities determined from thermodynamic linkage to protons by titration calorimetry. *Methods Enzymol*, 1995. **259**:183–94.
94. Sigurskjold, B.W., Exact analysis of competition ligand binding by displacement isothermal titration calorimetry. *Anal Biochem*, 2000. **277**(2): 260–6.
95. Zhang, Y.L. and Z.Y. Zhang, Low-affinity binding determined by titration calorimetry using a high-affinity coupling ligand: a thermodynamic study of ligand binding to protein tyrosine phosphatase 1B. *Anal Biochem*, 1998. **261**(2): 139–48.
96. Baker, B.M. and K.P. Murphy, Evaluation of linked protonation effects in protein binding reactions using isothermal titration calorimetry. *Biophys J*, 1996. **71**(4): 2049–55.
97. Parker, M.H., *et al.*, Analysis of the binding of hydroxamic acid and carboxylic acid inhibitors to the stromelysin-1 (matrix metalloproteinase-3) catalytic domain by isothermal titration calorimetry. *Biochemistry*, 1999. **38**(41): 13592–601.
98. Velazquez-Campoy, A., *et al.*, Thermodynamic dissection of the binding energetics of KNI-272, a potent HIV-1 protease inhibitor. *Protein Sci*, 2000. **9**(9): 1801–9.
99. Christensen, J.J., L.D. Hansen, and R.M. Izatt, *Proton Ionization Heats and Related Thermodynamic Quantities*. 1976, New York: Wiley.
100. Jelesarov, I. and H.R. Bosshard, Thermodynamics of ferredoxin binding to ferredoxin: NADP+ reductase and the role of water at the complex interface. *Biochemistry*, 1994. **33**(45): 13321–8.
101. Gilson, M.K. and H.X. Zhou, Calculation of protein-ligand binding affinities. *Annu Rev Biophys Biomol Struct*, 2007. **36**:21–42.
102. Murphy, K.P., Predicting binding energetics from structure: looking beyond DeltaG degrees. *Med Res Rev*, 1999. **19**(4): 333–9.
103. Luque, I. and E. Freire, Structure-based prediction of binding affinities and molecular design of peptide ligands. *Methods Enzymol*, 1998. **295**:100–27.
104. Luque, I., O.L. Mayorga, and E. Freire, Structure-based thermodynamic scale of alpha-helix propensities in amino acids. *Biochemistry*, 1996. **35**(42): 13681–8.
105. Pace, C.N. and J.M. Scholtz, A helix propensity scale based on experimental studies of peptides and proteins. *Biophys J*, 1998. **75**(1): 422–7.
106. Robertson, A.D. and K.P. Murphy, Protein Structure and the Energetics of Protein Stability. *Chem Rev*, 1997. **97**(5): 1251–68.

107. Xie, D. and E. Freire, Molecular basis of cooperativity in protein folding. V. Thermodynamic and structural conditions for the stabilization of compact denatured states. *Proteins*, 1994. **19**(4): 291–301.
108. Hilser, V.J. and E. Freire, Structure-based calculation of the equilibrium folding pathway of proteins. Correlation with hydrogen exchange protection factors. *J Mol Biol*, 1996. **262**(5): 756–72.
109. Lavigne, P., *et al.*, Structure-based thermodynamic analysis of the dissociation of protein phosphatase-1 catalytic subunit and microcystin-LR docked complexes. *Protein Sci*, 2000. **9**(2): 252–64.
110. Sundberg, E.J., *et al.*, Estimation of the hydrophobic effect in an antigen-antibody protein–protein interface. *Biochemistry*, 2000. **39**(50): 15375–87.
111. Gomez, J., N. Semo, and E. Freire, Structural thermodynamic study of the binding of renin inhibitors to endothiapepsin. *Adv Exp Med Biol*, 1998. **436**:325–8.
112. Gomez, J. and E. Freire, Thermodynamic mapping of the inhibitor site of the aspartic protease endothiapepsin. *J Mol Biol*, 1995. **252**(3): 337–50.
113. Luque, I., *et al.*, Structure-based thermodynamic design of peptide ligands: application to peptide inhibitors of the aspartic protease endothiapepsin. *Proteins*, 1998. **30**(1): 74–85.
114. Nezami, A., *et al.*, Identification and characterization of allophenylnorstatine-based inhibitors of plasmepsin II, an antimalarial target. *Biochemistry*, 2002. **41**(7): 2273–80.
115. Raschke, T.M., Water structure and interactions with protein surfaces. *Curr Opin Struct Biol*, 2006. **16**(2): 152–9.
116. Park, S. and J.G. Saven, Statistical and molecular dynamics studies of buried waters in globular proteins. *Proteins*, 2005. **60**(3): 450–63.
117. Bhat, T.N., *et al.*, Bound water molecules and conformational stabilization help mediate an antigen-antibody association. *Proc Natl Acad Sci USA*, 1994. **91**(3): 1089–93.
118. Wang, H. and A. Ben-Naim, A possible involvement of solvent-induced interactions in drug design. *J Med Chem*, 1996. **39**(7): 1531–9.
119. Janin, J., Wet and dry interfaces: the role of solvent in protein–protein and protein-DNA recognition. *Structure*, 1999. **7**(12): R277–9.
120. Ben-Naim, A., Molecular recognition–viewed through the eyes of the solvent. *Biophys Chem*, 2002. **101–2**:309–19.
121. Rodier, F., *et al.*, Hydration of protein–protein interfaces. *Proteins*, 2005. **60**(1): 36–45.
122. Denisov, V.P. and B. Halle, Hydrogen exchange and protein hydration: the deuteron spin relaxation dispersions of bovine pancreatic trypsin inhibitor and ubiquitin. *J Mol Biol*, 1995. **245** (5): 698–709.
123. Denisov, V.P. and B. Halle, Protein hydration dynamics in aqueous solution. *Faraday Discuss*, 1996 (103): 227–44.
124. Pal, S.K., *et al.*, Biological water: femtosecond dynamics of macromolecular hydration. *J. Phys. Chem. B*, 2002. **106**(48): 12376–95.
125. Pal, S.K., J. Peon, and A.H. Zewail, Biological water at the protein surface: dynamical solvation probed directly with femtosecond resolution. *Proc Natl Acad Sci USA*, 2002. **99**(4): 1763–8.
126. Teyra, J. and M. Teresa Pisabarro, Characterization of interfacial solvent in protein complexes and contribution of wet spots to the interface description. *Proteins*, 2007.
127. Palencia, A., *et al.*, Key role for interfacial water molecules in Abl-SH3 complexes. Towards a new paradigm for proline-rich ligand recognition by SH3 domains. Submitted, 2009.
128. Martin-Garcia, J.M., C. Corbi, and I. Luque,In preparation.
129. Teyra, J., *et al.*, SCOWLP: a web-based database for detailed characterization and visualization of protein interfaces. *BMC Bioinformatics*, 2006. **7**(1): 104.
130. Ladbury, J.E., Just add water! The effect of water on the specificity of protein-ligand binding sites and its potential application to drug design. *Chem Biol*, 1996. **3**(12): 973–80.
131. Li, Z. and T. Lazaridis, Thermodynamic contributions of the ordered water molecule in HIV-1 protease. *J Am Chem Soc*, 2003. **125**(22): 6636–7.
132. Li, Z. and T. Lazaridis, Thermodynamics of buried water clusters at a protein-ligand binding interface. *J. Phys. Chem. B*, 2006. **110**(3): 1464–75.

133. Dunitz, J.D., Win some, lose some: enthalpy-entropy compensation in weak intermolecular interactions. *Chem Biol*, 1995. **2**(11): 709–12.
134. Cooper, A., Heat capacity effects in protein folding and ligand binding: a re-evaluation of the role of water in biomolecular thermodynamics. *Biophys Chem*, 2005. **115**(2–3): 89–97.
135. Li, Z. and T. Lazaridis, Water at biomolecular binding interfaces. *Phys Chem Chem Phys*, 2007. **9**(5): 573–81.
136. Minke, W.E., *et al.*, The role of waters in docking strategies with incremental flexibility for carbohydrate derivatives: heat-labile enterotoxin, a multivalent test case. *J Med Chem*, 1999. **42**(10): 1778–88.
137. Rarey, M., B. Kramer, and T. Lengauer, The particle concept: placing discrete water molecules during protein-ligand docking predictions. *Proteins*, 1999. **34**(1): 17–28.
138. Verdonk, M.L., J.C. Cole, and R. Taylor, SuperStar: a knowledge-based approach for identifying interaction sites in proteins. *J Mol Biol*, 1999. **289**(4): 1093–1108.
139. Mancera, R.L., De novo ligand design with explicit water molecules: an application to bacterial neuraminidase. *J Comput Aided Mol Des*, 2002. **16**(7): 479–99.
140. Garcia-Sosa, A.T., S. Firth-Clark, and R.L. Mancera, Including tightly-bound water molecules in de novo drug design. Exemplification through the in silico generation of poly(ADP-ribose) polymerase ligands. *J Chem Inf Model*, 2005. **45**(3): 624–33.
141. Garcia-Sosa, A.T., R.L. Mancera, and P.M. Dean, WaterScore: a novel method for distinguishing between bound and displaceable water molecules in the crystal structure of the binding site of protein-ligand complexes. *J Mol Model*, 2003. **9**(3): 172–82.
142. Jiang, L., *et al.*, A 'solvated rotamer' approach to modeling water-mediated hydrogen bonds at protein–protein interfaces. *Proteins*, 2005. **58**(4): 893–904.
143. Raymer, M.L., *et al.*, Predicting conserved water-mediated and polar ligand interactions in proteins using a K-nearest-neighbors genetic algorithm. *J Mol Biol*, 1997. **265**(4): 445–64.
144. Kellogg, G.E., S.F. Semus, and D.J. Abraham, HINT: a new method of empirical hydrophobic field calculation for CoMFA. *J Comput Aided Mol Des*, 1991. **5**(6): 545–52.
145. Chen, D.L. and G.E. Kellogg, A computational tool to optimize ligand selectivity between two similar biomacromolecular targets. *J Comput Aided Mol Des*, 2005. **19**(2): 69–82.
146. Samsonov, S., J. Teyra, and M.T. Pisabarro, A molecular dynamics approach to study the importance of solvent in protein interactions. *Proteins*, 2008. **73**(2): 515–25.
147. Freire, E., The thermodynamic linkage between protein structure, stability and function, in *Protein Structure, Stability and Folding*, K.P. Murphy,Editor, Humana Press Inc.: Totowa, NJ.
148. Englander, S.W., Protein folding intermediates and pathways studied by hydrogen exchange. *Annu Rev Biophys Biomol Struct*, 2000. **29**: 213–38.
149. Luque, I., S.A. Leavitt, and E. Freire, The linkage between protein folding and functional cooperativity: two sides of the same coin? *Annu Rev Biophys Biomol Struct*, 2002. **31**:235–56.
150. Sadqi, M., *et al.*, The native state conformational ensemble of the SH3 domain from alpha-spectrin. *Biochemistry*, 1999. **38**(28): 8899–8906.
151. Pan, H., J.C. Lee, and V.J. Hilser, Binding sites in Escherichia coli dihydrofolate reductase communicate by modulating the conformational ensemble. *Proc Natl Acad Sci USA*, 2000. **97**(22): 12020–5.
152. Hilser, V.J., *et al.*, The structural distribution of cooperative interactions in proteins: analysis of the native state ensemble. *Proc Natl Acad Sci USA*, 1998. **95**(17): 9903–8.
153. Hilser, V.J. and E. Freire, Predicting the equilibrium protein folding pathway: structure-based analysis of staphylococcal nuclease. *Proteins*, 1997. **27**(2): 171–83.
154. Hilser, V.J., B.D. Townsend, and E. Freire, Structure-based statistical thermodynamic analysis of T4 lysozyme mutants: structural mapping of cooperative interactions. *Biophys Chem*, 1997. **64**(1–3): 69–79.
155. Ferreon, J.C., *et al.*, Solution structure, dynamics, and thermodynamics of the native state ensemble of the Sem-5 C-terminal SH3 domain. *Biochemistry*, 2003. **42**(19): 5582–91.
156. Luque, I. and E. Freire, Structural stability of binding sites: consequences for binding affinity and allosteric effects. *Proteins*, 2000. **Suppl 4**: 63–71.

157. Clackson, T. and J.A. Wells, A hot spot of binding energy in a hormone-receptor interface. *Science*, 1995. **267**(5196): 383–6.
158. Bogan, A.A. and K.S. Thorn, Anatomy of hot spots in protein interfaces. *J Mol Biol*, 1998. **280**(1): 1–9.
159. Moreira, I.S., P.A. Fernandes, and M.J. Ramos, Hot spots--a review of the protein–protein interface determinant amino-acid residues. *Proteins*, 2007. **68**(4): 803–12.
160. Thorn, K.S. and A.A. Bogan, ASEdb: a database of alanine mutations and their effects on the free energy of binding in protein interactions. *Bioinformatics*, 2001. **17**(3): 284–5.
161. Keskin, O., B. Ma, and R. Nussinov, Hot regions in protein–protein interactions: the organization and contribution of structurally conserved hot spot residues. *J Mol Biol*, 2005. **345**(5): 1281–94.
162. Keskin, O., *et al.*, Protein–protein interactions: organization, cooperativity and mapping in a bottom-up Systems Biology approach. *Phys Biol*, 2005. **2**(2): S24–35.
163. Li, X., *et al.*, Protein–protein interactions: hot spots and structurally conserved residues often locate in complemented pockets that pre-organized in the unbound states: implications for docking. *J Mol Biol*, 2004. **344**(3): 781–95.
164. Freire, E., Statistical thermodynamic linkage between conformational and binding equilibria. *Adv Protein Chem*, 1998. **51**:255–79.
165. Freire, E., The propagation of binding interactions to remote sites in proteins: analysis of the binding of the monoclonal antibody D1.3 to lysozyme. *Proc Natl Acad Sci USA*, 1999. **96**(18): 10118–22.
166. Muzammil, S., *et al.*, Unique thermodynamic response of tipranavir to human immunodeficiency virus type 1 protease drug resistance mutations. *J Virol*, 2007. **81**(10): 5144–54.
167. Carbonell, T. and E. Freire, Binding thermodynamics of statins to HMG-CoA reductase. *Biochemistry*, 2005. **44**(35): 11741–8.

Part II

Approaches

3

On the Logic of Natural Product Binding in Protein–Protein Interactivity

James J. La Clair
Xenobe Research Institute, San Diego, CA, USA

3.1 Introduction

Often the introduction to small molecule protein interactions begins with Emil Fisher's depiction of an enzyme as the interaction between a lock and a key [1]. This discussion usually leads to a structural description of the interaction between the locks (usually ascribed as the proteins) and their keys (usually the substrates or ligands) [2]. In Fisher's description as well as subsequent reviews, the context rarely examines the 'door' that the lock and key go to. One explanation arises from the fact that many classical treatments of protein interactions describe structure ('the lock and key mechanism') and function ('the door') as two different subjects. Are the two really separate disciplines and if so what gain is obtained by distinguishing their information?

In part, the answers to these questions are derived from the way in which we inspect the interaction between a protein and its ligands. Historically, the focus of structural information has been placed on elucidation of the features of specific ligand binding sites on a given protein. Structures such as the tubulin–colchicine complex provide classical examples depicting the interaction between natural products and proteins. Once characterized, the function of these ligation events is often described with regards to their upstream and downstream signaling. In the context of Fisher's analogy, these signaling are no longer directly related to the lock and key mechanism.

In recent years, the advance of integrative biology has provided a complete genomic and proteomic descriptions of a cell [3]. Using a panel of new graphical user interfaces such as

Protein Surface Recognition: Approaches for Drug Discovery Ernest Giralt, Mark W. Peczuh and Xavier Salvatella. © 2011 John Wiley & Sons, Ltd

Osprey [4], VisANT [5], yEd [6], NAViGaTOR [7], Walrus [8], Grafta [9], Cytoscape [10], software now exists that depict the seemingly daunting complexity of protein networks. Here the interaction between individual proteins gains context in terms of its regulatory and functional interactions. Resources such as Mammalian Protein–protein Interaction Database (MIPS) [11], Biomolecular Interaction Network Database (BIND) [12] or Small Molecule Interaction Database (SMID) [13] now provide online databases that correlate protein–protein and small-molecule protein binding events with their positions within a protein network [14]. While a complete union between structure and function has yet to be established, these assignments provide a critical tool in establishing protein interactions.

Perhaps the best way to understand the complexity of protein–protein relations begins with evaluating their perturbation by small molecule ligands. In ecosystems, secondary metabolites are the predominant signals synthesized to regulate protein function and interactivity. While synthesized through common biosynthetic processes (i.e. nonribsomal peptide synthesis or polyketide synthesis), the mechanisms that make these materials undergo considerable post-synthase modifications allowing the production of a complex myriad of diverse structure. It is within this diversity that natural products access a robust and complex function as protein regulators.

Until recently, the focus on natural product activity was focused on the identification of protein targets. While these studies have contributed to ~400 targets of interest in drug discovery, less than ~2 % of them involve a small molecule mediated protein–protein interaction. The combination of this lack in materials and high specificity of protein–protein interactions has recently led to their increased awareness as therapeutic targets [15]. A series of excellent reviews now exists outlining the challenges in targeting protein–protein interactions, including excellent synopses on the proteomic [16], theoretical [17], *in silico* modelling [18], and high-throughput screening [19] studies.

This chapter provides an overview of the pending issues within the regulation of protein–protein interactions from the perspective of the natural product. The discussion is developed into three sections. Section 3.2 focuses on structural biology of natural product interactions at protein–protein interfaces. The discussion then moves to evaluate the functional role of natural product induced protein–protein interactions. The final section of the chapter provides an outline to the short-term requirements and culminates in the projection of an all-encompassing system to depict all natural product regulations of protein–protein interactions, in which both theoretical and practical considerations are addressed.

3.2 Structural Logic

One of the most important facets of protein–protein research has originated from the development of models to understand the interactions that guide protein association. While an understanding of the features that regulate protein–protein interactions is far from being solved at the systems-wide level, a number of key structural elements have evolved through the solution of NMR and X-ray crystallographic structures [20]. One of the most important elements of these discoveries arises from developing a system to classify modes in which proteins and natural products interact. Questions then arise as to the generality within the ascribed modes.

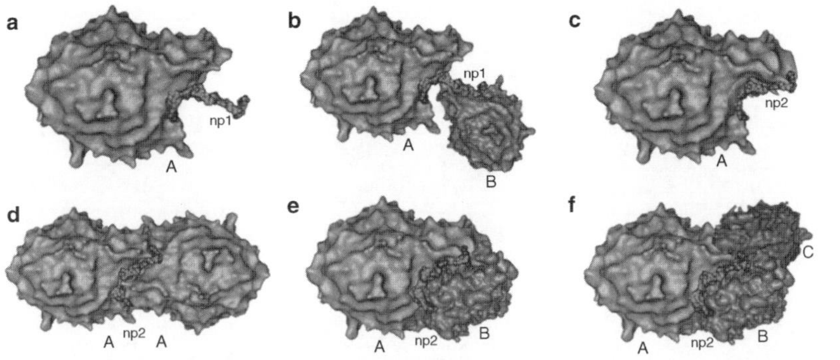

Figure 3.1 *Modes of interaction between natural products and proteins. A cartoon depiction of natural products (np1, np2) interacting with proteins A, B and C. (a) An exemplary mode in which the interaction between natural product np1 and protein A provides a second external binding domain. (b) This A•np1 complex can recruit a second protein B to form an A•np1•B complex. (c) Alternatively, natural product np2 can bind to the surface of a protein A. (d) The resulting A•np2 complex can then induce dimerization to form A• np2•A trimer. (e) The A•np2 complex can also recruit a different protein B to form a heterodimeric A•np2•B complex. (f) Higher order complexes as illustrated by the recruitment of a third protein C in formation of the A•np2•B•C complex. Each frame a-f depicts one mode of interaction as given by the formation of a unique type of complex (See Plate 5.)*

Figure 3.1 provides several theoretical examples depicting different modes of interaction between natural products and proteins. Natural products np1 or np2 can bind to the surface of protein A leaving a pendant tail in solution (as shown for np1 in Figure 3.1a) or completely absorbed on its surface (as shown for np2 in Figure 3.1c). The natural product tail in the A•np1 complex (Figure 3.1a) can then recruit a second protein B to form an A•np1•B trimer (Figure 3.1b). This interaction does not necessarily require that the surfaces of the two proteins interact.

More commonly, the natural product serves to create a new surface to link the two proteins together. This can occur either in the form of protein oligomerization as shown by the formation of an homodimeric A•np2•A complex (Figure 3.1d) or by the formation of heterodimeric A•np2•B complex (Figure 3.1e). Protein oligomerization is a well-established mechanism gated by the binding of a natural product. Natural products examined have been shown to partake both in the induction and inhibition of protein oligomerization. To date, these phenomena are commonly seen in cytosketal proteins. Examples as the modification of tubulin assembly by colchicine (**1**) [21], podophyllotoxin (**2**) [22] and combretastatin A4 (**3**) [23] or the stabilization of microtubule polymers with paclitaxel (**4**) [24] provide classical examples of this cytoskeletal protein–protein interactions [25] (see structures of **1-4** in Figure 3.2). In these examples, the interplay between the natural products and protein alters the natural oligomerization process required by cytoskeletal movement. For the cell, modification of these processes delivers an effective vehicle to block the cell cycle and movement.

The importance of natural product modifications of protein oligomerization events is not limited to the cytoskeleton. Other events such as the inhibition the biotin carboxylase domain of acetylcoenzyme A carboxylases by soraphen A (**5**) [26] or interactions between rapamycin (**6**) (see structure of **6** in Figure 3.2) and FRAP [27] provide examples of

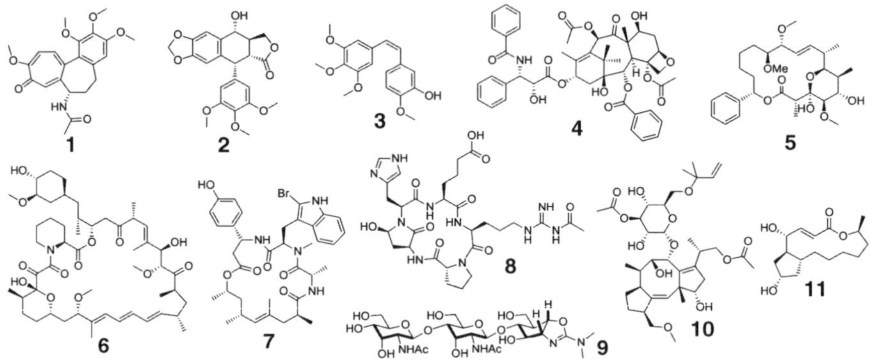

Figure 3.2 *Structures of several natural products shown to regulate protein-protein interactions*

regulation of enzymatic and signal transduction events. Moreover, the binding of small molecules such as soraphen A (**5**) to the surface of a protein can lead to modifications that induce dimerization, as illustrated by the formation of an A•np2•A trimer (Figure 3.1d). Other molecules such as rapamycin (**6**) recruit another second type of protein to generate an A•np2•B complex (Figure 3.1e). Even less is understood in terms of developing a structural understanding as to the formation of higher order complexes such as shown in Figure 3.1f. While some indication exists in terms of small molecule regulation of multiunit actin-cadherin-catenin complexes by jasplakinolide (**7**) (see structure of **7** in Figure 3.2) [28], the intricacy of these complexes complicates structure elucidation efforts.

An issue regarding the structure of natural product dimerization events arises in the methods used for their elucidation (Figure 3.3). In both NMR and X-ray structure elucidation efforts, the effective protein concentration is typically high and can therefore induce nonnatural oligomerization events [29]. In the crystal state, such associations are often unavoidable and therefore portray close associations that may not necessarily be apparent in solution. Comparable assumptions can also be made for the solution and solid-state NMR structures. One such example becomes apparent in the evaluation of the binding of the cyclic peptide argadin (**8**) (see structure of **8** in Figure 3.2) to the chitinase B (ChiB) from *Serratia marcescens* [30]. In these studies, a 2.0 Å (R = 0.204) resolution crystal structure presents interactions between the natural product and the targeted ChiB binding pocket with hydrogen bonds to Tyr214, Asp215, Arg294, Glu144, Asp142 and hydrophobic interactions with Tyr10, Try292, Met214, Trp403, Trp97, Ile 339, Asp316). As indicated in Figures 3.3a–b, these interactions are also accompanied by a hydrophobic interaction with Tyr481 on second molecule of ChiB. The question then exists as to the relevance of this interaction. Does the affinity between Tyr481 and argadin (**8**) induce the dimerization or were the proteins in a dimeric state prior to doping ChiB crystals with the natural product?

The answer to this question comes from examination of second structure of ligand bound ChiB (Figures 3.3c–d) [31]. In this structure, Tyr481 does not interact with the bound ligand, allosamadin (**9**) (see structure of **9** in Figure 3.2). Rather, it folds downward, contacting residues Trp220 and Glu221 in the adjacent protein molecule (Figure 3.3d). While the authors of both structures did not make any conclusions as to the role of their ligands in dimerization [3,31], this demonstration illustrates the potential for one to deduce the role of a natural product in enhancing protein–protein interfaces. The caution comes as to what level

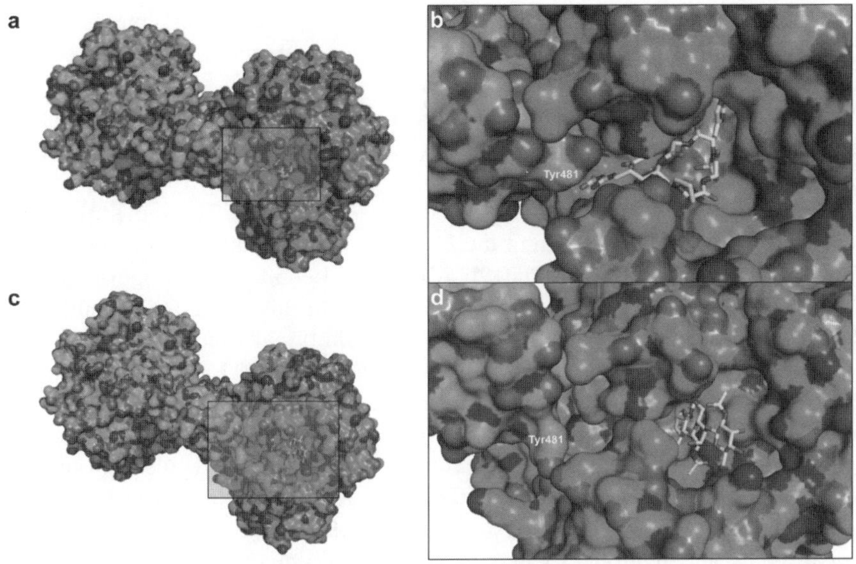

Figure 3.3 *False-positive pockets. Crystal structures of chitinase B (ChiB) from Serratia marcescens with bound ligands. (a) Structure of ChiB with argadin (**8**) bound. (b) Close-up of argadin (**8**) bound to ChiB. (c) Structure of mutant D142N ChiB with allosamadin bound. (d) Close-up of allosamadin (**9**) bound to ChiB. Each structure contains two ChiB proteins with a bound small molecule. Images were developed from structure files 1h0g and 1ogg.[31] (See Plate 6.)*

of interaction between natural products and protein interfaces is required to identify a clear effect. As illustrated by this example (Figure 3.3), overzealous interpretation can lead to the identification of association events that may not be critical for protein oligomerization, so called false-positive pockets.

While caution is warranted, strong structural evidence does exist depicting the role of natural products in recruiting protein–protein interactions. Clardy's structure of the FKBP12-rapamycin-FRB ternary complex provides an excellent example [27]. In their refined 2.0 Å structure, rapamycin (**6**) provides a two-faced binding wherein the C2–C12 region of **6** binds to FKBP12 and C16–C23 region interacts with FRB. The function of the natural product in these studies was verified by comparison to with the crystal structure of the FKBP12-rapamycin binary complex. Interestingly, while the conformation of the natural product remained the same, the formation of the binding of FRB to FKBP12-rapamycin delivered only a slight shift in the residues of FKBP12 within the rapamycin binding pocket and a more pronounced shift in the residues interacting with FRB with the most pronounced flexibility within two loops at residues 40–47 and 80–89.

A second example comes from the binding of the brefeldin A (**10**) (see structure of **10** in Figure 3.2) to a ternary complex containing a Ras-related GTPase (ARF), guanidine diphosphate (GDP) and a member of the Ser7 family of ARF exchange factors [32]. Brefeldin A (**10**) was shown to target the ARF•GDP•Sec7 complex and not target Sec7, ARF or ARF•GDP. The binding of brefeldin A (**10**) locked ARF and Sec7 in the ARF•GDP•Sec7 complex and thereby prevented its catalytic activity. The structure of the Brefeldin A (**10**)• ARF•GDP•Sec7 complex was solved by Goldberg [33] and depicted a tight interaction where **10** was sandwiched between two alpha helixes (Ile258-Ser265 and Ser198-Glu206) of Sec7 and an alpha helix (residues Gln71-Trp79) and beta-sheet (Val23-Thr55) on ARF. The structure contained 92 residues of Ser7 and ARF that were within ~4 Å of the natural product.

Examples also exist for the formation of higher order protein complexes. Recent studies on P-type H^+-ATPases provide an excellent example [34]. Fusicoccin (**11**) (see structure of **11** in Figure 3.2) [35], an known activator of plant H^+ pumps was shown to bind to a preformed complex between an H^+-ATPase and a dimer of a regulatory 14-3-3 protein [36]. The natural product was bound to the A-domain of the 14-3-3 protein with hydrophobic interactions to Ile225, Leu225, Pro174, Met130, Phe126, Val53 and Leu50 and hydrophilic interactions with Lys128, Asn49 and Asp222. The Lys128 residue bridged the fusicoccin(**11**)•14-3-3 complex to the CT52[YDI] peptide which was additionally held in the complex by hydrophylic interactions with Ile956 and His930 Interestingly, **11** also bound to the second 14-3-3 molecule with a comparable structure and associated residues. As shown by Oecking, natural products also serve as 'glue' to complete and lock large protein complexes.

Clearly, the breath of architectures in which natural products regulate protein–protein interactions has only been touched upon. The models provided in Figure 3.1 and demonstrated in Figures 3.3–3.6 provide only a few examples of the potential combinations of between proteins and natural products. The continued development of high-resolution structures of natural product-protein–protein complexes continues to provide vital information towards understanding the potential for regulating protein–protein affinity.

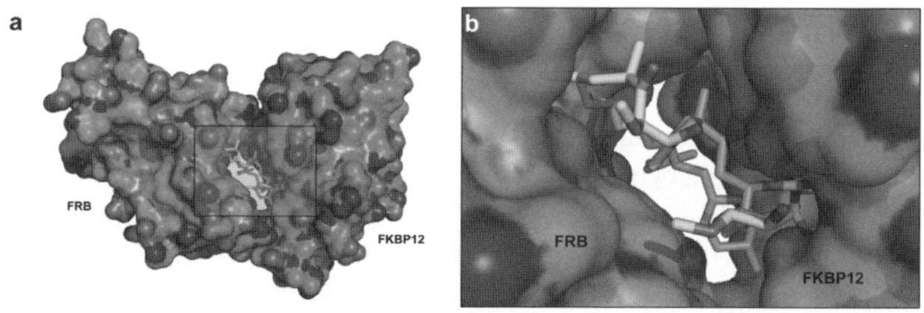

Figure 3.4 *The FKBP12•rapamycin(**6**)•FRB complex. A depiction of natural-product induced interface between FKBP12 and FRB. Each structure contains two proteins, FRB and FKBP12 with a bound rapamycin (**6**). Images were developed from structure file 4fap (27a) (See Plate 7.)*

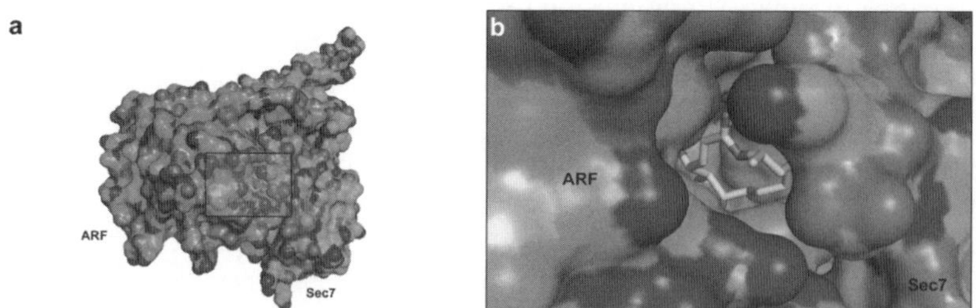

Figure 3.5 *The ARF•GDP•brefeldin(**10**)•Sec7 complex. A depiction of brefeldin A (**10**) interface between ARF and Sec7. The structure contains two proteins, ARF and Sec7 with a bound molecule of brefeldin A (**10**). Images were developed from structure file 1re0. [33a] (See Plate 8.)*

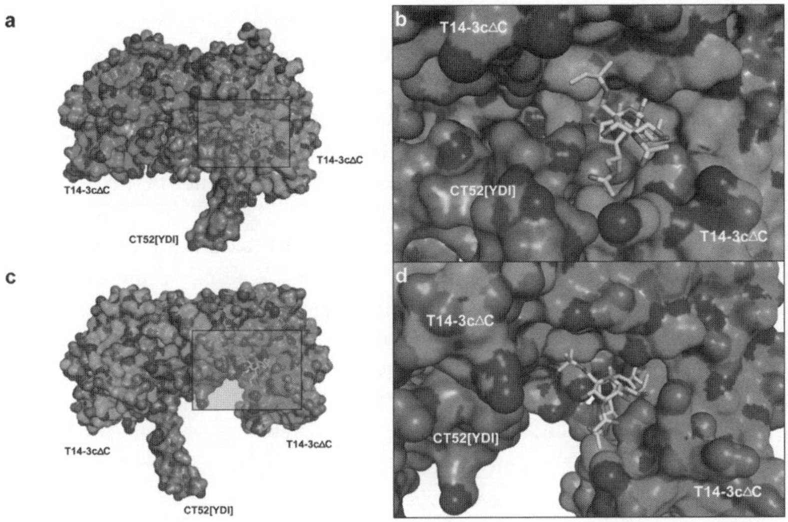

Figure 3.6 *The CT52[YDI]·fusicoccin(**11**)· 14-3-3 complex. (a) A depiction of fusicoccin (**11**) at the interface between the CT53[YDI] peptide and a dimer containing two 14-3-3 proteins. (b) close-up of the binding pocket in (a). (c) The back face of the image in (a). (d) A close-up of (c). Images were developed from structure file 1ia0 [36] (See Plate 9.)*

3.3 Functional Logic

Structure, however, is not the only facet of protein–protein interactions that requires detailed study. Natural products serve to induce and inhibit protein–protein interactions. Structures usually provide information only on systems that induce interactions and therefore one half of the function is therefore lost. While predictions can be made as to how a natural product blocks key binding sites for a given protein–protein interaction, the approximations must remain predictions. Moreover, structural information fails to predict the outcome of these events with regards to up and down stream regulatory networks. The nature of these regulations remains as complex as their structural potential.

The understanding of protein networks has recently accelerated in part by the development of tools to display and edit their structures [3–8]. One of the key advances in these tools has arisen within the ability to depict an entire database of interactions within one user-accessible map [8]. As these tools progress, the ability to correlate specific regulatory events to their upstream and downstream signal transduction will also advance as one can now visualize the entire protein system.

In order to build these networks there is a definitive need to identify each protein interaction within the cell. Without a global tool such as that used in genomic and proteomic analysis, the elucidation of protein associations is slow requiring the arduous development of individual research efforts. Over the last two decades, considerable detail has developed depicting a considerable portion of the interactions within the proteome [16], however, the information is far from complete. One area where there is an immediate need for discovery arises in the evaluation of metabolite mediated associations [37]. While studies evaluating specific members of primary [38] and secondary [39] metabolism have been described, we are far from developing a global perspective as to how small molecules regulate protein association. One of the first steps towards developing this understanding arises through the classification of the modes in which a metabolite regulates the interaction of proteins. Figure 3.7 provides a series of examples of these modes and defines them by evaluating their context with regards to a specific natural product.

Modifications can occur to the local protein network as a natural product associates multiple proteins. The induction of binding between proteins A and B leads to reconstruction of the network by creating a new interaction as shown by a yellow connection between proteins A and B in Figure 3.7a. The FKBP12•rapamycin(**6**)•FRB complex (Figure 3.4) provides an excellent example of this type of interaction. In this example, rapamycin serves to bind protein A (FKBP12) to second protein B (FRB) forming a new complex.

For many interactions, the formation of stable protein–protein complexes can lead to a net loss in the concentration of either protein. As depicted in Figure 3.7b, the loss in B by the formation of the terniary A•np•B complex (where np represents a natural product) now modifies the protein network by rendering the association between B and C is rendered inactive. The effect of this interaction leads to the inhibition of processes regulated by signal transduction through B and C. Studies on the activity of brefeldin A (**10**) provides an example for this event, as the binding of brefeldin A (**10**) locked the ARF and Sec7 proteins in an ARF•GDP•Sec7 complex (Figure 3.5). In this example, ARF and Sec7 represent proteins

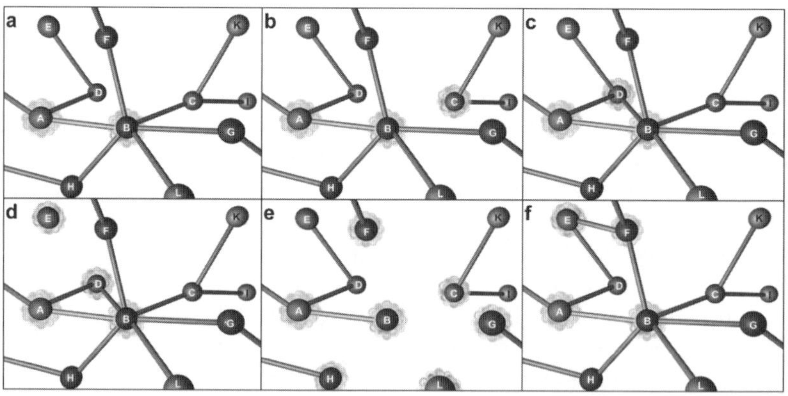

Figure 3.7 *Functional response. A hypothetical model depicting natural product derived regulation of protein-protein interactions. Each frame depicts a model network of proteins (A–L) in spheres and their interactions in tubes. This 3D network was modeled after protein network program Grafta 9. For each panel a-d, the natural product regulates the interaction between proteins A and B. The effects of this interaction as also indicated by tubes and proteins involved in this process are shaded with clouds. The proteins targeted by the natural product move about the cell (See Plate 10.)*

A and B, respectively and the lack of activity of Sec7 leads to the loss of its catalytic activity as depicted in Figure 3.7b by elimination of its connectivity to protein C.

The induced affinity between proteins A and B can also lead to secondary binding events and lead to the formation of new network partners. As depicted in Figure 3.7c, the induction of association between proteins A and B leads to the binding between protein A and protein D. This type of association typically occurs when the induced binding between proteins A and B arises at a site on protein A that is independent of its bind domain with protein D. Again, secondary events can also arise from the formation of the ternary complex between A•np•B•D. When the formation of the A•np•B•D complex leads to a stable state, the concentration of monomeric protein D in the cell can be reduced, resulting in a net loss in its affinity (as illustrated in Figure 3.7d by the loss of connectivity between proteins D and E). Again this restructuring of the normal protein network can have deleterious effects on the cell. For the case in Figure 3.7d, the formation of A•np•B•D complex ablates the contact between proteins D and E that subsequent blocks signal transduction.

A more profound example arises when induction of the A•np•B complex leads to a state that sequesters one of the two protein partners from its network. As illustrated in Figure 3.7e, the formation of the A•np•B complex sequesters B eliminating its normal associations with proteins C, F, G, H and L. Such events can occur in the cell when the formation of the A•np•B complex is accompanied by relocation of the proteins within the sub-cellular structure of the cell. A recent example of this event was demonstrated by studies on the phorboxazole B (**12**). In these studies, affinity analogs prepared from **12** recruited the cyclin dependent kinase 4 (cdk4) in the cytosol to the cytokeratin proteins (KRT10) within the intermediate filaments (IF) of the cell. Now bound to the surface of the IF, cytosolic cdk4 is no longer available for its delivery to the nucleus [40].

The final example (Figure 3.7f) was provided to indicate the potential complexity within the natural-product remodeling of protein networks. In this example, the formation of the A•np•B complex serves to stimulate the formation of a second nonnatural interaction between proteins E and F. The formation of such events can arise again by the shuttling of proteins throughout the cell. To demonstrate lets examine an example where proteins A and B appear in the cytosol bound as A•np•B and move into the Golgi apparatus. The trafficking of the A•np•B complex into the Golgi can also serve to direct other downstream cytosolic proteins such as E to the Golgi. This new location within the cell presents new contacts between proteins allowing the formation of secondary protein complexes such as E•F.

The events within the examples in Figure 3.7 depict several potential events that can arise upon induction of affinity between two proteins by a natural product. This mechanisms (Figure 3.7) are in no means comprehensive and only begin to touch upon the potential for small molecule regulation as a multitude of secondary events including post-translational modifications and expression profiles that can also lead to major restructurings within the cell.

3.4 The Need for Programmers

While the study of natural product regulation of protein–protein interactions may be in its early stages of development, it is key to develop a systematic approach that allows all fundamental modes of activity to be delineated within a single objective. The ultimate goal in the study of natural product regulation of protein–protein networks is to create a natural product-protein–protein interaction (NPPI) map that depicts all protein–protein interactions for a given cell or organism. This NPPI map would depict each natural products involvement within a global protein network and demonstrate how and by what means the small molecule regulates a given pathway. Creation of this map will take decades of effort and require individuals to program its code at many different levels of scientific investigation. This effort includes the unique association between natural product chemists, medicinal chemists, chemical biologists, geneticists, biochemists, molecular biologists, cell biologists and computer software designers and programmers. The following section provides a brief overview of the studies needed to complete the development of a global natural product-protein–protein (NPPI) network map.

While far from these requirements, Figure 3.8 provides a crude prototype for this map. One of the first steps required to complete this system originates in the development of graphical tools to depict the regulation of protein networks by small molecules. Current protein network display platforms (i.e. Walrus or Cytoscape) fail to provide such access. The central problem arises from the fact that each of the examples in Figure 3.7 required an individual frame for display. This would mean that any program designed to depict how a natural product altered a protein network would be required to generate a new screen display for each natural product. While effective, this approach would eliminate any potential for the user to compare the roles of two different natural product events. Moreover, it would require enormous data sets and complex algorithms for search and modification. A more direct approach is illustrated in Figure 3.8.

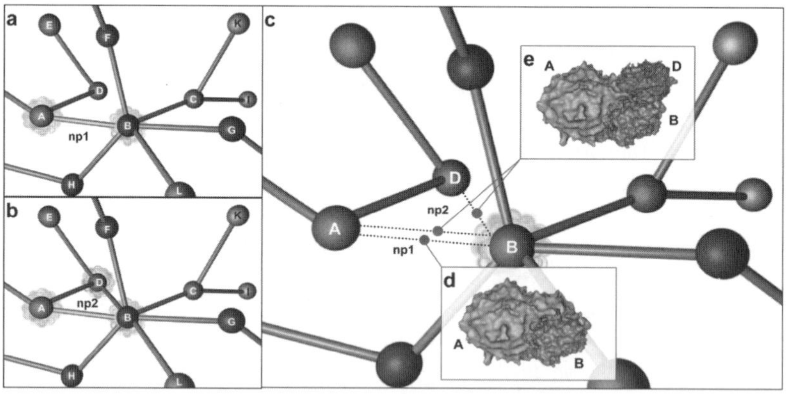

***Figure 3.8** Natural product-Protein-Protein Interaction (NPPI) map. (a) Example 1: Natural product (np1) induces affinity between proteins A and B form an A•np1•B complex. (b) Example 2: Natural product (np2) induces affinity between A and B which then recruits protein D to form a quaternary complex A•np2•B•C. (c) An model NPPI map. Proteins are depicted as spheres and interactions as tubes. Natural products are depicted as dots and their induced protein-protein interactions as dotted lines. (d) The structures of the A•np1•B complex can be provided in inset windows by simply clicking on the dot representing np1. (e) The display could host multiple structures including A•np2•B•C complex (See Plate 11.)*

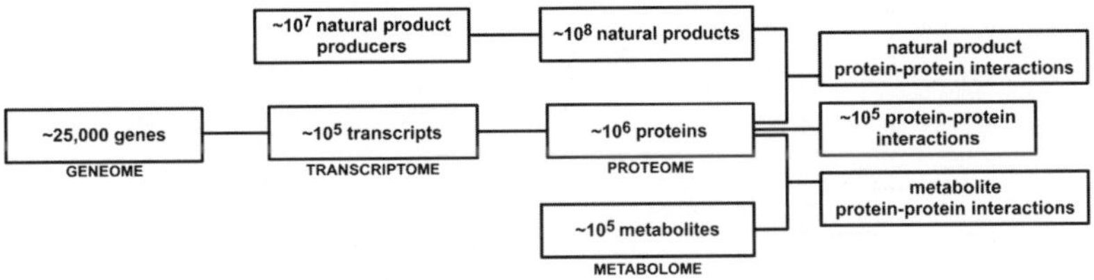

Figure 3.9 *Systems-wide estimate of potential natural product interactions. Estimates of members of the genome, transcriptome and proteome are developed based on current human databases.[45] The number of natural product producers was based on recent approximations of the number of microorganisms, fungi and plant species known to participate in secondary metabolism.[46] The number of natural products is based on the fact that the majority of producers, either contain more than one natural product synthase or produce analogs of the same family of natural products. An index of 10 natural products per producer organism was used. The number of protein-protein interactions was based on recent approximations in yeast.[41]*

Interactions due to two natural products, np1 and np2, be displayed an individual frames (Figures 3.8a–b) can be compressed into a single window (Figure 3.8c). In the latter window, a single 3D environment can be used to display the interaction of multiple natural products as represented by dots and dotted lines. In addition, this tool can also be developed such that clicking on each natural product opens a second window that displays the three-dimensional structure of the complex as suggested for np1 (Figure 3.8d) and np2 (Figure 3.8e). Software development is however perhaps the least of the many requirements for development of a comprehensive NPPI map as only a fraction of the potential structures and complexes have been identified to date. Using different search tools, the number of natural product-protein–protein interactions have been identified is a factor of 10^2. While it remains difficult to approximate the number of potential interactions, recent systems wide estimates (Figure 3.9) suggest that the number of natural product is $\sim$1000 fold greater than the number of proteins. Given the recent estimates of 10^5 protein–protein interactions [41], one can estimate that the number of natural product induced protein–protein complexes would be at a comparable magnitude with a population of $\sim 10^4$-10^5.

Given these estimations, considerable effort is required to advance the current number of natural product induced protein interactions ($\sim 10^2$) to its estimated potential (10^4-10^5). A massive multidisciplinary effort is required to achieve this goal, as with the development of genomic and proteomic efforts. Associated with this effort, critical developments will arise through the design and actualization of technologies that find and characterize natural product induced protein–protein interactions at high-throughput. Figure 3.10 provides a brief synopsis of several facets of this endeavor. This outline is in no means comprehensive.

As outlined in Figure 3.10, a number of factors are required to develop a functional NPPI map. This effort includes development of tools to search and mine the current proteomic [42], protein–protein interaction [43], and protein-ligand databases [44]. This information serves as the foundation for the development of the NPPI map. Experimental efforts are then needed to rapidly screen large sets of proteins and natural products. Hits identified through these screening efforts must then be characterized both at the structural and functional levels. Each interaction characterized through these studies can then be tabulated within a database.

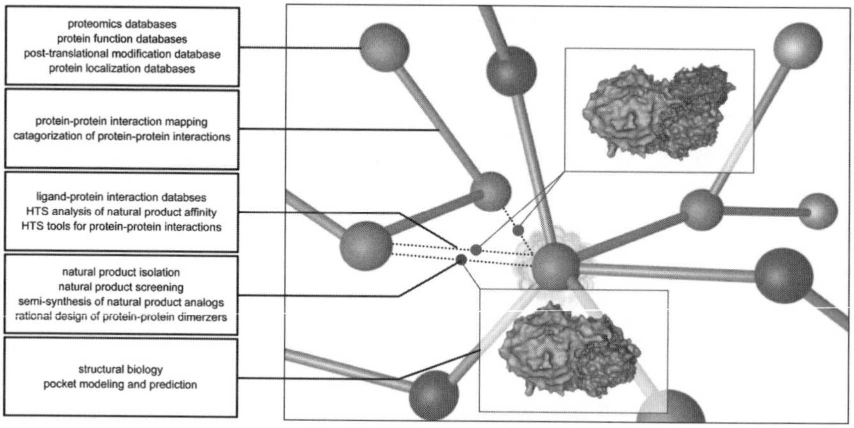

Figure 3.10 *Perspectives in the development of global Metabolite-Protein-Protein (NPPI) Networks. A depiction of five key aspects of research required for the development of a NPPI map. This effort includes research at (a) the level of the protein, (b) protein-protein interaction, (c) natural product protein interaction, (d) natural product and (e) the structural composite of natural product protein interactions (See Plate 12.)*

Complementary campaigns are also required to deliver sets of natural products for these studies. Traditional culture and isolation methods still continue to be a rich resource for material delivery. These methods can be enhanced through the adoption of high-throughput robotics. While yet to reach maturation, there is a critical need for technologies that can express, isolate and characterize secondary metabolites at high-throughput. Currently, leading natural products laboratories characterize between 1 and 100 novel materials by hand per year. The current bottleneck in this process arises due to the material and time requirements during structural characterization. The key to this development arises with the advance of low volume HT-NMR methods and LC-NMR technologies. In addition, techniques such as that used in protein crystallography can be applied to small molecule crystallography. While current small molecule crystallography centers can complete structures with 10–100 μm sized crystals at high throughput [47], few programs have developed as sophisticated tools as that established for protein crystallography. These facts aside, the advancement of HT methods provides an effective tool for expanding natural product science.

In addition, natural products can also be converted to materials that induce dimerization, so called dimerizers, through semisynthesis or biosynthesis. Dimeric materials based on compounds such as FK506 [48], coumermycin [49], amphotericin [50], vancomycin [51], and glycopeptides [52] are the established examples of this approach. As envisoned, FK1012, a synthetic dimer of FK506, induces the dimerization of FKBP12, an FK506-binding protein. Such studies can be further expanded to the synthesis of fusional natural product conjugates. The fusion of biotin perhaps is one of the first and most prevalent example of heterodimeric natural product fusions [53]. While used for protein target studies, biotinylated natural products formally consist of the fusion of two natural products, the molecule of interest and biotin. Its assembly delivers a dual activity which has been well documented to associate two unrelated proteins (i.e. streptavidin affinity purification of target proteins) [54]. More recent examples have focused on the elaboration of novel small molecule oligomers [55]. Such synthetic hybrids can then be used to create and study new protein–protein complexes.

3.5 Compiling the NPPI Mapper

To conclude, our understanding of the role of natural products at the protein–protein interface is in its infancy. The identification of new technologies both at the molecular level and *in silico* are required to enhance the throughput of their analyses. Techniques such as robotic natural product culturing and isolation are required to expand the number of natural products. Complementary systems are then required to screen these materials for their ability to induce the affinity between two or more proteins. Leads identified through the latter process must then be characterized both at the structural and functional level. Theoretically, this entire process could be automated using a single assembly line which when optimized could using modern methods to screen 10^6 binding events per day. Data collected from these studies can then be depicted using 2D or 3D maps thereby providing a global perspective as to how natural products regulate protein–protein interactions.

The building of this mapping tool in many ways is similar to the construction of a computer game as individual bits of code are assembled into a single 3D videographic application. No longer will individual protein association events be viewed the fitting of keys in locks but rather be described by a virtual object (i.e. spheres in Figure 3.8b) with functional and structural details. Each protein and natural product could become a character in this videogame. In essence, this NPPI mapping program becomes a realistic game in life. Access to this information can also be made as facile and enjoyable as a videogame. The question then exists as to the value of this information. The latter shall be left up to the reader. The fundamental question is as to whether you going to program the game or play it? Logout. This user has exited. Game over. It is now yours to build.

References

1. E. Fischer, Einfluss der configuration auf die wirkung der enzyme, *Ber. Dt. Chem. Ges.*, **27**, 2985–2993 (1894).
2. U. R. Lemieux and U. Spohr, How Emil Fischer was led to the lock and key concept for enzyme specificity, *Adv. Carbohydrate Chem. Biochem.*, **50**, 1–20 (1994).
3. (a) S. Ivakhno, From functional genomics to systems biology, *FEBS J.* **274**, 2439–48 (2007). (b) S. Huang and J. Wikswo, Dimensions of systems biology, *Rev. Physiol Biochem. Pharmacol.*, **157**, 81–104 (2006). (c) F. J. Bruggeman and H. V. Westerhoff, The nature of systems biology, *Trends Microbiol.*, **15**, 45–50 (2007). (d) A. Brazma, M. Krestyaninova and U. Sarkans, Standards for systems biology, *Nat. Rev. Genet.*, **7**, 593–605 (2006).
4. (a) G. W. Bell and F. Lewitter, Visualizing networks, *Methods Enzymol.*, **411**, 408–21 (2006). (b) P. M. Gordon and C. W. Sensen, Osprey: a comprehensive tool employing novel methods for the design of oligonucleotides for DNA sequencing and microarrays, *Nucleic Acids Res.*, **32**, e133 (2004). (c) C. Stark, B. J. Breitkreutz, T. Reguly, L. Boucher, A. Breitkreutz and M. Tyers, BioGRID: a general repository for interaction datasets, *Nucleic Acids Res.*, **34**, D535–D539 (2006).

5. (a) Z. Hu, J. Mellor, J. Wu, T. Yamada, D. Holloway and C. Delisi, VisANT: data-integrating visual framework for biological networks and modules, *Nucleic Acids Res.*, **33**, W352–W357 (2005). (b) Z. Hu, J. Mellor, J. Wu and C. DeLisi, VisANT: an online visualization and analysis tool for biological interaction data, *BMC Bioinformatics*, **5**, 17 (2004).
6. http://www.yworks.com/en/products_yed_about.htm
7. P. D. Karp, S. Paley and P. Romero, The Pathway Tools software, *Bioinformatics*, **1**, S225–S232 (2002).
8. D. Butler, Virtual globes: the web-wide world, *Nature*, **439**, 776–8 (2006).
9. G. W. Bell and F. Lewitter, Visualizing networks, *Methods Enzymol.*, **411**, 408–21 (2006).
10. (a) P. Shannon, A. Markiel, O. Ozier, *et al.*, Cytoscape: a software environment for integrated models of biomolecular interaction networks, *Genome Res.*, **13**, 2498–2504 (2003). (b) M. Albrecht, C. Huthmacher, S. C. Tosatto and T. Lengauer, Decomposing protein networks into domain-domain interactions, *Bioinformatics*, **21**, ii220–ii221 (2005). (c) M. Singhal and K. Domico, CABIN: Collective Analysis of Biological Interaction Networks, *Comput Biol Chem.*, **31**, 222–5 (2007).
11. (a) P. Pagel, S. Kovac, M. Oesterheld, *et al.*, The MIPS mammalian protein–protein interaction database, *Bioinformatics*, **21**, 832–4 (2006). (b) M. Costanzo, G. Giaever, C. Nislow and B. Andrews, Experimental approaches to identify genetic networks, *Curr. Opin. Biotechnol.*, **17**, 472–80 (2006). (c) G. Lubec, L. Afjehi-Sadat, J. W. Yang and J. P. John, Searching for hypothetical proteins: theory and practice based upon original data and literature, *Prog. Neurobiol.*, **77**, 90–127 (2005).
12. D. Gilbert, Biomolecular interaction network database, *Brief Bioinform.*, **6**, 194–8 (2005).
13. (a) K. A. Snyder, H. J. Feldman, M. Dumontier, J. J. Salama and C. W. Hogue, Domain-based small molecule binding site annotation, *BMC Bioinformatics*, **17**, 152 (2006). (b) H. J. Feldman, K. A. Snyder, A. Ticoll, G. Pintilie and C. W. Hogue, A complete small molecule dataset from the protein data bank, *FEBS Lett.*, **580**, 1649–53 (2006).
14. (a) B. A. Shoemaker and A. R. Panchenko. Deciphering protein–protein interactions. Part I. Experimental techniques and databases, *PLoS Comput Biol.*, **3**, e42 (2007). (b) S. Zhong, A. T. Macias and A. D. MacKerell Jr., Computational identification of inhibitors of protein–protein interactions, *Curr. Top. Med. Chem.* **7**, 63–82 (2007). (c) K. Kuroda, M. Kato, J. Mima and M. Ueda, Systems for the detection and analysis of protein–protein interactions. *Appl. Microbiol. Biotechnol.* **71**, 127–36 (2006).
15. (a) M. J. Vicent, E. Perez-Paya and M. Orzaez, Discovery of inhibitors of protein–protein interactions from combinatorial libraries, *Curr. Top. Med. Chem.* **7**, 83–95 (2007). (b) D. C. Fry, Protein–protein interactions as targets for small molecule drug discovery, *Biopolymers*, **84**, 535–52 (2006). (c) P. Chene, Drugs targeting protein–protein interactions, *ChemMedChem*, **4**, 400–11 (2006). (d) D. C. Fry and L. T. Vassilev, Targeting protein–protein interactions for cancer therapy, *J. Mol. Med.* **83**, 955–63 (2005). (e) H. Yin and A. D. Hamilton, Strategies for targeting protein–protein interactions with synthetic agents, *Angew. Chem. Int. Ed. Engl.*, **44**, 4130–63 (2005). (f) A. Loregian and G. Palu, Disruption of protein–protein interactions: towards new targets for chemotherapy, *J. Cell Physiol.*, **204**, 750–62 (2005). (g) L. O. Sillerud and R. S. Larson, Design and structure of peptide and peptidomimetic antagonists of protein–protein interaction, *Curr. Protein Pept. Sci.* **6**, 151–69 (2005).
16. (a) K. Kuroda, M. Kato, J. Mima and M. Ueda, Systems for the detection and analysis of protein–protein interactions, *Appl. Microbiol. Biotechnol.*, **71**, 127–36 (2006). (b) M. A. Trakselis, S. C. Alley and F. T. Ishmael, Identification and mapping of protein–protein interactions by a combination of cross-linking, cleavage and proteomics, *Bioconjug. Chem.*, **16**, 741–50 (2005). (c) S. Cho, S. G. Park, D. H. Lee and B. C. Park. Protein–protein interaction networks: from interactions to networks, *J. Biochem Mol. Biol.*, **37**, 45–52 (2004).
17. (a) J. F. Uhrig. Protein interaction networks in plants, *Planta*, **224**, 771–81 (2006). (b) I. G. Khalil and C. Hill. Systems biology for cancer, *Curr. Opin. Oncol.*, **17**, 44–8 (2005). (c) C. L. Vizcarra and S. L. Mayo. Electrostatics in computational protein design, *Curr. Opin. Chem. Biol.*, **9**, 622–6 (2005).

18. (a) B. A. Shoemaker and A. R. Panchenko, Deciphering protein–protein interactions. Part II. Computational methods to predict protein and domain interaction partners, *PLoS Comput Biol.*, **X**, X (2007). (b) B. A. Shoemaker and A. R. Panchenko, Deciphering protein–protein interactions. Part I. Experimental techniques and databases, *PLoS Comput Biol.*, **X**, X (2007). (c) S. Zhong, A. T. Macias and A. D. MacKerellJr., Computational identification of inhibitors of protein–protein interactions, *Curr. Top. Med. Chem.*, **7**, 63–82 (2007). (d) D. Gonzalez-Ruiz and H. Gohlke, Targeting protein–protein interactions with small molecules: challenges and perspectives for computational binding epitope detection and ligand finding, *Curr. Med. Chem.*, **13**, 2607–25 (2006). (e) T. L. Shi, Y. X. Li, Y. D. Cai and K. C. Chou. Computational methods for protein–protein interaction and their application, *Curr. Protein Pept. Sci.*, **6**, 443–9 (2005).
19. (a) S. Ivakhno, From functional genomics to systems biology, *FEBS J.*, **274**, 2439–48 (2007). (b) M. J. de Vega, M. Martin-Martinez and R. Gonzalez-Muniz. Modulation of protein–protein interactions by stabilizing/mimicking protein secondary structure elements. *Curr. Top. Med. Chem.*, **7**, 33–62 (2007). (c) L. Cekaite, E. Hovig and M. Sioud, Protein arrays: a versatile toolbox for target identification and monitoring of patient immune responses, *Methods Mo.l Biol.*, **360**, 335–48 (2007). (d) W. Huber and F. Mueller, Biomolecular interaction analysis in drug discovery using surface plasmon resonance technology, *Curr. Pharm. Des.*, **12**, 3999–4021 (2006). (e) M. Arkin, Protein–protein interactions and cancer: small molecules going in for the kill, *Curr. Opin. Chem. Biol.*, **9**, 317–24 (2005).
20. (a) J. A. Gerrard, C. A. Hutton and M. A. Perugini, Inhibiting protein–protein interactions as an emerging paradigm for drug discovery, *Mini Rev. Med. Chem.*, **7**, 151–7 (2007). (b) R. G. Efremov, A. O. Chugunov, T. V. Pyrkov, J. P. Priestle, A. S. Arseniev and E. Jacoby, Molecular lipophilicity in protein modeling and drug design, *Curr. Med. Chem.*, **14**, 393–415 (2007). (c) M. J. de Vega, M. Martin-Martinez and R. Gonzalez-Muniz, Modulation of protein–protein interactions by stabilizing/mimicking protein secondary structure elements, *Curr. Top. Med. Chem.*, **7**, 33–62 (2007). (d) S. Fletcher and A. D. Hamilton. Targeting protein–protein interactions by rational design: mimicry of protein surfaces, *J. R. Soc. Interface*, **3**, 215–33 (2006). (e) P. Chene, Drugs targeting protein–protein interactions, *Chem. Med. Chem.*, **4**, 400–11 (2006).
21. (a) K. H. Downing and E. Nogales, Crystallographic structure of tubulin: implications for dynamics and drug binding, *Cell Struct. Funct.*, **24**, 269–75 (1999). (b) S. Gupta and B. Bhattacharyya. Antimicrotubular drugs binding to vinca domain of tubulin, *Mol. Cell. Biochem.*, **253**, 41–7 (2003). (c) S. B. Hastie. Interactions of colchicine with tubulin, *Pharmacol. Ther.*, **51**, 377–401 (1991). (d) W. R. Mundy and H. A. Tilson, Neurotoxic effects of colchicines, *Neurotoxicology.*, **11**, 539–47 (1990). (e) K. H. Downing and Nogales E., Crystallographic structure of tubulin: implications for dynamics and drug binding, *Cell Struct. Funct.*, **24**, 269–75 (1999). (f) Y. Engelborghs, General features of the recognition by tubulin of colchicine and related compounds, *Eur. Biophys. J.*, **27**, 437–45 (1998).
22. (a) S. Desbene and S. Giorgi-Renault, Drugs that inhibit tubulin polymerization: the particular case of podophyllotoxin and analogues, *Curr. Med. Chem. Anticancer Agents*, **2**, 71–90 (2002). (b) Y. Damayanthi and J. W. Lown, Podophyllotoxins: current status and recent developments, *Curr. Med. Chem.*, **5**, 205–52 (1998). (c) D. L. Sackett, Podophyllotoxin, steganacin and combretastatin: natural products that bind at the colchicine site of tubulin, *Pharmacol. Ther.*, **59**, 163–228 (1993).
23. (a) M. Medarde, A. B. Maya and C. Perez-Melero. Naphthalene combretastatin analogues: synthesis, cytotoxicity and antitubulin activity, *J. Enzyme Inhib. Med. Chem.*, **19**, 521–40 (2004). (b) N. H. Nam, Combretastatin A-4 analogues as antimitotic antitumor agents, *Curr. Med. Chem.*, **10**, 1697–1722 (2003). (c) D. L. Sackett, Podophyllotoxin, steganacin and combretastatin: natural products that bind at the colchicine site of tubulin, *Pharmacol. Ther.*, **59**, 163–228 (1993).
24. (a) G. A. Orr, P. Verdier-Pinard, H. McDaid and S. B. Horwitz, Mechanisms of Taxol resistance related to microtubules, *Oncogene*, **22**, 7280–95 (2003). (b) E. A. Nogales, A structural view of microtubule dynamics, *Cell. Mol. Life Sci.*, **56**, 133–42 (1999). (c) L. A. Amos and J. Lowe, How taxol stabilizes microtubule structure, *Chem. Biol.* **6**, 65–9 (1999).
25. K. H. Downing and E. Nogales, Tubulin structure: insights into microtubule properties and functions, *Curr. Opin. Struct. Biol.*, **8**, 785–9 (1998).

26. Y. Shen, S. L. Volrath, S. C. Weatherly, T. D. Elich and L. Tong, A mechanism for the potent inhibition of eukaryotic acetyl-coenzyme A carboxylase by soraphen A, a macrocyclic polyketide natural product, *Mol. Cell.*, **16**, 881–91 (2004).
27. (a) J. Liang, J. Choi and J. Clardy, Refined structure of the FKBP12-rapamycin-FRB ternary complex at 2.2 A resolution, *Acta Crystallogr. D Biol. Crystallogr.*, **55**, 736–44 (1999). (b) J. Choi, J. Chen, S. L. Schreiber and J. Clardy, Structure of the FKBP12-rapamycin complex interacting with the binding domain of human FRAP, *Science*, **273**, 239–42 (1996).
28. (a) C. Higashida, T. Miyoshi, A. Fujita, *et al.*, Actin polymerization-driven molecular movement of mDia1 in living cells, *Science*, **303**, 2007–2010 (2004). (b) O. E. Christian, J. Compton, K. R. Christian, S. L. Mooberry, F. A. Valeriote and P. Crews. Using jasplakinolide to turn on pathways that enable the isolation of new chaetoglobosins from *Phomospis asparagi*, *J. Nat. Prod.*, **68**, 1592–7 (2005). (c) S. Yamada, S. Pokutta, F. Drees, W. I. Weis and W. J. Nelson, Deconstructing the cadherin-catenin-actin complex, *Cell.*, **123**, 889–901 (2005).
29. (a) J. Clarkson and I. D. Campbell, Studies of protein-ligand interactions by NMR, *Biochem. Soc. Trans.*, **31**, 1006–9 (2003). (b) R. A. Palmer and H. Niwa, X-ray crystallographic studies of protein-ligand interactions, *Biochem. Soc. Trans.*, **31**, 973–9 (2003). (c) A. M. Davis, S. J. Teague and G. J. Kleywegt, Application and limitations of X-ray crystallographic data in structure-based ligand and drug design, *Angew. Chem. Int. Ed. Engl.*, **42**, 2718–36 (2003).
30. (a) D. R. Houston, K. Shiomi, N. Arai, *et al.*, High-resolution structures of a chitinase complexed with natural product cyclopentapeptide inhibitors: mimicry of carbohydrate substrate, *Proc. Natl. Acad. Sci USA*, **99**, 9127–32 (2002). (b) D. R. Houston, B. Synstad, V. G. Eijsink, M. J. Stark, I. M. Eggleston and D. M. van Aalten. Structure-based exploration of cyclic dipeptide chitinase inhibitors, *J. Med. Chem.*, **47**, 5713–20 (2004). (c) N. Arai, K. Shiomi, Y. Yamaguchi, *et al.*, Argadin, a new chitinase inhibitor, produced by *Clonostachys sp.* FO-7314., *Chem. Pharm. Bull.*, **48**, 1442–6 (2000).
31. G. Vaaje-Kolstad, D. R. Houston, F. V. Rao, *et al.*, Structure of the D142N mutant of the family 18 chitinase ChiB from Serratia marcescens and its complex with allosamidin, *Biochim. Biophys. Acta.*, **1696**, 103–11 (2004).
32. (a) G. Pacheco-Rodriguez, J. Moss and M. Vaughan, Cytohesin-1: structure, function and ARF activation, *Methods Enzymol.*, **404**, 184–95 (2005). (b) J. Viaud, M. Zeghouf, H. Barelli, *et al.*, Structure-based discovery of an inhibitor of Arf activation by Sec7 domains through targeting of protein–protein complexes, *Proc. Natl. Acad. Sci. USA*, **104**, 10370–5 (2007).
33. (a) E. Mossessova, R. A. Corpina and J. Goldberg, Crystal structure of ARF1*Sec7 complexed with Brefeldin A and its implications for the guanine nucleotide exchange mechanism, *Mol. Cell.*, **12**, 1403–11 (2003). (b) J. C. Zeeh, M. Zeghouf, C. Grauffel, B. Guibert, E. Martin, A. Dejaegere and J. Cherfils. Dual specificity of the interfacial inhibitor brefeldin A for arf proteins and sec7 domains, *J. Biol. Chem.*, **281**, 11805–14 (2006).
34. (a) C. Ottmann, S. Marco, N. Jaspert, *et al.*, Structure of a 14-3-3 coordinated hexamer of the plant plasma membrane H+ -ATPase by combining X-ray crystallography and electron cryomicroscopy, *Mol. Cell.* **25**, 427–40 (2007). (b) M. Wurtele, C. Jelich-Ottmann, A. Wittinghofer and C. Oecking, Structural view of a fungal toxin acting on a 14-3-3 regulatory complex, *EMBO J.*, **22**, 987–94 (2003). (c) T. Jahn, J. Dietrich, B. Andersen, *et al.*, Large scale expression, purification and 2D crystallization of recombinant plant plasma membrane H+-ATPase, *J. Mol. Biol.*, **309**, 465–76 (2001).
35. (a) N. Tajima, M. Nukina, N. Kato and T. Sassa, Novel Fusicoccins R and S and the fusicoccin S aglycon (phomopsiol) from *Phomopsis amygdali niigata* 2-A and their seed germination-stimulating activity in the presence of abscisic acid, *Biosci. Biotechnol. Biochem.*, **68**, 1125–30 (2004). (b) M. Malerba, P. Crosti, R. Cerana and R. Bianchetti. Fusicoccin affects cytochrome c leakage and cytosolic 14-3-3 accumulation independent of H-ATPase activation, *Physiol. Plant.*, **120**, 386–94 (2004). (c) J. Singh and M. R. Roberts, Fusicoccin activates pathogen-responsive gene expression independently of common resistance signalling pathways, but increases disease symptoms in Pseudomonas syringae-infected tomato plants, *Planta*, **219**, 261–9 (2004).
36. (a) J. Borch, K. Bych, P. Roepstorff, M. G. Palmgren and A. T. Fuglsang. Phosphorylation-independent interaction between 14-3-3 protein and the plant plasma membrane H+-ATPase,

Biochem. Soc. Trans., **30**, 411–15 (2002). (b) M. Malerba and R. Bianchetti. 14-3-3 protein-activated and autoinhibited forms of plasma membrane H(+)-ATPase, *Biochem. Biophys. Res. Commun.*, **286**, 984–90 (2001). (c) H. A. Korthout and A. H. de Boer. A fusicoccin binding protein belongs to the family of 14-3-3 brain protein homologs, *Plant Cell*, **6**, 1681–92 (1994). (d) S. C. Huber, C. MacKintosh and W. M. Kaiser, Metabolic enzymes as targets for 14-3-3 proteins, *Plant Mol. Biol.*, **50**, 1053–63 (2002). (e) A. H. de Boer, Plant 14-3-3 proteins assist ion channels and pumps, *Biochem. Soc. Trans.*, **30**, 416–21 (2002).

37. T. O. Larsen, J. Smedsgaard, K. F. Nielsen, M. E. Hansen and J. C. Frisvad. Phenotypic taxonomy and metabolite profiling in microbial drug discovery, *Nat. Prod. Rep.*, **22**, 672–95 (2005).
38. (a) D. Steinhauser and J. Kopka, Methods, applications and concepts of metabolite profiling: primary metabolism, *EXS*. **97**, 171–94 (2007). (b) E. Fridman and E. Pichersky, Metabolomics, genomics, proteomics and the identification of enzymes and their substrates and products, *Curr. Opin. Plant Biol.*, **8**, 242–8 (2005). (c) J. Smedsgaard and J. Nielsen, Metabolite profiling of fungi and yeast: from phenotype to metabolome by MS and informatics, *J. Exp. Bot.*, **56**, 273–86 (2005).
39. (a) J. C. Lindon, E. Holmes and J. K. Nicholson, Metabonomics in pharmaceutical R&D, *FEBS J.*, **274**, 1140–51 (2007). (b) W. Schwab, Metabolome diversity: too few genes, too many metabolites?, *Phytochemistry*, **62**, 837–49 (2003).
40. C. J. Forsyth, L. Ying, J. Chen and J. J. La Clair, Phorboxazole analogues induce association of cdk4 with extranuclear cytokeratin intermediate filaments, *J. Am. Chem. Soc.*, **128**, 3858–9 (2006).
41. (a) N. Nariai, Y. Tamada, S. Imoto and S. Miyano, Estimating gene regulatory networks and protein–protein interactions of *Saccharomyces cerevisiae* from multiple genome-wide data, *Bioinformatics*, **21**, ii206–ii212 (2005). (b) A. Grigoriev, On the number of protein–protein interactions in the yeast proteome, *Nucleic Acids Res.*, **31**, 4157–61 (2003). (c) G. D. Bader and C. W. Hogue, Analyzing yeast protein–protein interaction data obtained from different sources, *Nat. Biotechnol.*, **20**, 991–7 (2002).
42. (a) Z. R. Yang and R. Hamer, Bio-basis function neural networks in protein data mining, *Curr. Pharm. Des.*, **13**, 1403–13 (2007). (b) X. Hu and D. D. Wu, Data mining and predictive modeling of biomolecular network from biomedical literature databases, *IEEE/ACM Trans. Comput. Biol. Bioinform.*, **4**, 251–63 (2007). (c) P. M. Roberts, Mining literature for systems biology, *Brief Bioinform.*, **7**, 399–406 (2006). (d) H. Li, J. Li and L. Wong, Discovering motif pairs at interaction sites from protein sequences on a proteome-wide scale, *Bioinformatics*, **22**, 989–96 (2006).
43. (a) J. Xu and Y. Li, Discovering disease-genes by topological features in human protein–protein interaction network, *Bioinformatics*, **22**, 2800–5 (2006). (b) X. L. Li, S. H. Tan, C. S. Foo and S. K. Ng, Interaction graph mining for protein complexes using local clique merging, *Genome Inform.*, **16**, 260–9 (2005). (c) M. Kirac, G. Ozsoyoglu and J. Yang, Annotating proteins by mining protein interaction networks, *Bioinformatics*, **22**, e260–e270 (2006). (d) T. Shlomi, D. Segal, E. Ruppin and R. Sharan, QPath: a method for querying pathways in a protein–protein interaction network, *BMC Bioinformatics*, **7**, 199 (2006). (e) M. A. van Driel, J. Bruggeman, G. Vriend, H. G. Brunner and J. A. Leunissen, A text-mining analysis of the human phenome, *Eur. J. Hum. Genet.*, **14**, 535–42 (2006). (f) J. J. Kim, Z. Zhang, J. C. Park and S. K Ng, BioContrasts: extracting and exploiting protein–protein contrastive relations from biomedical literature, *Bioinformatics*, **22**, 597–605 (2006). (g) J. Fang, R. J. Haasl, Y. Dong and G. H. Lushington, Discover protein sequence signatures from protein–protein interaction data, *BMC Bioinformatics*, **6**, 277 (2005). (h) S. D. Buckingham, Data mining for protein–protein interactions in invertebrate model organisms, *Invert. Neurosci.* **5**, 183–7 (2005).
44. (a) O. Sperandio, M. A. Miteva, F. Delfaud and B. O. Villoutreix. Receptor-based computational screening of compound databases: the main docking-scoring engines, *Curr. Protein Pept Sci.*, **7**, 369–93 (2006). (b) G. Scapin. Structural biology and drug discovery, *Curr. Pharm. Des.*, **12**, 2087–97 (2006). (c) S. J. Potts, D. J. Edwards and R. Hoffman. Challenges of target/compound data integration from disease to chemistry: a case study of dihydrofolate reductase inhibitors, *Curr. Drug Discov. Technol.*, **2**, 75–87 (2005). (d) J. C. Cole, C. W. Murray, J. W. Nissink, R. D. Taylor and R. Taylor, Comparing protein-ligand docking programs is difficult, *Proteins*, **60**, 325–

32 (2005). (e) R. Paulini, K. Muller and F. Diederich, Orthogonal multipolar interactions in structural chemistry and biology, *Angew. Chem. Int. Ed. Engl.*, **44**, 1788–1805 (2005). (f) M. Hendlich, A. Bergner, J. Gunther and G. Klebe, Relibase: design and development of a database for comprehensive analysis of protein-ligand interactions, *J. Mol. Biol.*, **326**, 607–20 (2003).

45. (a) S. J. Galbraith, L. M. Tran and J. C. Liao, Transcriptome network component analysis with limited microarray data, *Bioinformatics*, **22**, 1886–94 (2006). (b) P. Block, N. Weskamp, A. Wolf and G. Klebe, Strategies to search and design stabilizers of protein–protein interactions: a feasibility study, *Proteins*, **68**, 170–86 (2007). (c) P. S. Hegde, I. R. White and C. Debouck. Interplay of transcriptomics and proteomics, *Curr. Opin. Biotechnol.*, **14**, 647–51 (2003).
46. (a) I. Nobeli and J. M. Thornton, A bioinformatician's view of the metabolome, *Bioessays*, **28**, 534–45 (2006). (b) C. E. Thomas and G. Ganji, Integration of genomic and metabonomic data in systems biology –are we 'there' yet?, *Curr. Opin. Drug Discov. Devel.*, **9**, 92–100 (2006). (c) A. M. Richard, L. S. Gold and M. C. Nicklaus, Chemical structure indexing of toxicity data on the internet: moving toward a flat world, *Curr. Opin. Drug Discov. Devel.*, **9**, 314–25 (2006). (d) D. S. Wishart, C. Knox, A. C. Guo, *et al.*, DrugBank: a comprehensive resource for in silico drug discovery and exploration, *Nucleic Acids Res.*, **34**, D668–D672 (2006).
47. (a) A. Jain and V. Stojanoff, Are you centered? An automatic crystal-centering method for high-throughput macromolecular crystallography, *J. Synchrotron Radiat.*, **14**, 355–60 (2007). (b) B. A. Manjasetty, W. Shi, C. Zhan, A. Fiser and M. R. Chance, A high-throughput approach to protein structure analysis, *Genet. Eng.* **28**, 105–28 (2007). (c) G. Scapin, Structural biology and drug discovery, *Curr. Pharm. Des.*, **12**, 2087–97 (2007). (d) L. W. Tari, M. Rosenberg,and A. B. Schryvers. Structural proteomics in drug discovery, *Expert Rev. Proteomics*, **2**, 511–19 (2005). (e) M. L. Pusey, Z. J. Liu, W. Tempel, *et al.*, Life in the fast lane for protein crystallization and X-ray crystallography, *Prog. Biophys. Mol. Biol.*, **88**, 359–86 (2005). (f) T. L. Blundell and S. Patel, High-throughput X-ray crystallography for drug discovery, *Curr. Opin. Pharmacol.*, **4**, 490–6 (2004).
48. (a) L. W. Schultz and J. Clardy, Chemical inducers of dimerization: the atomic structure of FKBP12-FK1012A-FKBP12, *Bioorg. Med. Chem. Lett.*, **8**, 1–6 (1998). (b) M. N. Pruschy, D. M. Spencer, T. M. Kapoor, H. Miyake, G. R. Crabtree and S. L. Schreiber. Mechanistic studies of a signaling pathway activated by the organic dimerizer FK1012, *Chem. Biol.* **1**, 163–72 (1994). (c) A. Liakatos and H. Kunz, Synthetic glycopeptides for the development of cancer vaccines, *Curr. Opin. Mol. Ther.*, **9**, 35–44 (2007). (d) N. Yamaji, N. Matsumori, S. Matsuoka, T. Oishi and M. Murata. Amphotericin B dimers with bisamide linkage bearing powerful membrane-permeabilizing activity, *Org Lett.*, **4**, 2087–9 (2002).
49. C. L. Freel Meyers, M. Oberthur, L. Heide, D. Kahne and C. T. Walsh. Assembly of dimeric variants of coumermycins by tandem action of the four biosynthetic enzymes CouL, CouM, CouP and NovN, *Biochemistry*, **43**, 15022–36 (2004).
50. N. Yamaji, N. Matsumori, S. Matsuoka, T. Oishi and M. Murata, Amphotericin B dimers with bisamide linkage bearing powerful membrane-permeabilizing activity, *Org. Lett.* **4**, 2087–9 (2002).
51. L. Li and B. Xu, Multivalent vancomycins and related antibiotics against infectious diseases., *Curr. Pharm. Des.*, **11**, 3111–24 (2005).
52. (a) A. Liakatos and H. Kunz, Synthetic glycopeptides for the development of cancer vaccines, *Curr. Opin. Mol. Ther.*, **9**, 35–44 (2007). (b) J. Huskens, Multivalent interactions at interfaces, *Curr. Opin. Chem. Biol.*, **10**, 537–43 (2006). (c) N. V. Bovin, A. B. Tuzikov, A. A. Chinarev and A. S. Gambaryan, Multimeric glycotherapeutics: new paradigm, *Glycoconj. J.*, **21**, 471–8 (2004).
53. (a) G. Guizzunti, T. P. Brady, V. Malhotra and E. A. Theodorakis, Trifunctional norrisolide probes for the study of Golgi vesiculation, *Bioorg. Med. Chem. Lett.*, **17**, 320–5 (2007). (b) W. W. Qiu, J. Xu, D. Z. Liu, *et al.*, Design and synthesis of a biotin-tagged photoaffinity probe of paeoniflorin, *Bioorg. Med. Chem. Lett.*, **16**, 3306–9 (2006). (c) N. Kanoh, S. Kumashiro, S. Simizu, *et al.*, Immobilization of natural products on glass slides by using a photoaffinity reaction and the

detection of protein-small-molecule interactions, *Angew. Chem. Int. Ed. Engl.*, **24**, 5584–7 (2003).
54. M. D. Alexander, M. D. Burkart, M. S. Leonard, *et al.*, A central strategy for converting natural products into fluorescent probes. *Chembiochem.*, **7**, 409–16 (2006).
55. (a) D. L. Boger, J. Desharnais and K. Capps, Solution-phase combinatorial libraries: modulating cellular signaling by targeting protein–protein or protein-DNA interactions, *Angew. Chem. Int. Ed. Engl.*, **42**, 4138–76 (2003). (b) D. L. Boger, Solution-phase synthesis of combinatorial libraries designed to modulate protein–protein or protein-DNA interactions, *Bioorg. Med. Chem.*, **11**, 1607–13 (2003).

4

Interface Peptides

Mark W. Peczuh and Richard T. Desmond
Department of Chemistry, University of Connecticut, Storrs, CT, USA,

4.1 Interface Peptides Defined

In a protein–protein interaction (PPI), it is often only a fraction of the surface on the interacting partners that is involved in binding. The complementary regions of each define an interface. It follows that a peptide corresponding to one side or the other of this interface region should be able to bind its complementary partner in competition with the native protein itself (Figure 4.1). An interface peptide could be broadly considered in terms of the structure of the interface rather than the exact sequence. That is, the interface peptide may be composed of fragments defined by residues that are close in three dimensional space but are far apart in sequence space. More frequently, it is a contiguous sequence, as part of a loop or helix for example, that defines this interface region. Interface peptides are perhaps the most straightforward examples of rational inhibitors of PPIs.

Some properties of short peptides put limits on their overall effectiveness as inhibitors of PPIs. First amongst these is the fact that an isolated peptide's conformation is often poorly defined. The binding conformation of an interface peptide may be only one of many low energy conformations it can adopt. The lack of preorganization would then be reflected in a lower affinity of the interface peptide for the target protein. In a cellular context, flexible interface peptides may be especially susceptible to degradation by native proteases. This too would attenuate their activity. Therefore, not surprisingly, much of the work on interface peptide strategies for inhibiting PPIs has focused on countering these interrelated issues. The advantage of interface peptides is that they are easily amenable to investigation of structure activity relationships. They also serve as a starting point for the development of

Protein Surface Recognition: Approaches for Drug Discovery Ernest Giralt, Mark W. Peczuh and Xavier Salvatella. © 2011 John Wiley & Sons, Ltd

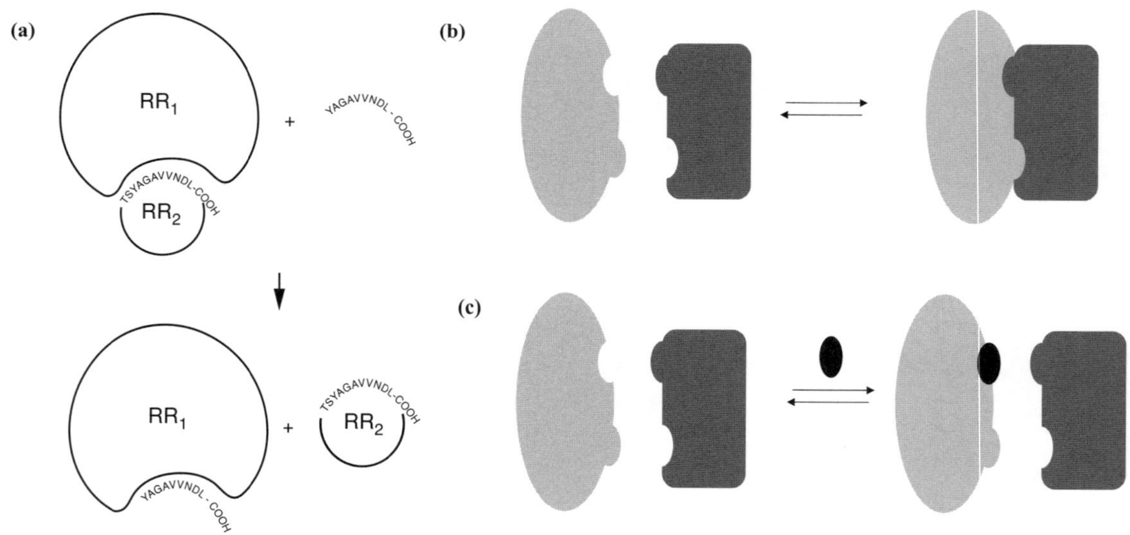

Figure 4.1 *Schematic representations of interface peptide inhibition of protein-protein interactions (PPIs). (a) Interface peptide inhibition of herpesvirus ribonucleotide reductase. A peptide from subunit 2 (RR_2) prevents RR_1/RR_2 association, thereby inhibiting enzyme activity. Reprinted by permission from Macmillan Publishers Ltd [23], copyright 1986. (b) A generalized scheme showing a PPI. (c) Inhibition of the association in (b) by an interface peptide*

small molecule peptidomimetics that would have the same activity and may prove to be therapeutic candidates. The ease in preparation of short peptides by either solid-phase peptide synthesis (SPPS) or expression in bacterial hosts is also a contributing factor to their use. Incorporation of unnatural amino acids that may increase the affinity of the interface peptide for its target is possible by both methods, although significantly easier via the SPPS route. Expression in bacteria or other hosts allows for facile preparation of large numbers of variants by genetic engineering. Overall, interface peptides present a readily available first-pass approach to inhibiting PPIs.

The main objectives of this chapter are to collect several illustrations of interface peptides and to detail reported strategies that address the shortcomings of natural peptides. Recent reviews of interface peptides in a number of contexts offer further reading. The cited reviews often place interface peptides in a broader context of inhibiting PPIs generally [4–12]. Information gleaned from the examples of interface peptides presented here will be elaborated in subsequent chapters. This fact emphasizes the starting point nature of interface peptides as inhibitors of PPIs. The information provided by interface peptides helps to propel biological investigations and also to develop therapeutic small molecules. Major divisions in this chapter have been made based on the nature of the interface peptides themselves – unmodified or modified. Unmodified peptides, as the name suggests, are simply oligopeptides of varying lengths. Their sequence may be from one protein of the PPI or a naïve/random sequence. Mini-proteins, whose sequence includes portions dedicated to inhibiting a given PPI and other portions dedicated to maintaining its overall fold, are treated in this section. Modified peptides are peptide sequences that have been manipulated or altered in some way. Common modifications included in this section are peptides that have been conformationally constrained by cyclization and sequences that contain unnatural amino acids. A special group of peptides derived in large part from unnatural β-amino acids, known as β-peptides, is considered as a subset of the modified peptides.

4.2 Unmodified Peptides

The rudimentary exercise of using short oligopeptides to inhibit PPIs was already an important tool for cellular biology by the mid 1980s. In these experiments, fragments from proteins of the extracellular milieu were used to inhibit cell/cell-matrix or cell/virus interactions. For example, fibronectin [13–17] based peptides or collagen based peptides [18] were used to inhibit cell binding to the matrix. The arginine-glycine-aspartate (RGD) tripeptide sequence of fibronectin interacts with cell surface receptors. In addition to being essential for the maintence of cell shape, the RGD motif has been widely used for artificially anchoring cells on surfaces for biochemical applications [19]. Other early studies include a loop from TGFa inhibiting viral infection [20] and a peptide from a tryptic digest inhibiting diazepam binding to its receptor [21].

Although there are earlier reports of similar phenomena [22], two back-to-back papers reporting the inhibition of herpesvirus ribonucleotide reductase using short, synthetic peptides derived from one of the enzyme's two subunits are noteworthy [23–24]. The subunits are unequal, RR_1 being relatively large (ca 135–140 kDa) and RR_2 small (ca 40 kDa). It was known that both subunits were necessary for activity, and that the carboxy terminus of the smaller subunit was bound by the larger one [25]. Both reports used a nonapeptide (9 aa) with a sequence that matched the C-terminal residudes of the RR_2 subunit as an inhibitor of the reductase activity. The papers demonstrated that enzymes consisting of multiple subunits may yield to inhibitors that were not targeted at the active site or an allosteric site. Instead, the short peptides from one subunit (RR_2) acted by binding the complimentary subunit (RR_1) thereby preventing the organization of a competent, active enzyme. Figure 4.1 shows the model schematically and is also a clear representation of the interface peptide approach in a general sense. Kinetics measurements [24] of the enzyme in the presence of varying concentrations of the interface peptide showed noncompetitive inhibition, consistent with the model described. A high degree of selectivity between viral and cellular ribonucleotide reductases was also demonstrated [23].

The perspective offered in these papers also anticipated the future and further application of the interface peptide strategy for the inhibition of PPIs. The 1986 Langelier paper [24] is broad in speculating on the application of interface peptides. It states:

> We predict that the use of such synthetic peptides, derived from amino-acid sequences implicated in the association of proteins with receptors, membranes, DNA, RNA or enzyme subunits, will be of prime importance in elucidating macromolecular interactions.

The foresight of these comments are remarkable twenty years on. As mentioned previously, the importance of interface peptides stems from their use as tool for molecular and cellular biology, and as starting point for the development of nonpeptide inhibitors of PPIs. The two sections on unmodified peptides below cover different aspects of interface peptide inhibition of PPIs. The first section focuses on listing examples of the strategy in action and then tries to draw some generalities out of these. The second section focuses in on the use of small folded mini-proteins to effectively present functionality in interface peptides.

4.2.1 Examples of Interface Peptides

Application of the interface peptide strategy is best illustrated by providing some specific examples. Table 4.1 lists several of these along with a number of characteristics of the PPI that are potentially important for the development of small molecule inhibitors of the interactions. The parameters listed are: (a) the origin of the interface peptide; (b) its size, in terms of the number of amino acids; (c) a secondary structural motif, if applicable; (d) the dissociation constant of the protein-interface peptide interaction.

Considering first the origin of the peptide, it is reasonable to expect that in a PPI, the two protein faces that comprise the interface are not equal. Selecting from which protein the interface peptide should be derived is an important design consideration. Secondly, the number of amino acids of the interface peptide gives an indication about the size of the interface itself. The facility in simplifying a given interface peptide to a small molecule is also a practical implication of the interface peptide size. Identification of a structural motif integrates information about the first two parameters. Lastly, the individual dissociation constant values provide information about the magnitude of a particular interaction and considering the K_d values as a whole gives insight into the strength of the PPIs in general. In some cases, IC_{50} values are given rather than dissociation constants. These are marked

Q1 **Table 4.1** *Examples of interface peptide inhibition of protein-protein interactions (PPIs)*

Viral/Bacterial/Pathogenic

Entry	Proteins of PPI	Origin	Size	Motif	K_d/IC_{50}*	Ref.
HIV gp120/CD4	CD4/gp41	phage(CD4)	12 aa	extended or β-sheet	330 nM*	26
		CD4	27 aa	β-hairpin	40 μM*	27
HIV gp41	gp41 six-helical bundle	gp41	34 aa	α -helix	2 ng/mL*	28
	gp41 six-helical bundle	gp41	34 aa	α -helix	2–5000 nM*	29
	gp41 six-helical bundle	phage	7 aa	extended	9 μg/mL*	30
HIV p24	HIV p24 dimerization	p24	10 aa	α -helix	37 μM	31
HIV RT	HIV RT dimerization	HIV RT	10 aa	W repeat	0.24 μM	32
HIV protease	HIV protease dimerization	protease	27 aa	β -sheet	0.23–0.32 μM*	33
	HIV protease dimerization	protease	11 aa	β -sheet	2 μM*	34
	HIV protease dimerization		13 aa	β -sheet	20 μM*	35
HIV integrase	HIV integrase dimerization		29 aa	α –helix/coiled coil		36
	HIV integrase dimerization		21 aa	α –helix	80 nM*	37
	HIV integrase dimerization	SPLC	6 aa		2 μM*	38
HIV Vif proteins	HIV Vif oligomerization	Vif	12 aa	poly-Pro (PP)	4.8 μM*	39
HIV assembly	Gag capsid domain	phage	12 aa		15 μM	40
HIV Vpr/HHR23a	HHR23 UBA/HIV-Vpr	UBA(2)	44 aa	helix/loop		41
HSV ribonucleotide reductase	reductase subunits RR_1/RR_2	HSV-R2	9 aa (5 aa)		24 μM*	23, 24
HSV-1 VP16	VP16/HCF	VP16	8 aa or 14 aa		60 μM	42
HSV-1 DNA polymerase	HSV Pol-UL42	pol	36 aa	α-helix	2.8 μM*	43
	HSV Pol-UL42	pol	18 aa	α -helix	11 μM*	44
RSV	HR-N/HR-C	HR-C	13 aa	α -helix	36 nM*	45
SARS-CoV	Spike Protein	HRN/HRC	18 aa	α –helix/coiled coil	5–12 μM	46
Rb/HPV E7		HPV E7	9 aa	extended	110 nM	47
HeV F	Hendra virus Fusion/Human ParaInflu prot	HPIV	45 aa	α –helix/coiled coil	7.5–20 nM*	48
HCMV DNA Polymerase	DNA Pol (UL54)/UL44	DNA Pol	22 aa		11 μM*	49
L. casei thymidylte synthase	Thymidylate synthase-homodimer	TS	20 aa	β-hairpin		50

(continued)

Table 4.1 *(Continued)*

Viral/Bacterial/Pathogenic

Entry	Proteins of PPI	Origin	Size	Motif	K_d/IC_{50}^*	Ref.
EcoR1	EcoRI homodimer	EcoRI	16 aa	α -helix	27 μM*	51
IgG Fc	IgG Fc/protein A of Staphylococcus aureus		13 aa	β-hairpin		52
PfTIM	Triosephosphate Isomerase Dimerization	PfTIM	12 aa	β -sheet	600 nM*	53

Extra-Cellular

Entry	Proteins of PPI	Origin	Size	Motif	K_d/IC_{50}^*	Ref.
Aβ	Abeta protein oligomerization	phage	11 aa		40–85 μM*	54
	Abeta protein oligomerization	phage	56 aa	β-sheet	50 μM*	55
hER	hER dimerization	hER	12 aa	SH2 domain	50 μM	56
	hER dimerization	hER	16 aa	α-helix		57
p185(neu)	Homodimerization	Tneu	115 kDa	α -helix		58
p185(neu)/EGF	p185/EGF heterodimer	p185	46 aa			59
VEGF/VEGFR	VEGFmimic/VEGFR	VEGF	15 aa	helix	1 nM	60
VEGF/KDR	VEGF/KDR receptor	phage	20 aa		700 nM*	61
erbB2	erbB2 dimerization	erbB2	12 aa	β-turn/sheet	300 nM	62
	erbB2/mAb	erbB2	55 aa			63
PDGFb	PDGFb/PI3K	PDGFb	4 aa (P*)	extended	0.45 μM*	64
IGF-1/IGFBP-1	IGF-1/IGFBP	phage	14 aa	α-helix	50 nM	65
IGF	IGF/peptide	phage	17 aa		1 μM*	66
(TNF)BAFF/BR3	TNFlike ligand BAFF/BR3	BR3	12 aa	β-hairpin	70 nM*	67
C5a/C5a receptor	C5a/C5a receptor	C5a	6 aa	β-turn	20 nM*	68
ZAP-70/T-cell receptor	ZAP-70/T-cell receptor	TAM	17 aa	SH2 domain		69

TCR	T-cell receptor/MHC peptide		9 aa	extended	3 μM	70
IL-1	IL-1/IL-1 receptor	phage	21 aa		45 μM*	71
IL-4	IL4/IL4-R	IL4		α-helix	5 μM	72
gpVI/collagen	collagen/gpVI	collagen	15 aa	helix		73
α2β1/collagen	collagen/integrin α2β1	collagen	60 aa	collagen helix		74
GPCR	G-protein heptahelical receptor		83 aa	helix		75
G-protein (heterotrimer)	beta-gamma subunits	phage	14 aa	helix	470 nM	76
G-protein/βARK	beta-gamma/βARK	βARK	28 aa		76 μM*	77
Rhodopsin/arrestin	Rhodopsin/arrestin	rhodopsin	23 aa		34 μM*	78
Rhodopsin	Rhodopsin/Gt	G-prot.C-term.	15 aa		200 μM	79
AchR/α-bungarotoxin	AchR/α-bungarotoxin	phage	13 aa	β-sheet	2 nM*	80, 81

Intra-Cellular

Entry	Proteins of PPI	Origin	Size	Motif	K_d/IC_{50}*	Ref.
Jun	Jun dimerization		46 aa	helix (colied-coil)		82
	Jun dimerization		32/15 aa	helix (colied-coil)	11 μM	83
bHLH TF E47	E47 dimerization		20 aa	β -sheet	3.7 μM	84
Ras-Raf	Ras-Raf	Ras effector	23 aa	β -sheet	4.4–28 μM	85
DR5	DR5/pro-apopoptic peptide	phage	19 aa*		300 nM	86
BakBH3/Bcl	Bak BH3 domain with Bcl-2/Bcl-X_L	BakBH3	34 aa	α -helix (mixed)	52 nM	87
	Bak BH3 domain with Bcl-2/Bcl-X_L	BakBH3	35 aa	α -helix (mixed)	505 nM	88
p53/MDM2	p53-MDM2	p53	12 aa	α -helix	5 nM*	89
	p53-MDM2	p53	15 aa	α -helix	600 nM	90
	p53-MDM2	p53	15 aa	α -helix (retroinverso)	15 μM*	91
p53/hDM2	p53/hDM2	p53	37 aa	α -helix (mixed)	35 nM	92
	p53/hDM2	p53	10 aa	β-hairpin	140 nM*	93

(*continued*)

Table 4.1 (*Continued*)

Entry	Proteins of PPI	Origin	Size	Motif	K_d/IC_{50}^*	Ref.
Intra-Cellular						
	p53/hDM2	phage	11 aa	extended	313 nM*	94
Cdk4-p16	cdk4-cyclin D1	p16	26 aa	α -helix		95
	cdk4-cyclin D1	p16	20 aa			96
Cdk2	cdk2-cyclinA	cyclinA	22 aa	α -helix		97
	cdk2/p53	p53	20 aa			98
Cdk2/CyclinE	cdk2/cylinE	HIV Tat	10 aa			99
p21 cyclinE/Cdk	p21-cyclinE/Cdk	p21	65 aa	β -sheet		100
p21 PCNA	p21–PCNA	p21	16 aa		100 nM	101
Src-SH2 domain	Src-SH2 domain		4 aa	SH2 domain	5.6 μM	102
Grb2 SH2	Grb2 SH2		3 aa	type I h-turn	8.6 μM*	103
Grb2-SOS	Grb2SH3/SOS	SOS	10 aa	PP helix	3.5 μM	104
CK2	protein kinase CK2 a/b subunits	CK2b	60 aa			105
Abl-SH3/PPII	Abl SH3 domain/APP-PPII hybrid	APP-PPII	39 aa	PPII helix	20 μM	106
AMAP1/Cortactin	AMAP1-PP/Cortactin SH3	AMAP1	32 aa	PP helix	10 μM*	107
Stat3	Stat3 dimerization	Stat3	6 aa (P*)	extended	0.15 μM*	108
EVH1 domains	ActA with EVH1 domain of Mena	ActA	31 aa	PP II helix	290 nM	109
Ubiquitin/Proteasome	HHR23 UBA/S5a	S5a	26 aa	helix		110
gluthatione reductase	homodimer	GR	24 aa	helix		111
guanyl cyclase	dimer	GC-A	42 aa	α -helix		112
RNAse	RNAse/S-peptide	phage(S-peptide)	19 aa	helix	5 nM	113
PKC	PKC/RACK1	RACK1	8 aa		4 nM*	114
γ-secretase	γ-secretase/amyloid-β precursor protein (APP)	APP		helix		115
FVIIa/FactorX	Factor VIIa/X	phage	17 aa	β-sheet	1 nM	116
Gt a subunit	Gt a subunit/PDE	Py	11 aa		2.5 μM	117

PKA/MAP2	Prot. Kin. A/Microtub.-assoc. Prot. 2	MAP2	24 aa	α-helix	4 nM	118
ADR	homodimerization	ADR	20 aa		2.1-3.2 μM	119
Smac/IAPs	Smac/IAP	Smac	9 aa	α-helix	2.0 μM	120
CaM	CaM/CaM-dependent kinases	phage	10 aa			121
BEACH/PH	Beach domain of FAN/PH domain	PH	115 aa	β-sandwich	1 μM	122
NEMO/IKK	NEMO/IKK-Nemo binding domain (NBD)	NEMO	44 aa	helix		123
KIX/KID	KIX domain of CBP with KID domain of CREB	KID	34 aa	α-helix (mixed)	0.5–10 μM	124
APC	Coiled coil dimer	anti-APCp1	54 aa	α -helix		125

with an asterisk in the table. These numbers are not as accurate as K_d values in quantifying the affinity of an interface peptide/protein interaction, but they do provide a reference point for the general value. Note that, where there are blank entries in the table, the parameters were unable to be culled from the manuscript based on the given citation.

4.2.2 Guiding Concepts

An analysis of the information in Table 4.1 provides some insights into the underlying themes that characterize interface peptide/protein interactions. This information may be useful for the implementation of the interface peptide strategy in a new system. Observations include: (i) Interface peptides are of general applicability. The table is segregated into PPIs from viral/bacterial/pathogenic systems to extracellular and intracellular systems, with nearly equal representation in each category. (ii) The average size of an interface peptide is 24 aa. The definition of interface peptide here is intentionally broad. Nonetheless, the average value reported is reasonable if not slightly high. An analysis of crystallographic data from protein–protein complexes indicates that approximately 150–400 atoms comprise an average PPI [126]. This translates to 15–40 amino acids. While the identity or nature of the amino acids at the interfaces in Table 4.1 was not analysed, these trends have been presented elsewhere [127]. (iii) Secondary structures from those listed show that of the 98 total entries, 39 involve an α-helix and 25 involve β-sheets, hairpins, or related extended structures. The similar distribution of these motifs agrees with an early analysis of PPIs by Thornton [128]. As will become apparent, secondary structures are attractive targets for development of interface peptides and peptidomimetics. They allow for the concentration of interacting residues in a small amount of sequence space. Knowledge of the interface topology, including whether or not it is part of a secondary structure, is imperative for the further development of modified interface peptides and peptidomimetics. (iv) K_d/IC_{50} values in Table 4.1 range from 10^{-9} to 10^{-4} M with an average value of 12.5×10^{-6} M (12.5 μM) for an interface peptide/protein interaction. Dividing the K_d by the number of residues of the interface peptide allows for a measure of the efficiency of the interaction. Combining secondary structure information with the K_d values may also provide an estimate of the magnitude of binding for a specific motif. (v) Interface peptides may come about by design, by combinatorial biology, or a melding of the two. The most direct method for designing an interface peptide is by utilizing a sequence from one of the constituent proteins of the PPI. The retro-iverso approach [129] is not presented here. Methods such as phage display [11], bacterial display [130], yeast display [131], and mRNA display [132] used to identify ligands for a target protein are a powerful way to develop interface peptides [10]. Several examples from phage display libraries are in Table 4.1. Overall, consideration of these parameters could facilitate implementation of an interface peptide strategy for a new targeted PPI.

Another consideration when targeting a protein protein interaction is deciding which side or face to inhibit of the interface. Binding to either effector or receptor will disrupt the binding process, but one face may be a more favorable target in terms synthesis, secondary structure and physical accessibility (cleft vs projection). Figure 4.1 demonstrates how each side of the interface can be bound by a peptide to disrupt binding. For example, a secondary structural element could fit into the cleft of one face. Alternatively, an extended peptide could conform to the projection on the complimentary protein of the interface.

4.2.3 Folded Interface Peptides – Protein Grafting

Combining the power of both the design and the combinatorial approaches is an attractive way to develop active inhibitors of PPIs. Miniature proteins arising from natural sequences [133, 134] or that have been developed *de novo* [135, 136] have a clear application as scaffolds for the development of interface peptides. Usually a sequence that takes up a well defined fold is manipulated so that individual residues that are not essential for structure are substituted with residues that will facilitate binding to a target protein surface. The residues used in the replacement are said to be 'grafted' onto the mini-protein scaffold [137, 138]. The grafted mini-protein strategy is perhaps related to or a limiting case of engineered PPIs [139–143]. A need to maintain a folded state in the mini-protein strategy often leads to a longer peptide which in some respects is contrary to the developmental aspect of interface peptides. The mini-protein is typically smaller than the parent interface protein but larger than the interface segment itself. However, the idea of a temporally expressed gene whose product could inhibit a given PPI with high affinity and specificity is attractive for molecular biology applications.

One grafting strategy was originally utilized in the preparation of novel DNA major-groove binders where key residues of the α-helix from GCN4 were grafted onto a avian pancreatic polypeptide (aPP) scaffold [144–146]. This natural polypeptide has a 36 residue amino acid sequence which has an eight residue polyproline type II (PPII) stretch stacked against eighteen residue α-helix (Figure 4.2). The PPII helix/α-helix interaction stabilizes the fold and allows the exposed face of the α-helix to interact with its target. The technique has since been adopted to develop ligands for protein targets [147]. In the PPI, one protein becomes the target while the interface, or epitope, of the other is mimicked through the grafted miniature protein. Proteins most amenable to mimicry are those that use structural motifs inherent in the scaffold as part of the protein–protein interface. For example, binding residues from α-helices have been grafted onto the α-helix of the aPP scaffold. Alternatively, the PPII helix of aPP has also been used to bind SH3 and EVH1domains, respectively. Details on both of these applications are considered below.

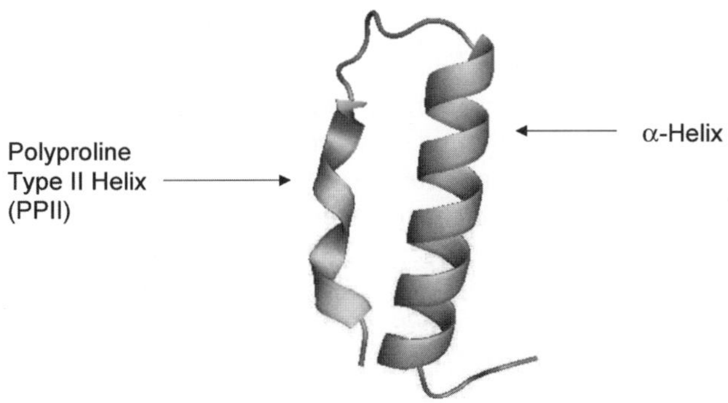

__Figure 4.2__ Avian Pancreatic Polypeptide (aPP) has a 36 residue amino acid sequence which has an 8 residue polyproline type II (PPII) stretch stacked against an 18 residue α-helix. PDB code 1PPT[148] viewed using PyMol

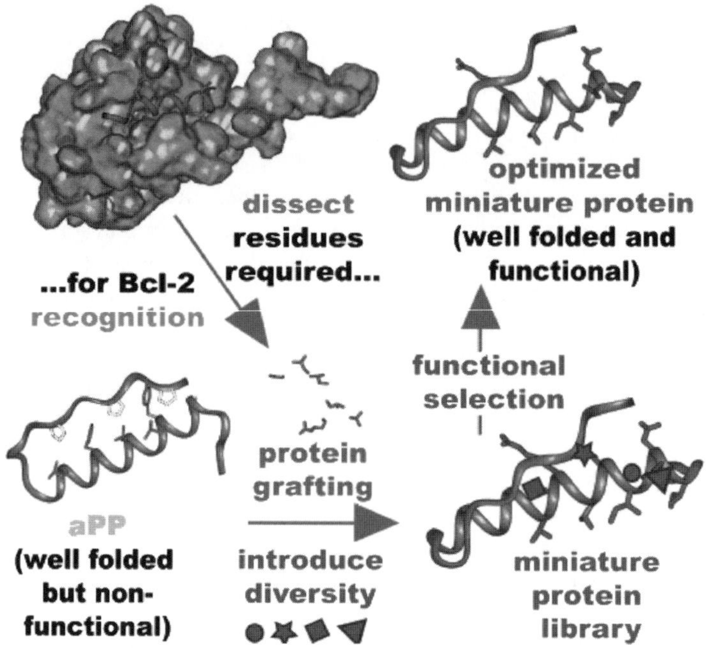

Figure 4.3 *Protein grafting using the aPP scaffold. Residues important for the fold are maintained and residues for binding are grafted onto the scaffold. Randomization at positions where these roles conflict provides a folded and functional mini-protein. Reprinted with permission from (87). Copyright Wiley-VCH Verlag GmbH & Co. KGaA*

The aPP scaffold has been applied to protein–protein interactions where an α-helix of one partner interacts with a deep cleft of another. Protein pairs that exemplify this type of interaction are the p53 helix with M/hDM2 [92] and Bak BH3 helix with Bcl-2 [87]. Focusing on the Bak-Bcl-2(Bcl-X_L) work (Figure 4.3), the first step was to identify the binding sequence of Bak responsible for Bcl-2 recognition. The grafting of a primary binding sequence to a scaffold is a balance of affinity to the target and a well folded structure. A key element of the design process is the alignment of the binding sequence with the aPP scaffold structure. Schepartz defined a three step scoring procedure that was used for the Bcl-2 binding mini-proteins. It can, however, be considered a general procedure. First, conflicts in alignment between residues which are crucial to maintaining the aPP structure or binding the target protein are considered. Alignments with multiple conflicts are then eliminated. Second, sequence alignments are ranked based on potentially unfavorable steric interactions in the protein/mini-protein complex. The third scoring step looks at the backbone RMSD of each alignment. At the conclusion of the scoring they defined an alignment of 14 residues that include amino acids essential to expressing both binding affinity and well folded structural properties. The mini-proteins included six Bak BH3 residues that contribute to binding Bcl-2 protein and four aPP residues that contribute to structure. A phage display library was developed and subsequently screened (using Bcl-2 for affinity maturation) with these ten specific residues and a random combination of all 20 amino acids across the remaining four residues. A hybrid aPP-Bak BH3 mini-protein was identified by this process that had a 100-fold higher affinity than the native Bak sequence for the Bcl-2 protein (52 nM vs 4.9 μM).

The authors note that affinities in the nM regime may be able to demonstrate significant selectivity, even amongst related proteins. Similar to the native Bak peptide, the new peptide identified showed a greater affinity for a related protein, Bcl-X_L. In subsequent work [88], the original Bcl-2, Bcl-X_L binding mini-protein was subjected to a positive/negative selection protocol. The affinity of the new aPP-Bak mini-protein to Bcl-2 was increased over that of Bcl-X_L. This new mini-protein *reversed* the selectivity of the natural peptide, binding Bcl-2 with a K_d of 505 nM and Bcl-X_L with a K_d of 2.7 μM. The switch in selectivity was influenced by subtle electrostatic differences in the binding grooves to the target proteins. Overall, it underscores the power of phage display maturation when combined with the protein grafting design strategy [124].

In turn, the PPII helix of aPP can also be optimized to target protein partners. Polyproline II (PPII) helices are known to be bound by WW, SH3, and EVH1 domains [109]. A report on an aPP based ligand of the Abelson tyrosine kinase SH3 domain (Abl-SH3) indicates a generality in the strategy [106]. SH3 domains have been shown to interact with synthetic proline rich sequences, but the PPII conformation required to bind is maintained by Pro residues which cannot be mutated because of entropic destabilization [149, 150]. This problem was overcome by grafting the PPII helical epitope onto the aPP scaffold. The sequence introduced on the PPII helix of this miniature protein was shown to bind to the Abl-SH3 domain with a K_d of 20 μM while the α-helix aided in maintaining the aPP fold. Similarly, a designed mini-protein based on the aPP scaffold showed selectivities up to 120:1 ($K_d = 290$ nM) for the EVH1 domains from Mena, VASP, and Ev1 [109]. An interesting observation is the relative magnitude of binding in the SH3 and EVH1 examples above. The K_d of the mini-protein ligand for Abl-SH3 was essentially the same as the truncated PPII sequence, whereas the affinity of the mini-protein ligand for the EVH1 domains was approximately 100–1000 times greater than the PPII sequences alone. The origin in these differences is unclear but may reflect differences in the domains themselves.

A coiled-coil motif [151] has also been used as a scaffold for mini-protein interface peptides. In the same way that the α-helix/PPII helix interactions stabilize the aPP structure, the leucine zipper helps to define the overall structure in these systems. The heptad repeat of coiled-coils is itself a unique and useful PPI. In the systems described, the individual helices of the coiled-coil unit are modified so that they are covelantly linked by a disulfide linkage. With a defined fold in place, the solvent exposed residues of the leucine-zipper can be grafted to produce an active epitope. GCN4 is a formidable scaffold candidate because it is a stable dimer and extensive study has rendered it easy to manipulate. Mimicry of both interleukin-4 (IL-4) [72] and the HIV gp41 C-peptide [152] using GCN4 has been reported.

Interaction between the cytokine IL-4 and its receptor IL-4Rα is an important event in T-cell mediated immune responses. Due to this fact, the IL-4/IL-4Rα interaction is integral to the allergic response and is a target of drug design. IL-4 exists as a four-helix bundle and binds to IL-4Rα on one face, via two helices of the bundle. Residues on IL-4 that contribute to the binding of IL-4Rα had previously been identified. The authors then introduced binding residues onto the GCN4 scaffold in a stepwise fashion. Alignments were initially partitioned into two models. Proper folding of the grafted proteins based on either alignment was assessed by molecular dynamics simulations prior to the synthesis of the proteins. Peptides from the first alignment group showed only weak binding to IL-4Rα. The second group were better ligands, and were optimized by addition of the disulfide to eventually reach a K_d value of 5 μM for their interaction with IL-4Rα, in comparison to a K_d of 1.4 nM for IL-4 itself.

The discrepancy in K_d values was attributed to three residues that were shown to be significant for binding to IL-4Rα, but could not be mutated because they would also compromise the GCN4 fold. This introduces a limitation in the mini-protein strategy. Depending on the fold of the mini-protein and the match at the interface, overall affinity may be refractory to optimization due the need for balance between structure and function. This may be partially alleviated by turning to display techniques to increase affinity in nonintuitive ways while maintaining the integrity of the scaffold.

4.3 Modified Peptides

The disulfide linkage introduced into the GCN4 scaffolded mini-protein above formally made it a modified peptide by our earlier definition. Unmodified interface peptides are those that bind to a target protein without any modification to the primary sequence (other than folding). In general, interface peptides may be modified in at least three ways. First, reactive functional groups such as the N-terminal amine, the C-terminal acid, or sidechain groups may be derivatized in the process of its synthesis. The second main modification is to constrain the peptide's conformation, usually by cyclization. Most common among the cyclization strategies are the formation of disulfides from cysteine thiol groups, by amide bond formation through lysine/aspartate/glutamate sidechains or hydrazone linkages formed by reaction of acetal and hydrazine derivatives. Last, unnatural amino acids may be incorporated into the sequence. Obviously, the three types of modification are related and a given example could be counted in both categories.

4.3.1 Peptides Constrained to an α-Helical Conformation

Fuzeon® (enfuvirtide) is a prime example of the potential therapeutic and commercial success promised by interface peptides [28, 153, 154]. Worldwide sales of the drug were 249 million (USD) in 2006 [155]. Fuzeon® is an α-helical polypeptide (36 aa) that inhibits HIV infection by preventing viral and host cell fusion [156]. In this process, a trimeric coiled-coil fusion protein (gp41) reorganizes into a six helix hairpin complex by folding the C-terminal domain of the protein back upon the N-terminal domain. The peptide drug, which corresponds to residues 643–678 of the fusion protein (the C-terminal domain), acts by preventing N- and C-peptide reorganization by binding the central N-terminal trimeric coiled coil [157]. This mode of action is unique amongst current anti-HIV drugs. Importantly, the targeted step is fundamental to the life cycle of the virus and is unlikely therefore to develop resistance to this type of inhibition.

The Fuzeon® story and its mechanism of action are fairly well-known. It is presented as paradigm for the potential significance of this strategy. The only modification of this peptide relative to a natural peptide is that the C-terminus is capped as a primary amide and the N-terminus is acetylated during its preparation by solid phase peptide synthesis (Figure 4.4). This modification stabilizes the α-helical conformation of the peptide by partially alleviating the unfavorable fixed charge/helix macrodipole interaction. In a strict interpretation, then, this may be considered a modified peptide although the modifications made are minor and better approximate the amide linkages that would be present in the native protein from which the interface peptide was derived.

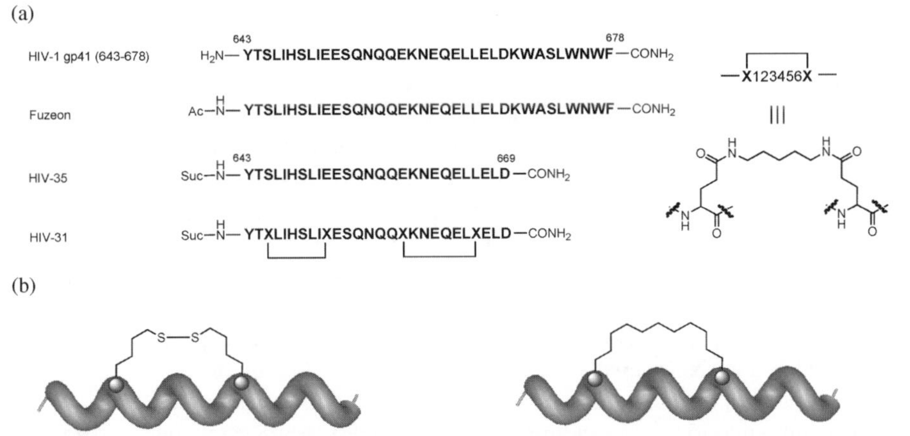

Figure 4.4 *(a) Native and modified sequences of HIV gp41 C-domain. On the right is the i, i + 7 relationsip between the modified glutamate residues. (b) Disulfide stabilized α-helix (left) and hydrocarbon 'stapled' helix (right), both in an i, i + 7 relationship*

A peptide based on the Fuzeon® sequence was subjected to more extensive modification [158]. First, the sequence was reduced to 27 amino acids rather than the original 36 by removal of the C-terminal nine amino acids from Fuzeon (HIV-35). Truncation of this sort usually leads to a less conformationally stable peptide. The sequence was also capped as a primary amide at the C-terminus, but the N-terminus was derivatized as as a succinate as opposed to the acetate in Fuzeon®. The designed peptide HIV-31 shared the N- and C-terminal caps, but also incorporated diamide linkages between engineered glutamate residues positioned along the same face of the helix (in an i, i + 7 relationship – Figure 4.4a). The two diamides were positioned so that the HIV-31 peptide could properly present its binding epitope and still associate with the N-terminal trimeric coiled-coil. The constrained peptide (HIV-31) presents nearly twice as much helical character in solution relative to the parent peptide HIV-35 as determined by CD spectroscopy. Similarly, the HIV-31 peptide was approximately 100 times more effective as HIV-35 at inhibiting viral entry in a primary infectivity assay. Further, HIV-31 was as potent as Fuzeon® itself in the same assay.

The theme underpinning the process in the Fuzeon® example is consistent with the general concept of minimization. Interface peptides are the first step; this includes the iterative process of evaluating truncated peptides to find a minimum epitope. Modification of the shortened interface peptide is then the next step toward development of small molecule inhibitors of PPIs. The ability of HIV-31 to match the activity of Fuzeon® is telling to this end. Fuzeon® is administered as a subcutaneous injection because of its poor bioavailability and susceptibility to proteases. Although a direct comparison is not able to be made, the protease issue is also a significant one at this point. By rigidifiying the helical conformation by the sidechain constraints, the peptide is likely to be more resistant to peptidases [159], and have greater bioavailability. A disulfide based strategy for stabilizing the gp41 C-peptide has also been reported [160].

Among other examples of amide [115, 161–164], hydrazone [165] or disulfide [166, 167] linkages stabilizing α-helical peptides are hydrocarbon linked peptides (Figure 4.4b) [159, 168, 169]. The hydrocarbon links have been constructed by Ring-Closing Metathesis (RCM) of allyl functionalized serine or homoserine residues, or by incorporation of α,α-disubstituted amino acids that contain a terminal alkene of varying lengths [159]. The initial work incorporating hydrocarbon linkers was systematic in evaluating the disubstituted amino acid configuration (*R vs. S*) and relative spacing on the helix (i, i + 4 versus i, i + 7) so that the resulting RCM-derived macrocycle was efficient in stabilizing the helix conformation. A near doubling of helicity was observed (to >80% helicity) by derivatizing the ribonuclease S-peptide in this study. Moreover, the newly prepared peptides were over 40 fold more resistant to proteolysis by trypsin relative to the parent peptide. The formation of side-chain derived macrocycles prepared by this process has been termed 'hydrocarbon stapling' [170].

A hydrocarbon stapled BH3 peptide from the pro-apoptotic BID protein was used in the development of a BAX agonist [170]. The α-helical BH3 domain of BID binds to BAX and triggers apoptosis [171]. The 23 residue natural BID BH3 exists with 16% α-helicity in solution and the majority as a random coil. An all-hydrocarbon staple was inserted into the BID BH3 peptide sequence based on the earlier model work [159], increasing helicity to 87 %. Binding assays show a considerable improvement of K_d from 269 nM for the natural peptide to 39 nM for the 'stabilized alpha-helix of the Bcl-2 domain' (SAHB). Note that, relative to the mini-protein approach, the K_d values for these shortened, modified peptides are comparable or slightly better for their target Bcl-2 family members (52 nM for the mini-protein). Normalized for affinity per residue, these helices perform well. The authors also enhanced peptide half-life in a protease assay and improved cell permeability relative to the control, or unmodified, peptide. Importantly, the BH3 domain stapled helix peptides have been utilized to probe specific questions about the biological role of other Bcl-2 members in apoptosis. This work emphasizes the utility of interface peptides as tools for biochemical investigations [172].

4.3.2 Peptides Constrained to a β-Hairpin Conformation

A continuing theme in the application of the interface peptide strategy is minimalism. From a given PPI an interface peptide from one of the partners is taken to then inhibit the interaction. As has been hinted at and will become more apparent throughout this text, the jump from protein to interface peptide to peptidomimetic/small molecule is the process by which new therapeutics that act by inhibiting PPIs is likely to come about. In the α-helix examples just presented, the helix often uses one of its faces to bind with a target protein of a PPI. From one perspective this is an inefficient way to present properly oriented functionality in terms of the number of amino acids needed to bind the target protein with a given affinity. A similar presentation of important side-chain functionality can be achieved in peptide sequences which form other secondary structures such as β-hairpins.

Cyclic β-hairpin sequences that mimic α-helices have been reported [93, 173, 174]. An example, which will also be explained in more detail in Chapter 5, covers the p53/(h)MDM2 interaction. The original p53 15 amino acid α-helical peptide inserts a Phe, Trp, and Leu (i, i + 4, i + 7) that are situated along the same face of the helix into hydrophobic pockets of the surface of MDM2. Looking at the spacing of these key binding residues, the authors identified a β-hairpin as a scaffold for a mimetic that required only 8 amino acids in a cyclic array. The loop was templated in a β-hairpin conformation by a D-Pro, L-Pro dipeptide. The binding functionality is positioned along the backbone in a similar orientation relative

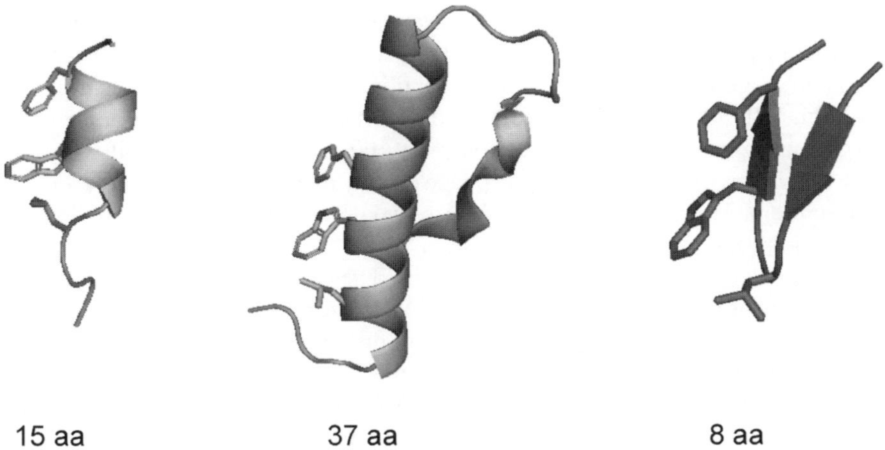

Figure 4.5 *(h)MDM2 ligands shown with key binding ligands Phe, Trp and Leu. The key binding residues are situated along one face of each peptide and insert into hydrophobic pockets of the surface of hDM2/MDM2. (a) original p53 epitope(1YCR) (b) aPP based miniprotein (1PPT) and (c) β hairpin mimetic(2AXI). PDB viewed and modified using PyMol*

to the natural epitope and other inhibitors such as the aPP based mini-protien (Figure 4.5). The first α-helical interface peptide (15 aa) bound with a K_d of 600 nM. The initial β-hairpin mimic produced an inhibitor with an IC_{50} of 125 μM which was optimized to an IC_{50} of 140 nM, a 1000 fold increase. The optimized β-hairpin incorporated unnatural amino acids in the hairpin motif. For comparison, the aPP based mini-protien (37 aa) had a K_d of 35 nM. Although it is impossible to compare IC_{50} to K_d values, consideration of the ratio of binding affinity per residue shows that the 8 residue cyclic β-hairpin is a respectable inhibitor of the p53/hDM2 interaction. The loop is smaller than the natural epitope as it eliminates the residues 'wasted' in the turn of a helix while displaying the 'hotspot' residues in the proper orientation for binding. Further, the optimized peptide benefits from the incorporation of unnatural amino acids and cyclization to rigidify its structure. In total these modifications demonstrate the power of the strategy.

The β-hairpin is also an independent motif that mediates PPIs in its own right. Examples include chemokine/CXCR4 interactions [175] and the gp120/CD4 interaction [176]. A D-Pro, L-Pro dipeptide templated cyclic β-hairpin related to the system just described (a 14 aa loop instead of the 8 aa loop above) was used to inhibit the chemokine/CXCR4 interaction [177]. CXCR4 is a cofactor to CD4 [178] and the CD4/SDR-1 (chemokine) interface holds great potential in the development of future therapeutics for treatment of diseases such as AIDS, cancer, inflammation, and arthritis.

A mini-protein based strategy was used in another study aimed at inhibiting the CD4 gp120 interaction [27]. The strategy fits into the modified section because the mini-protein used as the template contains three disulfides that constrain the peptide in an α/β-structural motif. The authors found that the scorpion toxin, scyllatoxin, presents the β-hairpin motif in a manner equivalent to the CD4 CDR2-like loop. Earlier work had shown that a mini-protein based on charbydotoxin (33 aa) from scorpion inhibited the gp120/CD4 interaction in the 10–100 μM range [179]. The scyllatoxin scaffold was desirable because it reduced the number of amino acids in the mini-protein to 27 and it also had superior overlap with the

CDR2 loop. The initial mini-protein where the key residues for binding gp120 were mapped onto the toxin gave IC_{50} values of approximately 40 μM. Alanine scanning on this initial mini-protein suggested other sites for modification, and an overlay was then optimized relative to the natural gp120 ligand in a rational fashion. The resulting optimized mini-protein gave an IC_{50} in the same assay of 400 nM, a 100-fold increase in affinity. The alanine scanning step in this process is akin to the phage display maturation presented earlier for the aPP mini-protein. The uniting element of these two processes is the added randomization step in addition to the initial design exercise.

β-Sheets make up 30% of all protein structures. They are primarily scaffolding but also play a key role in protein/protein and protein/DNA surface recognition. The motif is difficult to mimic as the residues that make up a sheet are distant in sequence [180] and the sheet structures tend to aggregate due to the nature of their hydrophilic and hydrophobic faces [135]. There are currently no human therapeutics involving β-sheet mimicry [181]. However, there are over 30 diseases that involve amyloid fibrils composed of proteins 'misfolded' into β-sheets including Alzheimer's, Parkinson's, Creutzfeldt-Jakob and Huntington's. Focus has been directed to inhibiting this type of protein aggregation using β-strands and miniature proteins with β-sheets. Ghosh has established that a 'miniature protein which presents a well-folded, aromatic residue-enriched, β-sheet can abrogate Aβ fibrillization at stoichometric concentrations' [55]. Also, Meredith [182, 183], Doig [184] and Kessling [185, 186] have done work toward disrupting amyloid aggregation. Real value in the use of β-sheets comes in the application of information from their PPI interactions to the development of β-strand peptidomimetics and small molecules [187].

4.3.3 β-Peptides as Interface Peptides: Foldamers

'A foldamer is defined as an oligomer with a characteristic tendency to fold into a specific structure in solution that is stabilized by noncovalent interactions between nonadjacent subunits [188].' Proteins, RNA, and even the mini-proteins discussed previously could be considered foldamers. A defining feature of designed foldamers is that the oligomers consist at least in part of unnatural monomers. β-Peptides, vinylogous peptides [189, 190], α-sulfonopeptides [191], vinylogous sulfonopeptide [192, 193] and m-phenylene ethynylene oligomers [194] represent the variations in monomeric units and linkages that have been explored. Many of these motifs have been shown to adopt discrete and predictable secondary structures. Gellman defined three steps in creating foldamers that hold true for foldamer design in general. They are: (i) identification of new polymeric backbones with suitable folding propensities, (ii) introduction of chemical function to the resulting foldamers; and (iii) efficient production of foldamers [195]. The extensive research on β-peptides and their growing application for inhibiting PPIs required that they be treated separately in the modified peptide section.

Oligopeptides that contain β-amino acids are commonly referred to as β-peptides. The connection between interface peptides and β-peptides is direct – the only difference is the nature of the amino acid monomers. Oligomers containing cyclic or acyclic β-amino acids, or mixtures of β- and α-amino acids have been shown to fold into novel helices that resemble the α-helices of natural peptides. Additionally, β-peptides are resistant to protease enzymes and provide an opportunity to investigate the importance of individual side-chains toward the complexation of a target protein surface. Identifying or designing a foldamer

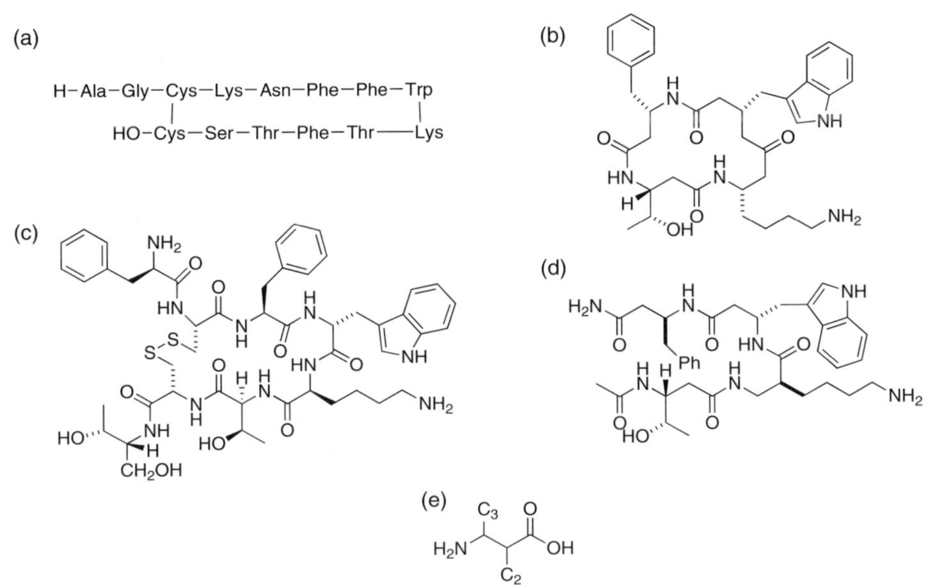

Figure 4.6 *(a) Somatostatin and analogs (b) Octreotide, (c) cyclic β-peptide, (d) acyclic β-peptide. The acyclic β-peptide takes up a turn conformation due to intramolecular H-bonding. (e) β-amino acid showing the C2 and C3 positions*

that can bind tightly and selectively to the surface of a protein is an exciting avenue of PPI inhibition.

The preponderance of PPIs that involve an α-helix has made them good model systems to test the β-peptide strategy for mimicking α-helices. Example systems include the Bak/Bcl-2 [196], p53/(h)MDM2 [197], HIV/gp41 [198] and gB/HCMV [199] interactions. A detailed discussion of the utility of β-peptides as a tool for inhibiting PPIs is presented in Chapter 5. Its treatment there also demonstrates the importance of β-peptides to motivate the design of small molecule inhibitors.

β-Peptides can also mimic other α-amino acid secondary structures such as β-turns and β-sheets. The most developed example is the β-peptide mimicry of the peptide hormone somatostatin. Somatostatin is a 14 α-amino acid cyclic (disulfide) peptide that regulates the release of growth hormone and insulin by binding to cell surface receptors (Figure 4.6). Studies on a cyclic octapeptide, octreotide, showed that the hormone took up a β-turn conformation in the binding site of the receptor. Initially, a cyclic four amino acid β-peptide was designed to mimic this motif [200]. The peptide showed micromolar affinity for the human somatostatin receptors. An acyclic β-peptide, based on the same sequence, was able to fold through a ten-membered hydrogen bonded ring into the β-turn epitope [201]. Control of the β-peptide conformation was governed by the substitution pattern (R groups at either C2 or C3) of the constituent β-amino acids. The acyclic β-peptide showed increased affinity and also selectivity for the somatostatin receptor with a K_d of 83 nM, much closer to the low nanomolar affinity of the natural ligand. Once again through this foldamer example the theme of efficiency as measured by the observed affinity per amino acid residue is demonstrated.

4.4 Summary/Perspective

The interface peptide strategy is an exciting avenue of PPI inhibition that represents valuable tools for biological research and is a key step in the development of drug-like targets. Interface peptides are the simplest designed inhibitors of PPIs that may be envisioned. Several design motifs were discussed including grafting, constraining modifications, mimics and foldamers. Table 4.1 emphasizes the broad applicability of interface peptides in therapeutically significant systems. The application of this strategy develops the simple concept of interface peptides into fuller principles that deal with minimizing the epitope, considering secondary structure of the peptides, controlling conformations, and measuring efficiency as affinity per amino acid residue. Additionally, the tools of modern biochemical research (display techniques, alanine scanning, etc.) can contribute to the optimization of an interface peptide in ways that are beyond the rational design stage. Finally, having arrived at a sufficiently active interface peptide allows for further development into the small molecule arena. The goal of these inhibitory studies is to strive toward molecular minimization while maintaining activity. Overall, interface peptides occupy an important and accessible position for the inhibition of PPIs.

References

1. J.J. Osterhout, Understanding protein folding through peptide models, *Prot. Pept. Lett.*, **12**, 159–64 (2005).
2. C.M. Dobson, Protein folding and misfolding, *Nature*, **426**, 884–90 (2003).
3. V. Daggett, and A. Fersht, The present view of the mechanism of protein folding, *Nat. Rev. Mol. Cell Bio.*, **4**, 497–502 (2003).
4. M.W. Peczuh and A.D. Hamilton, Peptide and protein recognition by designed molecules, *Chem. Rev.*, **100**, 2479–92 (2000).
5. A. Loregian, H.S. Marsden and G. Palù, Protein–protein interactions as targets for antiviral chemotherapy, *Rev. Med. Virol.*, **12**, 239–62 (2002).
6. R. Zutshi, M. Brickner and J. Chmielewski, Inhibiting the assembly of protein–protein interfaces, *Curr. Op. Chem. Biol.*, **2**, 62–6 (1998).
7. S.K. Sharma, T.M. Ramsey and K.W. Bair, Protein–protein interactions: lessons learned, *Curr. Med. Chem. – Anti-Cancer Agents*, **2**, 311–30 (2002).
8. A. Loregian and G. Palù, Disruption of protein–protein interactions towards new targets for chemotherapy, *J. Cell. Phys.*, **204**, 750–62 (2005).
9. A.G. Cochran, Antagonists of protein–protein interactions, *Chem. Biol.*, **7**, R85–R94 (2000).
10. A.M. Levin and G.A. Weiss, Optimizing the affinity and specificity of proteins with molecular display, *Mol. Biosystems*, **2**, 49–57 (2006).
11. S.S. Sachdev, W.J. Fairbrother and K. Deshayes, Exploring protein–protein interactions with phage display, *Chem. Bio. Chem.*, **4**, 14–25 (2003).
12. K. Machida and B.J. Mayer, The SH2 domain: Versatile module and pharmaceutical target, *Biochim. Biophys. Acta*, **1747**, 1–25 (2005).
13. S.K. Akiyama and K.M. Yamada, Synthetic peptides competitively inhibit both direct binding to fibroblasts and functional biological assays for the purified cell-binding domain of fibronectin, *J. Biol. Chem.*, **260**, 10402–5 (1985).
14. M.D. Pierschbacher and E. Ruoslahti, Cell attachment activity of fibronectin can be duplicated by small synthetic fragments of the molecule, *Nature*, **309**, 30–9 (1984).
15. M.D. Pierschbacher, E.G. Hayman and E. Ruoslahti, Location of the cell-attachment site in fibronectin with monoclonal antibodies and proteolytic fragments of the molecule, *Cell*, **26**, 259–67 (1981).

16. T.K. Gartner and J.S. Bennett, Platelet aggregation inhibitors having high specificity for GPIIBIIIA, *J. Biol. Chem.*, **260**, 11891–4 (1985).
17. E.F. Plow, M.D. Pierschbacher and E. Ruoslahti, The effect of Arg-Gly-Asp-containing peptides on fibrinogen and von Willebrand factor binding to platelets, *Proc. Natl. Acad. Sci. U.S.A.*, **82**, 8057–61 (1985).
18. E.C. Tsilibary, L.A. Reger, A.M. Vogel, G.G. Koliakos, S.S. Anderson, A.S. Charonis, J.N. Alegre and L.T. Furcht, Identification of a multifunctional, cell-binding peptide sequence from the a1(NC1) of type IV collagen, *J. Cell. Biol.*, **111**, 1583–91 (1990).
19. D.L. Wilson, R. Martin, S. Hong, M. Cronin-Golomb, C.A. Mirkin and D.L. Kaplan, Surface organization and nanopatterning of collagen by dip-pen nanolithography, *Proc. Natl. Acad. Sci. U.S.A.*, **98**, 13660–4 (2001).
20. D.A. Eppstein, Y.V. Marsh and A.B. Schreiber, Epidermal growth factor occupancy inhibits vaccinia virus infection, *Nature*, **318**, 663–5 (1985).
21. P. Ferrero, A. Guidotti, B. Conti-Tronconi and E. Costa, A brain octadecaneuropeptide generated by tryptic digestion of DBI (diazepam binding inhibitor) functions as a proconflict ligand of benzodiazepine recognition sites, *Neuropharmacology*, **23**, 1359–62 (1984).
22. S. Mintogawa and P.J. Russell, Inhibition of enzyme-antienzyme interaction by tryptic digests of enzyme, *Biochem. Biophys. Res. Comm.*, **31**, 48–53 (1968).
23. B.M. Dutia, M.C. Frame, J.H. Subak-Sharpe, W.N. Clark and H.S. Marsden, Specific inhibition of herpesvirus ribonucleotide reductase by synthetic peptides, *Nature*, **321**, 439–41 (1986).
24. E.A. Cohen, P. Gaudreau, P. Brazeau and Y. Langelier, Specific inhibition of herpesvirus ribonucleotide reductase by a nonapeptide derived from the carboxy terminus of subunit 2, *Nature*, **321**, 441–3 (1986).
25. M.C. Frame, H.S. Marsden and B.M. Dutia, The ribonucleotide reductase induced by herpes simplex virus type 1 involves minimally a complex of two polypeptides (136K and 38K), *J. Gen. Virol.*, **66**, 1581–7 (1985).
26. M. Ferrer and S.C. Harrison, Peptide ligands to human immunodeficiency virus type 1 gp120 identified from phage display libraries, *J. Virol.*, **73**, 5795–5802 (1999).
27. C. Vita, E. Drakopoulou, J. Vizzavona, S. Rochette, L. Martin, A. Menez, C. Roumestand, Y.-S. Yang, L. Ylisastigui, A. Benjouad and J.C. Gluckman, Rational engineering of a miniprotein that reproduces the core of the CD4 site interacting with HIV-1 envelope glycoprotein, *Proc. Natl. Acad. Sci. U.S.A.*, **96**, 13091–6 (1999).
28. C.T. Wild, D.C. Shugars, T.K. Greenwell, C.B. Mcdanal and C.J. Mathews, Peptides corresponding to a predictive α-helical domain of human immunodeficiency virus type 1 gp41 are potent inhibitors of virus infection, *Proc. Natl. Acad. Sci. U.S.A.*, **91**, 9770–4 (1994).
29. J.K. Seo, H.K. Kim, T.Y. Lee, K.S. Hahm, K.L. Kim and M.K. Lee, Stronger anti-HIV-1 activity of C-peptide derived from HIV-$1_{89.6}$ gp41 C-terminal heptad repeated sequence, *Peptides*, **26**, 2175–81 (2005).
30. J.H. Huang, Z.Q. Liu, S.W. Liu, S.B. Jiang and Y.B. Chen, Identification of the HIV-1 gp41 core-binding motif-HXXNPF, *FEBS Lett.*, **580**, 4807–14 (2006).
31. K. Hilpert, J. Behlke, C. Scholz, R. Misselwitz, J. Schneider-Mergener and W. Hohne, Phenotype of HIV-1 lacking a functional nuclear localization signal in matrix protein of gag and Vpr is comparable to wild-type HIV-1 in primary macrophages, *Virology*, **254**, 6–10 (1999).
32. M.C. Morris, V. Robert-Hebmann, L. Chaloin, J. Mery, F. Heitz, C. Devaux, R.S. Goody and G. Divita, A new potent HIV-1 reverse transcriptase inhibitor – A synthetic peptide derived from the interface subunit domains, *J. Biol. Chem.*, **274**, 24941–6 (1999).
33. D.A. Davis, C.A. Brown, K.E. Singer, *et al.*, Inhibition of HIV-1 replication by a peptide dimerization inhibitor of HIV-1 protease, *Antiviral Res.*, **72**, 89–99 (2006).
34. R. Zutshi, J. Fraciskovich, M. Shultz, B. Schweitzer, P. Bishop, M. Wilson and J. Chmielewski, Targeting the dimerization interface of HIV-1 protease: Inhibition with cross-linked interfacial peptides, *J. Am. Chem. Soc.*, **119**, 4841–5 (1997).
35. L.M. Babe, J. Rose and C.S. Craik, Synthetic 'interface' peptides alter dimeric assembly of the HIV 1 and 2 proteases, *Protein Science*, **1**, 1244–53 (1992).
36. F. Sourgen, R.G. Maroun, V. Frere, M. Bouzaine, C. Auclair, F. Troalen, and S.A. Fermandjian, A synthetic peptide from the human immunodeficiency virus type-1 integrase exhibits

coiled-coil properties and interferes with the in vitro integration activity of the enzyme – Correlated biochemical and spectroscopic results, *Eur. J. Biochem.*, **240**, 765–73 (1996).
37. R.G. Maroun, S. Gayet, M.S. Benleulmi, *et al.*, Peptide inhibitors of HIV-I intergrase dissociate enzyme oligomers. *Biochemistry*, **40**, 13840–8 (2001).
38. R.A.P. Lutzke, N.A. Eppens, P.A. Weber, R.A. Houghton and R.H.A. Plasterk, Identification of a hexapeptide inhibitor of the human immunodeficiency virus integrase protein by using a combinatorial chemical library. *Proc. Natl. Acad. Sci. U.S.A.*, **92**, 11456–61 (1995).
39. B. Yang, L. Gao, L. Li, *et al.*, Potent suppression of viral infectivity by the peptides that inhibit multimerization of human immunodeficiency virus type 1 (HIV-1) Vif proteins, *J. Biol. Chem.*, **278**, 6596–6602 (2003).
40. J. Sticht, M. Humbert, S. Findlow, *et al.*, A peptide inhibitor of HIV-1 assembly in vitro, *Nat. Struct. Mol. Bio.*, **12**, 671–7 (2005).
41. E.S. Withers-Ward, T.D. Mueller, I.S.Y. Chen and J. Feigon, Biochemical and structural analysis of the interaction between the UBA(2) domain of the DNA repair protein HHR23A and HIV-1 Vpr, *Biochemistry*, **39**, 14103–12 (2000).
42. K.A. Simmen, A. Newell, M. Robinson, *et al.*, Protein interactions in the herpes simplex virus type 1 VP16-induced complex: VP16 peptide inhibition and mutational analysis of host cell factor requirements, *J. Virol.*, **71**, 3886–94 (1997).
43. P. Digard, K.P. Williams, P. Hensley, I.S. Brooks, C.E. Dahl and D.M. Coen, Specific inhibition of herpes simplex virus DNA polymerase by helical peptides corresponding to the subunit interface, *Proc. Natl. Acad. Sci. U.S.A.*, **92**, 1456–60 (1995).
44. K.G. Bridges, Q.X. Hua, M.R. Brigham-Burke, *et al.*, Secondary structure and structure-activity relationships of peptides corresponding to the subunit interface of herpes simplex virus DNA polymerase, *J. Biol. Chem.*, **275** 472–8 (2000).
45. N.E. Shepherd, H.N. Hoang, V.S. Desai, E. Letouze, P.R. Young and D.P. Fairlie, Modular alpha-helical mimetics with antiviral activity against respiratory syncitial virus, *J. Amer. Chem. Soc.*, **128**, 13284–9 (2006).
46. Z. Yan, B. Tripet and R.S. Hodges, Biophysical characterization of HRC peptide analogs interaction with heptad repeat regions of the SARS-coronavirus Spike fusion protein core, *J. Struc. Biol.*, **155**, 162–75 (2006).
47. J.O. Lee, A.A. Russo and N.P. Pavletich, Structure of the retinoblastoma tumour-suppressor pocket domain bound to a peptide from HPV E7, *Nature*, **391**, 859–65 (1998).
48. M. Porotto, L. Doctor, P. Carta, *et al.*, Inhibition of Hendra virus fusion, *J. Virol.*, **80**, 9837–49 (2006).
49. A. Loregian, R. Rigatti, M. Murphy, E. Schievano, G. Palu and H.S. Marsden, Inhibition of human cytomegalovirus DNA polymerase by C-terminal peptides from the UL54 subunit, *J. Virol.*, **77**, 8336–44 (2003).
50. V. Prasanna, S. Bhattacharjya and P. Balaram, Synthetic interface peptides as inactivators of multimeric enzymes: Inhibitory and conformational properties of three fragments from Lactobacillus casei thymidylate synthase, *Biochemistry*, **37**, 6883–93 (1998).
51. M. Brickner and J. Chmielewski, Inhibiting the dimeric restriction endonuclease EcoRI using interfacial helical peptides, *Chem. Biol.*, **5**, 339–43 (1998).
52. W.L. DeLano, M.H. Ultsch, A.M. de Vos and J.A. Wells, Convergent solutions to binding at a protein–protein interface, *Science*, **287**, 1279–83 (2000).
53. S.K. Singh, K. Maithal, H. Balaram and P. Balaram, Synthetic peptides as inactivators of multimeric enzymes: inhibition of Plasmodium falciparum triosephosphate isomerase by interface peptides, *FEBS Lett.*, **501**, 19–23 (2001).
54. B.P. Orner, L. Liu, R.M. Murphy and L.L. Kiessling, Phage display affords peptides that modulate β-amyloid aggregation, *J. Am. Chem. Soc.*, **128**, 11882–9 (2006).
55. T.J. Smith, C.I. Stains, S.C. Meyer and I.G. Ghosh, Inhibition of β-amyloid fibrillization by directed evolution of a β-sheet presenting miniature protein, *J. Am. Chem. Soc.*, **128**, 14456–7 (2006).
56. S.F. Arnold and A.C. Notides, An antiestrogen: A phosphotyrosyl peptide that blocks dimerization of the human estrogen receptor, *Proc. Natl. Acad. Sci. U.S.A.*, **92**, 7475–9 (1995).

57. M.R. Yudt and S. Koide, Preventing estrogen receptor action with dimer-interface peptides, *Steroids*, **66**, 549–58 (2001).
58. X.L. Qian, D.M. O'Rourke, H.Z. Zhao and M.I. Greene, Inhibition of p185(neu) kinase activity and cellular transformation by co-expression of a truncated neu protein, *Oncogene*, **13**, 2149–57 (1996).
59. T. Kumagai, J.G. Davis, T. Horie, D.M. O'Rourke and M.I. Greene, The role of distinct p185(neu) extracellular subdomains for dimerization with the epidermal growth factor (EGF) receptor and EGF-mediated signaling, *Proc. Natl. Acad. Sci. U.S.A.*, **98**, 5526–31 (2001).
60. L.D. D'Andrea, G. Iaccarino, R. Fattorusso, D. Sorriento, C. Carannante, D. Capasso, B. Trimarco and C. Pedone, Targeting angiogenesis: Structural characterization and biological properties of a de novo engineered VEGF mimicking peptide, *Proc. Natl. Acad. Sci. U.S.A.*, **102**, 14215–20 (2005).
61. W.J. Fairbrother, H.W. Christinger, A.G. Cochran, *et al.*, Novel peptides selected to bind vascular endothelial growth factor target the receptor-binding site, *Biochemistry*, **37**, 17754–64 (1998).
62. A. Berezov, J.Q. Chen, Q.D. Liu, H.T. Zhang, M.I. Greene and R. Murali, Disabling receptor ensembles with rationally designed interface peptidomimetics, *J. Biol. Chem.*, **277**, 28330–9 (2002).
63. Y.L. Yip, G. Smith, J. Koch, S. Dubel and R.L. Ward, Identification of epitope regions recognized by tumor inhibitory and stimulatory anti-ErbB-2 monoclonal antibodies: Implications for vaccine design, *J Immunol.*, **166**, 5271–8 (2001).
64. S.R. Eaton, W.L. Cody, A.M. Doherty, *et al.*, Design of peptidomimetics that inhibit the association of phosphatidylinositol 3-kinase with platelet-derived growth factor-beta receptor and possess cellular activity, *J. Med. Chem.*, **41**, 4329–42 (1998).
65. H.B. Lowman, Y.M. Chen, N.J. Skelton, *et al.*, Molecular mimics of insulin-like growth factor 1 (IGF-1) for inhibiting IGF-1: IGF-binding protein interactions, *Biochemistry*, **37**, 8870–8 (1998).
66. K. Deshayes, M.L. Schaffer, N.J. Skelton, G.R. Nakamura, S. Kadkhodayan and S.S. Sidhu, Rapid identification of small binding motifs with high-throughput phage display: Discovery of peptidic antagonists of IGF-1 function, *Chemistry& Biology*, **9**, 495–505 (2002).
67. N. Kayagaki, M.H. Yan, D. Seshasayee, *et al.*, BAFF/BLyS receptor 3 binds the B cell survival factor BAFF ligand through a discrete surface loop and promotes processing of NF-kappa B2, *Immunity*, **10**, 515–24 (2002).
68. A.M. Finch, A.K. Wong, N.J. Paczkowski, *et al.*, Low-molecular-weight peptidic and cyclic antagonists of the receptor for the complement factor C5a, *J. Med. Chem.*, **42**, 1965–74 (1999).
69. R.L. Wange, N. Isakov, T.R. Jr.*et al.*, F2(Pmp)2-TAM zeta 3, a novel competitive inhibitor of the binding of ZAP-70 to the T cell antigen receptor, blocks early T cell signaling, *J. Biol. Chem.*, **270**, 944–8 (1995).
70. S.A. Richman, S.J. Healan, K.S. Weber, *et al.*, Development of a novel strategy for engineering high-affinity proteins by yeast display, *Prot. Eng. Des. Sel.*, **19**, 255–64 (2006).
71. S.D. Yanofsky, D.N. Baldwin, J.H. Butler, *et al.*, High affinity type I interleukin 1 receptor antagonists discovered by screening recombinant peptide libraries, *Proc. Natl. Acad. Sci. U.S.A.*, **93**, 7381–6 (1996).
72. H. Domingues, D. Cregut, W. Sebald, H. Oschkinat and L. Serrano, Rational design of a GCN4-derived mimetic of interleukin-4, *Nat. Struct. Biol.*, **6**, 652–6 (1999).
73. M.N. O'Connor, P.A. Smethurst, L.W. Davies, *et al.*, Selective blockade of glycoprotein VI clustering on collagen helices, *J. Biol. Chem.*, **281**, 33505–10 (2006).
74. N. Raynal, S.W. Hamaia, P.R.M. Siljander, *et al.*, Use of synthetic peptides to locate novel integrin alpha(2)beta(1)-binding motifs in human collagen III, *J. Biol. Chem.*, **281**, 3821–31 (2006).
75. D.S. Feldman, A.M. Zamah, K.L. Pierce, *et al.*, Selective inhibition of Heterotrimeric G(s) signaling – Targeting the receptor-G protein interface using a peptide minigene encoding the G alpha(s) carboxyl terminus, *J. Biol. Chem.*, **277**, 28631–40 (2002).

76. T.L. Davis, T.M. Bonacci, S.R. Sprang and A.V. Smrcka, Structural and molecular characterization of a preferred protein interaction surface on G protein beta gamma subunits, *Biochemistry*, **44**, 10593–10604 (2005).
77. W.J. Koch, J. Inglese, W.C. Stone and R.J. Lefkowitz, The binding site for the beta gamma subunits of heterotrimeric G proteins on the beta-adrenergic receptor kinase, *J. Biol. Chem.*, **268**, 8256–60 (1993).
78. J.G. Krupnick, V.V. Gurevich, T. Schepers, H.E. Hamm and J.L. Benovic, Arrestin-rhodopsin interaction-multisite binding delineated by peptide inhibition, *J. Biol. Chem.*, **269**, 3226–32 (1994).
79. E.L. Martin, S. RensDomiano, P.J. Schatz and H.E. Hamm, Potent peptide analogues of a G protein receptor-binding region obtained with a combinatorial library, *J. Biol. Chem.*, **271**, 361–6 (1996).
80. T. Scherf, R. Kasher, M. Balass, M. Fridkin, S. Fuchs and E. Katchalski-Katzir, A beta-hairpin structure in a 13-mer peptide that binds alpha-bungarotoxin with high affinity and neutralizes its toxicity, *Proc. Natl. Acad. Sci. U.S.A.*, **98**, 6629–34 (2001).
81. S. Fuchs, R. Kasher, M. Balass, *et al.*, The binding site of acetylcholine receptor: From synthetic peptides to solution and crystal structure, *Ann. N. Y. Acad. Sci.*, **998**, 93–100 (2003).
82. N.P.S. Bains, J.A. Wilce, K.H. Heuer, *et al.*, Zipping up transcription factors: Rational design of anti-Jun and anti-Fos peptides, *Lett. Pept. Sci.*, **4**, 67–77 (1997).
83. S. Yao, M. Brickner, E.I. Pares-Matos and J. Chmielewski, Uncoiling c-Jun coiled coils: Inhibitory effects of truncated fos peptides on jun dimerization and DNA binding in vitro, *Biopolymers*, **47**, 277–83 (1998).
84. I. Ghosh and J. Chmielewski, A beta-sheet peptide inhibitor of E47 dimerization and DNA binding, *Chem. Biol.*, **5**, 439–45 (1998).
85. M. Ohnishi, Y. Yamawaki-Kataoka, K. Kariya, M. Tamada, C.D. Hu and T. Kataoka, Selective inhibition of ras interaction with its particular effector by synthetic peptides corresponding to the ras effector region, *J. Biol. Chem.*, **273**, 10210–15 (1998).
86. B. Li, S.J. Russell, D.M. Compaan, *et al.*, Activation of the proapoptotic death receptor DR5 by oligomeric peptide and antibody agonists, *J. Mol. Biol.*, **361**, 522–36 (2006).
87. J.W. Chin and A. Schepartz, Design and evolution of a miniature bcl-2 binding protein, *Agnew. Chem. Int. Ed.*, **40**, 3806–9 (2001).
88. A.C. Gemperli, S.E. Rutledge, A. Maranda, and A. Schepartz, Paralog-selective ligands for Bcl-2 proteins, *J. Am. Chem. Soc.*, **127**, 1596–7 (2005).
89. P. Chene, Inhibition of the p53-MDM2 interaction: Targeting a protein–protein interface, *Mol. Can. Res.*, **2**, 20–8 (2004).
90. P.H. Kussie, S. Gorina, V. Marechal, *et al.*, Structure of the MDM2 Oncoprotein Bound to the p53 Tumor Suppressor Transactivation Domain, *Science*, **274**, 948–53 (1996).
91. K. Sakurai, H.S. Chung and D. Kahne, Use of a retroinverso p53 peptide as an inhibitor of MDM2, *J. Am. Chem. Soc.*, **126**, 16288–9 (2004).
92. J.A. Kritzer, R. Zutshi, M. Cheah, F.A. Ran, R. Webman, T.M. Wongjirad and A. Schepartz, Miniature protein inhibitors of the p53-hDM2 interaction, *ChemBioChem.*, **7**, 29–31 (2006).
93. R. Fasan, R.L.A. Dias, K. Moehle, *et al.*, Using a β-hairpin to mimic an α-helix: Cyclic peptidomimetics inhibitors of the p53-HDM2 protein–protein interaction, *Angew. Chem. Int. Ed.*, **43**, 2109–12 (2004).
94. C. Garcia-Echeverria, P. Chene, M.J.J. Blommers and P. Furet, Discovery of potent antagonists of the interaction between human double minute 2 and tumor suppressor p53. *J. Med. Chem.* **43**, 3205–8 (2000).
95. R. Fahraeus, S. Lain, K.L. Ball and D.P. Lane, Characterization of the cyclin-dependent kinase inhibitory domain of the INK4 family as a model for a synthetic tumour suppressor molecule, *Oncogene*, **16**, 587–96 (1998).
96. R. Fahraeus, J.M. Paramio, K.L. Ball, S. Lain and D.P. Lane, Inhibition of pRb phosphorylation and cell-cycle progression by a 20-residue peptide derived from p16(CDKN2/INK4A), *Curr. Biol.*, **6**, 81–7 (1996).
97. C. Gondeau, S. Gerbal-Chaloin, P. Bello, G. Aldrian-Herrada, M.C. Morris, and G. Divita, Design of a novel class of peptide inhibitors of cyclin-dependent kinase/cyclin activation, *J. Biol. Chem.*, **280**, 13793–13800 (2005).

98. M. Ferguson, M.G. Luciani, L. Finlan, *et al.*, The development of a CDK2-docking site peptide that inhibits p53 and sensitizes cells to death, *Cell Cycle*, **3**, 80–9 (2004).
99. A. Emmanuel, Z. B. Naigong, S. Dadgar, *et al.*, Inhibition of HIV-1 virus replication using small soluble Tat peptides, *Virology*, **345**, 373–89 (2006).
100. I.T. Chen, M. Akamatsu, M.L. Smith, *et al.*, Characterization of p21(Cip1/Waf1) peptide domains required for cyclin E/Cdk2 and PCNA interaction, *Oncogene*, **12**, 595–607 (1996).
101. G. Kontopidis, S.Y. Wu, D.I. Zheleva, *et al.*, Structural and biochemical studies of human proliferating cell nuclear antigen complexes provide a rationale for cyclin association and inhibitor design, *Proc. Natl. Acad. Sci. U.S.A.*, **102**, 1871–6 (2005).
102. S.M. Violette, W.C. Shakespeare, C. Bartlett, *et al.*, A Src SH2 selective binding compound inhibits osteoclast-mediated resorption, *Chem. Biol.*, **7**, 225–35 (2000).
103. P. Furet, B. Gay, C. GarciaEcheverria, *et al.*, Discovery of 3-aminobenzyloxycarbonyl as an N-terminal group conferring high affinity to the minimal phosphopeptide sequence recognized by the Grb2-SH2 domain, *J. Med. Chem.* **40**, 3551–6 (1997).
104. M. Wittekind, C. Mapelli, B.T. Farmer, *et al.*, Orientation of peptide-fragments from SOS proteins bound to the N-terminal SH3 domain of GRB2 determined by NMR-Spectroscopy, *Biochemistry*, **33**, 13531–9 (1994).
105. F. Meggio, O. Marin, S. Sarno and L.A. Pinna, Functional analysis of CK2 beta-derived synthetic fragments, *Mol. Cell Biochem.*, **191**, 35–42 (1999).
106. E.S. Cobos, M.T. Pisabarro, M.C. Vega, *et al.*, A miniprotein scaffold used to assemble the polyproline II binding epitope recognized by SH3 domains, *J. Mol. Biol.*, **342**, 355–65 (2004).
107. S. Hashimoto, M. Hiroso, A. Hashimoto, *et al.*, Targeting AMAP1 and cortactin binding bearing an atypical src horology 3/proline interface for prevention of breast cancer invasion and metastasis, *Proc. Natl. Acad. Sci. U.S.A.*, **103**, 7036–41 (2006).
108. Z.Y. Ren, L.A. Cabell, T.S. Schaefer and J.S. McMurray, Identification of a high-affinity phosphopeptide inhibitor of Stat3, *Bioorg. Med. Chem. Lett.*, **13**, 633–6 (2003).
109. D. Golemi-Kotra, R. Mahaffy, M.J. Footer, *et al.*, High affinity, paralog-specific recognition of the Mena EVH1 domain by a miniature protein, *J. Am. Chem. Soc.*, **126**, 4–5 (2004).
110. T.D. Mueller and J. Feigon, Structural determinants for the binding of ubiquitin-like domains to the proteasome, *EMBO J.*, **22**, 4634–5 (2003).
111. A. Nordhoff, C. Tziatzios, J.A. vandenBroek, *et al.*, Denaturation and reactivation of dimeric human glutathione reductase – An assay for folding inhibitors, *Eur. J. Biochem.*, **245**, 273–82 (1997).
112. E.M. Wilson and M. Chinkers, Identification of sequences mediating guanylyl cyclase dimerization, *Biochemistry*, **34**, 4696–4701 (1995).
113. J.J. Dwyer, M.A. Dwyer and A.A. Kossiakoff, High affinity RNase S-peptide variants obtained by phage display have a novel 'hot-spot' of binding energy, *Biochemistry*, **40**, 13491–13500 (2001).
114. D. Ron and D. Mochly-Rosen, Agonists and antagonists of protein kinase C function, derived from its binding proteins, *J. Biol. Chem.*, **269**, 21395–8 (1994).
115. M.S. Wolfe, W.P. Esler and C. Das, Continuing strategies for inhibiting Alzheimer's γ-secretase, *J. Mol. Neurosc.*, **19**, 83–7 (2002).
116. M.S. Dennis, C. Eigenbrot, N.J. Skelton, *et al.*, Peptide exosite inhibitors of factor VIIa as anticoagulants, *Nature*, **404**, 465–70 (2000).
117. N.P. Skiba, N.O. Artemyev and H.E. Hahm, The Carboxyl-terminus of the gamma-subunit of Rod CGMP Phosphodiesterase Contains Distinct Sites of Interaction with the Enzyme Catalytic Subunits and alpha-subunit of Transduction, *J. Biol. Chem.*, **270**, 13210–15 (1995).
118. D.W. Carr, Z.E. Hausken, I.D. Fraser, R.E. Stofko-Hahn and J.D. Scott, Association of the type II cAMP-dependent protein kinase with a human thyroid RII-anchoring protein. Cloning and characterization of the RII-binding domain, *J. Biol. Chem.*, **267**, 13376–82 (1992).
119. T.E. Hebert, S. Moffett, J.P. Morello, *et al.*, A peptide derived from a beta(2)-adrenergic receptor transmembrane domain inhibits both receptor dimerization and activation, *J. Biol. Chem.*, **271**, 16384–92 (1996).

120. Z.H. Liu, C.H. Sun, E.T. Olejniczak, *et al.*, Structural basis for binding of Smac/DIABLO to the XIAP BIR3 domain, *Nature*, **408**, 1004–8 (2000).
121. L.T. Nevalainen, T. Aoyama, M. Ikura, *et al.*, Characterization of novel calmodulin-binding peptides with distinct inhibitory effects on calmodulin-dependent enzymes, *Biochem. J.*, **321**, 107–15 (1997).
122. G. Jogl, Y. Shen, D. Gebauer, *et al.*, Crystal structure of the BEACH domain reveals an unusual fold and extensive association with a novel PH domain, *EMBO J.*, **21**, 4785–95 (2002).
123. M.J. May, F. D'Acquisto, L.A. Madge, J. Glockner, J.S. Pober and S. Ghosh, Selective inhibition of NF-kappa B activation by a peptide that blocks the interaction of NEMO with the I kappa B kinase complex, *Science*, **289**, 1550–4 (2000).
124. S.E. Rutledge, H.M. Volkman and A. Schepartz, Molecular recognition of protein surfaces: High affinity ligands for the CBPKIX domain, *J. Am. Chem. Soc.*, **125**, 14336–7 (2003).
125. V.A. Sharma, J. Logan, D.S. King, R. White and T. Alber, Sequence-based design of a peptide probe for the APC tumor suppressor protein, *Curr. Bio.*, **8**, 823–30 (1998).
126. A. Lo Conte, C. Chothia, and J. Janin, The atomic structure of protein–protein recognition sites, *J. Mol. Biol.*, **285**, 2177–98 (1999).
127. A.A. Bogan and K.S. Thorn, Anatomy of hot spots in protein interfaces, *J. Mol. Biol.* **280**, 1–9 (1998).
128. S, Jones and J. Thornton, Principles of protein–protein interactions, *Proc. Natl. Acad. Sci. U.S.A.*, **93**, 13–20 (1996).
129. M. Chorev and M. Goodman, Recent developments in retro peptides and proteins – An ongoing topochemical exploration, *Trends Biotechnol.*, **15**, 438–45 (1997).
130. M. Little, P. Fuchs, F. Breitling, and S. Dubel, Bacterial surface presentation of proteins and peptides: an alternative to phage technology?, *Trends Biotechnol.*, **11**, 3–5 (1993).
131. E.T. Boder and K.D. Wittrup, Yeast surface display for screening combinatorial polypeptide libraries, *Nat. Biotech.* **15**, 553–7 (1997).
132. R.W. Roberts and J.W. Szostak, RNA-peptide fusions for the in vitro selection of peptides and proteins, *Proc. Natl. Acad. Sci. U.S.A.*, **94**, 12297–12302 (1997).
133. K. Tonan, Y. Kawata and K. Hamaguchi, Conformations of isolated fragments of pancreatic polypeptide, *Biochemistry*, **29**, 4424–9 (1990).
134. H. Darbon, J.-M. Bernassau, C. Deleuze, J. Chenu, A. Roussel and C. Cambillau, Solution confromation of human neuropeptide Y by ^{1}H nuclear magnetic resonance and restrained molecular dynamics, *Eur. J. Biochem.*, **209**, 765–71 (1992).
135. B. Imperiali and J.J. Ottesen, Uniquely folded mini-protein motifs, *J. Pept. Res.*, **54**, 177–84 (1999).
136. B. Imperiali and J.J. Ottesen, Design strategies for the construction of independently folded polypeptide motifs, *Biopolymers*, **47**, 23–9 (1998).
137. P.A. Nygren and M. Uhlèn, Scaffolds for engineering novel binding sites in proteins, *Curr. Op. Struct. Biol.*, **7**, 463–9 (1997).
138. P.A. Nygren and A. Skerra, Binding proteins from alternative scaffolds, *J. Immunol. Methods*, **290**, 3–28 (2004).
139. S. Liu S. Liu, X. Zhu, H. Liang, A. Cao, Z. Chang, L. Lai, Nonnatural protein–protein inateraction pair design by key residues grafting, *Proc. Natl. Acad. Sci. U.S.A.*, **104**, 5330–5 (2007).
140. J. Reina, E. Lacroix, S.D. Hobson, *et al.*, Computer-aided design of a PDZ domain to recognize new target sequences, *Nat. Struct. Biol.*, **9**, 621–7 (2002).
141. J. M. Shifman, and S. L. Mayo, Modulating Calmodulin Binding Specificity through Computational Protein Design, *J. Mol. Biol.*, **323**, 417–23 (2002).
142. J. J. Havranek, and P. B. Harbury, Automated design of specificity in molecular recognition, *Nat. Struct. Biol.*, **10**, 45–52 (2003).
143. T. Kortemme, L. A. Joachimiak, A. N. Bullock, A. D. Schuler, B. L. Stoddard, and D. Baker, Computational redesign of protein–protein interaction specificity, *Nat. Struct. Mol. Biol.*, **11**, 371–9 (2004).
144. N. J. Zondlo, and A. Schepartz, Highly specific DNA recognition by a designed miniature protein, *J. Am. Chem. Soc.*, **121**, 6938–9 (1999).

145. J. W. Chin, and A. Schepartz, Concerted evolution of structure and function in a miniature protein, *J. Am. Chem. Soc.*, **123**, 2929–30 (2001).
146. J. K. Montclare, and A. Schepartz, Miniature homeodomains: high specificity without an N-terminal arm, *J. Am. Chem. Soc.*, **125**, 3416–17 (2003).
147. N. Shimba, A. M. Nomura, A. B. Marnett, and C. S. Craik, Herpesvirus protease inhibition by dimer disruption, *J. Virol.*, **78**, 6657–65 (2004).
148. T. L. Blundell, J. E. Pitts, I. J. Tickle, S. P. Wood, and C. W. Wu, X-ray analysis (1.4 Å resolution) of avian pancreatic polypeptide: small globular protein hormone, *Proc. Natl. Acad. Sci. U.S.A.*, **78**, 4175–9 (1981).
149. M. T. Pisabarro, and L. Serrano, Rational design of specific high-affinity peptide ligands for the Abl-SH3 domain, *Biochemistry*, **35**, 10634–40 (1996).
150. A. Palencia, E. S. Cobos, P. L. Mateo, J. C. Martinez, and I. Luque, Thermodynamic dissection of the binding energetics of proline-rich peptides to the Abl-SH3 domain: Implications for rational ligand design, *J. Mol. Biol.*, **336**, 527–37 (2006).
151. Branden, C., Tooze, B. *Introduction to Protein Structure*. Garland Publishing Inc. New York, NY (1991).
152. S. K. Sia, P. S. Kim, Protein grafting of an HIV-1 inhibiting epitope, *Proc. Natl. Acad. Sci. U.S.A.* **100**, 9756–61 (2003).
153. C. Wild, T. Oas, C. McDanal, D. Bolognesi, and T. Matthews, A synthetic peptide inhibitor of human immunodeficiency virus replication: Correlation between solution structure and viral inhibition, *Proc. Natl. Acad. Sci. U.S.A.*, **89**, 10537–41 (1992).
154. C.-H. Chen, T. J. Matthews, C. B. McDanal, D. P. Bolognesi, and M. L. Greenberg, A Molecular Clasp in the Human Immunodeficiency Virus (HIV) Type 1 TM Protein Determines the Anti-HIV Activity of gp41 Derivatives: Implication for Viral Fusion, *J. Virol.*, **69**, 3771–7 (1995).
155. Trimeris, Inc. (Press Release) Trimeris Reports 2006 FUZEON Sales Results Morrisville, N.C. (2007).
156. D. C. Chan, and P. S. Kim, HIV entry and its inhibition, *Cell*, **93**, 681–4 (1998).
157. C. Wild, T. Greenwell, and T. Matthews, A synthetic peptide from HIV-1 gp41 is a potent inhibitor of virus-mediated cell-cell fusion, *AIDS Res. and Human Retrovir.*, **9**, 1051–3 (1993).
158. J. K. Judice, J. Y. K. Tom, W. Huang, *et al.*, Inhibition of HIV type 1 infectivity by constrained a-helical peptides: Implications for the viral fusion mechanism, *Proc. Natl. Acad. Sci. U.S.A.* **94**, 1326–30 (1997).
159. C. E. Schafmeister, J. Po, G. L. Verdine, An all-hydrocarbon cross-linking system for enhancing the helicity and metabolic stability of peptides, *J. Am. Chem. Soc.*, **122**, 5891–2 (2000).
160. K. L. Myung, K. K. Hee, Y. L. Tae, K.-S. Hahm, and L. K. Kil, Structure-activity relationships of anti-HIV-1 peptides with disulfide linkage between D- and L-cysteine at positions i and $i+3$, respectively, derived from HIV-1 gp41 C-peptide, *Exp. Mol. Med.* **38**, 18–26 (2006).
161. C. Blaines-Mira, M. T. Pastor, E. Valera, *et al.*, Identification of SNARE complex modulators that inhibit exocytosis from an α-helix constrained combinatorial library, *Biochem. J.*, **375**, 159–66 (2003).
162. J. C. Phelan, N. J. Skelton, A. C. Braisted, and R. S. McDowell, A general method for constraining short peptides to an α-helical conformation, *J. Am. Chem. Soc.*, **119**, 455–60 (1997).
163. C. Bracken, J. Gulyas, J. W. Taylor, and J. Baum, Synthesis and nuclear magnetic resonance structure determination of an α-helical, bicyclic, lactam-bridged hexapeptide, *J. Am. Chem. Soc.*, **116**, 6431–2 (1994).
164. G. Udugamasooriya, D. Saro and M.R. Spaller, Bridged peptide macrocycles as ligands for PDZ domain proteins. *Org. Lett.* **7**, 1203–6 (2005).
165. J.C. Calvo, K.C. Choconta, D. Daiz, *et al.*, An alpha helix conformationally restricted peptide is recognized by cervical carcinoma patients' sera. *J. Med. Chem.* **46**, 5389–94 (2003).
166. D. Y. Jackson, D. S. King, J. Chmielewski, S. Singh, and P. G. Schultz, General Approach to the Synthesis of Short α-Helical Peptides, *J. Am. Chem. Soc.*, **113**, 9391–9392 (1991).
167. A.M. Leduc, J.O. Trent, J.L. Wittliff, K.S. Bramlett, S.L. Briggs, N.Y. Chirgadze, Y. Wang, T.P. Burris and A.F. Spatola, Helix-stabilized cyclic peptides as selective inhibitors of steroid receptor-coactivator interactions. *Proc. Natl. Acad. Sci. U.S.A.*, **100**, 11273–11278 (2003).

168. H. E. Blackwell, and R. H. Grubbs, Highly Efficient Synthesis of Covalently Cross-Linked Peptide Helices by Ring-Closing Metathesis, *Angew. Chem. Int. Ed.*, **37**, 3281–3284 (1998).
169. D. Wang, K. Chen, J. L Kulp and P.S. Arora, Evaluation of biologically relevant short alpha-helices stabilized by a main-chain hydrogen-bond surrogate. *J. Amer. Chem. Soc.* **128**, 9248–9256 (2006).
170. L. D. Walensky, A. L. Kung, I. Escher, *et al.*, Activation of Apoptosis in Vivo by a Hydrocarbon-Stapled BH3 Helix, *Science*, **305**, 1466–70 (2004).
171. L. D. Walensky, BCL-2 in the crosshairs: tipping the balance of life and death, *Cell Death Differ.*, **13**, 1339–50 (2006).
172. L. D. Walensky, K. Pitter, J. Morash, *et al.*, A Stapled BID BH3 Helix Directly Binds and Activates BAX, *Mol. Cell*, **24**, 199–210 (2006).
173. R. Fasan, R. L. A. Dias, K. Moehle, *et al.*, Structure-Activity Studies in a Family of β-Hairpin Protein Epitope Mimetic Inhibitors of the p53-HDM2 Protein–protein Interaction, *Chem. BioChem.*, **7**, 515–26 (2006).
174. S. Krause, H.-U. Schmoldt, A. Wentzel, M. Ballmaier, K. Friedrich, and H. Kolmar, Grafting of thrombopoietin-mimetic peptides into cysteine knot miniproteins yields high-affinity thrombopoietin antagonists and agonists, *FEBS Journal* **274**, 86–95 (2007).
175. C. C. Bleul, M. Farzan, H. Choe, *et al.*, The lymphocyte chemoattractant SDF-1 is a ligand for LESTR/fusin and blocks HIV-1 entry, *Nature*, **382**, 829–33 (1996).
176. A. G. Dalgleish, A. C. Beverly, P. R. Clapham, D. H. Crawford, M. F. Greaves, and R. A. Weiss, The CD4 (T4) antigen is an essential component of the receptor for the AIDS retrovirus, *Nature*, **312**, 763–7 (1984).
177. S. J. DeMarco, H. Henze, A. Lederer, *et al.*, Discovery of novel, highly potent and selective beta-hairpin mimetic CXCR4 inhibitors with excellent anti-HIV activity and pharmacokinetic profiles, *Bioorg. Med. Chem.*, **14**, 8396–8404 (2006).
178. B. A. Jameson, J. M. McDonnell, J. C. Marini, and R. Korngold, A rationally designed CD4 analogue inhibits experimental allergic encephalomyelitis, *Nature*, **368**, 744–6 (1994).
179. E. Drakopoulou, J. Vizzanova, and C. Vita, Engineering a CD4 mimetic inhibiting the binding of the human immunodeficiency virus-1 (HIV-1) envelope glycoprotein gp120 to human lymphocyte CD4 by the transfer of a CD4 functional site to a small natural scaffold, *Lett. Pept. Sci.*, **5**, 241–5 (1998).
180. C. K. Smith, and L. Regan, Construction and design of β-Sheets. *Acc. Chem. Res.*, **30**, 153–61 (1997).
181. W. A. Loughlin, J. D. A. Tyndall, M. P. Glenn, and D. P. Fairlie, Beta-Strand Mimetics, *Chem. Rev.*, **104**, 6085–6117 (2004).
182. K.L. Sciarretta, A. Biore, D.J. Gordon and S.C. Meredith, Spatial separation of beta-sheet domains of beta-amyloid: Disruption of each beta-sheet by N-methyl amino acids. *Biochemistry*, **45**, 9485–95 (2006).
183. K.L. Sciarretta, D.J. Gordon and S.C. Meredith, Peptide-based inhibitors of amyloid assembly. *Methods in Enzymology*, **413**, 273–312 (2006).
184. N. Kokkoni, K. Scott, H. Amijee, J.M. Mason and A.J. Doig, N-methylated peptide inhibitors of beta-amyloid aggregation and toxicity. *Optimization of the inhibitor structure. Biochemistry*, **45**, 9906–18 (2006).
185. C.W. Cairo, A. Strzelec, R.M. Murphy and L.L. Kiessling, Affinity-based inhibition of beta-amyloid toxicity. *Biochemistry* **41**, 8620–9 (2002).
186. T.L. Lowe, A. Strzelec, L.L. Kiessling and R.M. Murphy, Structure-function relationships for inhibitors of beta-amyloid toxicity containing the recognition sequence KLVFF. *Biochemistry* **40**, 7882–9 (2001).
187. S. Rajagopal, S. C. Meyer, A. Goldman, M. Zhou, and I. Ghosh, A minimalist approach toward protein recognition by epitope transfer from functionally evolved β-sheet surfaces, *J. Am. Chem. Soc.*, **128**, 14356–63 (2006).
188. G. Licini, L. J. Prins, and P. Scrimin, Oligopeptide foldamers: From structure to function, *Eur. J. Org. Chem.*, **6**, 969–77 (2005).

189. C. Grison, P. Coutrot, S. Geneve, C. Didierjean, and M. Marraud, Structural investigation of 'cis' and 'trans' vinylogous peptides: cis-Vinylog turn in folded cis-vinylogous peptides, an excellent mimic of the natural beta-turn, *J. Org. Chem.*, **70**, 10753–64 (2005).
190. M. Hagihara, N. J. Anthony, T. J. Stout, J. Clardy, and S. L. Schreiber, Vinylogous polypeptides: an alternative peptide backbone, *J. Am. Chem. Soc.*, **114**, 6568–70 (1992).
191. M. Gude, U. Piarulli, D. Potenza, B. Salom, and C. Gennari, A new method for the solution and solid phase synthesis of chiral beta-sulfonopeptides under mild conditions, *Tetrahedron Lett.*, **37**, 8589–2 (1996).
192. C. Gennari, B. Salom, D. Potenza, and A. Williams, Synthesis of sulfonamido-pseudopeptides – new chiral unnatural oligomers, *Angew. Chem. Int. Ed. Engl.* **33**, 2067–9 (1994).
193. C. Gennari, B. Salom, D. Potenza, *et al.*, Conformational studies of chiral vinylogous sulfonamidopeptides, *Chem. Eur. J.*, **2**, 644–55 (1996).
194. R. B. Prince, R. B. Barnes, and J. S. Moore, Foldamer-based molecular recognition, *J. Am. Chem. Soc.*, **122**, 2758–62 (2000).
195. S. H. Gellman, Foldamers: A Manifesto, *Acc. Chem. Res.*, **31**, 173–80 (1997).
196. J. D. Sadowsky, W. D. Fairlie, E. B. Hadley, *et al.*, (alpha/beta + alpha)-Peptide antagonists of BH3 Domain/Bcl-x(L) recognition: Toward general strategies for foldamer-based inhibition of protein–protein interactions, *J. Am. Chem. Soc.*, **129**, 139–54 (2007).
197. J. A. Kritzer, M. E. Hodsdon, and A. Schepartz, Solution structure of a beta-peptide ligand for hDM2, *J. Am. Chem. Soc.* **127**, 4118–19 (2005).
198. O. M. Stephens, S. Kim, B. D. Welch, M. E. Hodsdon, M. S. Kay, and A. Schepartz, Inhibiting HIV fusion with a beta-peptide foldamer, *J. Am. Chem. Soc.*, **127**, 13126–7 (2005).
199. E.P. English, R.S. Chumanov, S.H. Gellman and T. Compton, Rational development of beta-peptide inhibitors of human cytomegalovirus entry. *J. Bio. Chem.* **281**, 2661–7 (2006).
200. K. Gademann, M. Ernst, D. Hoyer, and D. Seebach, Synthesis and biological evaluation of a cyclo-β-tetrapeptide as a somatostatin analogue, *Angew. Chem. Int. Ed.*, **38**, 1223–6 (1999).
201. K. Gademann, T. Kimmerlin, D. Hoyer, and D. Seebach, Peptide folding induces high and selective affinity of a linear and small β-peptide to the human somatostatin receptor 4, *J. Med. Chem.*, **44**, 2460–8 (2001).

5

Inhibition of Protein–Protein Interactions by Peptide Mimics

Jorge Becerril, Johanna M. Rodriguez, Pauline N. Wyrembak and Andrew D. Hamilton
Department of Chemistry, Yale University, New Haven, CT, USA

5.1 Introduction

Virtually all biological processes are controlled by the communication between proteins through specific surface interactions. Protein–protein interactions (PPIs) are crucial for normal cell function and regulate processes as diverse as cell growth, signal transduction, and apoptosis. Because of their central role in cellular biochemistry, the disturbance of PPIs lies at the heart of many diseases.

Recent advances in protein structure determination have made possible the characterization of an increasing number of important protein–protein complexes. The insight obtained from mutational and thermodynamic studies has led to the elucidation of the critical binding regions and the structural requirements necessary for complex formation. The wealth of new information available to scientists has allowed the design of molecules that can potentially target a specific protein surface by mimicking the binding epitope of its natural protein partner. A subgroup among the studied PPIs includes those in which one of the protein partners interacts through residues that are localized within a small region on its surface. These key residues are usually found as part of common secondary structural motifs such as α-helices, β-turns and sheets, and serve as binding epitopes for the interaction.

As the structural details of a growing number of protein complexes have been elucidated, it has become more apparent that α-helices play a significant role in numerous protein–protein interactions. As many as 40% of the amino acids in an average protein form part of a chain

Protein Surface Recognition: Approaches for Drug Discovery Ernest Giralt, Mark W. Peczuh and Xavier Salvatella. © 2011 John Wiley & Sons, Ltd

folded into this configuration [1]. Typically, a series of residues on the solvent-exposed face of the helix of one protein is projected into a binding pocket present on the surface of its corresponding protein partner. In many instances, it is the interaction of these amino acids in consecutive turns of the helix that accounts for most of the binding energy. The abundance of α-helices, as well as their characteristic structural features, make them suitable for mediating protein–protein interactions that involve binding to an elongated hot-spot.

The β-sheet is the second major structural element found in globular proteins. β-strands consist of five to ten amino acids that are in an almost fully extended conformation where the side chains alternate above and below the plane of the peptide backbone. The majority of β-strands are aligned adjacent to each other, forming a planar hydrogen-bonding network. Successive β-strands can alternate directions to form an antiparallel β-sheet or extend in the same direction to form a parallel β-sheet. Two antiparallel β-strands connected by a loop of two to five amino acids forms a β-hairpin, a simple supersecondary structure.

Turns are another common structural element found in many PPIs that are often sites for receptor binding, post-translational modifications and antibody recognition [2]. β-turns contain four amino acid residues in the loop designated as $i, i+1, i+2$, and $i+3$ and may be stabilized by a hydrogen bond between the carbonyl and amide N-H of the i and $i+3$ residues, respectively.

The prevalence of these motifs has prompted academic as well as industrial groups to pursue the development of molecules that mimic the key residues of α-helices, β-sheets and β-turns as potential therapeutic leads. In the following section, we discuss some examples of molecules reported in the literature that have been designed to explicitly mimic the amino acid residues of these secondary motifs and have been shown to disrupt PPIs. We have grouped these examples by target, illustrating the various approaches taken by different groups to inhibit the same PPI.

5.2 Inhibition of Calmodulin

5.2.1 Introduction

Calmodulin (CaM) is a calcium-binding protein that mediates a variety of processes including metabolism, muscle contraction, inflammation and the immune response [3, 4]. The ability of CaM to affect many cellular pathways stems from its affinity for the amphiphilic α-helical domains of an array of different protein targets [4]. The tendency of CaM to bind α-helical structures makes it a useful model system for the evaluation of scaffolds that attempt to mimic peptidic α-helices.

5.2.2 Small-Molecule Inhibitors

Hamilton and coworkers first demonstrated the validity and versatility of a terphenyl scaffold as an α-helix mimetic by successfully targeting the interaction between calmodulin and a helical domain of smooth myosin light-chain kinase (smMLCK) [5, 6]. The terphenyl scaffold was developed by Hamilton *et al.* as a general strategy to mimic key amino acid residues along one face of an α-helix while eliminating the complexity of the peptidic

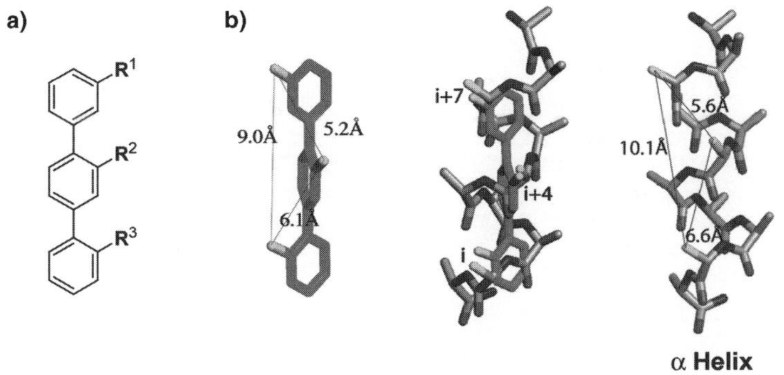

Figure 5.1 *(a) Structure of the trisubstituted 3,2′,2″-terphenyl scaffold designed by Hamilton and coworkers; (b) staggered conformation of trimethyl-substituted terphenyl (left), idealized poly-alanine α-helix (right), and overlay of trimethyl-substituted terphenyl and α-helix (center)*

backbone (Figure 5.1a). The initial design intended to mimic three residues of an α-helix with a trisubstituted 3,2′,2″-terphenyl derivative [7].

Steric repulsion between the aromatic hydrogens of consecutive rings forces a staggered conformation that positions the R groups at similar distances and angular relations to those of the residues in the i, $i+3$ (or $i+4$) and $i+7$ positions of an α-helix (Figure 5.1b). The terphenyl scaffold possesses inherent rigidity while maintaining some degree of rotational freedom that ensures that conformations that best mimic the specific α-helix are easily accessible.

Mutational studies concluded that Trp800 (i), Thr803 ($i+3$), and Val807 ($i+7$) of a 20-mer fragment of the smMLCK helical domain are vital for the stability of the complex with CaM [6]. The terphenyl derivatives **1** and **2** intended to mimic these key hydrophobic residues (Figure 5.2a). In order to simplify the synthesis, the indole of Trp800 was replaced by a phenyl group and the hydroxyl of Thr803 was removed. Improved water solubility was achieved by incorporation of a carboxylic acid.

Polyacrylamide gel permeation chromatography showed that terphenyl **1** formed a complex with CaM; the retention time of the terphenyl was significantly shorter when mixed with CaM. Inhibition of PDE (3′-5-cyclic nucleotide phosphodiesterase), an enzyme that requires binding of CaM to be functional, was observed upon addition of **1** ($IC_{50} = 800$ nM). Since binding of CaM to PDE occurs through the same hydrophobic region that recognizes α-helical motifs, inhibition of the enzyme suggests that **1** binds to CaM in this region. Additional competition experiments with fluorescently labeled C20W led to the same conclusion. Dansylated peptide C20W binds selectively to the C-terminal domain of CaM and the fluorescence of the dansyl group increases upon formation of the complex. Addition of compound **1** resulted in a decrease in the fluorescence intensity, suggesting competition for the same binding site. Structural optimization led to compound **2** incorporating a naphthyl group that more closely resembles the indole of Trp800. In fact, this compound showed inhibition of PDE activity with an IC_{50} value of 9 nM.

Degrado *et al.* targeted CaM employing an arylamide scaffold that had been previously used to mimic amphipathic antibacterial peptides [8]. The presence of intramolecular hydrogen bonds in this scaffold induces a conformation in which the hydrophobic *tert*-butyl and the positively charged ammonium groups are segregated on different areas of the

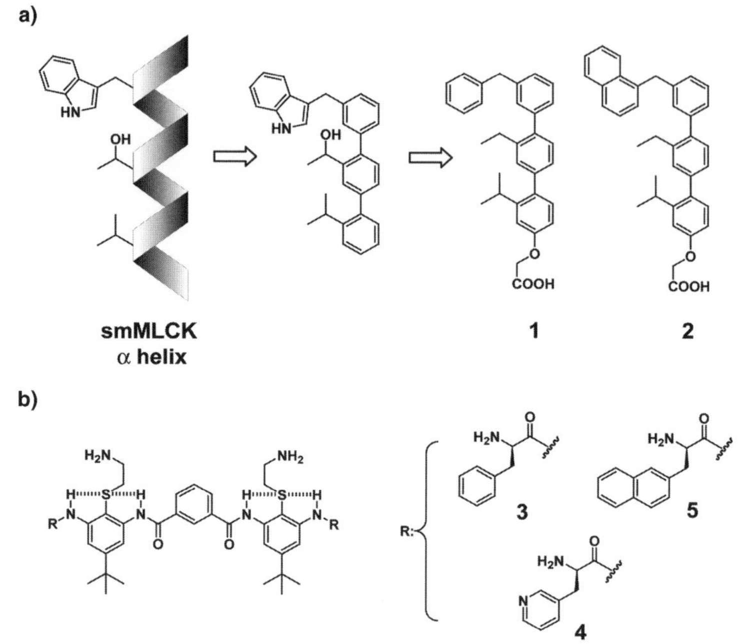

Figure 5.2 (a) Design strategy used by Hamilton and coworkers to mimic key residues Trp800, Thr803, and Val807 from the smMLCK peptide; (b) structure of DeGrado's arylamide foldamers

molecule (Figure 5.2b). The lateral aryl amines were functionalized with hydrophobic D-amino acids bearing phenyl, pyridyl, and naphthyl groups, **3**, **4**, and **5**, respectively.

Competition FP assays showed that these compounds were able to bind CaM with high affinity; inhibitor **3** had a K_i value of 7 nM. The pyridyl and naphthyl derivatives exhibited slightly weaker affinity, but maintained K_i values in the low nanomolar range. [^{15}N, ^{1}H]-HSQC experiments suggested that **3** binds CaM in a similar manner to that of smMLCK.

5.3 Inhibition of HIV-1 Fusion

5.3.1 Introduction

Inhibitors of protein–protein interactions could become important allies in the fight against the HIV-1 epidemic. The initial step of the HIV-1 infection pathway involves attachment of the viral particle to the human cell followed by fusion and delivery of the viral genetic and enzymatic cargo. Proteins involved in this process, such as gp41 and gp120, have become important targets for prevention of HIV-1 infection [9].

Ectodomain gp41 undergoes a series of conformational changes in response to the interaction of viral gp120 with cell membrane-bound CD4 glycoprotein, ultimately leading to fusion of the viral and target cell membranes. Crystallographic studies have shown that gp41 exists as a trimeric helical bundle composed of the N-terminal and C-terminal helices

a)

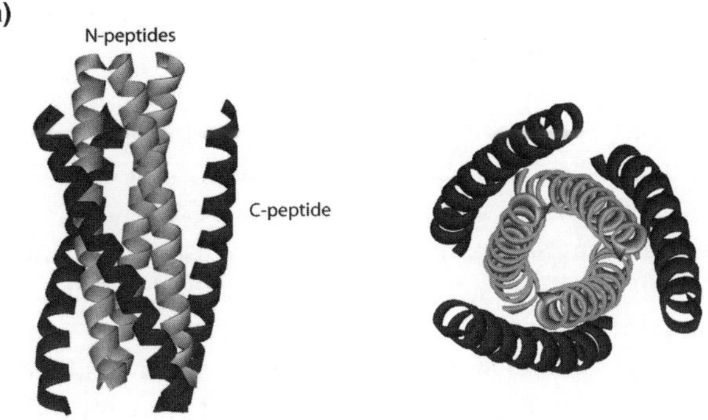

b)

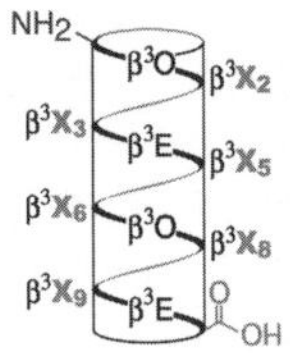

	X3	X6	X9	X2	X5	X8
βWWI-1	I	W	W	V	V	V
βWWI-2	V	V	V	I	W	W
βWWI-3	V	V	V	W	W	I
βWWI-4	W	W	I	V	V	V
βWAI-1	I	A	W	V	V	V

Figure 5.3 *(a) Structure of the trimeric helical bundle formed by the gp41 N-terminal and C-terminal peptides side-view (left) and top-view (right); (b) β-peptide inhibitors designed by Schepartz and coworkers for the inhibition of HIV fusion*

of each of the subunits. In this bundle, the N-terminal helices form a coiled-coil that is surrounded by the C-terminal helices (Figure 5.3a) [10, 11]. During the membrane fusion process, gp41 undergoes a conformational change that exposes the hydrophobic N-terminal bundle, facilitating its insertion into the cell membrane. Refolding of the transient intermediate and reassembly of the C-terminal helices bring the virus into proximity of the cell and induce fusion of their membranes.

Extensive hydrophobic contacts between the residues of the C-terminus and the hydrophobic grooves of the N-terminal coiled-coil account for the bundle stability. The amino acids in the *a* and *d* positions of an *abcdefg* heptad repeat of the C-helical peptides are key for the stability of the complex. Nonnatural oligomers and synthetic small molecules that mimic these hydrophobic repeats can bind to the N-terminal coiled-coil region and prevent reassembly of the C-terminal helices. Antagonists that target either gp120 binding to membrane-bound CD4 or gp41 bundle formation hold great promise as anti-HIV therapeutics [9].

5.3.2 Nonnatural Oligomers: *β*-peptides

β-peptides are a logical progression from *α*-peptides due to the minor modification that results from introducing an additional methylene unit, and have been reviewed extensively in the literature [12]. Significant effort has been made to study the different structural variations available to *β*-peptides [13–16]. A wealth of information is now available about how the presence and positioning of different *β*-amino acids affect the overall secondary structure. The understanding of the factors that govern *β*-peptide folding has facilitated the design of β-peptides that mimic α-peptide motifs and led to the development of several active inhibitors of protein–protein interactions.

Schepartz and coworkers have developed a small series of β^3-decapeptides in order to inhibit gp41-cell fusion [17]. These short β-peptides adopt a 14-helical conformation with three distinct faces. A salt-bridging network between ornithine and β^3-Glu along one of the three faces of the 14-helix and a stabilized macrodipole confer overall structural stability in water [18]. For their design, Schepartz and coworkers used the second face of the helix to present the Trp-Trp-Ile (WWI) residues that mimic the hydrophobic repeats of C-terminal gp41 peptides. The third face of the 14-helix was functionalized with hydrophobic β^3-Val residues. In addition, they interchanged these two faces to allow all possible orientations of the WWI epitope resulting in β-decapeptides **βWWI-1** through **βWWI-4** (Figure 5.3b).

All of the decapeptides were fluorescently labeled at the N-terminus and used directly in FP experiments *in vitro* to measure their affinity for gp41 in model system IZN17 [19]. All four decapeptides showed affinity for this target independent of the WWI epitope orientation to the 14-helix macrodipole or the salt-bridging face. The best inhibitor **βWWI-1** was found to have a K_d of 750 nM. Additional FP competition assays showed that the peptides competed well for the IZN17 binding pocket. A derivative of **βWWI-1**, where the central β^3-Trp derivative was replaced by β^3-Ala resulting in **βWAI-1**, showed no binding affinity for gp41 illustrating the importance of the indole group for activity. Cell-based studies showed that the β-decapeptides could inhibit cell-cell fusion with low micromolar EC_{50} values.

5.3.3 β-Turn Mimetics

As mentioned previously, HIV infection begins with attachment of viral gp120 to membrane protein CD4. Extensive mutagenesis studies [20, 21] and peptide mapping experiments [22, 23] have indicated that amino acid residues 40–55 in the CDR2 region of the CD4 protein are essential for binding of viral gp120, and that Phe43 plays a significant role in this interaction. An X-ray structure of an N-terminal fragment of CD4 shows that amino acids Gln40 through Phe43 are placed in a surface-exposed β-turn [24, 25]. Designed peptides derived from this loop region failed to show potent binding affinity to gp120, suggesting that both the sequence and the structural arrangement of a β-turn are important for binding [22, 26]. Kahn and coworkers reported the development of **6** that mimics residues Gln40-Thr45 and incorporates the β-turn region that extends from Gln40 to Phe43 (Figure 5.4a) [27]. The affinity of fluorescently labeled gp120 for cell-surface CD4 in the presence of peptidomimetic **6** was measured in a fluorescence intensity assay. The inhibitor was able to disrupt binding of gp120 to cells expressing CD4 with an IC_{50} in the low micromolar range.

5.3.4 Small-Molecule Inhibitors

Encouraged by the results obtained from the calmodulin studies, Hamilton *et al.* used the terphenyl approach to target gp41 [28]. They synthesized the terphenyl derivatives shown in Figure 5.4b that mimic the hydrophobic residues usually found in the 3–4 heptad repeats. Terminal carboxylic acids were incorporated to mimic the anionic character of the C-terminal helices while improving water solubility.

The activity of terphenyls **7–9** was studied by circular dichroism (CD) with a model system that consisted of two peptides, N36 and C34, from the N- and C-heptad repeat regions

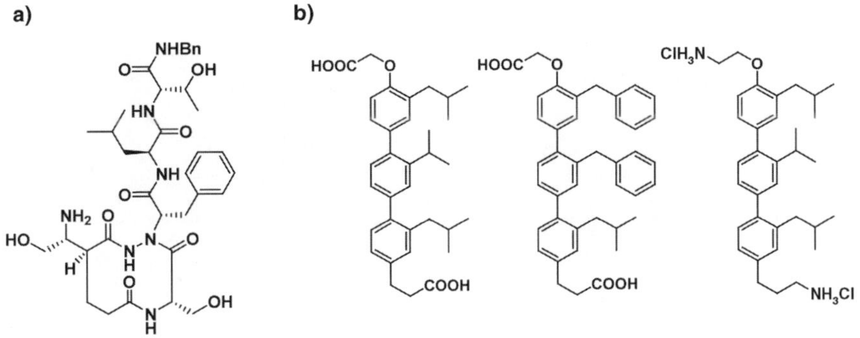

6 7 8 9

Figure 5.4 (a) β-turn mimetic **6** designed by Kahn and coworkers for the inhibition of the gp120/CD4 interaction; (b) structures of terphenyls **7–9**

of gp41. Individually, these peptides are unstructured but when combined, they associate into a six-helix bundle analogous to the gp41 core. Addition of inhibitors **7** and **8** led to a decrease in helicity which indicated a destabilization of the bundle. Moreover, ELISA experiments using an antibody that binds to the folded bundle, but not to the disordered peptides, gave an IC_{50} value of 13 μM for compound **7**. Finally, the results obtained in a dye-transfer cell fusion assay showed that the fusion mechanism of HIV-1 was indeed inhibited by **7** with an IC_{50} value of 16 μM. Substitution of the carboxylic acids by alkyl ammonium groups in **9** completely abolished affinity, underscoring the fact that both electrostatic and hydrophobic factors are required to achieve acceptable potency.

5.4 Inhibition of the Nuclear Estrogen Receptor

5.4.1 Introduction

The members of the nuclear receptor family are involved in a great variety of disease processes and include the estrogen receptor (ER), the androgen receptor (AR), and the thyroid receptor (TR) among others [29, 30]. The nuclear receptors mediate the expression of specific genes upon reception of the appropriate signal (e.g. hormone). Recognition and binding to the cognate DNA sequence is followed by recruitment of coactivator proteins that help assemble the transcriptional machinery (Figure 5.5a) [31].

The coactivator proteins possess multiple repeats of a conserved Leu-X-X-Leu-Leu (LXXLL) motif that is necessary and sufficient for binding [32]. Additional residues flanking this key sequence provide specificity between the different nuclear receptors [33]. Structural information obtained from crystallographic studies has provided evidence that bound coactivator peptides adopt an α-helical conformation where the leucine side chains in positions i, $i + 3$, and $i + 4$ are projected into a hydrophobic area on the surface of the nuclear receptor (Figure 5.5b) [34]. During the last few years, a number of reports have demonstrated that disruption of the ER/coactivator interaction can be achieved by appropriately mimicking the structural features of the LXXLL sequence.

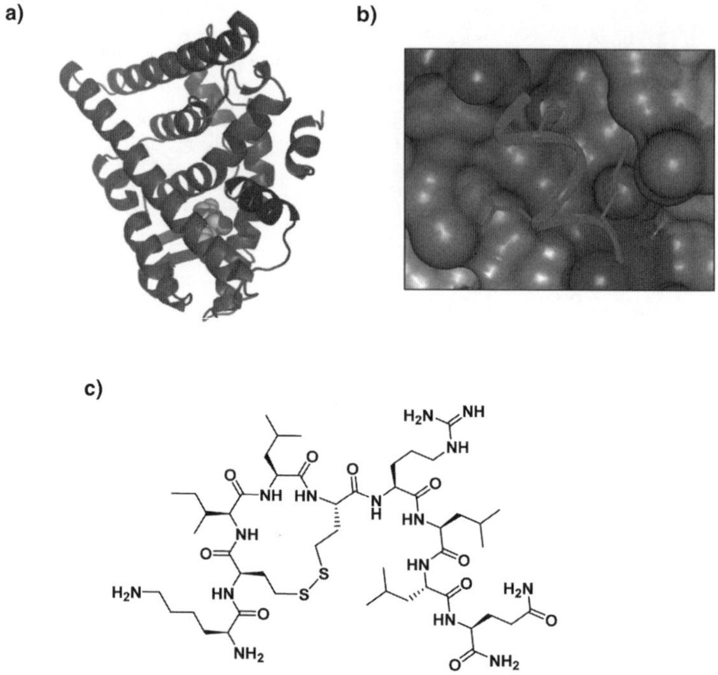

10

Figure 5.5 *(a) X-ray structure of the ER ligand binding domain bound to an agonist molecule and GRIP1 coactivator peptide; (b) view of the GRIP1 peptide on the surface of ER; (c) structure of disulfide-bridged inhibitor* ***10*** *of the ER receptor developed by Spatola and coworkers*

5.4.2 Nonnatural Oligomers: Cyclic Peptides

Spatola and coworkers have recently reported potent inhibitors of the ER/coactivator interaction based on short helices that are stabilized by side-chain to side-chain covalent linkages [35, 36]. The X-ray structure of the coactivator peptide GRIP1 bound to the ER was used to identify residues that could be modified and covalently linked without significant disturbance of the binding epitope. Molecular modeling studies showed that a disulfide-bridged peptide, connected *via* a D-Cys and L-Cys at positions X_{-2} and X_1 of the sequence $X_{-2}X_{-1}\mathbf{L}X_1X_2\mathbf{LL}$, had the best structural resemblance to the bound GRIP1 NR Box II peptide. Other strategies to increase helical propensity within the cyclic peptides included introduction of amide bridges and helix-stabilizing Aib amino acids.

A fluorescence-based binding assay was employed to evaluate the inhibitory activity of the compounds. It was found that the disulfide-bridged compound **10** in Figure 5.5c, that had the greatest helical content in CD studies, also had the highest affinity for the ERα isoform with a K_i value of 25 nM. Interestingly, the affinity for the ERβ isoform was 15 times lower, with a K_i value of 390 nM. In order to investigate the importance of preorganization, several linear peptide counterparts of the cyclic peptides were also studied. Compounds from this series displayed weaker inhibition illustrating the importance of conformational constraints that promote α-helical character. The authors were able to confirm the predicted binding mode of **10** through an X-ray structure of the ligand bound to ERα.

Other researchers have exploited similar methods for the stabilization of short α-peptide secondary structures including introduction of hydrogen bond surrogates [37, 38], covalent linkages [39, 40] and side-chain to side-chain interactions [41, 42].

5.4.3 Small-Molecule Inhibitors

5.4.3.1 Trisubstituted Heterocycles

Katzenellenbogen and coworkers demonstrated that the ER/coactivator complex can be successfully targeted with small molecules [43]. They noted that the three key leucine residues of the LXXLL box lie in a triangular arrangement (Figure 5.6a) and that a trisubstituted aromatic core could project hydrophobic functionality in a similar fashion.

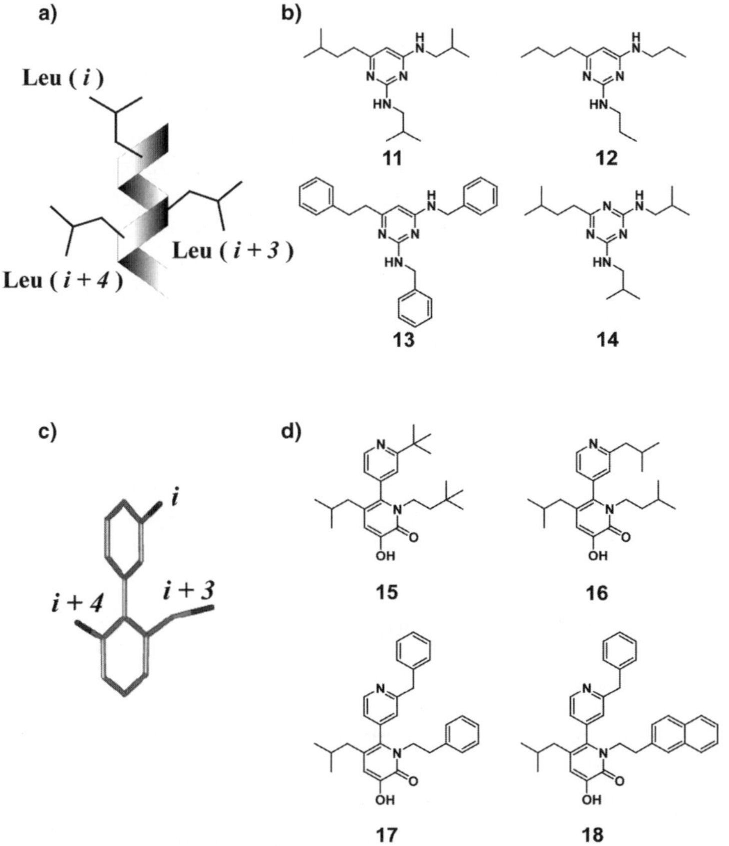

Figure 5.6 *(a) Schematic representation of the leucine residues in the i, i + 3, and i + 4 positions of the LXXLL nuclear receptor box; (b) structures of inhibitors* ***11–14*** *designed by Katzenellenbogen and coworkers; (c) general biaryl scaffold for the mimicry of the LXXLL motif; (d) structures of ER inhibitors* ***15–18*** *based on the biaryl scaffold*

A series of derivatives containing triazene, pyrimidine and trithiane heterocyclic cores were studied by molecular modeling and synthesized for affinity evaluation, some of which are shown in Figure 5.6b.

The ability of these small molecules to displace the coactivator peptide from the surface of the ER was assessed through an FP assay. Most of the compounds showed weak binding with K_i values above 200 μM. However, derivatives **11–13** that were based on the pyrimidine core, showed more promising activities with the best compound **11** having a K_i of 29 μM (Figure 5.6b). Interestingly, **14**, which closely resembles the most potent inhibitor **11**, had low affinity ($K_i = 590$ μM) for the ER. Even though higher affinity inhibitors are desired, these results are the proof of concept that small molecules can indeed be used to target the ER and potentially other nuclear receptors as well.

5.4.3.2 Pyridylpyridones

Hamilton and colleagues targeted the ER receptor by incorporating the leucine-mimicking groups into a biaryl scaffold [44]. They and others [45] have shown that the i, $i+3$ and $i+4$ residues of an α-helix can be mimicked by simply adding an additional *ortho*-substituent to the bottom ring of a substituted biaryl derivative (Figure 5.6c). Incorporation of a single methylene in the $i+3$ mimicking side chain permits the adoption of a conformation that closely mimics the distances and angular projections of the i, $i+3$, and $i+4$ groups of an α-helix.

The use of aromatic heterocyclic rings yielded derivatives with good water solubility while facilitating the introduction of substituents into the 2-pyridyl and 1,5-pyridone positions (Figure 5.6d). An X-ray structure of **15** showed a nonplanar conformation with an aryl-aryl dihedral angle of 82° and distances of 5.6, 5.4 and 5.7 Å between the atoms that mimic the β-carbons of the key amino acid residues. The β-carbon atoms of the i, $i+3$, and $i+4$ leucine residues of the GRIP1 coactivator peptide superimposed with the corresponding carbon atoms of **15** showed good spatial matching, thereby validating this design as an α-helix mimetic.

In vitro FP assays showed that derivatives **16**, **17**, and **18** effectively bound to the surface of the ER displacing the coactivator peptide (Figure 5.6d). The best K_i value reported was 4.2 μM for the most hydrophobic compound **18**. Remarkably, the reported values are within the same order of magnitude as those found for natural isolated Leu-X-X-Leu-Leu motifs which lie in the high nanomolar to low micromolar range.

5.5 Inhibition of the Bcl-x_L/Bak Interaction

5.5.1 Introduction

Apoptosis, or programmed cell death, is tightly regulated by the B-cell lymphoma-2 (Bcl-2) family of proteins that include both anti-apoptotic and pro-apoptotic proteins [46]. Anti-apoptotic members, such as Bcl-x_L and Bcl-w, form stable complexes with pro-apoptotic proteins neutralizing their activity and preventing them from initiating the apoptotic cascade [47]. A balance between these two subfamilies is required for cell function [48]. In normal cells, a series of factors such as DNA damage or death signals, cause up-regulation and activation of pro-apoptotic proteins such as Bak and Bad and ultimately result in the initiation of apoptosis [49]. In cancerous cells, however, over-expressed anti-apoptotic proteins override the pro-apoptotic signals abrogating the cell's ability to initiate apoptosis [50]. Disruption of the Bcl-x$_L$/Bak complex, enabling Bak to carry out its apoptotic function, is an attractive strategy for inducing apoptosis in cancerous cells. Several low molecular weight inhibitors of Bcl-x$_L$ have been identified by screening various chemical libraries [51, 52].

Alanine-scanning experiments have identified Val74, Leu78, Ile81, and Ile85 as the critical residues, found on the Bak BH3 α-helix, that provide most of the complex stability [53]. Additional evidence for the importance of these residues has been provided by an NMR structure of the Bcl-x$_L$/Bak complex that revealed that these key residues make direct contact with a hydrophobic pocket on the surface of Bcl-x$_L$. The availability of detailed structural information has prompted the development of various α-helix mimicking compounds to inhibit the Bcl-x$_L$/Bak interaction [54, 55].

5.5.2 Nonnatural Oligomers: β-peptides

Recently, Gellman *et al.* have used chimeric ($\alpha/\beta + \alpha$)-peptide ligands to disrupt the Bak BH3 domain interaction with Bcl-x$_L$ [56, 57]. Initial efforts from the Gellman group to develop inhibitors based on 12- and 14-helical β-peptides were unsuccessful. However, a 14/15-helical scaffold derived from α/β-peptide oligomers was met with more success and a potent inhibitor **19** was identified by FP (Figure 5.7a).

This chimeric ($\alpha/\beta + \alpha$) oligomer was composed of two segments: a 1:1 alternation of α- and β-amino acids and a sequence of only α-amino acids. The residues projected on this

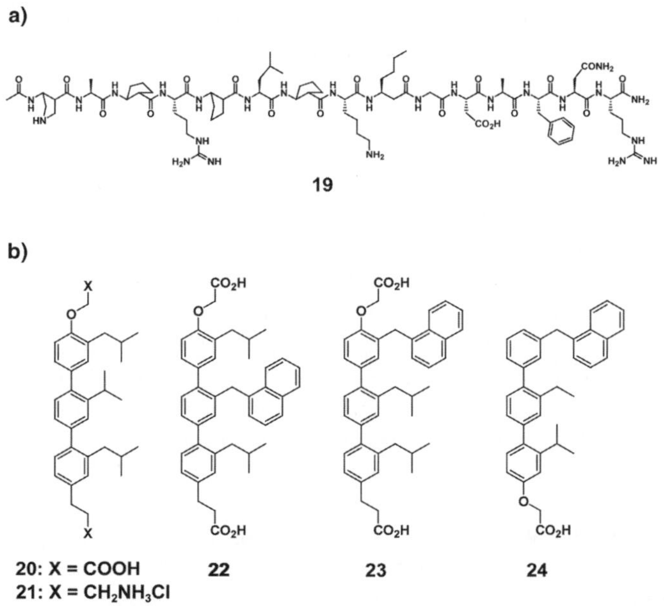

Figure 5.7 *(a) Structure of Gellman's ($\alpha/\beta + \alpha$) chimeric peptide* ***19****; (b) structure of terphenyls* ***20–24*** *designed by Hamilton and coworkers for the inhibition of the Bcl-x$_L$/Bak protein-protein interaction*

oligomer correspond to the sequence of the BH3 domain of Bak with the exception of β-Phe13 in **19** that was used instead of Ile84. To discard nonspecific hydrophobic interactions as the driving force for the binding, the authors assessed the affinity of a fluorescein-labeled derivative of **19** for bovine γ-globulin and bovine serum albumin proteins. Weak binding to these proteins, which preferentially bind to hydrophobic ligands, ruled out nonspecific hydrophobic effects as the cause of the affinity of **19** for Bcl-x_L.

5.5.3 Small-Molecule Inhibitors

5.5.3.1 Terphenyls

Hamilton's terphenyl strategy was also employed to mimic the pro-apoptotic protein Bak. Terphenyls **20–24** were evaluated for their ability to displace the Bak peptide from the surface of Bcl-x_L (Figure 5.7b) [58]. An FP assay provided K_i values in the low micromolar range for the different derivatives, except for **22** that was more potent with a K_i value of 114 nM. Scrambling of the naphthyl residue (**23**) or the use of a 2-naphthyl group led to less potent inhibitors which demonstrated that hydrophobicity alone was not responsible for the potent activity of **22** but that shape complementarity was also an important factor. [^{15}N, ^{1}H]-HSQC experiments showed that the residues around the binding cleft on the Bcl-x_L surface experienced significant chemical shift changes upon addition of **22**, which confirmed that the inhibitors bind in the same area as the natural Bak peptide.

Additional studies were carried out to determine whether the terphenyl strategy could be applied to disrupt the p53/MDM2 interaction. Hamilton *et al.* synthesized terphenyl derivatives that inhibited this complex with low nanomolar K_i values and more importantly found compounds that had different selectivity for Bcl-x_L over MDM2. A more detailed description of this important protein–protein complex and attempts to inhibit it with these and other small molecules are described in a later chapter.

5.5.3.2 Pyridyl Oligoamides

Structures that adopt preferred specific conformations are useful starting points in the design of scaffolds from which to project functionality in a controlled fashion. Hamilton and coworkers reported the proteomimetic foldamer **25** that mimics the i, $i + 4$ and $i + 7$ residues of an α-helix (Figure 5.8a) [59]. The pyridyl-oligoamide scaffold was accessible through a convenient iterative synthesis and incorporated a series of intramolecular hydrogen bonds that stabilize a conformation in which the R groups are all projected from the same side of the molecule. Indeed, the X-ray structure of one of the derivatives showed that the foldamer adopts this planar conformation in the solid state. The network of bifurcated hydrogen bonds stabilizes the desired conformer and favors the presence of a significant population of active molecules even in hydrogen bond-competing solvents such as DMSO.

The oligoamide derivatives were tested for their ability to disrupt the Bcl-x_L/Bak interaction in an FP assay and found to effectively displace the Bak peptide with K_i values in the low micromolar range. Compound **25** (where R^1, R^2 = benzyl, R^3 = isopropyl) was the most potent with a K_i value of 1.6 μM. Surprisingly, extended derivatives that mimicked the fourth key residue ($i + 11$ position) of the α-helix failed to improve the binding affinity. Recently, Guy and coworkers have reported an analogous oligoamide scaffold that lacks an

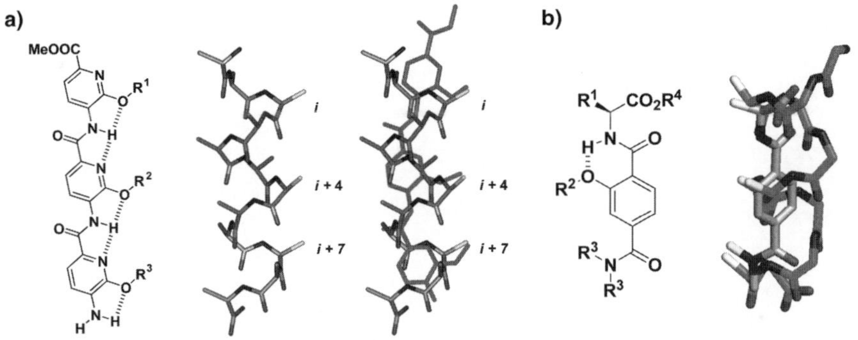

25 26

Figure 5.8 (a) Hamilton's oligoamide foldamer **25** (left), poly-alanine α-helix (center), and overlay of **25** (where $R^1=R^2=R^3=$ -Me) and poly-alanine α-helix (right); (b) structure of terephthalamide scaffold **26** (left), and overlay of a poly-alanine α-helix and **26** (where $R^1=R^2=R^3=$ -Me, $R^4=$H, right)

intramolecular hydrogen-bonding network to target the p53/MDM2 protein–protein interaction [60].

5.5.3.3 Terephthalamides

In the search for simpler, more drug-like proteomimetics than the terphenyl and the oligoamide structures, the Hamilton laboratory developed the terephthalamide scaffold **26** (Figure 5.8b) [61]. The new design combined the basic structural features of the terphenyls along with the use of intramolecular hydrogen bonds to preorganize the structure and yield a scaffold that can mimic the i, $i + 4$ and $i + 7$ residues of an α-helix. The top and bottom phenyl rings of the terphenyl scaffold were replaced by two functionalized carboxamide groups that could be easily introduced and varied while retaining part of the structural rigidity. Additionally, an intramolecular hydrogen bond was used to project the R^1 group in an optimal position.

FP titrations were employed to assess the ability of the terephthalamide derivatives to disrupt the Bcl-x_L/Bak complex. Compound **26** (where $R^1 =$ isobutyl, R^2 and $R^3 =$ isopropyl, $R^4 =$ methyl) demonstrated the highest activity with a K_i value of 0.78 nM. Two-dimensional [^{15}N, ^1H]-HSQC experiments showed that addition of the inhibitors affected the chemical shift of residues located in the shallow hydrophobic cleft where the Bak peptide binds. More importantly, both the terephthalamide and the terphenyl scaffolds perturb the chemical shift of the same amino acids on the Bcl-x_L surface which suggests a similar binding mode for the two proteomimetics.

5.5.3.4 Trisubstituted Imidazoles

Following a similar strategy, Antuch and coworkers developed trisubstituted imidazole scaffolds **27** and **28** that mimic the i, $i + 4$ and $i + 7$ hydrophobic residues of Bak and that can be synthesized in a couple of simple steps (Figure 5.9) [62]. Compound **29** showed promising micromolar activity in cell proliferation assays in HL-60 cells and induced apoptosis in a dose dependent manner. *In vitro* FP assays showed that **29** was the best

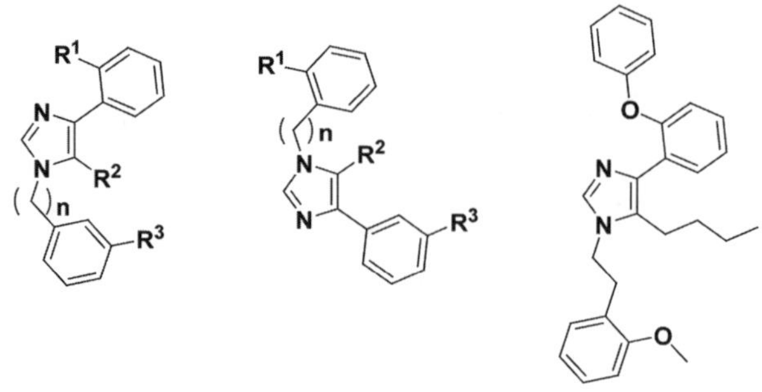

27 28 29

Figure 5.9 Structures of trisubstituted imidazoles **27–29** designed by Antuch and coworkers as inhibitors of Bcl-w

inhibitor and that it can indeed bind to Bcl-w from the Bcl-2 family displacing the Bak BH3 peptide ($K_i = 8\,\mu M$).

5.6 Inhibition of the p53/MDM2 Interaction

5.6.1 Introduction

The p53/MDM2 interaction is another protein–protein interaction involved in the apoptotic pathway [63]. In unstressed cells, p53 is present in low concentrations due to tight regulation by MDM2 [64, 65]. Over-expression of p53 results in cycle arrest and apoptosis. In many cancers, MDM2 is over-expressed attenuating p53's ability to induce apoptosis. Three key residues of the p53 helix project into a hydrophobic cleft on the surface of MDM2 and contribute to most of the binding energy [66]. Extensive structural characterization of the p53/MDM2 interaction has allowed the design of molecules capable of inhibiting this interaction that can be used to force cancer cells to undergo apoptosis.

Small-molecule inhibitors of the p53/MDM2 interaction are not described in the following section since they are discussed in a later chapter.

5.6.2 Nonnatural Oligomers

5.6.2.1 β-Peptides

Schepartz and coworkers have reported the use of a family of β^3-peptides that form 14-helices as inhibitors of the p53/HDM2 interaction [67, 68]. Residues Phe19, Trp23, and Leu26, from the p53 binding epitope were grafted onto the i, $i+3$ and $i+6$ positions of a 14-helix β-peptide scaffold (Figure 5.10). This arrangement positions these residues on one face of the β-helix.

Competition FP assays were performed to measure the ability of these oligomers to inhibit the p53/HDM2 interaction. ***β*53-1** showed the best inhibition with an IC_{50} value of 94.5 μM. Direct binding of N- and C-labeled fluorescein conjugates of ***β*53-1** to HDM2 showed that these derivatives have an affinity similar to that of p53. Further optimization of the nonbinding face of the β-peptide led to an increase in affinity ($IC_{50} = 13\,\mu M$).

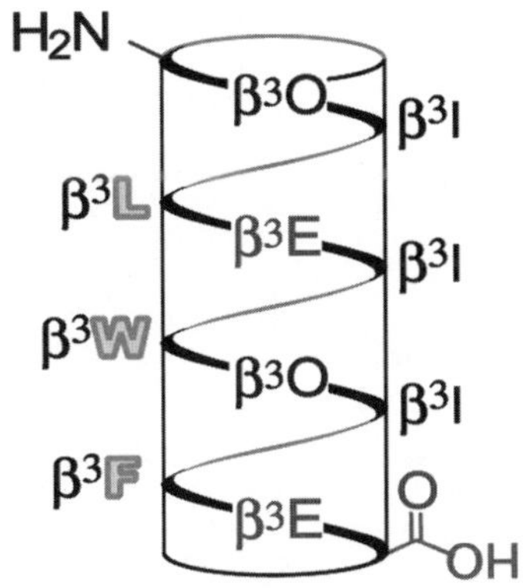

β53-1

Figure 5.10 *Structure of β-peptide **β53-1**, developed by Schepartz and coworkers for the inhibition of the p53/HDM2 interaction*

5.6.2.2 Peptoids

Peptoids are another class of peptidomimetic oligomers that have been used to mimic α-helices [69]. In peptoids, the amino acid side chain is appended to the amide nitrogen which eliminates stereogenic carbons and modifies the conformational preferences of the oligomer. Appella and coworkers have recently developed peptoid inhibitors of the p53/HDM2 interaction [70]. In order to mimic the relative positions of the Phe19, Trp23, and Leu26 residues corresponding to the i, $i + 4$, and $i + 7$ position of the p53 helix, the tryptophan residue was moved from the $i + 4$ to the $i + 3$ position on the peptoid mimetic. This minor modification accounts for the difference in helicity between an α-helix (3.6 residues per turn) and the more tightly wound peptoid helix (3.0 residues per turn) facilitating better alignment of the p53 helix with the peptoid residues. The peptoid type-1 poly-proline helix also projects side chains at different angles from those of an α-helix. This is overcome by addition of extra methylene groups at both the phenylalanine and tryptophan side chains which allows for a greater range of possible orientations.

Appella *et al.* began with a peptoid helix that contained chiral side chains at every position, and found that even though this promoted helix stability in water, the solubility was diminished. Incorporation of an increasing number of achiral side chains led to better peptoid inhibitors. The stepwise substitution with more hydrophilic side chains resulted in their best peptoid inhibitor **30** having an IC_{50} value of 6.6 μM (Figure 5.11a).

5.6.3 *β*-Hairpin Mimetics

Robinson *et al.* proved that a β-hairpin scaffold can be used to mimic the spatial arrangement of critical groups found along one face of the helical p53 peptide and disrupt the p53/HDM2 interaction. The distance expected between the Cα atoms of residues i and $i + 2$ along one strand of a β-hairpin is similar to that found between the Cα

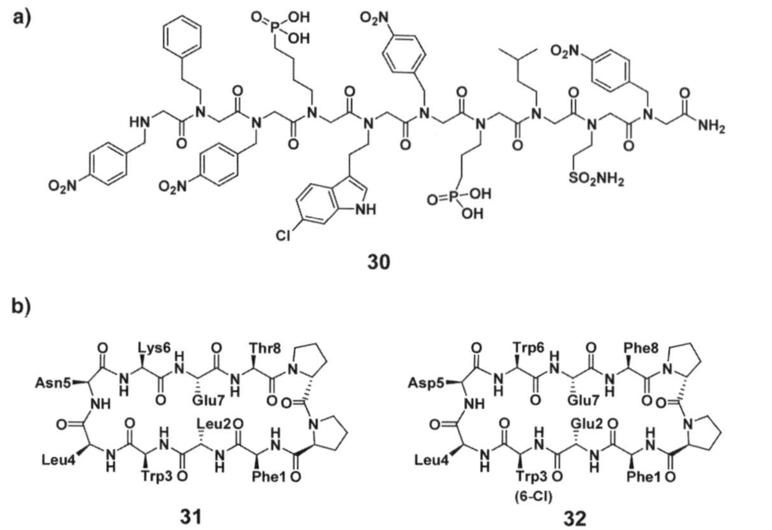

Figure 5.11 (a) Structure of peptoid inhibitor **30** designed by Appella and coworkers for the inhibition of the p53/HDM2 interaction; (b) β-hairpin mimetics **31** and **32** designed by Robinson and coworkers

atoms of the key binding residues Phe19 and Trp23. The authors argued that a β-hairpin mimetic scaffold could thus be used to preorganize the side chains of phenylalanine and tryptophan (and possibly also leucine), providing a geometry similar to the p53 α-helix binding epitope. For their design, a D-Pro-L-Pro dipeptide template was introduced to stabilize the β-hairpin conformation of compound **31** (Figure 5.11b) [71, 72]. Residues Phe1, Trp3, and Leu4 in **31** mimic residues Phe19, Trp23, and Leu26 in p53, as seen from computational modeling studies. A BIAcore solution-phase competition assay was used to study binding of **31** to HDM2, and the IC_{50} value was found to be 125 μM. Two-dimensional $[^{15}N, {}^{1}H]$-HSQC experiments confirmed that mimetic **31** interacts with the p53-binding site, however, further NMR studies suggested that the peptide was not populating a β-hairpin conformation in solution, which may explain the weak affinity for HDM2.

Subsequent optimization steps resulted in several libraries in which the residues in **31** were exchanged for other proteinogenic and nonproteinogenic amino acid building blocks. The potency of lead **31** was improved by a factor of 1000 to give **32** with an IC_{50} value of 140 nM (Figure 5.11b). A 1.4 Å resolution crystal structure of a HDM2/**32** complex confirmed that the β-hairpin conformation of **32** exists in the bound state and showed that interaction with HDM2 mainly occurs through aromatic contacts. Residues Phe1, (6-Cl) Trp3, and Leu4 are projected into the hydrophobic p53-binding site on the surface of HDM2. Trp6 and Phe8 participate in π-stacking interactions with Phe55 on HDM2 while the side chain of Asp5 seems to interact with the N-terminal region of HDM2. Interestingly, these contacts have not been previously seen in complexes of HDM2 with p53. Overall, the structural and functional binding data demonstrate that a β-hairpin can be used to successfully mimic an α-helix.

5.7 Miscellaneous Protein Targets

5.7.1 Inhibition of Neurotrophins

Neurotrophins are dimeric growth factors that regulate the differentiation and survival of select neurons in the peripheral and central nervous system [73]. Nerve growth factor (NGF), brain-derived neurotrophic factor (BDNF) and neurotrophin-3 (NT-3) are members of this protein family which bind to tyrosine kinase (Trk) receptors TrkA, TrkB, and TrkC, respectively [74]. Binding of NGF to TrkA induces receptor dimerization and phosphorylation of tyrosine residues which sets off a signaling cascade that promotes cell survival and differentiation. Several neuronal cell types implicated in pathologies including neuropathies, pain, and neurodegeneration, express the TrkA receptor and therefore could benefit from treatment with neurotrophins [75]. Clinical trials with NGF, however, have not been successful due to its *in vivo* instability and lack of selectivity [76]. Compounds that can mimic the binding mode of NGF to TrkA and elicit neurotrophic activity are highly desirable.

Several studies suggest that the β-turns on NGF are putative hot spots that mediate binding to TrkA [77, 78]. As a result, there has been a significant effort focused on the development of compounds that mimic these regions. Preliminary work in this field led to the preparation of disulfide-bridged cyclic peptides that adopt β-turn conformations corresponding to regions of NGF [77, 79]. Building upon this research, Burgess and co-workers designed a library of small β-turn peptidomimetics with decreased peptidic character [80]. Compound **33** was identified from this library and shown to be proteolytically stable and a selective partial agonist of TrkA (Figure 5.12) [75]. β-turn mimetic **33** induced tyrosine phosphorylation, receptor dimerization, and acted as a synergistic potentiator of NGF.

The same strategy was used to design peptidomimetic compounds that bind to TrkC by mimicking the binding epitope of NT-3. In this case, the more rigid inhibitor **34** was found to be a partial agonist that induced phosphorylation of TrkC (Figure 5.12) [81]. Recently, the authors constructed a mini-library that incorporated the β-turn residues of both NT-3 and NGF into their bisphenyl macrocyclic scaffold [82]. These peptidomimetics showed cell survival and/or neuronal differentiation activities, depending on the compound.

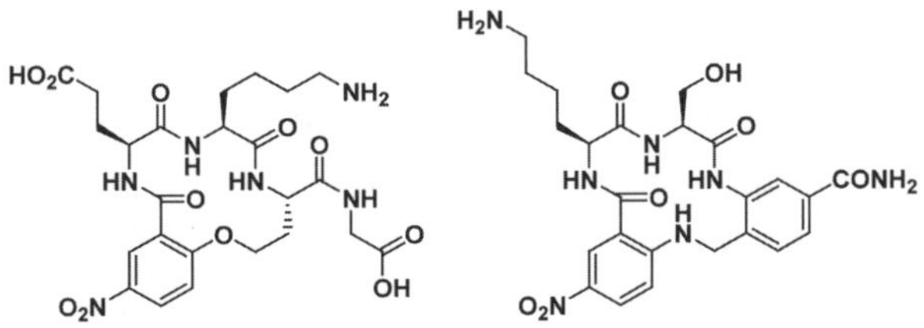

33 **34**

Figure 5.12 *β-turn analogues **33** and **34** that bind the NGF receptor TrkA and TrkC, respectively*

5.7.2 Inhibition of the Grb2 SH2 Domain

Growth factor receptor bound protein 2 (Grb2) is a critical adaptor protein involved in intracellular signal transduction pathways [83]. It is composed of two Src homology 3 (SH3) domains flanking a single Src homology 2 (SH2) domain. The SH2 domain interacts with protein-tyrosine kinases (PTKs) ultimately leading to Ras activation. SH2 domain binding antagonists offer an alternative to PTK inhibitors as therapeutics for the treatment of cancer.

The preferred binding peptide sequence for SH2 domains is pTyr-X-Asn-X where pTyr represents phosphotyrosine. An X-ray structure of a Grb2 SH2 domain complexed with a pTyr-containing peptide shows that the ligand binds in a β-turn conformation in contrast to many other SH2 domains where the peptide binds in an extended conformation [84]. The pTyr and Asn residues of the peptide are placed at the i and $i + 2$ positions of the β-turn, respectively.

In an effort to develop antagonists of the SH2 domain, researchers from Novartis optimized the minimal Grb2 SH2 domain binding peptide Ac-pTyr-Ile-Asn-NH_2 [85]. They noted that the binding of the phosphopeptide in a β-turn configuration forces a local right-handed 3_{10} helical conformation at the Ile residue. In order to increase the β-turn propensity of their peptidomimetics, the Ile residue was replaced with 1-aminocyclohexane carboxylic acid (Ac_6c), a nonnatural amino acid that stabilizes a 3_{10} helical conformation [86]. Additionally, a 3-naphthalen-1-yl-propyl group was introduced at the C-terminus to further increase contacts with a hydrophobic surface on the SH2 domain, near the C-terminus of the bound phosphopeptide. The resulting peptide mimetic **35** was significantly more potent ($IC_{50} = 47$ nM) than the starting minimal tripeptide Ac-pTyr-Ile-Asn-NH_2 ($IC_{50} = 8.9$ µM) in competitive binding assays (Figure 5.13).

The pTyr residues are necessary for ligand recognition and binding by the SH2 domains, however, they are highly susceptible to hydrolysis by phosphatases. Burke and coworkers set out to develop inhibitors that incorporated pTyr mimetics. They replaced the pTyr residue

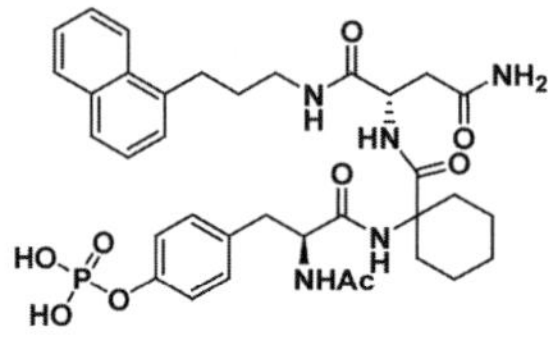

35

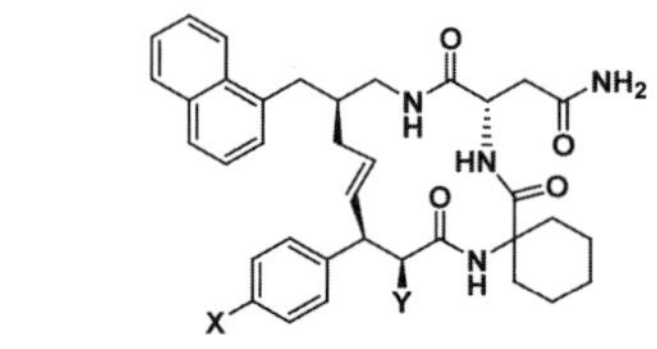

36: X = $CH_2P(O)(OH)_2$; Y = CH_2COOH
37: X = $CH(COOH)_2$; Y = CH_2COOH

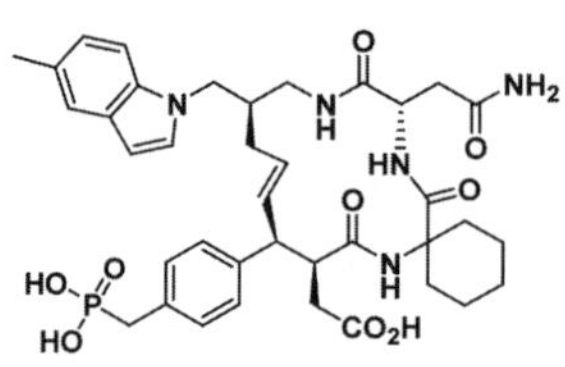

38

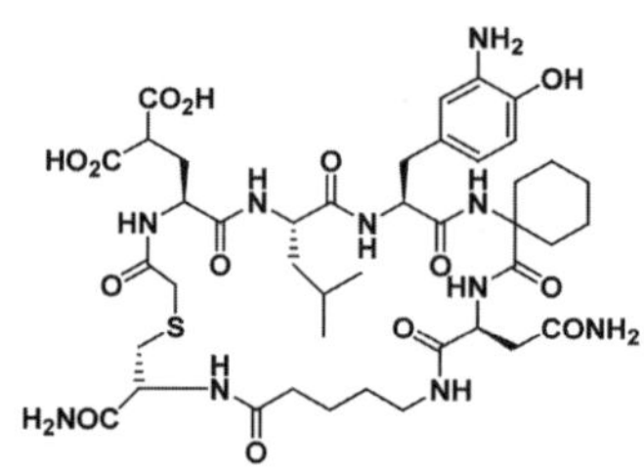

39

Figure 5.13 *Inhibitors of the Grb2 SH2 domain based on a β-turn*

in **35** with phosphonomethyl phenylalanine (PmP) [87] and *p*-(2-malonlyl)phenylalanine (Pmf) [88]. Further optimizations led to a more rigid macrocyclic scaffold that incorporated both the bend-inducing Ac_6c residue and negatively charged but nonhydrolyzable groups, resulting in compounds **36** [89] and **37** [90] (Figure 5.13). These inhibitors showed increased potency with respect to their open-chain congeners and had improved cellular efficacy. A crystal structure of **37** complexed with the Grb2 SH2 domain showed that the compound bound to the pTyr target site in a similar mode to that of the phosphopeptide [91]. Finally, the related 5-methylindolyl-containing macrocycle **38** had a K_d value of 75 pM as determined by surface plasmon resonance experiments (Figure 5.13) [92]. In whole cell assays, **38** blocked binding of Grb2 to erbB-2 and exhibited anti-mitogenic effects against erbB-2-dependent breast cancers at noncytotoxic concentrations.

Roller and coworkers have also developed nonphosphorylated ligands for the Grb2 SH2 domain based on a disulfide-bridged cyclic decapeptide discovered from a phage-display library [93]. A series of thioether-cyclized derivatives designed based on this lead showed potency and promising cellular activity [94, 95]. Further efforts were focused on the development of a new scaffold with only five amino acids in the hope of decreasing the peptidic character and improving affinity. Compound **39**, which also contains the Ac_6c β-turn inducing residue, exhibited an IC_{50} value of 58 nM and showed antiproliferative activity against breast cancer cells with an IC_{50} value of 19 nM (Figure 5.13) [96]. This series of compounds serves as a lead in the development of Grb2-SH2 domain antagonists that lack a hydrolyzable pTyr group.

In addition to the previous examples, β-turn based peptidomimetics have also been used as antagonists of the integrins and have been reviewed elsewhere [97].

5.7.3 Inhibition of the Myd88/IL-1RI Interaction

Toll-like receptors (TLRs) and the superfamily of interleukin-1 receptors (IL-1Rs) are necessary for proper innate and adaptive immune system function [98, 99]. Both families share a conserved Toll/IL-1 resistance (TIR) domain. Myeloid differentiation primary response protein 88 (MyD88) is an adaptor protein that binds via its TIR domain to the TIR domain of IL-1RI in complex with IL-1 receptor accessory protein (IL-1RAcp) [100]. Myd88 then facilitates the recruitment of signaling proteins that ultimately lead to the activation of mitogen-activated protein (MAP) kinases and transcription factors [101].

Crystal structures of the TIR domains of the related TLR1 and TLR2 combined with mutational and functional studies provided insight about the basis of the TIR homotypic protein/protein complex [102]. A loop that links one of the α-helices to a β-sheet motif, the BB loop, has the consensus sequence (Phe/Tyr)-(Val/Leu/Ile)-(Pro/Gly) and comprises the majority of a large surface patch that is critical for stability of the TIR-TIR domain-mediated interaction.

Rebek and coworkers prepared derivatives of the central three amino acid residues of the BB-loop and studied their ability to disrupt the TIR mediated IL-1RI/Myd88 interaction *in vitro* and *in vivo* [103]. The compound hydrocinnamoyl-L-valyl pyrrolidine **40** was found to inhibit IL-1β mediated phosphorylation of p38 MAP in EL4 thymoma cells (Figure 5.14a). In addition, sandwich ELISA assays showed that **40** prevents the IL-1β mediated association of IL-1RI and MyD88 in EL4 cells and lymphocytes from mouse spleen. The disruption of the IL-1RI/MyD88 interaction is specific over other members of

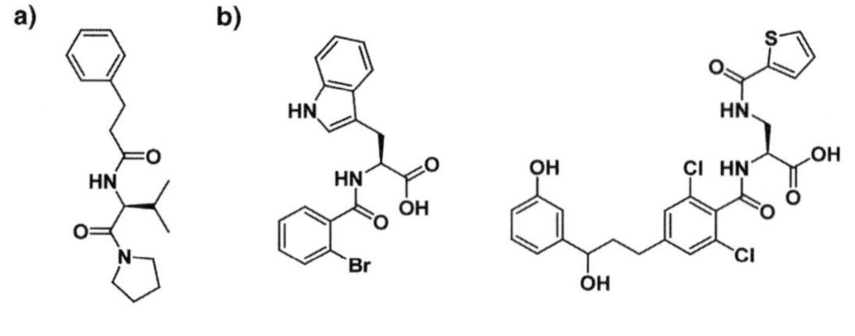

40 **41** **42**

Figure 5.14 *(a) β-turn peptidomimetic **40** that disrupts the IL-1RI/Myd88 interaction; (b) inhibitors of the ICAM-1/LFA-1 interaction **41** and **42***

the Toll receptor superfamily. Attenuation of the IL-1β-induced fever response in mice confirmed that compound **40** exerts an inhibitory effect *in vivo*. Recently, modifications of this lead resulted in bifunctional TIR domain mimetics that also disrupt the interaction of MyD88 with the Il-1RI/IL-1RAcP complex [104].

5.7.4 Inhibition of the ICAM-1/LFA-1 Interaction

The interaction between leukocyte functional antigen-1 (LFA-1) and intercellular adhesion molecule-1 (ICAM-1) is necessary for proper immunological and inflammatory reactions [105]. The epitope comprising ICAM-1's residues Glu34, Lys39, Met64, Tyr66, Asn68 and Gln73, is necessary for its interaction with the αL subunit of LFA-1 [106]. These amino acids are located on three different β-strands across the face of the protein [107]. Rather than using a small molecule as a lead in their development of an inhibitor, Gadek and coworkers used the epitope of ICAM-1, the native ligand of LFA-1, as the starting point for their design [108]. Mutagenesis studies had shown that Glu34 and Lys39 are required for LFA-1 binding [106, 109]. The authors found that kistrin, a disintegrin protein containing an Arg-Gly-Asp sequence, inhibited the binding of LFA-1 and ICAM-1 *in vitro*, most likely due to its ability to mimic Glu34 and Lys39 from ICAM-1. The linear epitope Arg-Gly-Asp-Met-Pro from kistrin was identified through alanine mutagenesis studies [110], and was subsequently grafted into cyclic peptides that ultimately furnished the disulfide H_2N-Cys-Gly-Tyr$^{(m)}$-Asp-Met-Pro-Cys-CO_2H (Tyr$^{(m)}$ = *meta*-tyrosine) with an IC_{50} value of 1.6 μM in an ELISA assay.

Simultaneous, related efforts identified *ortho*-bromobenzoyl tryptophan **41** as an inhibitor of the ICAM-1/LFA-1 interaction (IC_{50} = 1.4 μM) with structure and potency similar to those of peptide H_2N-Cys-Gly-Tyr$^{(m)}$-Asp-Met-Pro-Cys-CO_2H (Figure 5.14b). Incorporation of the *meta*-phenol group of the peptide in a subsequent inhibitor gave a 30-fold improvement in potency. Further optimization yielded compound **42** which had the best IC_{50} value of 1.4 nM (Figure 5.14b). An MLR assay, which measures lymphocyte function, showed that compound **42** inhibits ICAM-1/LFA-1 binding, LFA-1 mediated lymphocyte proliferation *in vitro* and the immune response *in vivo*. The structural similarities found among the various active inhibitors led Gadek and colleagues to propose that these compounds successfully mimic the ICAM-1 epitope.

Several authors have proposed an alternative allosteric mechanism by which these inhibitors might bind to an alternative site located on the $\beta 2$ subunit I-like domain of LFA-1 [111–114]. This binding event could significantly alter the β-propeller region of the α-subunit of LFA-1 preventing ICAM-1 binding. Recent data, however, suggests that there is a high-affinity binding site in the αL subunit of LFA-1 that overlaps with the ICAM-1 binding site, but also a lower-affinity binding site in the β-subunit [115]. The presence of these two binding regions could explain both observations.

5.7.5 Inhibition of PDZ Domains

PDZ protein-interaction domains act in tandem or in concert with other interaction domains to form scaffold proteins onto which large molecular complexes can be assembled [116, 117]. Although they were first discovered as key components of synaptic signaling complexes, it is now known that they also function as recognition elements in other membrane assemblies and in protein trafficking networks. The structure of PDZ domains is characterized by six β-strands and two α-helices folded into a six-stranded β-sandwich. A groove formed between one β-sheet and an α-helix serves as the interacting site with the protein ligand, capturing the ligand's C-terminal tail as an extension of the β-sheet in a mechanism known as β-strand addition [118]. The sequence of the C-terminal peptide determines the specificity of the domain.

Bartlett and coworkers designed β-strand mimetics using a solution structure of the PDZ domain of α1-syntrophin bound to heptapeptide Gly-Val-Lys-Glu-Ser-Leu-Val, derived from the conserved C-terminus of the voltage-gated sodium channel [119]. The structure of the complex of α1-syntrophin shows that the backbone of the heptapeptide is hydrogen-bonded to the PDZ cleft [120]. They developed a series of @-tides, protease-resistant peptidomimetics containing cyclic amino acid surrogates @-unit **43** and aza-@-unit **44** (Figure 5.15) [121]. When placed at alternating positions of a β-strand, both units maintain the hydrogen-bonding pattern and extended conformation typically found in this secondary structure element [122]. In addition, the versatility of this approach is increased by using **44** that allows the incorporation of amino acid side chains.

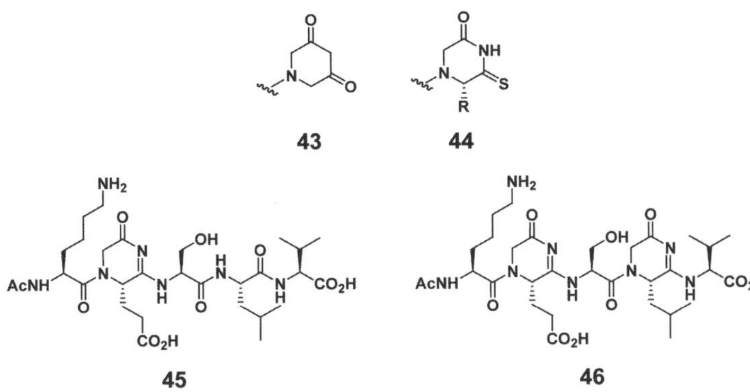

*Figure 5.15 Unsubstituted @-unit **43** and substituted aza-@-unit **44** (top), and substituted @-tides, **45** and **46** (bottom), designed by Bartlett and coworkers as PDZ domain ligands*

The authors replaced the Glu and Leu residues of the pentapeptide Ac-Lys-Glu-Ser-Leu-Val with **43** and **44** (where R = $-CH_2CH_2CO_2H$ or $-CH_2CH(CH_3)_2$) at either or both positions, since these residues do not make significant interactions with the PDZ cleft. When Glu and Leu were simultaneously substituted with their analogous aza-@-units, the resulting peptidomimetic **45** had a K_d of 1 μM, in comparison to a K_d of 6.4 μM for the more flexible pentapeptide (Figure 5.15). Surprisingly, the single substitution of Glu with $@^{Glu}$ (where the superscript indicates the amino acid side chain present on the aza-@-unit) in **46** showed the best binding with a K_d value of 320 nM (Figure 5.15).

The stability of several derivatives was also evaluated in aqueous phosphate buffer at pH 4,7, and 10. From these studies, it was determined that the acyl amidine linkage is fairly stable at neutral or basic pH, but labile under acidic conditions. Peptidomimetic Ac-Lys-$@^{Leu}$-Val was resistant to both trypsin and carboxypeptidase A in contrast with the peptide Ac-Lys-Leu-Val that was rapidly degraded. The stability of the @-units along with their ability to stabilize an extended β-sheet conformation makes them useful β-strand peptidomimetics.

In addition to the previous examples, Ghosh and coworkers have described an elegant method to transfer necessary chemical information from a β-sheet to a small molecule mimetic that could be used to target thrombin [123].

5.8 Conclusion

Many protein–protein interactions are mediated by a binding epitope composed of residues belonging to a single structural motif such as an α-helix, β-turn or a β-sheet. Detailed structural analysis of these binding interfaces has facilitated the development of inhibitors that mimic the key elements necessary for complexation. The examples highlighted in this chapter demonstrate that the use of such peptidomimetics to target protein surfaces is an effective strategy for modulating protein function. These encouraging results will surely prompt the development of novel peptidomimetic scaffolds with increased affinity and versatility applicable to a broader range of biologically relevant systems.

References

1. W. Kabsch, C. Sander, Dictionary of protein secondary structure – pattern-recognition of hydrogen-bonded and geometrical features, *Biopolymers* **1983**, *22*, 2577–2637.
2. G. D. Rose, L. M. Gierasch, J. A. Smith, Turns in peptides and proteins, *Adv. Protein Chem.* **1985**, *37*, 1–109.
3. C. B. Klee, T. C. Vanaman, Calmodulin, *Adv. Protein Chem.* **1982**, *35*, 213–321.
4. A. Crivici, M. Ikura, Molecular and structural basis of target recognition by calmodulin, *Annual Review of Biophysics and Biomolecular Structure* **1995**, *24*, 85–116.
5. R. Chattopadhyaya, W. E. Meador, A. R. Means, F. A. Quiocho, Calmodulin structure refined at 1.7 angstrom resolution, *J. Mol. Biol.* **1992**, *228*, 1177–92.
6. W. E. Meador, A. R. Means, F. A. Quiocho, Target enzyme recognition by calmodulin – 2.4-angstrom structure of a calmodulin-peptide complex, *Science* **1992**, *257*, 1251–5.
7. B. P. Orner, J. T. Ernst, A. D. Hamilton, Toward proteomimetics: Terphenyl derivatives as structural and functional mimics of extended regions of an alpha-helix, *J. Am. Chem. Soc.* **2001**, *123*, 5382–3.

8. H. Yin, K. K. Frederick, D. H. Liu, A. J. Wand, W. F. DeGrado, Arylamide derivatives as peptidomimetic inhibitors of calmodulin, *Org. Lett.* **2006**, *8*, 223–5.
9. D. C. Chan, P. S. Kim, HIV entry and its inhibition, *Cell* **1998**, *93*, 681–4.
10. D. C. Chan, D. Fass, J. M. Berger, P. S. Kim, Core structure of gp41 from the HIV envelope glycoprotein, *Cell* **1997**, *89*, 263–73.
11. M. Lu, P. S. Kim, A trimeric structural subdomain of the HIV-1 transmembrane glycoprotein, *J. Biomol. Struct. Dyn.* **1997**, *15*, 465–71.
12. P. E. Nielsen, *Pseudo-Peptides in Drug Discovery*, WILEY-VCH, Weinheim (Germany), **2004**.
13. W. F. DeGrado, Introduction: Protein design, *Chemical Reviews* **2001**, *101*, 3025–6.
14. D. Seebach, J. L. Matthews, beta-peptides: a surprise at every turn, *Chemical Communications* **1997**, 2015–22.
15. S. H. Gellman, Foldamers: A manifesto, *Acc. Chem. Res.* **1998**, *31*, 173–80.
16. J. A. Kritzer, O. M. Stephens, D. A. Guarracino, S. K. Reznik, A. Schepartz, beta-Peptides as inhibitors of protein–protein interactions, *Bioorg. Med. Chem.* **2005**, *13*, 11–16.
17. O. M. Stephens, S. Kim, B. D. Welch, M. E. Hodsdon, M. S. Kay, A. Schepartz, Inhibiting HIV fusion with a beta-peptide foldamer, *J. Am. Chem. Soc.* **2005**, *127*, 13126–7.
18. S. A. Hart, A. B. F. Bahadoor, E. E. Matthews, X. Y. J. Qiu, A. Schepartz, Helix macrodipole control of beta(3)-peptide 14-helix stability in water, *J. Am. Chem. Soc.* **2003**, *125*, 4022–3.
19. D. M. Eckert, P. S. Kim, Design of potent inhibitors of HIV-1 entry from the gp41 N-peptide region, *Proc. Natl. Acad. Sci. U. S. A.* **2001**, *98*, 11187–92.
20. N. R. Landau, M. Warton, D. R. Littman, The Envelope Glycoprotein of the human immunodeficiency virus binds to the immunoglobulin-like domain of Cd4, *Nature* **1988**, *334*, 159–62.
21. A. Ashkenazi, L. G. Presta, S. A. Marsters, T. R. Camerato, K. A. Rosenthal, B. M. Fendly, D. J. Capon, Mapping the Cd4 binding-site for human-immunodeficiency-virus by alanine-scanning mutagenesis, *Proc. Natl. Acad. Sci. U. S. A.* **1990**, *87*, 7150–4.
22. B. A. Jameson, P. E. Rao, L. I. Kong, B. H. Hahn, G. M. Shaw, L. E. Hood, S. B. H. Kent, Location and chemical synthesis of a binding-site for Hiv-1 on the Cd4 protein, *Science* **1988**, *240*, 1335–9.
23. J. D. Lifson, K. M. Hwang, P. L. Nara, *et al.*, Synthetic Cd4 peptide derivatives that inhibit Hiv infection and cytopathicity, *Science* **1988**, *241*, 712–16.
24. J. H. Wang, Y. W. Yan, T. P. J. Garrett, *et al.*, Atomic-structure of a fragment of human Cd4 containing 2 immunoglobulin-like domains, *Nature* **1990**, *348*, 411–18.
25. S. E. Ryu, P. D. Kwong, A. Truneh, *et al.*, Crystal structure of an Hiv-binding recombinant fragment of human Cd4, *Nature* **1990**, *348*, 419–26.
26. M. H. Brodsky, M. Warton, R. M. Myers, D. R. Littman, Analysis of the site in Cd4 that binds to the Hiv envelope glycoprotein, *J. Immunol.* **1990**, *144*, 3078–6.
27. S. X. Chen, R. A. Chrusciel, H. Nakanishi, *et al.*, Design and synthesis of a Cd4 beta-turn mimetic that inhibits human-immunodeficiency-virus envelope glycoprotein Gp120 binding and infection of human-lymphocytes, *Proc. Natl. Acad. Sci. U. S. A.* **1992**, *89*, 5872–6.
28. J. T. Ernst, O. Kutzki, A. K. Debnath, S. Jiang, H. Lu, A. D. Hamilton, Design of a protein surface antagonist based on alpha-helix mimicry: Inhibition of gp41 assembly and viral fusion, *Angew. Chem. Int. Ed.* **2001**, *41*, 278–81.
29. J. M. Hall, J. F. Couse, K. S. Korach, The multifaceted mechanisms of estradiol and estrogen receptor signaling, *J. Biol. Chem.* **2001**, *276*, 36869–72.
30. J. M. Olefsky, Nuclear receptor minireview series, *J. Biol. Chem.* **2001**, *276*, 36863–4.
31. M. G. Rosenfeld, C. K. Glass, Coregulator codes of transcriptional regulation by nuclear receptors, *J. Biol. Chem.* **2001**, *276*, 36865–8.
32. D. M. Heery, E. Kalkhoven, S. Hoare, M. G. Parker, A signature motif in transcriptional co-activators mediates binding to nuclear receptor, *Nature* **1997**, *387*, 733–6.
33. C. Y. Chang, J. D. Norris, H. Gron, L. A. Paige, P. T. Hamilton, D. J. Kenan, D. Fowlkes, D. P. McDonnell, Dissection of the LXXLL nuclear receptor-coactivator interaction motif using combinatorial peptide libraries: Discovery of peptide antagonists of estrogen receptors alpha and beta, *Mol. Cell. Biol.* **1999**, *19*, 8226–39.

34. A. K. Shiau, D. Barstad, P. M. Loria, *et al.*, The structural basis of estrogen receptor/coactivator recognition and the antagonism of this interaction by tamoxifen, *Cell* **1998**, *95*, 927–37.
35. A. K. Galande, K. S. Bramlett, J. O. Trent, T. P. Burris, J. L. Wittliff, A. F. Spatola, Potent inhibitors of LXXLL-based protein–protein interactions, *Chembiochem* **2005**, *6*, 1991–8.
36. A. M. Leduc, J. O. Trent, J. L. Wittliff, *et al.*, Helix-stabilized cyclic peptides as selective inhibitors of steroid receptor-coactivator interactions, *Proc. Natl. Acad. Sci. U. S. A.* **2003**, *100*, 11273–8.
37. R. N. Chapman, G. Dimartino, P. S. Arora, A highly stable short alpha-helix constrained by a main-chain hydrogen-bond surrogate, *J. Am. Chem. Soc.* **2004**, *126*, 12252–3.
38. D. Wang, K. Chen, J. L. Kulp, P. S. Arora, Evaluation of biologically relevant short alpha-helices stabilized by a main-chain hydrogen-bond surrogate, *J. Am. Chem. Soc.* **2006**, *128*, 9248–56.
39. C. E. Schafmeister, J. Po, G. L. Verdine, An all-hydrocarbon cross-linking system for enhancing the helicity and metabolic stability of peptides, *J. Am. Chem. Soc.* **2000**, *122*, 5891–2.
40. S. K. Sia, P. A. Carr, A. G. Cochran, V. N. Malashkevich, P. S. Kim, Short constrained peptides that inhibit HIV-1 entry, *Proc. Natl. Acad. Sci. U. S. A.* **2002**, *99*, 14664–9.
41. V. Esteve, S. Blondelle, B. Celda, E. Perez-Paya, Stabilization of an alpha-helical conformation in an isolated hexapeptide inhibitor of calmodulin, *Biopolymers* **2001**, *59*, 467–76.
42. K. Ramalingam, T. L. Gururaja, N. Ramasubbu, M. J. Levine, Stabilization of helix by side-chain interactions in histatin-derived peptides: Role in candidacidal activity, *Biochem. Biophys. Res. Commun.* **1996**, *225*, 47–53.
43. A. L. Rodriguez, A. Tamrazi, M. L. Collins, J. A. Katzenellenbogen, Design, synthesis, and *in vitro* biological evaluation of small molecule inhibitors of estrogen receptor a coactivator binding, *J. Med. Chem.* **2004**, *47*, 600–11.
44. J. Becerril, A. D. Hamilton, Helix mimetics as inhibitors of the interaction of the estrogen receptor with coactivator peptides, *Angew. Chem. Int. Ed.* **2007**, vol. *46* (24) pp. 4471–3.
45. E. Jacoby, Biphenyls as potential mimetics of protein alpha-helix, *Bioorg. Med. Chem. Lett.* **2002**, *12*, 891–3.
46. A. Burlacu, Regulation of apoptosis by Bcl-2 family proteins, *Journal of Cellular and Molecular Medicine* **2003**, *7*, 249–57.
47. T. W. Sedlak, Z. N. Oltvai, E. Yang, K. Wang, L. H. Boise, C. B. Thompson, S. J. Korsmeyer, Multiple Bcl-2 family members demonstrate selective dimerizations with bax, *Proc. Natl. Acad. Sci. U. S. A.* **1995**, *92*, 7834–8.
48. M. C. Raff, Social controls on cell-survival and cell-death, *Nature* **1992**, *356*, 397–400.
49. C. M. Rudin, C. B. Thompson, Apoptosis and disease: Regulation and clinical relevance of programmed cell death, *Annual Review of Medicine* **1997**, *48*, 267–81.
50. D. E. Fisher, Apoptosis in cancer-therapy – crossing the threshold, *Cell* **1994**, *78*, 539–42.
51. A. G. Cochran, Protein–protein interfaces: mimics and inhibitors, *Curr. Opin. Chem. Biol.* **2001**, *5*, 654–9.
52. U. Fischer, K. Schulze-Osthoff, Apoptosis-based therapies and drug targets, *Cell Death and Differentiation* **2005**, 1–20.
53. M. Sattler, H. Liang, D. Nettesheim, *et al.*, Structure of Bcl-x(L)-Bak peptide complex: Recognition between regulators of apoptosis, *Science* **1997**, *275*, 983–6.
54. J. Qian, M. J. Voorbach, J. R. Huth, *et al.*, Discovery of novel inhibitors of Bcl-xL using multiple high-throughput screening platforms, *Anal. Biochem.* **2004**, *328*, 131–8.
55. I. J. Enyedy, Y. Ling, K. Nacro, *et al.*, Discovery of small-molecule inhibitors of bcl-2 through structure-based computer screening, *J. Med. Chem.* **2001**, *44*, 4313–24.
56. J. D. Sadowsky, W. D. Fairlie, E. B. Hadley, *et al.*, (alpha/beta + alpha)-Peptide antagonists of BH3 Domain/Bcl-x(L) recognition: Toward general strategies for foldamer-based inhibition of protein–protein interactions, *J. Am. Chem. Soc.* **2007**, *129*, 139–54.
57. J. D. Sadowsky, M. A. Schmitt, H. S. Lee, *et al.*, Chimeric (alpha/beta plus alpha)-peptide ligands for the BH3-recognition cleft of Bcl-x(L): Critical role of the molecular scaffold in protein surface recognition, *J. Am. Chem. Soc.* **2005**, *127*, 11966–8.

58. O. Kutzki, H. S. Park, J. T. Ernst, B. P. Orner, H. Yin, A. D. Hamilton, Development of a potent Bcl-x(L) antagonist based on alpha-helix mimicry, *J. Am. Chem. Soc.* **2002**, *124*, 11838–9.
59. J. T. Ernst, J. Becerril, H. S. Park, H. Yin, A. D. Hamilton, Design and application of an alpha-helix-mimetic scaffold based on an oligoamide-foldamer strategy: Antagonism of the bak BH3/Bcl-xL complex, *Angew. Chem. Int. Ed.* **2003**, *42*, *535*- +
60. F. Lu, S. W. Chi, D. H. Kim, K. H. Han, I. D. Kuntz, R. K. Guy, Proteomimetic libraries: Design, synthesis, and evaluation of p53-MDM2 interaction inhibitors, *J. Comb. Chem.* **2006**, *8*, 315–25.
61. H. Yin, G. I. Lee, K. A. Sedey, *et al.*, Terephthalamide derivatives as mimetics of helical peptides: Disruption of the Bcl-x(L)/Bak interaction, *J. Am. Chem. Soc.* **2005**, *127*, 5463–8.
62. W. Antuch, S. Menon, Q. Z. Chen, *et al.*, Design and modular parallel synthesis of a MCR derived alpha-helix mimetic protein–protein interaction inhibitor scaffold, *Bioorg. Med. Chem. Lett.* **2006**, *16*, 1740–3.
63. D. A. Vargas, S. Takahashi, Z. Ronai,in *Advances in Cancer Research,* Vol 89 **2003**, pp. 1–34.
64. Y. Haupt, R. Maya, A. Kazaz, M. Oren, Mdm2 promotes the rapid degradation of p53, *Nature* **1997**, *387*, 296–9.
65. J. Momand, H. H. Wu, G. Dasgupta, MDM2 – master regulator of the p53 tumor suppressor protein, *Gene* **2000**, *242*, 15–29.
66. P. H. Kussie, S. Gorina, V. Marechal, *et al.*, Structure of the MDM2 oncoprotein bound to the p53 tumor suppressor transactivation domain, *Science* **1996**, *274*, 948–53.
67. J. A. Kritzer, J. D. Lear, M. E. Hodsdon, A. Schepartz, Helical beta-peptide inhibitors of the p53-HDM2 interaction, *J. Am. Chem. Soc.* **2004**, *126*, 9468–9.
68. J. A. Kritzer, N. W. Luedtke, E. A. Harker, A. Schepartz, A rapid library screen for tailoring beta-peptide structure and function, *J. Am. Chem. Soc.* **2005**, *127*, 14584–5.
69. C. W. Wu, S. L. Seurynck, K. Y. C. Lee, A. E. Barron, Helical peptoid mimics of lung surfactant protein C, *Chemistry & Biology* **2003**, *10*, 1057–63.
70. T. Hara, S. R. Durell, M. C. Myers, D. H. Appella, Probing the structural requirements of peptoids that inhibit HDM2-p53 interactions, *J. Am. Chem. Soc.* **2006**, *128*, 1995–2004.
71. R. Fasan, R. L. A. Dias, K. Moehle, O. Zerbe, J. W. Vrijbloed, D. Obrecht, J. A. Robinson, Using a beta-hairpin to mimic an alpha-helix: Cyclic peptidomimetic inhibitors of the p53-HDM2 protein–protein interaction, *Angew. Chem. Int. Ed.* **2004**, *43*, 2109–12.
72. R. Fasan, R. L. A. Dias, K. Moehle, *et al.*, Structure-activity studies in a family of beta-hairpin protein epitope mimetic inhibitors of the p53-HDM2 protein–protein interaction, *Chembiochem* **2006**, *7*, 515–26.
73. G. R. Lewin, Y. A. Barde, Physiology of the neurotrophins, *Annual Review of Neuroscience* **1996**, *19*, 289–317.
74. M. Barbacid, The Trk Family of Neurotrophin Receptors, *Journal of Neurobiology* **1994**, *25*, 1386–1403.
75. S. Maliartchouk, Y. B. Feng, L. Ivanisevic, *et al.*, A designed peptidomimetic agonistic ligand of TrkA nerve growth factor receptors, *Mol. Pharmacol.* **2000**, *57*, 385–91.
76. H. U. Saragovi, K. Burgess, Small molecule and protein-based neurotrophic ligands: agonists and antagonists as therapeutic agents, *Expert Opinion on Therapeutic Patents* **1999**, *9*, 737–51.
77. L. Lesauteur, L. Wei, B. F. Gibbs, H. U. Saragovi, Small peptide mimics of nerve growth-factor bind Trka receptors and affect biological responses, *J. Biol. Chem.* **1995**, *270*, 6564–9.
78. N. Beglova, L. LeSauteur, I. Ekiel, H. U. Saragovi, K. Gehring, Solution structure and internal motion of a bioactive peptide derived from nerve growth factor, *J. Biol. Chem.* **1998**, *273*, 23652–8.
79. F. M. Longo, M. Manthorpe, Y. M. Xie, S. Varon, Synthetic NGF peptide derivatives prevent neuronal death via a p75 receptor-dependent mechanism, *J. Neurosci. Res.* **1997**, *48*, 1–17.
80. H. B. Lee, M. C. Zaccaro, M. Pattarawarapan, S. Roy, H. U. Saragovi, K. Burgess, Syntheses and activities of new C-10 beta-turn peptidomimetics, *J. Org. Chem.* **2004**, *69*, 701–13.
81. M. Pattarawarapan, M. C. Zaccaro, U. H. Saragovi, K. Burgess, New templates for syntheses of ring-fused, C-10 beta-turn peptidomimetics leading to the first reported small-molecule mimic of neurotrophin-3, *J. Med. Chem.* **2002**, *45*, 4387–90.

82. M. C. Zaccaro, H. B. Lee, M. Pattarawarapan, Z. B. Xia, A. Caron, P. J. L'Heureux, Y. Bengio, K. Burgess, H. U. Saragovi, Selective small molecule peptidomimetic ligands of TrkC and TrkA receptors afford discrete or complete neurotrophic activities, *Chemistry & Biology* **2005**, *12*, 1015–28.
83. T. K. Sawyer, Src homology-2 domains: Structure, mechanisms, and drug discovery, *Biopolymers* **1998**, *47*, 243–61.
84. J. Rahuel, C. Garcia-Echeverria, P. Furet, *et al.*, Structural basis for the high affinity of amino-aromatic SH2 phosphopeptide ligands, *J. Mol. Biol.* **1998**, *279*, 1013–22.
85. H. Fretz, P. Furet, C. Garcia-Echeverria, J. Rahuel, J. Schoepfer, Structure-based design of compounds inhibiting Grb2-SH2 mediated protein–protein interactions in signal transduction pathways, *Current Pharmaceutical Design* **2000**, *6*, 1777–96.
86. P. Furet, B. Gay, G. Caravatti, *et al.*, Structure-based design and synthesis of nigh affinity tripeptide ligands of the Grb2-SH2 domain, *J. Med. Chem.* **1998**, *41*, 3442–9.
87. J. Yao, C. R. King, T. Cao, *et al.*, Potent inhibition of Grb2 SH2 domain binding by non-phosphate-containing ligands, *J. Med. Chem.* **1999**, *42*, 25–35.
88. Y. Gao, J. Luo, Z. J. Yao, *et al.*, Inhibition of Grb2 SH2 domain binding by non-phosphate-containing ligands. 2. 4-(2-malollyl)phenylalanine as a potent phosphotyrosyl mimetic, *J. Med. Chem.* **2000**, *43*, 911–20.
89. C. Q. Wei, Y. Gao, K. Lee, *et al.*, Macrocyclization in the design of Grb2 SH2 domain-binding ligands exhibiting high potency in whole-cell systems, *J. Med. Chem.* **2003**, *46*, 244–54.
90. Z. D. Shi, C. Q. Wei, K. O. Lee, *et al.*, Macrocyclization in the design of non-phosphorus-containing Grb2 SH2 domain-binding ligands, *J. Med. Chem.* **2004**, *47*, 2166–9.
91. J. Phan, Z. D. Shi, T. R. Burke, D. S. Waugh, Crystal structures of a high-affinity macrocyclic peptide mimetic in complex with the Grb2 SH2 domain, *J. Mol. Biol.* **2005**, *353*, 104–15.
92. Z. D. Shi, K. Lee, H. P. Liu, *et al.*, A novel macrocyclic tetrapeptide mimetic that exhibits low-picomolar Grb2 SH2 domain-binding affinity, *Biochem. Biophys. Res. Commun.* **2003**, *310*, 378–83.
93. L. Oligino, F. D. T. Lung, L. Sastry, *et al.*, Nonphosphorylated peptide ligands for the Grb2 Src homology 2 domain, *J. Biol. Chem.* **1997**, *272*, 29046–52.
94. Y. Q. Long, F. D. T. Lung, P. P. Roller, Global optimization of conformational constraint on non-phosphorylated cyclic peptide antagonists of the Grb2-SH2 domain, *Bioorg. Med. Chem.* **2003**, *11*, 3929–36.
95. P. Li, M. C. Zhang, Y. Q. Long, *et al.*, Potent Grb2-SH2 domain antagonists not relying on phosphotyrosine mimics, *Bioorg. Med. Chem. Lett.* **2003**, *13*, 2173–7.
96. Y. L. Song, M. L. Peach, P. P. Roller, S. Qiu, S. M. Wang, Y. Q. Long, Discovery of a novel nonphosphorylated pentapeptide motif displaying high affinity for Grb2-SH2 domain by the utilization of 3′-substituted tyrosine derivatives, *J. Med. Chem.* **2006**, *49*, 1585–96.
97. M. J. P. de Vega, M. Martin-Martinez, R. Gonzalez-Muniz, Modulation of protein–protein interactions by stabilizing/mimicking protein secondary structure elements, *Curr. Top. Med. Chem.* **2007**, *7*, 33–62.
98. K. V. Anderson, Toll signaling pathways in the innate immune response, *Curr. Opin. Immunol.* **2000**, *12*, 13–19.
99. L. O'Neill, The Toll/interleukin-1 receptor domain: a molecular switch for inflammation and host defence, *Biochem. Soc. Trans.* **2000**, *28*, 557–63.
100. R. Medzhitov, P. Preston-Hurlburt, E. Kopp, *et al.*, MyD88 is an adaptor protein in the hToll/IL-1 receptor family signaling pathways, *Molecular Cell* **1998**, *2*, 253–8.
101. S. Janssens, R. Beyaert, A universal role for MyD88 in TLR/IL-1R-mediated signaling, *Trends Biochem. Sci.* **2002**, *27*, 474–82.
102. Y. W. Xu, X. Tao, B. H. Shen, *et al.*, Structural basis for signal transduction by the Toll/interleukin-1 receptor domains, *Nature* **2000**, *408*, 111–15.
103. T. Bartfai, M. M. Behrens, S. Gaidarova, J. Pemberton, A. Shivanyuk, J. Rebek, A low molecular weight mimic of the Toll/IL-1 receptor/resistance domain inhibits IL-1 receptor-mediated responses, *Proc. Natl. Acad. Sci. U. S. A.* **2003**, *100*, 7971–6.

104. C. N. Davis, E. Mann, M. M. Behrens, *et al.*, MyD88-dependent and -independent signaling by IL-1 in neurons probed by bifunctional toll/IL-1 receptor domain/BB-loop mimetics, *Proc. Natl. Acad. Sci. U. S. A.* **2006**, *103*, 2953–8.
105. N. OppenheimerMarks, P. E. Lipsky, Adhesion molecules as targets for the treatment of autoimmune diseases, *Clin. Immunol. Immunopathol.* **1996**, *79*, 203–10.
106. K. L. Fisher, J. Lu, L. Riddle, K. J. Kim, L. G. Presta, S. C. Bodary, Identification of the binding site in intercellular adhesion molecule 1 for its receptor, leukocyte function-associated antigen 1, *Molecular Biology of the Cell* **1997**, *8*, 501–15.
107. J. M. Casasnovas, T. Stehle, J. H. Liu, J. H. Wang, T. A. Springer, A dimeric crystal structure for the N-terminal two domains of intercellular adhesion molecule-1, *Proc. Natl. Acad. Sci. U. S. A.* **1998**, *95*, 4134–9.
108. T. R. Gadek, D. J. Burdick, R. S. McDowell, *et al.*, Generation of an LFA-1 antagonist by the transfer of the ICAM-1 immunoregulatory epitope to a small molecule, *Science* **2002**, *295*, 1086–9.
109. D. E. Staunton, M. L. Dustin, H. P. Erickson, T. A. Springer, The arrangement of the immunoglobulin-like domains of Icam-1 and the binding-sites for Lfa-1 and rhinovirus, *Cell* **1990**, *61*, 243–54.
110. M. S. Dennis, P. Carter, R. A. Lazarus, binding interactions of kistrin with platelet glycoprotein-Iib-Iiia – analysis by site-directed mutagenesis, *Proteins-Structure Function and Genetics* **1993**, *15*, 312–21.
111. K. Welzenbach, U. Hommel, G. Weitz-Schmidt, Small molecule inhibitors induce conformational changes in the I domain and the I-like domain of lymphocyte function-associated antigen-1 – Molecular insights into integrin inhibition, *J. Biol. Chem.* **2002**, *277*, 10590–8.
112. W. Yang, M. Shimaoka, A. Salas, J. Takagi, T. A. S Springer, Intersubunit signal transmission in integrins by a receptor-like interaction with a pull spring, *Proc. Natl. Acad. Sci. U. S. A.* **2004**, *101*, 2906–11.
113. A. Salas, M. Shimaoka, A. N. Kogan, C. Harwood, U. H. von Andrian, T. A. Springer, Rolling adhesion through an extended conformation of integrin alpha(L)beta(2) and relation to alpha I and beta I-like domain interaction, *Immunity* **2004**, *20*, 393–406.
114. M. Shimaoka, A. Salas, W. Yang, G. Weitz-Schmidt, T. A. Springer, Small molecule integrin antagonists that bind to the beta(2) subunit I-like domain and activate signals in one direction and block them in the other, *Immunity* **2003**, *19*, 391–402.
115. S. M. Keating, K. R. Clark, L. D. Stefanich, *et al.*, Competition between intercellular adhesion molecule-1 and a small-molecule antagonist for a common binding site on the alpha 1 subunit of lymphocyte function-associated antigen-1, *Protein Science* **2006**, *15*, 290–303.
116. B. Z. Harris, W. A. Lim, Mechanism and role of PDZ domains in signaling complex assembly, *J. Cell Sci.* **2001**, *114*, 3219–31.
117. E. J. Kim, M. Sheng, PDZ domain proteins of synapses, *Nature Reviews Neuroscience* **2004**, *5*, 771–81.
118. S. C. Harrison, Peptide-surface association: The case of PDZ and PTB domains, *Cell* **1996**, *86*, 341–3.
119. M. C. Hammond, B. Z. Harris, W. A. Lim, P. A. Bartlett, beta strand peptidomimetics as potent PDZ domain ligands, *Chemistry & Biology* **2006**, *13*, 1247–51.
120. J. Schultz, U. Hoffmuller, G. Krause, *et al.*, Specific interactions between the syntrophin PDZ domain and voltage-gated sodium channels, *Nature Structural Biology* **1998**, *5*, 19–24.
121. S. T. Phillips, G. Piersanti, M. Ruth, N. Gubernator, B. van Lengerich, P. A. Bartlett, Facile synthesis of @-tide beta-strand peptidomimetics: Improved assembly in solution and on solid phase, *Org. Lett.* **2004**, *6*, 4483–5.
122. S. T. Phillips, L. K. Blasdel, P. A. Bartlett, @-tide-stabilized beta-hairpins, *J. Org. Chem.* **2005**, *70*, 1865–71.
123. S. Rajagopal, S. C. Meyer, A. Goldman, M. Zhou, I. Ghosh, A minimalist approach toward protein recognition by epitope transfer from functionally evolved beta-sheet surfaces, *J. Am. Chem. Soc.* **2006**, *128*, 14356–63.

6

Discovery of Inhibitors of Protein–Protein Interactions by Screening Chemical Libraries

Carlos García-Echeverría
Novartis Institutes for Biomedical Research, Basel, Switzerland

6.1 Introduction

Protein–protein interactions (PPIs) are involved in basically all biological processes, and over the past few years a set of proteins have been selected as candidates for drug discovery activities based on their association with human malignancies. Until recently, the identification of antagonists of PPIs has been largely confined to biologic therapeutics that inhibit the physical interactions between surface receptors and their cognate ligands. PPIs often span large surface areas and, unlike enzyme/substrate interactions, they lack deep, well-defined binding pockets. Affinity is often achieved by summing up a large number of weak, widely spaced interactions. Due to the unique structural characteristics of the protein binding sites and interactions, and in spite of major improvements in bioassay-based high-throughput screening (HTS) capabilities, the identification and development of low molecular mass compounds capable of modulating specific PPIs has proven highly challenging, and consequently forced medicinal chemistry departments to rethink drug discovery strategies for this family of therapeutic targets. In this context, fragment based or virtual screening methodologies, tailored libraries and structure-based design approaches have led recently to true mechanism-based tool compounds to further explore biology, and drug candidates that have provided proof-of-concept for a limited set of PPIs in preclinical and clinical settings.

Protein Surface Recognition: Approaches for Drug Discovery Ernest Giralt, Mark W. Peczuh and Xavier Salvatella. © 2011 John Wiley & Sons, Ltd

This chapter covers salient achievement in the identification and development of modulators of PPIs, and highlight different screening techniques and structure-based design approaches that may be brought to bear on the discovery of inhibitors of this family of therapeutic targets. Due to space constraints, and the broad scope of this area of drug discovery, the intent of this chapter is to provide a comprehensive rather than an exhaustive overview of the pertinent literature. Several review articles on individual topics are available and the reader is referred to the most recent publications in the text (for recent reviews on this subject, see [1–7]). A brief introduction to the therapeutic target is often included to help the reader to understand the scientific rationale behind the drug discovery activities and the peculiar structural features of the protein–protein binding interactions.

6.2 Screening Strategies to Identify and Develop Antagonists of Protein–Protein Interactions

6.2.1 Phage Display Libraries, Peptides and Unnatural Biopolymers – Mapping Protein Surfaces

Recombinant monoclonal antibodies have been successfully used over the past few years to target extracellular PPIs. These proteins posses high target affinity and specificity, and are well-suitable therapeutic agents, particularly in targeted anticancer therapy. At this moment, greater than 400 mAbs are currently in clinical trials [8], and a few them have received marketing approval (e.g. trastuzumab, an anti-erbB2 mAb). In addition to antibodies, smaller and more compact proteins or peptides have also been investigated, and a few of them have been developed into clinical therapeutics. Thus, enfuvirtide, which is a 36-amino acid peptide -Ac-Tyr-Thr-Ser-Leu-Ile-His-Ser-Leu-Ile-Glu-Glu-Ser-Gln-Asn-Gln-Gln-Glu-Lys-Asn-Glu-Gln-Glu-Leu-Leu-Glu-Leu-Asp-Lys-Trp-Ala-Ser-Leu-Trp-Asn-Trp-Phe-NH_2- derived from the human immunodeficiency virus glycoprotein gp41, was the first marketed viral fusion antagonists for the treatment of HIV [9].

Another example of a peptide antagonist is CD4M9, which inhibits the CD4-gp120 interaction with an IC_{50} value of 40 μM. In this case, the drug discovery strategy for this target exploits the rigidity of Scylla toxin –a 31-residue scorpion protein with a double-stranded β sheet- to graft a functional epitope [10]. The chimeric protein is expected to effectively present the key residues of the epitope and have a greater proteolytic stability than the linear peptide. For other examples of the use of protein grafting [11] in the screening of antagonist of PPIs see, [12, 13].

Although the appeal of polypeptides to target intracellular targets is limited due to their poor membrane permeability and physiological stability, peptides and derivatives thereof can provide important information about the ligand binding sites on a target protein. In this context, combinatorial biological libraries such as bacteriophage-displayed peptide libraries have been extensively used to map PPIs and to develop tool compounds to further explore biology. Target-binding peptides can be affinity selected from complex mixtures of displayed peptides on phage and further enriched through the biopanning process. A clear example of the value of this technique to map PPIs and obtain potent and selective tool compounds is illustrated with the identification of hdm2 antagonists [14]. The HDM2 gene encodes a 491-amino acid residues polypeptide that binds to the transactivation domain of p53 and downregulates its ability to activate transcription; in turn, p53 activates the expression of the HDM2 gene in an autoregulatory feedback loop. Over-expression of hdm2 is an efficient way that tumor cells use to inactive p53 and block tumor cell death. The disruption of the p53/hdm2 PPI is therefore an attractive approach for targeted anticancer therapy because it provides the possibility to activate the p53 responsive reported genes in tumor cells.

Hdm2 contains a hydrophobic groove that binds to an α-helix in the transactivation domain of p53. The p52-hdm2 interface buries a total of 1498 Å^2 of surface area, 690 Å^2 thereof in hdm2. Screening of 12- and 15-mer phage display libraries resulted in the identification of a peptide –Ac-Met-Pro-Arg-Phe-Met-Asp-Tyr-Trp-Glu-Gly-Leu-Asn-NH_2- that had sub-micromolar affinity for hdm2 ($IC_{50} = 313$ nM), and was 28-fold more potent than the corresponding wild-type p53-derived peptide [15]. Truncation studies revealed that the octapeptide -H-Phe-Met-Asp-Tyr-Trp-Glu-Gly-Leu-NH_2- was the minimal sequence retaining micromolar activity for hdm2 ($IC_{50} = 8.9\,\mu$M). The iterative optimization of this recognition motif resulted in the identification of a peptide (Ac-Phe-Met-Aib-Pmp-6-*Cl*-Trp-Glu-Ac_3c-Leu-NH_2) that inhibited full-length p53 binding to GST-hdm2 with an IC_{50} value of 5 nM [16]. During this study, a significant affinity gain was obtained when chlorine was incorporated at the six position of the indole moiety of tryptophan. In accordance with the prediction that motivated this chemical replacement, the X-ray structure of the modified peptide bound to hdm2 has confirmed that the 6-Cl-Trp residue protrudes into the Trp-23 binding site of hdm2 and optimizes the steric complementarity of the protein-peptide interface by establishing additional van der Waals contacts [17]. These interactions have been exploited by several groups in the design of nonpeptidic, low-molecular mass hdm2 antagonists [14, 18]. Although the final compound did not qualify as a lead, it provided proof-of-concept for this therapeutic approach in cellular settings. Thus, the untagged peptide was capable of inducing p53 activation and apoptosis only in cells expressing wild-type p53 and high endogenous levels of hdm2 protein [19–21]. Overall, this example illustrates in a succinct manner how phage-display libraries and structural information can be used to map PPIs and provide a pharmacophore model suitable for virtual screening or structure-based design approaches [18].

In addition to the use of random phage display libraries, computational methods have been established to design biased peptide libraries to target a specific target antigen [22] or a protein surface (e.g. the Ras-Raf protein interface) [23]. For a recent review on this topic, see [24]

Polypeptides composed of homologues of natural α-amino acids –β-or γ-amino acids- and β-hairpins are of great interest due to their conformational vesatility, rigidity and *in vivo* stability compared to natural sequences. Such compounds have now been used as scaffolds to inhibit a set of PPIs, including CXCR4 modulators [25], p53-hdm2 [26, 27], and more recently helical peptide-based foldamers that bind to the anti-apoptotic Bcl-x_L protein [28].

6.2.2 Synthetic and Natural Modulators of Protein–Protein Interactions from High-throughput Data-Generation Techniques

Conventional bioassay-based HTS of corporate chemical libraries still remains a mainstream approach for the identification of potential lead candidates for PPIs. In any case, and independently of the source of the compound obtained by HTS, it is important to verify by

NMR or X-ray structural studies that the identified hit sits in the primary interaction site and fills the pocket(s) that are normally occupied by amino acid side chains from the natural binding partner. Representative examples have been selected in this section to illustrate the successful identification and further development of compounds identified by classical HTS approaches.

Imidazolines were originally identified as hits in a p53/hdm2 HTS campaign and were subsequently optimized to afford the nutlins.* These molecules – e.g. compounds 1 and 2, Figure 6.1 – displaced p53 from its complex with hdm2 with IC_{50} values in the 100 to 300 nM range [29, 30]. The hdm2 binding modes of these antagonists have been determined by X-ray crystallography and NMR [30]. The cis-imidazoline scaffold reproduces features of the helical backbone of the p53 protein and directs its substituents to the critical binding pockets on hdm2. As previously observed with peptide-like molecules (see Section 1.1.1), compound 1 increases in a dose-dependent manner the levels of hdm2, p53 and $p21^{waf1}$ in HCT116 and RKO tumor cells. Animal studies demonstrated that compound 2 is effective in decreasing tumor growth in nude mice bearing s.c. SJSA-1 tumor xenograft (200 mg/kg bid) over a 20-day period with no apparent toxicity.

Another successful example of HTS for the identification of p53-hdm2 antagonists entailed the use of the so-called ThermoFluor® screening method. This approach exploits fluorescent dyes to monitor thermal protein unfolding, allowing the detection of test compounds binding to target protein. Upon binding of a ligand, the conformation of the target protein is 'locked'. Fluorescence-based thermal shifts arise as a result of this conformational stabilization and energetic coupling of ligand binding and protein unfolding. Hdm2 screening of around 338 000 compounds provided 1216 hits, of which 116 originated from a benzodiazepinedione library (e.g. compound 3, racemic mixture; Figure 6.1). As in the case of nutlins, a lead optimization programme provided compounds with improved *in vitro* inhibitory activity (compound 4, Figure 6.1) [31] and established SARs, including the importance of stereochemistry for biological activity.

Another example of a HTS hit is compound 5 (FP-21399, Figure 6.1), which is a hit that was identified in an antiviral drug screen [32]. This bis(disulfonaphthalene) derivative inhibits gp120-mediated spreading of T-cell-topic strains and interacts in a specific manner with the HIV gp120-gp41 complex during virus entry [33]. Animal studies showed that FP-21399 decreases the infected population of mice from 71% (control) to 13–21% at a dosage of 10–50 mg/kg. This inhibitor was under Phase II clinical trials as a potential anti-HIV drug that operates trough the disruption of viral fusion. It is unclear at this moment if this compound is still under clinical development.

Parallel to the PPI disruptors obtained by screening proprietary compound collections, targeted combinatorial libraries have been designed and prepared with the intention to expand the chemical space cover by the 'classical' pharmaceutical targets (e.g. enzymes or GPCRs), and provide potential hits for challenging targets like PPIs. Representative examples of hits identified by screening targeted combinatorial libraries are: compound 6 (Figure 6.1; inhibitor of mmp2-$\alpha_v\beta_3$) [34], and compound 7 (Figure 6.1; inhibitor of c-Myc-Max) [35]. Overall, the size and linearity of these molecules are more excessive that what is normally expected for a lead candidate, but they provided proof-of-concept of the possibility to disrupt therapeutic relevant PPIs.

* The nutlins are named geographically after the research site in Nutley, N.J., USA, where they were discovered.

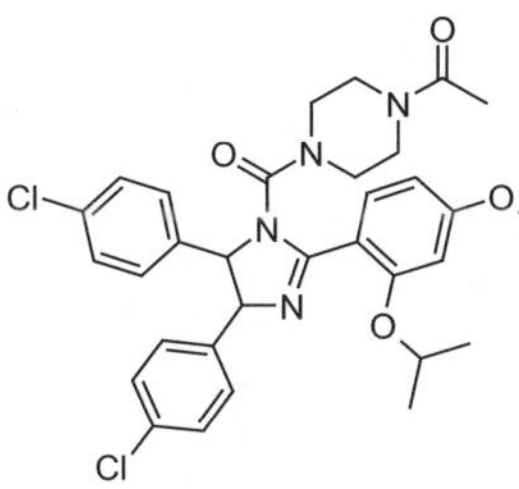

1
Nutlin 1

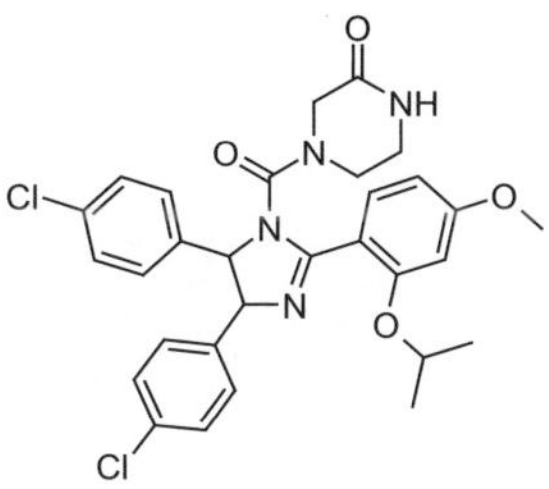

2
Nutlin 3

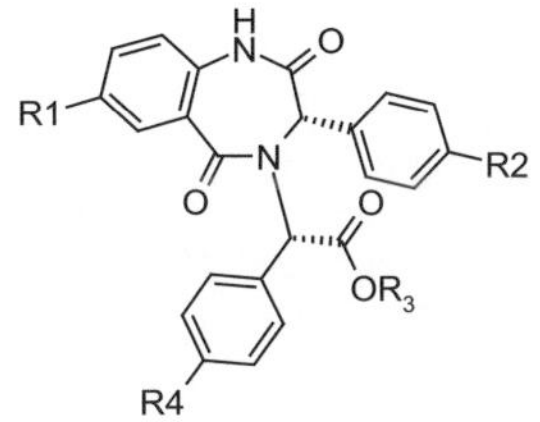

R_1= I, R_2= CF_3, R_3= Me, R_4= H, **3**
R_1= I, R_2= CF_3, R_3= H, R_4= H, **4**

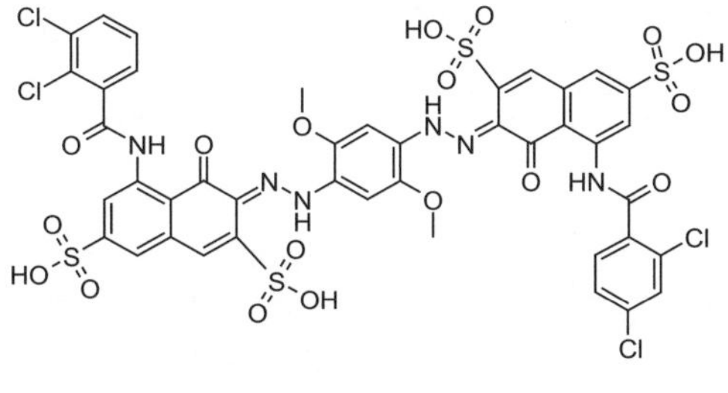

5
FP-21399

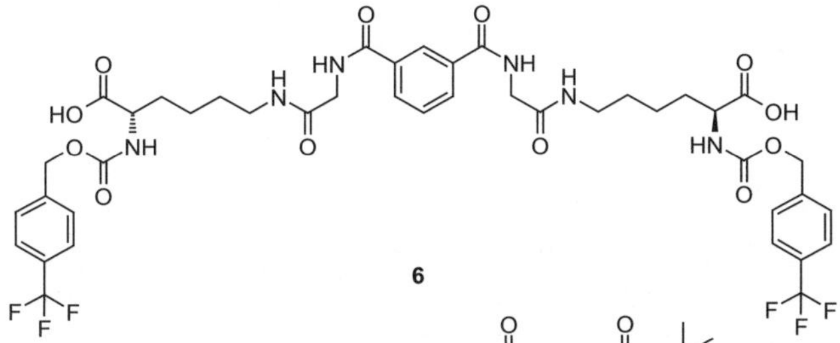

6

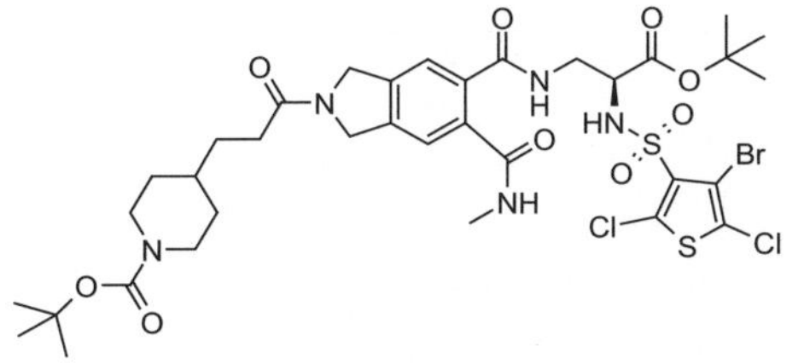

7

Figure 6.1

As for other families of therapeutic targets, the screening of compounds from natural products has been an important source of hits in the identification of inhibitors of PPIs, and, as illustrated below, a few of them can inhibit relevant targets in a clinically and commercially relevant manner.

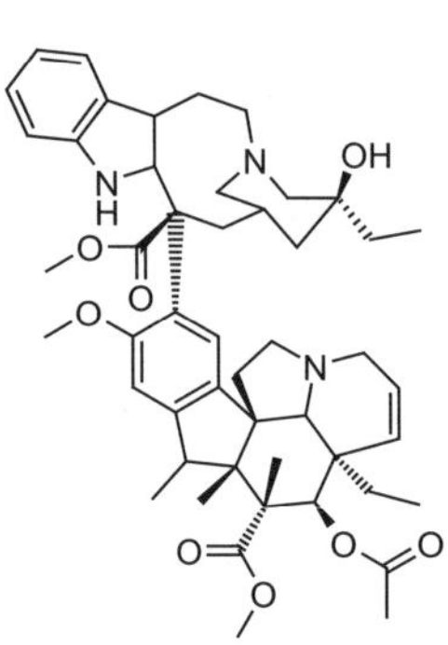

8
Vinblastine

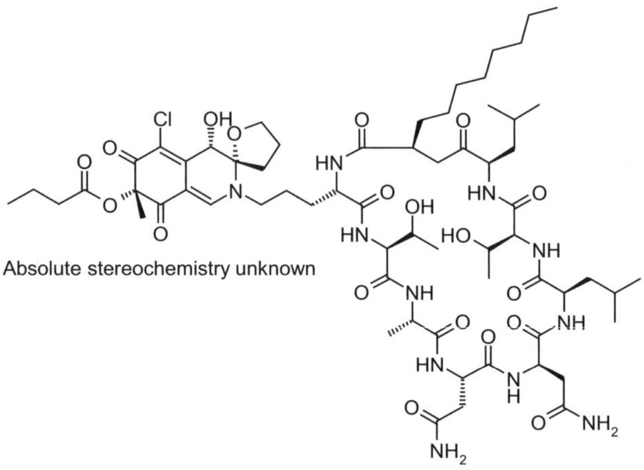

9
Chlorofusin

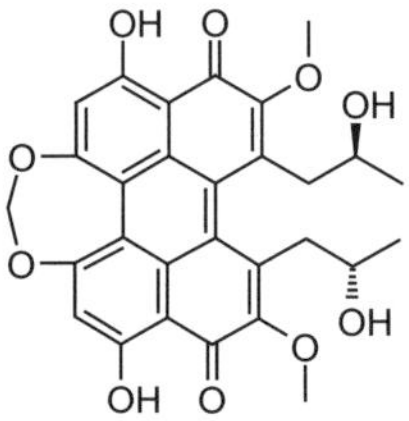

10
CGP049090

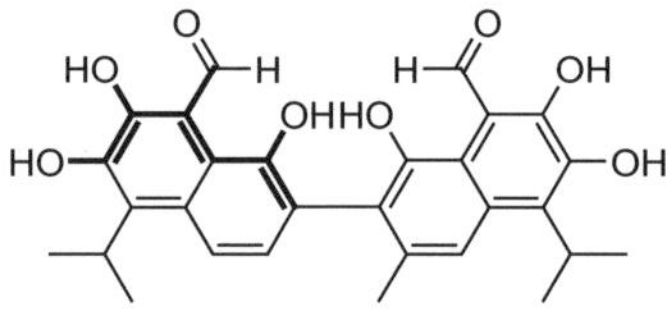

11
R-(-)-Gossypol

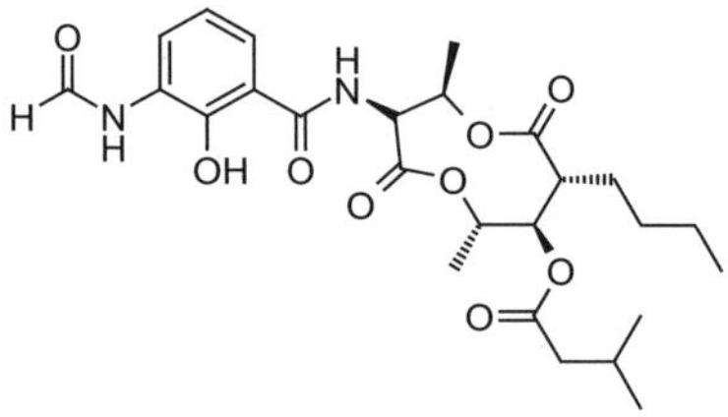

12
Antimycin A

Figure 6.2

The tubulin polymerization inhibitor vinblastine (compound 8; Figure 6.2) can be considered as one of the first PPI modulators reported in the literature. This vinca alkaloid and derivatives thereof were identified in the late 1950s as potential antineoplastic agents by traditional natural product screening methods. The elucidation of the mechanism of action of these antineoplastic agents [36] lead to further screening and medicinal chemistry efforts to identify other allosteric modulators of tubulin polymerization or depolymerization (e.g. taxanes and epothilone derivatives) with improved pharmacological and safety profiles [37–39].

Another recent example of a natural antagonists of PPIs is chlorofusin (compound 9; Figure 6.2). This fungal metabolite was discovered as a result of a screening programme that involved testing over 53 000 microbial extracts for the presence of inhibitors of the binding of p53 to hdm2 [40]. This natural product binds to the *N*-terminal domain of hdm2 with a K_d of 4.7 μM. Modifications of the target compound have shown that derivatives lacking the densely functionalized chromophore do not inhibit the p53-hdm2 interaction [41].

Other recent examples of PPI antagonist identified by screening of natural-product libraries are CGP049090 (compound 10; Figure 6.2), the first reported small-molecular mass inhibitor of the β-catenin-Tcf protein complex [42], R-(−)-gossypol (compound 11; Figure 6.2), which binds to different members of the anti-apoptotic Bcl-2 family with binding affinities in the low μM range ($K_i = 320$, 480 and 180 nM for Bcl-2, Bcl-xL and Mcl-1, respectively) [43], and antimycin A (compound 12; Figure 6.2), which antagonizes Bak binding to Bcl-x_L [44, 45].

Interestingly, a common feature of some of these natural products is the presence of multiple rings in succession. The cyclic, rigid scaffold seems to limit the conformational freedom of its functional substituents reducing the energetic penalty associated with loss of entropy upon binding and, eventually, contributing to the creation of additional interactions by displacing flexible portions of the target protein.

6.2.3 Virtual Database Screening Strategies

Computational 3-D database screening starts to be extensively employed as a complement to HTS in drug lead discovery for PPIs. The availability of large chemical databases and relative inexpensive high-performance computing platforms has transformed this rapidly evolving screening methodology. Two main approaches have been reported: (i) pharmacophore and (ii) structure-based screening. In the former, a pharmacophore model is defined consisting of functional groups critical for target binding, and their 3-D geometrical relationship(s). Using the pharmacophore model, a computational search is then performed to identify building blocks whose previously estimated 3-D structures fulfilled the requirements specified in the pharmacophore model. Although this approach requires relatively short computing times and allows fast elimination of compounds that lack the critical binding elements, it is qualitative in nature and may potentially provide many inactive hits.

In structure-based database screening, each compound in a chemical database is computationally docked into the binding site in the target protein surface and its potential binding affinity is calculated using a scoring function. A clear limitation of this strategy is that the computational time available for each compound is limited and this can result in inaccurate predictions of the binding mode(s), especially for flexible molecules. Moreover, the current scoring functions have severe limitations in estimating predictable binding affinities. The number of successful examples of virtual screening is still limited, but we can expect in the near future major advances in the development of new docking algorithms and score functions [3]. Representative examples have been selected to illustrate the progress made over the past few years in the establishment of in-silico screening approaches to search for antagonists of PPIs. For a recent review on this approach, see [46].

Compound 13 (Figure 6.3) was identified by performing a 3-D database search of the National Cancer Institute database using a computationally derived pharmacophore model for hdm2 binding [47]. This model was gauged by analysis with the HINT molecular modeling program [48]. Most of the selected compounds did not demonstrate dose-responses in p53-hdm2 binding assays with the exception of compound 13, which inhibits

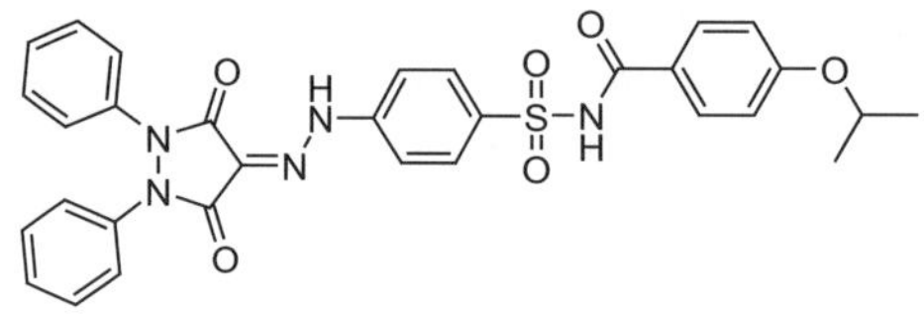

13

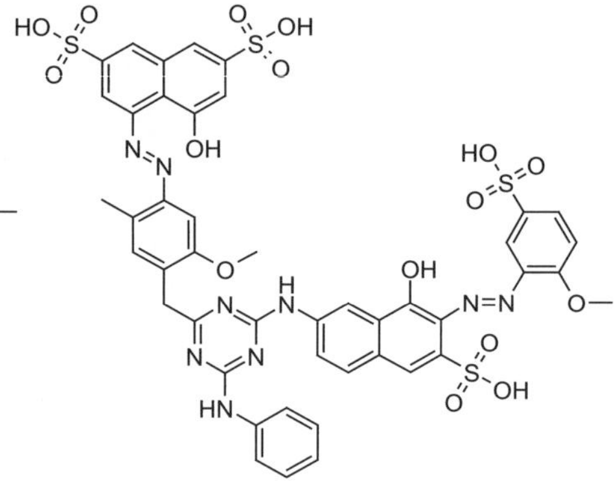

14
ADS-J1

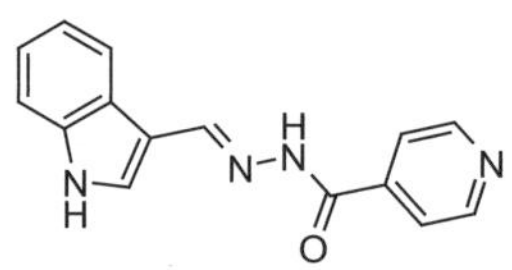

15
TJU103

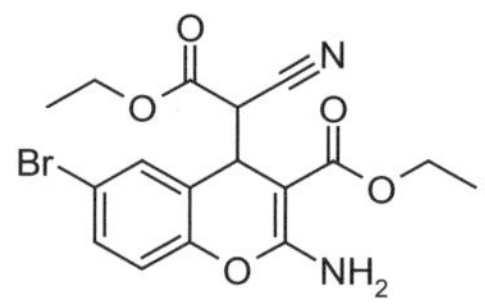

16
HA14-1

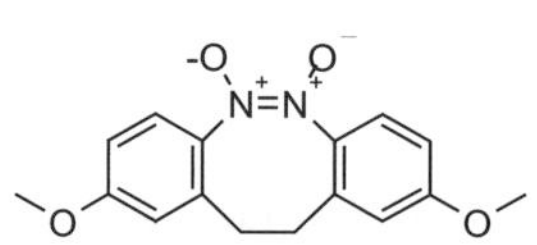

17

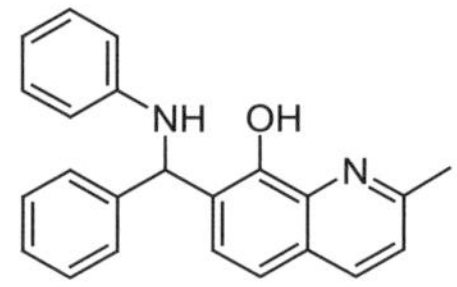

18

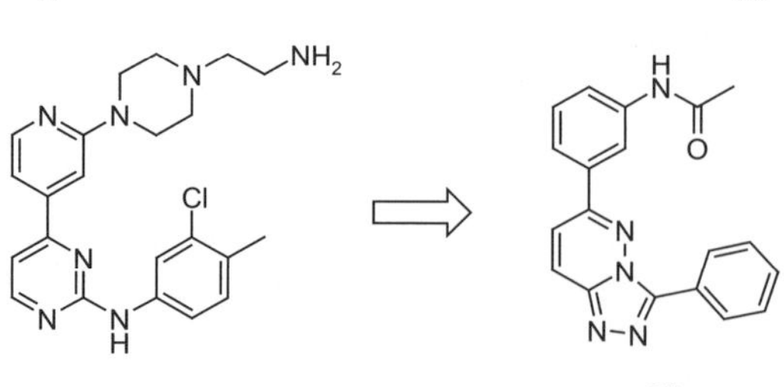

19 **20**

Figure 6.3

the interaction of recombinant p53 and mdm2 in biochemical assays ($IC_{50} = 31.8\,\mu M$) and modestly increases p53-dependent transcription in an hdm2-overexpressing cell line.

X-ray crystallographic studies demonstrated that the structure of the fusion-active conformation of the gp41 core of the HIV-1 virus is a six-stranded helical bundle [49]. Three N helices associate to form the internal coiled-coil trimmer, while three C helices pack against this trimer in an antiparallel fashion into the hydrophobic grooves formed on the surface of the trimer. Each of the grooves has a deep cavity that accommodates three highly conserved hydrophobic residues (Trp628, Trp631, and Ile635) in three C helices. Structure-activity studies demonstrated that mutation of these residues drastically reduced association of C-peptides with N-peptides to form the gp41 core and the antiviral activity of C-peptide, suggesting that these coiled-coil cavities are critical for gp41-mediated membrane fusion and are attractive targets for designing new anti-HIV drugs preventing the early fusion events [50]. Following this strategy, a database of 20 000 organic compounds was screened using molecular docking techniques to identify molecules that target the hydrophobic grooves of gp41 [51]. From this molecular docking study, 16 compounds showed the best fit for docking into the previously identified hydrophobic cavity within the gp41 core. These compounds were then tested in an enzyme-linked immunosorbent assay (disruption of the N-36/C-34 complex) and virus inhibition assays. Compoud 14 (ADS-J1, Figure 6.3) showed potent activity with an IC_{50} value of 0.73 μg/mL, and inhibited HIV-1 mediated cell fusion *in vivo*. Interestingly, this compound was ranked first in the computational scoring.

A computer screening was also successfully used to identify compounds that inhibit CD4 activation in T cells [52] by disrupting the oligomerization and stabilization of the CD4-MHC class II interaction. CD4 binds to nonpolymorphic regions of the MHC class II $\beta 2$ domain, resulting in the co-oligomerization of CD4, MHC class II and the T-cell receptor to initiate signaling in T-cells. From a library of 150 000 organic compounds, four molecules were found to have inhibitory effects on alloreactive T-cell proliferation, and one of them (compound 15, TJU103: Figure 6.3) showed *in vivo* activity in an experimental animal model for multiple sclerosis.

In-silico screenings have also been used by several groups to assist in the search of antagonist of Bcl-2 and Bcl-x_L. Proteins in the Bcl-2 family are central regulators of programmed cell death. All act by forming protein–protein complexes with other members of the group, and proteins that inhibit apoptosis, such as Bcl-x_L and Bcl-2, are overexpressed in many cancers. Inhibiting their interactions with their respective partners has emerged as a promising strategy for drug anticancer development. The molecular nature of this physical interaction has been revealed by the solving of the NMR structure of a complex in which Bcl-x_L is bound to a peptide fragment of Bak [53]. As Bcl-2 and Bcl-x_L have around 47 % sequence identity, a homology model of Bcl-2 was constructed based on the preceding structural information. Using the Bak-binding surface pocket in the homology model of Bcl-2 as the targeted active site, virtual screening of the ACD database of 193,833 compounds was carried out using the program DOCK3.5 [54]. From this screening, 28 candidate molecules that showed relatively lower binding energy, favorable shape complementarity, and/or potential of forming hydrogen bonds with the Bcl-2 protein residues were chosen for biological testing. Subsequent fluorescence polarization assays verified that the most potent compound (compound 16, HA14-1; Figure 6.3) competed with Bak BH3 peptide for binding to Bcl-2 ($IC_{50} \approx 9\,\mu M$). Another molecular inhibitor of Bcl-2 (compound 17, Figure 6.3) was also discovered by virtual screening the NCI database of 206,876 compounds against the modeled structure of Bcl-2 using DOCK3.5 [55]. The top 500 compounds with the best scores were further filtered to include only those with nonpeptidic character. 35 compounds were tested in *in vitro* biological assays (IC_{50} values from 1.6 to 14.0 μM). It was found that compound 17 inhibited tumor cell growth ($IC_{50} = 4$ μM), and induced apoptosis in cancer cells expressing high levels of Bcl-2. Furthermore, using NMR methods, compound 17 was shown to bind the BH3 binding site in Bcl-X_L with a binding constant of 7 μM.

As an alternative to the preceding methods, an integrated computational database screening strategy was employed to identify modulators of the p53/hdm2 protein–protein interaction. In this mixed strategy, a pharmacophore model was used to identify the initial hit list, and the preselected compounds were docked and scored using the GOLD program with the ChemScore fitness function [56]. The integrated virtual screening of 110.000 compounds from the NCI database led to the identification of compound 18 (Figure 6.3). This quinolinol binds hdm2 with a Ki value of 120 nM and dose-dependently activates p53 function in cellular settings.

Halfway between high-throughput and virtual screening, the shape-comparison program ROCS (Rapid Overlay of Chemical Structures) provides an interesting alternative to find antagonists of PPIs. This new shape-based computational procedure was used to identify antagonists of the ZipA-FtsZ protein–protein interaction [57]. The bacterial protein ZipA appears to play an essential role in the formation of a membrane-associated organelle that drives the formation of new cell walls during bacterial cell division. This biological effects requires the interaction of ZipA with FtsZ. To find novel inhibitors of this critical PPI interaction, a HTS of the 250 000 corporate compound collection was performed and compound 19 (Figure 6.3) was identified and selected for further chemical modifications. To this end, the X-ray structure of ZipA in complex with this lead was solved. Unfortunately, it was found that compound 19 had strong, nonspecific activity in both bacterial and yeast-based assay, and ROCS was used to identify alternative scaffolds. This computational method 'perceives' similarity between molecules based on their three-dimensional shape and uses this information for *in silico* searches of compound collections. This effort lead to the identification of two new and significantly different inhibitor scaffolds (e.g. compound 20, Figure 6.3) with similar target binding interactions. One important feature of this computational method is its speed: it processes conformers at a rate of one thousand per second on a single processor. However, and contrary to the methods described above, it does not give energies of interactions, and the cut-off for selecting the list of candidates for further profiling is based on Shape Tanimoto values.

6.2.4 Fragment Libraries – Screening for Weak Interactions

Fragment screening has become a popular technique for finding new chemical starting points for drug discovery programmes, in particular for the identification of antagonists of PPIs. In general terms, fragments are low molecular mass building blocks that interact weakly with the selected protein, but have the possibility of synthetic elaboration on the basis of structural information [58, 59]. The elucidation of its binding mode by X-ray crystallography or NMR provides insight into its mode of interaction, then a structure-guided medicinal chemistry approach can be undertaken to optimize the interactions and fine-tuning the pharmacological properties of the selected fragment(s).

The obvious challenge of all approaches to fragment screening is to first select a suitable fragment collection that is sufficiently diverse in chemical space and provides potential starting points for optimization of potency, target selectivity and pharmacological properties. Overall, it is reasonable to say that fragment library selection strategies have been focused on molecular property ranges, chemical space and suitability for medicinal chemistry elaboration to optimize potency, selectivity and drug-like properties [59, 60].

There are several fragment screening strategies available, including functional screening, nuclear magnetic resonance (NMR), mass spectrometry (MS), X-ray crystallography and surface plasmon resonance [58]. This section briefly covers the successful identification and development of hit candidates by fragment screening using NMR [61, 62] and tethering [63].

NMR-based strategies for hit identification from fragment libraries (e.g. heteronuclear-NMR-based screening) [62] rely of the ability of NMR spectroscopy techniques to detect and characterize the binding of weakly interacting molecules. If one or several of such fragments is found to interact with the protein of interest, focused libraries around the fragment(s) can be prepared and screen again. Then, a tighter chemical binder can be obtained by optimal linkage of the optimized fragment(s). This strategy was successfully used in the discovery of a class of biarylacylsulfonamide antagonists of the antiapoptotic protein Bcl-xL [64–67]. An NMR-based screen of a 10 000 compound fragment library was conducted using a truncated form of Bcl-x_L. This screen yielded a fluoro biary acid derivate (compound 21, Figure 6.4) that binds Bcl-x_L with a dissociation constant (K_d) of 300 μM [65]. Additional NMR studies showed that this weak ligand binds in a hydrophobic groove and occupies the same position as the key leucine amino acid of the bound Bak peptide [53]. Exploiting

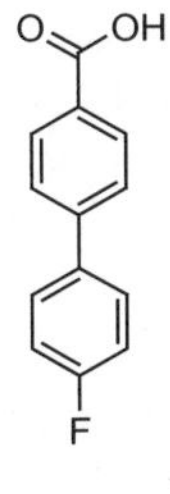

21

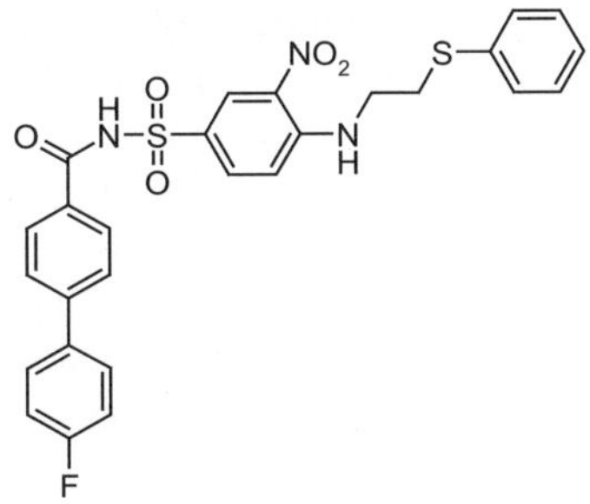

22

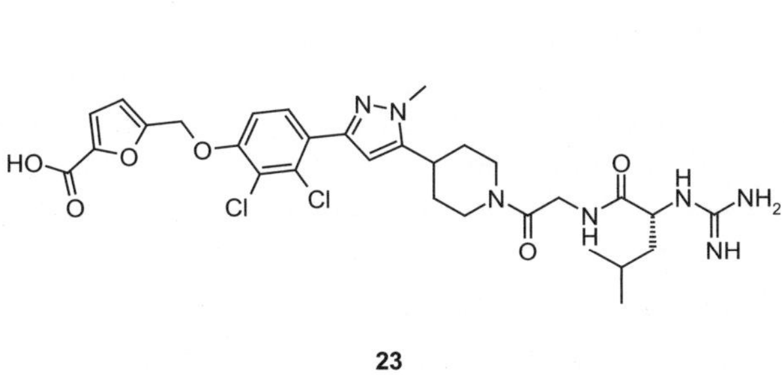

23

Figure 6.4

structure-activity relationships obtained from NMR studies [62], a second-site ligand was discovered by screening a 3500 compound library in the presence of an excess (c = 2 mM) of compound 21. From this second screen, several napthol analogues and a biary phenol were identified. Using a structure-based design strategy and parallel synthesis to identify the best linker and fine-tuning the interactions of the modified compound with the protein, a potent ligand (compound 22, Figure 6.4) was identified, which binds to Bcl-x_L with an inhibition constant (Ki) of 36 nM. This initial Bcl-x_L antagonist had limited pharmacological utility due to its poor aqueous solubility and tight biding to serum albumin. Through subsequent rounds of medicinal chemistry optimization the pharmacological limitations of this initial head were overcome, and a drug candidate was identified for futher clinical development (more details in the next section).

The tethering approach relies on the formation of a covalent bond between the fragment and the protein of interest [63]. To use this method, the protein must have a cysteine residue within or near the targeted site** – 5 to 10 Å-, and the fragments must contain a disulfide moeity (e.g. aminoethanethiol has been extensively used to form the disulfide bridge with each fragment of the library). The protein is reacted with the library of disulfide-containing fragments under partially reducing conditions. Assuming that the disulfide moiety in each fragment has similar reactivity, the thiol-disulfide equilibrium will be shifted in favor of the formation of a covalent disulfide bond with the fragment that has higher affinity for the targeted pocket/cleft of the protein. The identification of the selected fragment is then carried out by MS. Although limited by the thiol needs, this screening methodology requires less amount of protein compared to other fragment-based approaches, and is not limited by the size of the protein or its crystallization properties.

Fragment tethering was succesfully applied in the identification of IL2-inhibitors. This cytokine triggers T-cell proliferation by recognition of a heterotrimeric receptor complex on the T-cell surface, and disruption of this physical interaction can result in an effective immunosuppressive activity. A compound with millimolar affinity for IL-2 was identified by fragment screening [68, 69], and it was advanced into a lead series with nanomolar affinity (e.g. compound 23, $IC_{50} = 0.06\,\mu M$, disruption of the IL2/IL-2Rαcomplex) by using focused libraries and a modular chemical approach. The X-ray structures of compound 23 and derivatives thereof bound to IL-2 confirmed the design strategy and the existence of an adaptable binding site in the surface of the protein that undergoes conformational rearrangements upon binding of different ligands.

6.3 Mimetics of Common Protein Structure Motifs and Structure-based Design of Peptidomimetics

A long-standing goal of medicinal chemists has been to develop approaches for the *de novo* design of inhibitors based on the knowledge of a protein's 3D-structure. In this context, and following the path of early efforts to identify peptidomimetics,*** different groups have tried

** If cysteine is not in the natural sequence, it can be introduced at the desired placed by site-directed mutagenesis.
*** For an interesting and useful information about the term peptidomimetic please see [70, 71]. Under Rich and coworkers' scheme, inhibitors or ligands are classified as Type I peptidomimetics if the analogy is at the atomic level, whereas Type III peptidomimetics are topologically similar, but there is no clear analogy at the atom-by-atom level. Type II peptidomimetics appear to bind at different sites than the original partner, and are best described as functional mimetics.

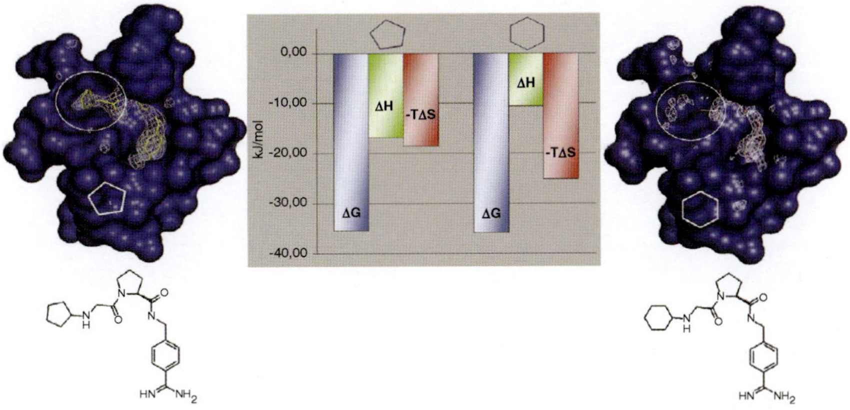

Plate 1 *The same binding affinity can arise from different combinations of enthalpic and entropic contributions. The central panel summarizes the binding energetics of two closely related thrombin inhibitors bearing a cyclopentyl or cyclohexyl moiety as terminal substituent that binds at the S3/S4 pocket in the thrombin binding site. Even though both molecules bind with the same binding affinity, significant differences are observed in the entropic and enthalpic contributions to the Gibbs energy. The corresponding crystal structures are shown in which the structure of the binding pocket is shown as a blue surface and the electron density maps for both ligands are shown as white chicken-wire contours. (Reprinted from [52] with permission from Elsevier)*

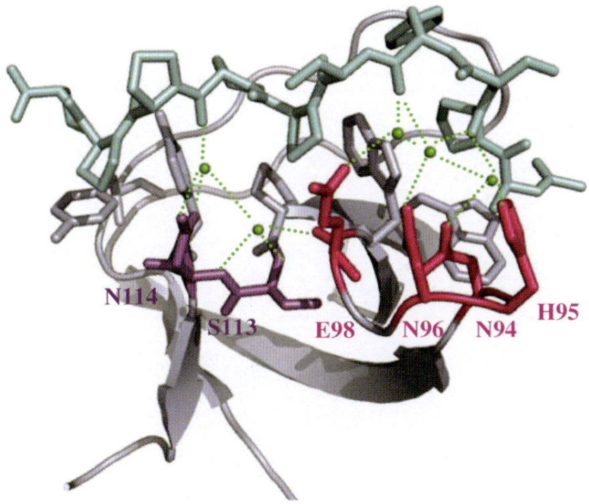

Plate 2 *Water molecules at the Abl-SH3/p41 binding interface. The structure of the Abl-SH3 domain is shown in a grey cartoon representation. Residues defining the canonical binding site for polyproline recognition are shown as grey sticks. The structure of the p41 peptide is shown as cyan sticks. Fully buried water molecules at the binding interface are shown as green spheres. Peripheral water-coordinating residues in the* 3_{10} *and n-Src regions are shown as purple and dark pink sticks respectively. Water-mediated hydrogen bonds are depicted as dotted green lines*

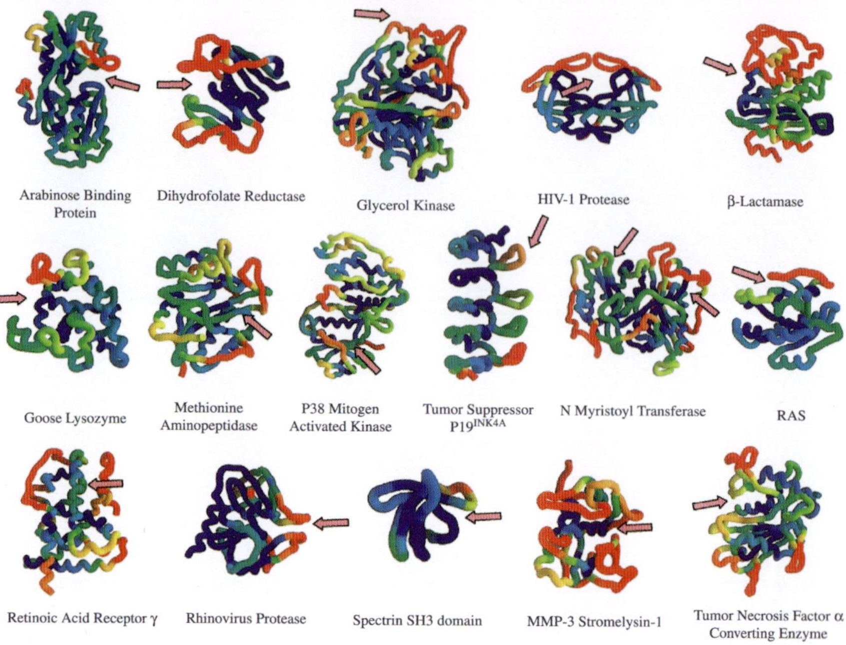

Plate 3 *The distribution of structural stability for 16 different proteins. The structures have been color-coded according to the magnitude of the individual stability constants per residue using a normalized rainbow scale ranging from 1 to 100 according to the overall Gibbs energy of stabilization for each protein. In this scheme, red regions correspond to residues with stability constants smaller than 25 and dark blue regions to residues with stability constants greater than 75. Green residues define the middle of the scale and correspond to residues with stability constants of 50. The arrows indicate the location of the binding sites. (Reprinted with permission from [156]. Copyright 2000 John Wiley & Sons., Inc)*

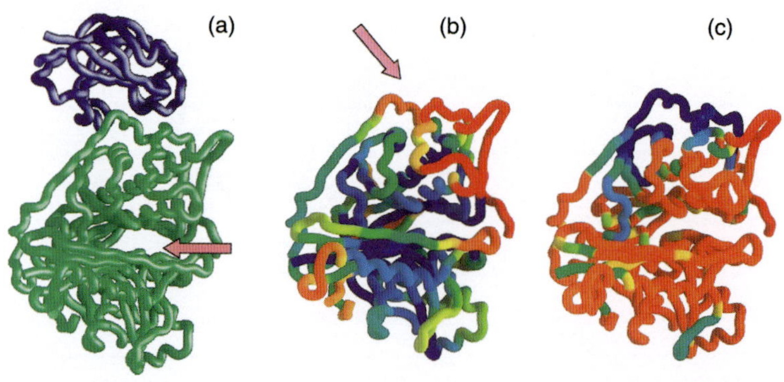

Plate 4 *(A) The complex between Glycerol Kinase (green) and the allosteric regulator IIAGlc (blue). The arrow indicates the location of the catalytic site. (B) The structural distribution of the stability of unbound Glycerol Kinase. The residues at the binding site for IIAGlc are intrinsically unstable and do not interact strongly with the rest of the molecule in the absence of the regulator. (C) Effect on the structural stability of the binding of IIAGlc. The structure of Glycerol Kinase has been color coded in a rainbow scheme according to the changes in stability constants induced by the binding of IIAGlc (dark blue regions are the most affected and red regions are unafected). Binding of the regulator triggers the propagation of cooperative interactions trough a stretch of residues that connect the regulatory and catalytic domains. (Reprinted with permission from [156]. Copyright 2000 John Wiley & Sons., Inc)*

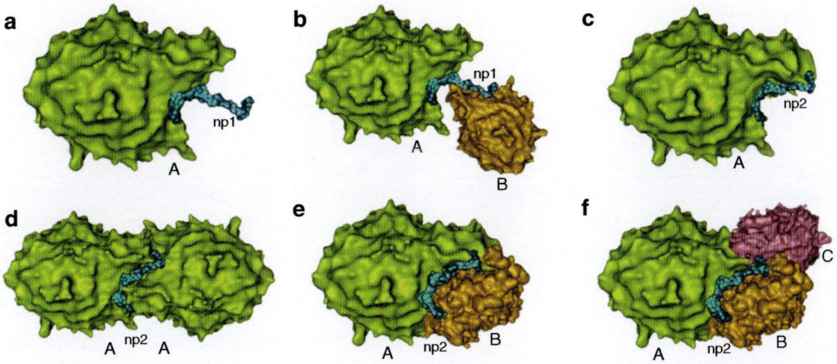

Plate 5 *Modes of interaction between natural products and proteins. A cartoon depiction of natural products (np1, np2) interacting with proteins A, B and C. (a) An exemplary mode in which the interaction between natural product np1 and protein A provides a second external binding domain. (b) This A•np1 complex can recruit a second protein B to form an A•np1•B complex. (c) Alternatively, natural product np2 can bind to the surface of a protein A. (d) The resulting A•np2 complex can then induce dimerization to form A• np2•A trimer. (e) The A•np2 complex can also recruit a different protein B to form a heterodimeric A•np2•B complex. (f) Higher order complexes as illustrated by the recruitment of a third protein C in formation of the A•np2•B•C complex. Each frame a-f depicts one mode of interaction as given by the formation of a unique type of complex*

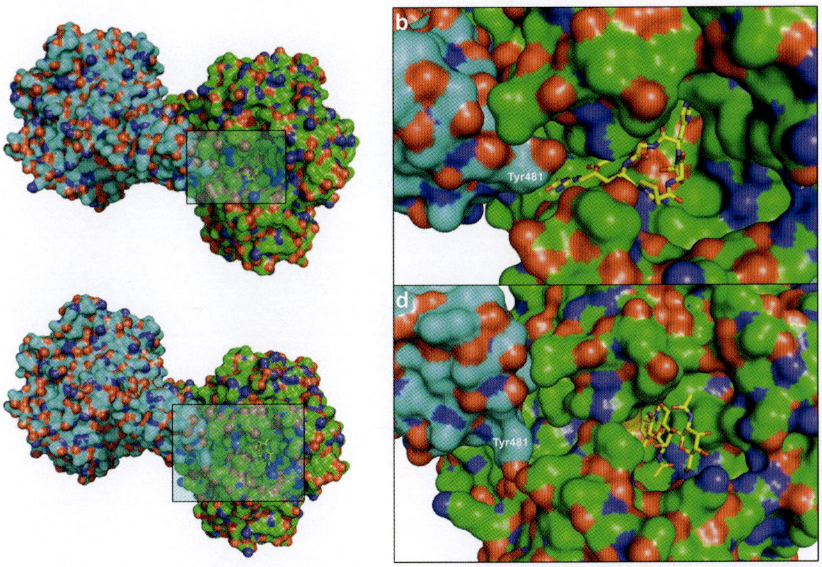

Plate 6 *False-positive pockets. Crystal structures of chitinase B (ChiB) from Serratia marcescens with bound ligands. (a) Structure of ChiB with argadin (**8**) bound. (b) Close-up of argadin (**8**) bound to ChiB. (c) Structure of mutant D142N ChiB with allosamadin bound. (d) Close-up of allosamadin (**9**) bound to ChiB. Each structure contains two ChiB proteins (cyan and green) with a bound small molecule (yellow). Images were developed from structure files 1h0g and 1ogg.[31]*

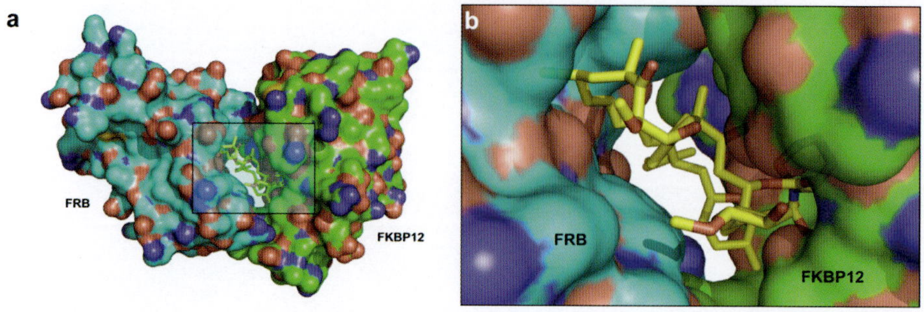

Plate 7 *The FKBP12•rapamycin(**6**)•FRB complex. A depiction of natural-product induced interface between FKBP12 and FRB. Each structure contains two proteins, FRB (cyan) and FKBP12 (green) with a bound rapamycin (**6**) (yellow). Images were developed from structure file 4fap (27a)*

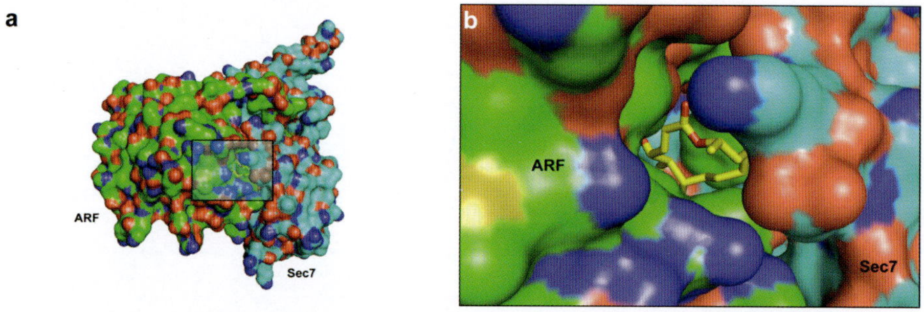

Plate 8 *The ARF•GDP•brefeldin(**10**)•Sec7 complex. A depiction of brefeldin A (**10**) interface between ARF and Sec7. The structure contains two proteins, ARF (green) and Sec7 (cyan) with a bound molecule of brefeldin A (**10**) (yellow). Images were developed from structure file 1re0. [33a]*

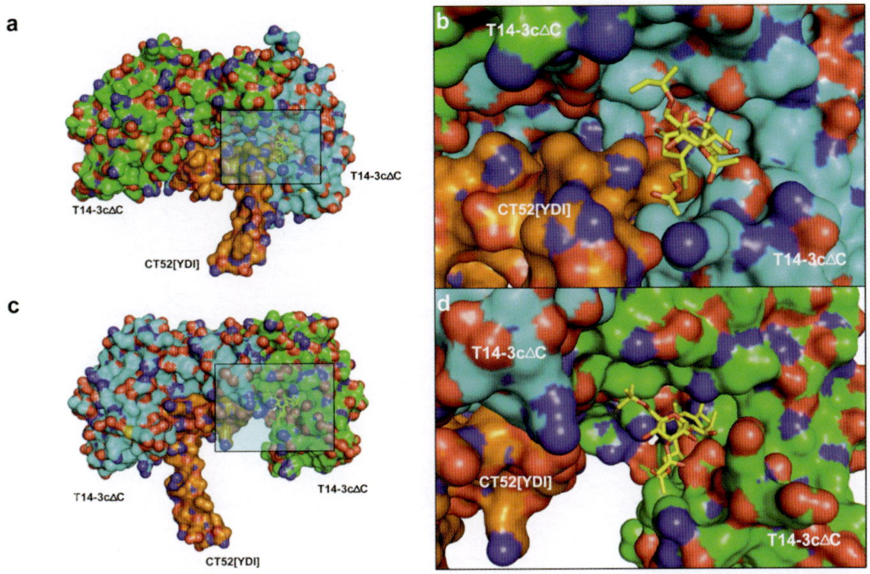

Plate 9 *The CT52[YDI]•fusicoccin(**11**)• 14-3-3 complex. (a) A depiction of fusicoccin (**11**) (yellow) at the interface between the CT53[YDI] peptide (orange) and a dimer containing two 14-3-3 proteins (cyan and green). (b) close-up of the binding pocket in (a). (c) The back face of the image in (a). (d) A close-up of (c). Images were developed from structure file 1ia0 [36]*

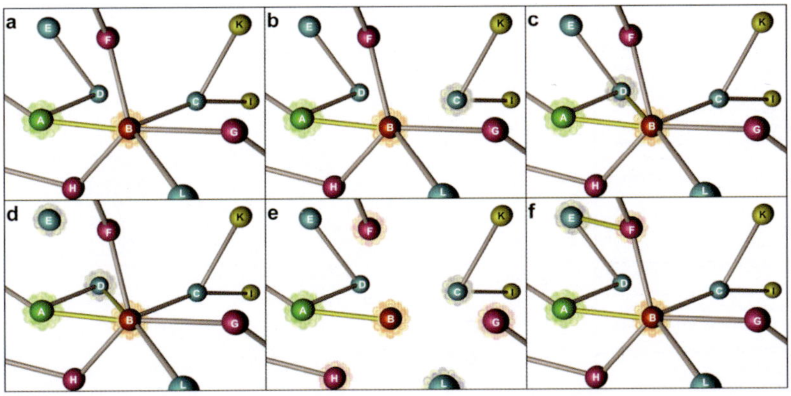

Plate 10 *Functional response. A hypothetical model depicting natural product derived regulation of protein-protein interactions. Each frame depicts a model network of proteins (A–L) in spheres and their interactions in tubes. This 3D network was modeled after protein network program Grafta 9. For each panel a-d, the natural product regulates the interaction between proteins A and B as indicated by a yellow highlighted tube. The effects of this interaction as also indicated by yellow tubes and proteins involved in this process are shaded with clouds. The color of the spheres indicate their location in the cell, as given by cytosol in cyan, the Golgi apparatus in magenta and the nucleus in yellow. The proteins targeted by the natural product are provided in green and red as they move about the cell*

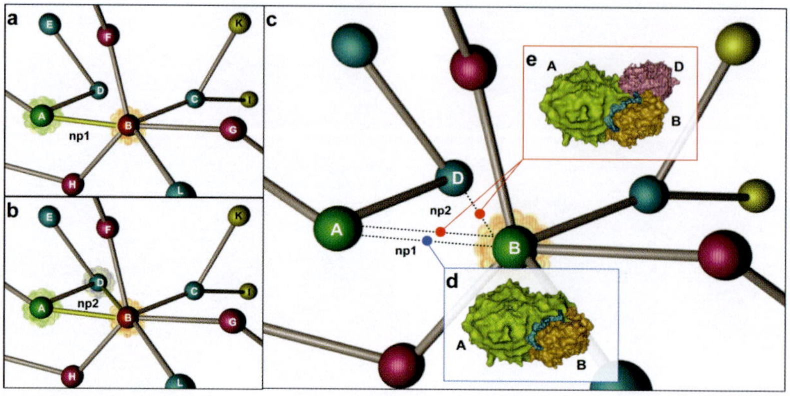

Plate 11 *Natural product-Protein-Protein Interaction (NPPI) map. (a) Example 1: Natural product (np1) induces affinity between proteins A and B form an A•np1•B complex. (b) Example 2: Natural product (np2) induces affinity between A and B which then recruits protein D to form a quaternary complex A•np2•B•C. (c) An model NPPI map. Proteins are depicted as spheres and interactions as tubes. Natural products are depicted as dots and their induced protein-protein interactions as dotted lines. (d) The structures of the A•np1•B complex can be provided in inset windows by simply clicking on the dot representing np1. (e) The display could host multiple structures including A•np2•B•C complex*

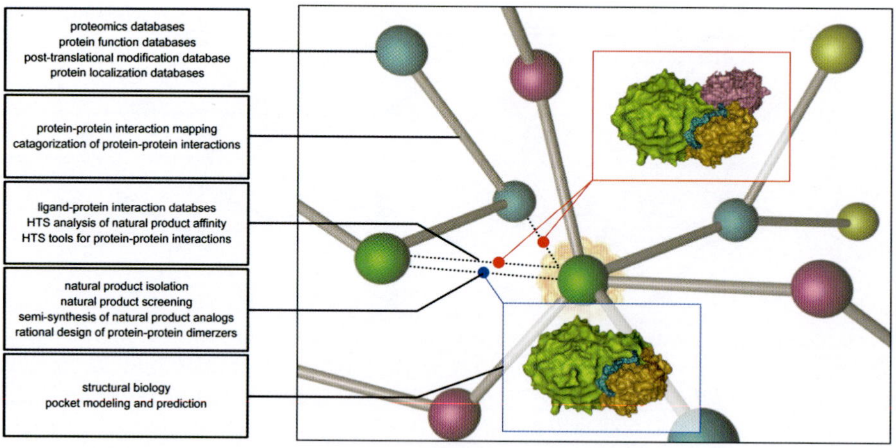

Plate 12 *Perspectives in the development of global Metabolite-Protein-Protein (NPPI) Networks. A depiction of five key aspects of research required for the development of a NPPI map. This effort includes research at (a) the level of the protein, (b) protein-protein interaction, (c) natural product protein interaction, (d) natural product and (e) the structural composite of natural product protein interactions*

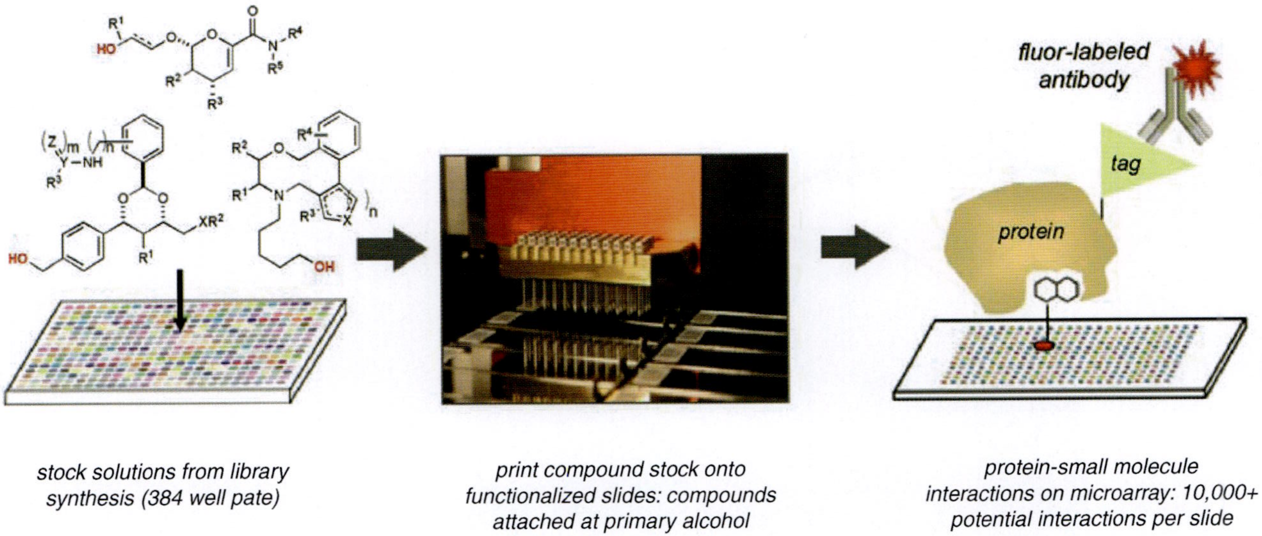

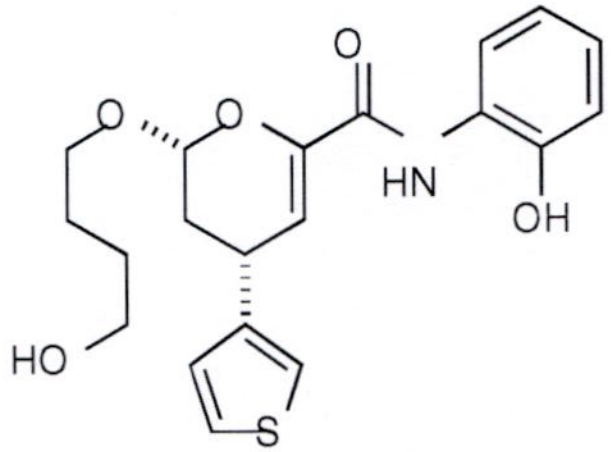

176, haptamide A
IC_{50} = 42.1 μM; K_D = 5.03 μM

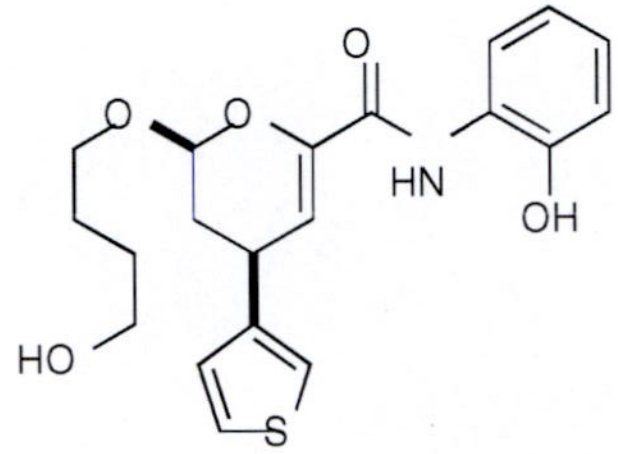

177, *ent*- haptamide A
IC_{50} = 30.6 μM; K_D = 0.66 μM

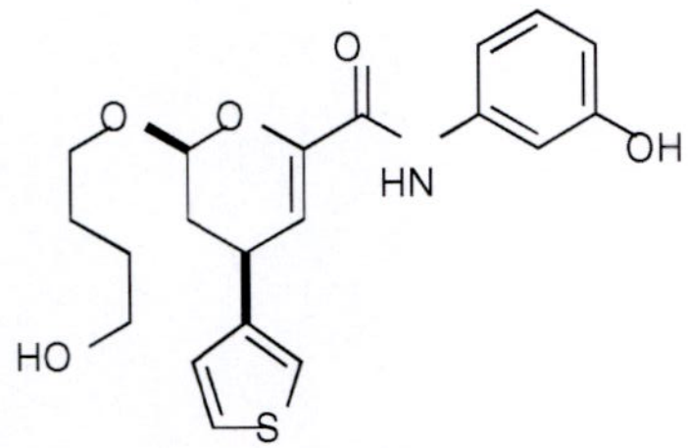

178, haptamide B
IC_{50} = 23.8 μM; K_D = 0.33 μM

Plate 13 *Preparation of small molecule microarrays (SMMs) from stock solutions. Detection of protein-small molecule interactions using SMM; structures of compounds that disrupt the Hap2/3/4/5p transcription factor complex (haptamides)*

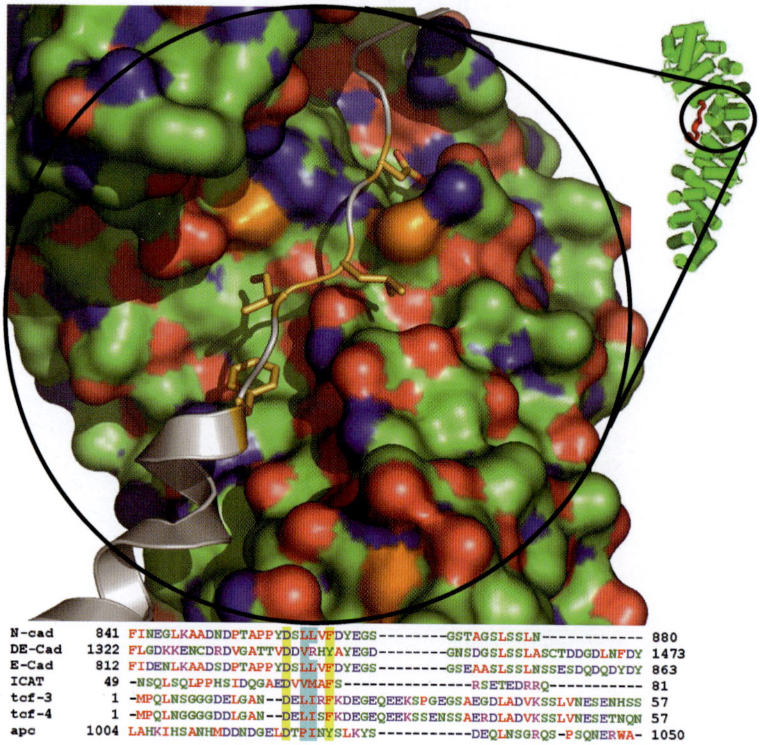

Plate 14 *Structure of the complex between β-catenin and Tcf-4 (PDB code: 1JDH) and structural alignment of the sequences of other β-catenin-binding proteins. The residues forming the common binding motif are shown in orange in the structure and are highlighted in the sequence alignment. All figures were prepared using PyMOL (http://www.pymol.org)*

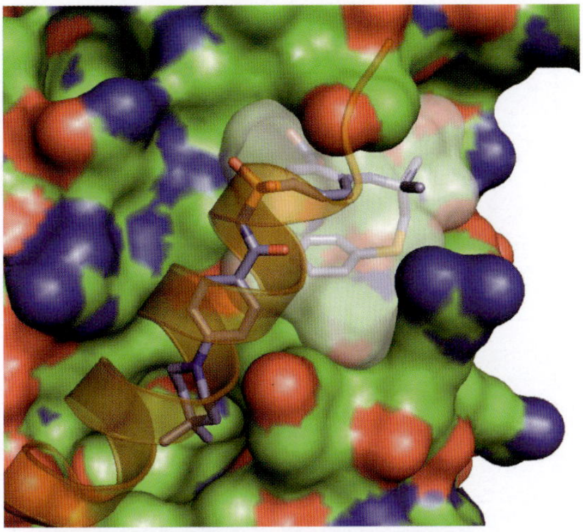

Plate 15 *The interaction between Bcl-xL and BAD (Bcl-2 antagonist of cell death) can be disrupted with drug-like compounds. The inhibitor (shown in sticks) binds to a cavity on the surface of Bcl-xL (white transparent surface) that can be predicted as druggable using computational methods. The PDB code of the ligand-protein complex is 2O22. The backbone of BAD is shown as a semi-transparent orange ribbon (PDB code 2BZW)*

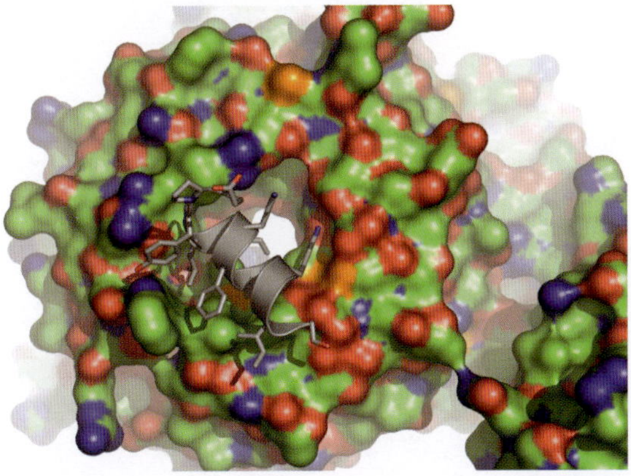

Plate 16 *SIGK peptide bound to Gβ (PDB: 1XHM). The peptide binds to a hydrophobic patch located near the axis of this β-propeller structure. Helix 9 of Gα and the C-terminus of GPCR Kinase 2 also bind to the same site (PDB codes: 1GP2 & 2BCJ, respectively)*

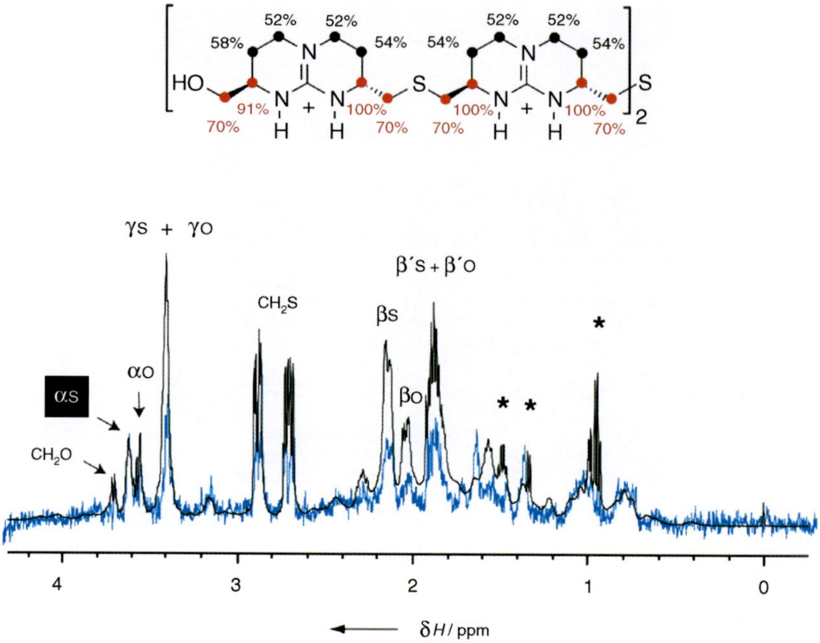

Plate 17 *STD experiment (blue) on a mixture of p53TD and α-tetraguanidinium ligand. A ^{1}H-1D-NMR spectrum of the sample is shown in black. The relative saturation transferred to each proton is shown on the chemical structure of the ligand.[11] Reproduced with permission from [11]. Copyright Wiley-VCH Verlag GmbH & Co. KGaA*

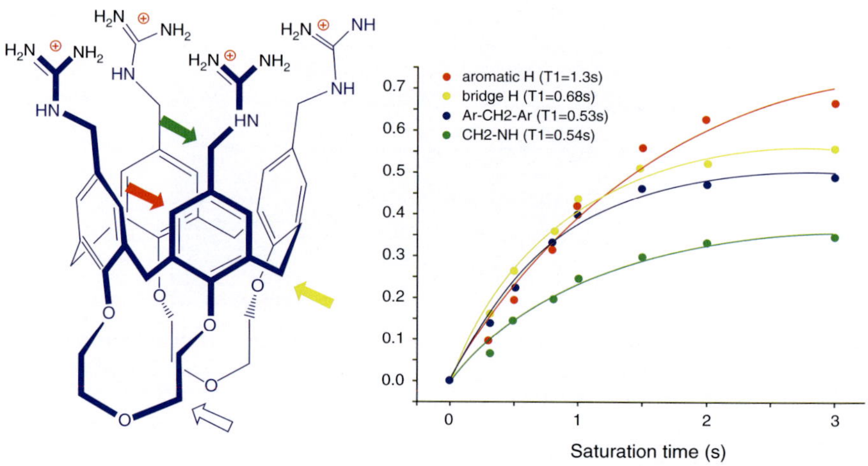

Plate 18 *STD build-up curves from the interaction between the tetramerization domain of protein p53 and a tetraguanidiniumcalixarene.[14] Copyright (2008) National Academy of Sciences, U.S.A*

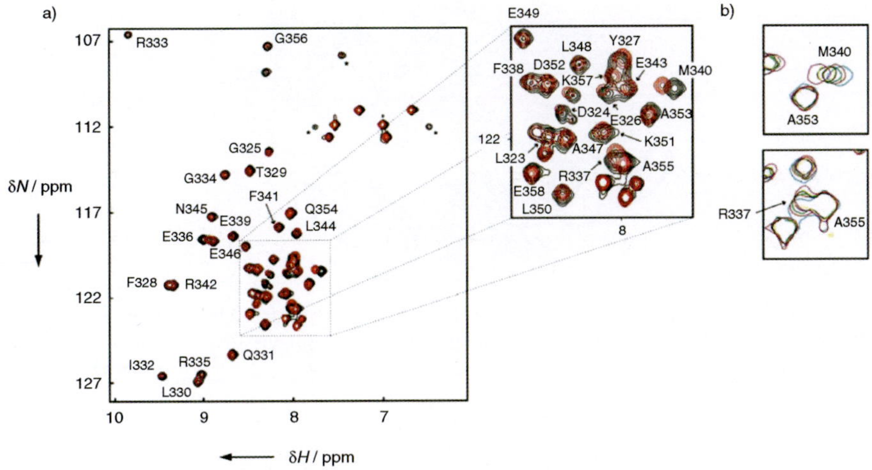

Plate 19 *(A) [^{1}H, ^{15}N]-HSQC spectrum of the tetramerization domain of protein p53 in the absence (black contours) and in the presence (red contours) of Ligand* **1** *(4 equiv.); and (B) titration results for residues Ala 353, Met 340, Arg 337 and Ala 355. Reprinted with Permission from [11]. Copyright Wiley-VCH Verlag GmbH & Co. KGaA*

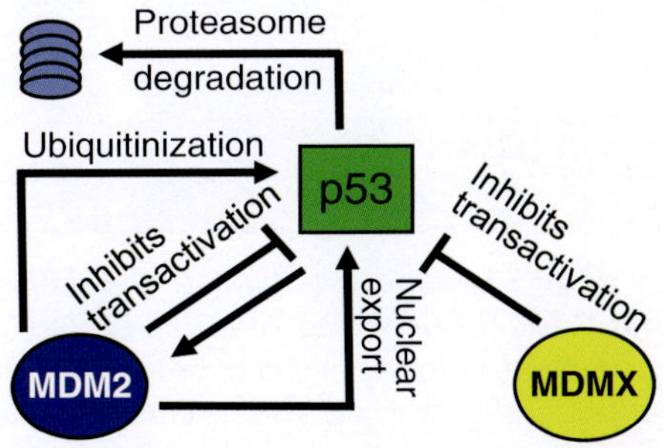

Plate 20 *Regulation of p53 by MDM2 and MDMX. Upon activation, p53 induces transcriptional upregulation of MDM2 and in turn, MDM2 inhibits p53 through an autoregulatory feedback loop. MDM2 directly binds to the transactivation domain of p53 and inhibits p53 transcriptional activity, facilitating nuclear export of p53. Through its E3 ligase activity, MDM2 causes the ubiquitinization and proteasomal degradation of p53. MDMX, a homologue of MDM2 also binds and conceals the transactivation domain of p53, inhibiting p53 activity*

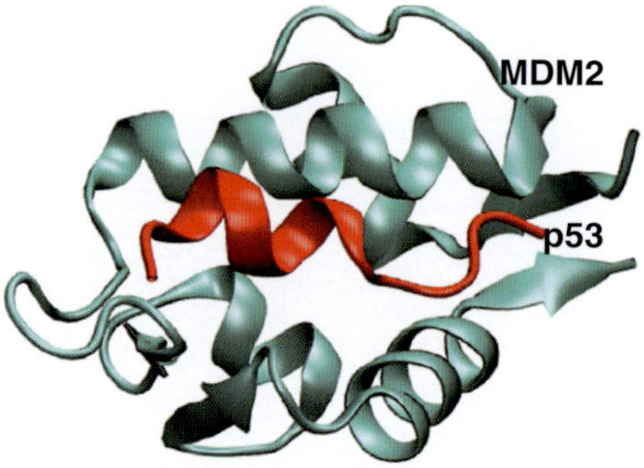

Plate 21 *Binding mode of p53 peptide (residues 15–29) in red to MDM2 (residues 25–109) in cyan [PDB ID:1YCR]. This figure was generated by the program VMD*

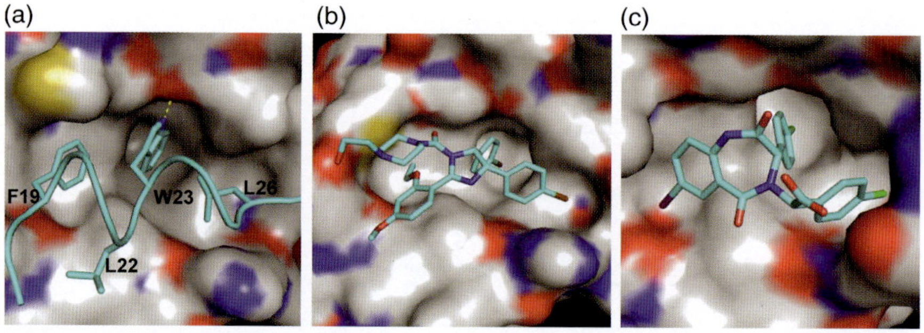

Plate 22 *Binding mode of (a) p53 peptide, (b) Nutlin-2 and (c) benzodiazepinedione compound to MDM2 [PDB ID: (a) 1YCR, (b) 1RV1 and (c) 1T4E]. Ligands are shown with carbons in cyan, nitrogen in blue, oxygen in red, chlorine in green, bromine in brown and iodine in purple. Key residues in the p53 peptide are shown in stick representation. The surface representation of MDM2 is shown with carbons in grey, nitrogen in blue, oxygen in red and sulfur in yellow. Hydrogen atoms are excluded for clarity with hydrogen bonds depicted by yellow dashed lines. This figure was generated by the program Pymol*

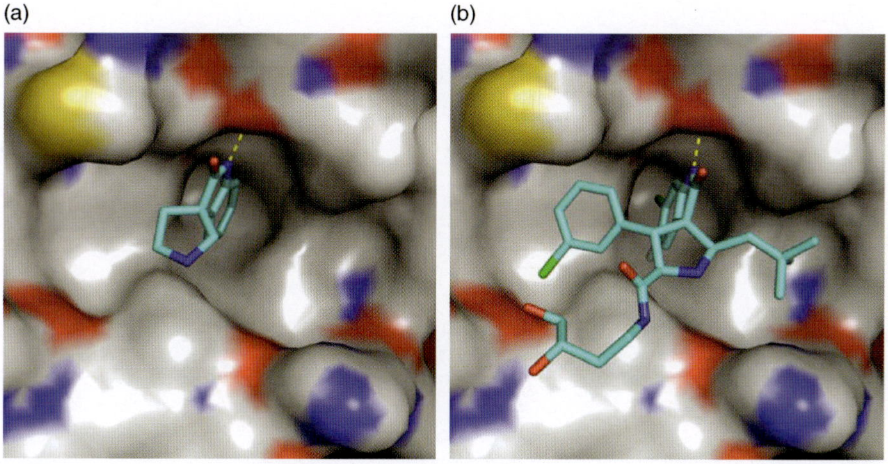

Plate 23 *Predicted binding mode of (a) spiro(oxindole-3,3′-pyrrolidine) and (b) MI-219 to MDM2 by GOLD. Ligands are shown with carbons in cyan, nitrogen in blue, oxygen in red and chlorine in green. The surface representation of MDM2 is shown with carbons in grey, nitrogen in blue, oxygen in red and sulfur in yellow. Hydrogen atoms are excluded for clarity with hydrogen bonds depicted by yellow dashed lines. This figure was generated by the program Pymol*

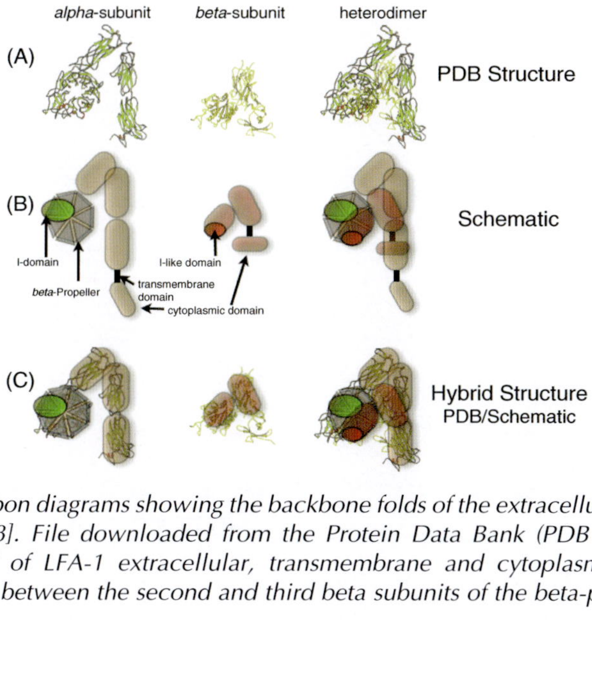

Plate 24 *(A) Ribbon diagrams showing the backbone folds of the extracellular domains of the integrin $\alpha V\beta 3$ [18]. File downloaded from the Protein Data Bank (PDB 1l5g). (B and C). Schematic model of LFA-1 extracellular, transmembrane and cytoplasmic domains with I-domain inserted between the second and third beta subunits of the beta-propeller*

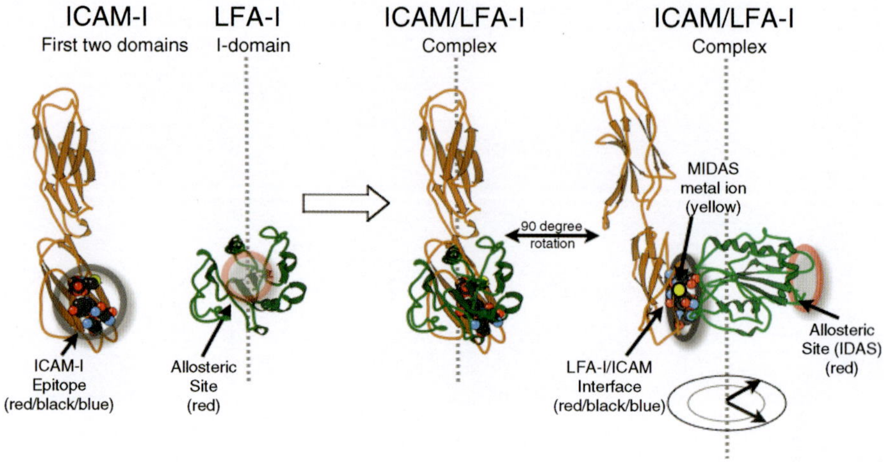

Plate 25 *Ribbon diagrams tracing the backbone folds in the protein-protein interaction between the first domain of ICAM-1 and the I-domain of LFA-1[20]. File downloaded from the Protein Data Bank (PDB 1MQ8). Note that the sidechains of amino acid residues E-34, M-64, Y-66, N68 and Q73, comprising the epitope of ICAM-1 responsible for its binding to LFA-1, have been displayed. The I-domain allosteric site (IDAS) binding site for allosteric antagonists of LFA-1 is proximal to the ICAM-1 binding site on the I-domain and is denoted by a red circle*

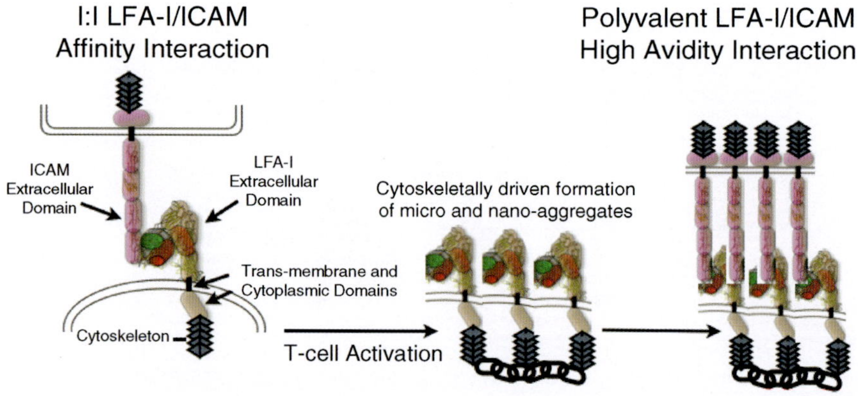

Plate 26 *Schematic representation of monovalent and polyvalent cell-cell interaction. Cytoskeletal engagement by the short cytoplasmic domains of LFA-1 can form nano/micro-aggregates of LFA-1 on the cell surface with enhanced avidity for ICAM. Similar surface aggregation of ICAM on the opposing cell surface will augment this avidity effect*

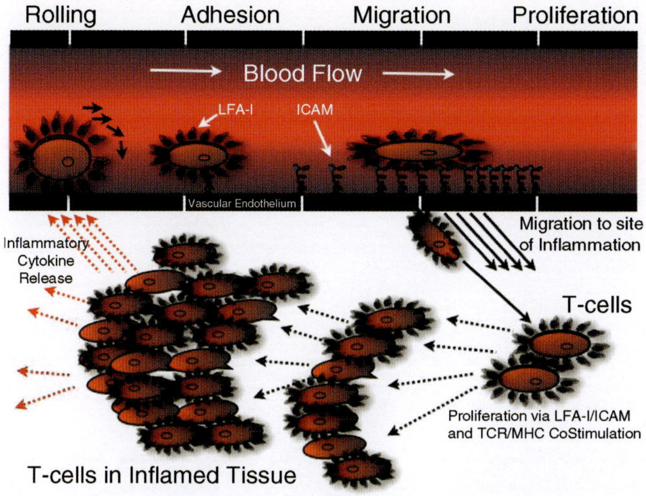

Plate 27 *T-cells involved in an inflammatory process. Cells rolling along the wall of a blood vessel encounter a locally high level of ICAM expression on the surface of vascular endothelial cells which has been induced by a local inflammatory signal. The adhesion of these cells to ICAM allows them to change their morphology from round to a more flattened adherent shape. Cells migrate out of the vessel and follow a cytokine concentration gradient to a site of inflammation. T-cells then self associate and proliferate in tissue via costimulation of LFA-1/ICAM and T-cell receptor/MHC expressed on their surface. These clonally expanding T-cells upregulate the expression of cytokines and cytokine receptors which continues a self propagating cycle of inflammation, attracting more cells from the local vasculature and stimulating their proliferation in the inflamed tissue*

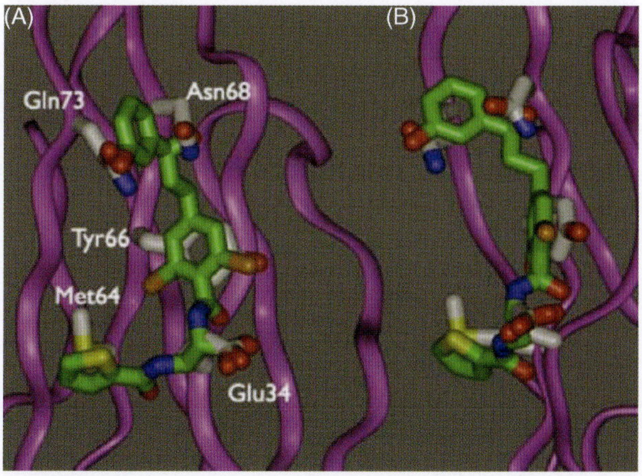

Plate 28 *Topdown (A) and Sideon (B) view of the structure of Genentech's optimized small molecule LFA-1 antagonist (Figure 11.5) overlaid on the epitope of the first domain of ICAM-1 (modeled from PDB1MQ8, adapted from Gadek et al. [40])*

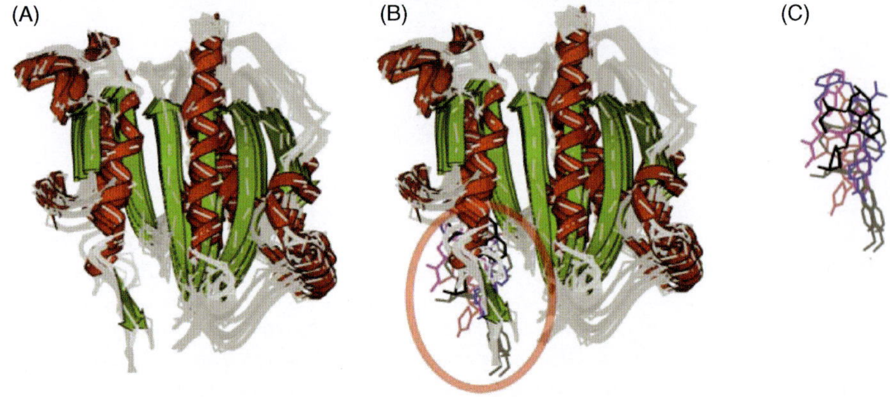

Plate 29 *Overlap of 5 crystal structures of I-domain and 5 different IDAS LFA-1 antagonists from BMS and Novartis. Note gross similarities in I domain folds (A) and diversity in antagonist bound structures (B/C). See text for references to structures*

to design nonpeptidic small molecule protein structural motif mimetics that might serve as biochemical tools or drug leads [2]. Development of a general strategy based on a scaffold that could be specifically tailored for different protein–protein therapeutic targets might yield insights into may different therapeutic targets. The focus of this section is on compounds that mimic the structural features of different protein secondary structures. In general, these templates, which were originally designed to allow facile synthetic chemical modifications, do not cover large surface areas, and therefore they do not really represent the intended secondary peptidic structure. Rather, they display the critical pharmacophores in a suitable 3D-spatial orientation for binding to the target protein.

Indanes, ter-phenyls, trisubstituted imidazoles and oligo-amide foldamers have been reported by several groups to mimic α-helical surface binding domains and disrupt PPIs (for a recent review, see [72]). A representative example of the potential use of these templates in the design of PPIs is illustrated with compound 24 (Figure 6.5), which blocks the formation of the Bcl-x_L-Bak complex. The NMR solution structure of Bcl-x_L bound to the BH3 domain of Bak [53] shows that the Bak derived peptide adopts an amphipathic α-helix, and that residues Val-74, Leu-78, Ile-81, and Ile-85 –corresponding to i, $i + 4$, $i + 7$ and $i + 11$– make van der Waals contacts with the hydrophobic binding cleft of Bcl-x_L. A series of terphenyl derivatives were prepared to mimic the α-helical structure of the BH3 peptide and the interactions mediated by the side-chains of the preceding amino acids [73]. From this synthetic effort and *in vitro* biological evaluation using a fluorescence polarization assay, compound 24 was identified as a potent antagonist of this PPI ($K_D = 114$ nM). NMR experiments with ^{15}N-labeled Bcl-x_L protein suggested that the binding cleft on the surface of Bcl-x_L known to interact with BH3 domain of the Bak peptide is the target area for this compound.

One important limitation of the terphenyl scaffold is that it requires relative difficult C-C bond reactions to attach the side chain mimetics to the template, and to assemble functionalized phenyl building blocks. This potential synthetic limitation has been overcome by using terephthalamide derivatives, which can be easily functionalized by O-alkylation and amide formations (e.g. compound 25, Figure 6.5) [74]. Interestingly, the intramolecular H-bond between the amide NH and the alkoxy oxygen atom of this template ensures that the O-substituent and the side chain of the amino acid are positioned on the same side of the terephathalamide group.

Benzodiazepine derivatives, D-glucose, spirocyclic β-lactams and polypyrrolinones have been used among other no peptide molecules [5], [75] as privileged scaffolds to mimic β-turn structures in peptide/protein recognition motifs. A common feature of these molecules is a rigid conformation that imposes an appropriate projection of several substituents. In terms of drug discovery activities, some of these scaffolds (e.g. pyrrolinones like compound 26, Figure 6.5) have been applied to the development of antagonists of HIV-1 protease [76].

Although β-sheets account for over 30% of all protein secondary structures, the examples of synthetic nonpeptidic templates that mimic β-strand/β-sheets motifs and disrupt PPIs are scanty. Compound 27 (Figure 6.5) was designed to inhibit the HIV-1 protease dimerization, which is necessary to exhibit the enzymatic activity of this aspartyl protease [77]. The enzymatic characterization of compound 27 showed that it inhibited the dimerization of HIV-1 protease by a dissociative mechanism with a K_{id} value of 5.4 μM. However, additional kinetic analyses suggested that this β-strand mimetic also inhibited partly the enzyme activity though an active-site directed mechanism.

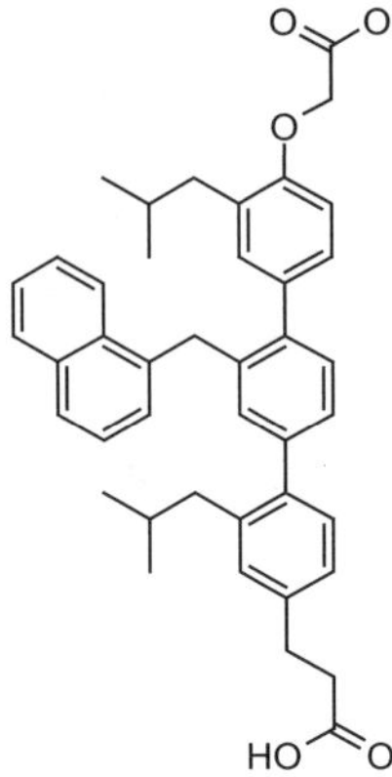

24

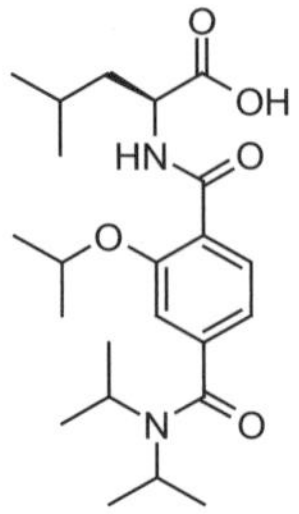

25

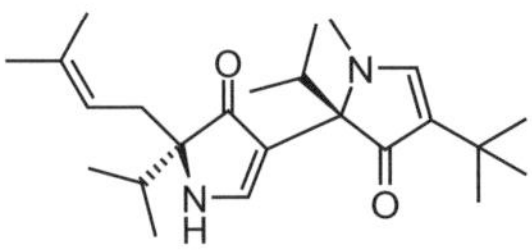

26

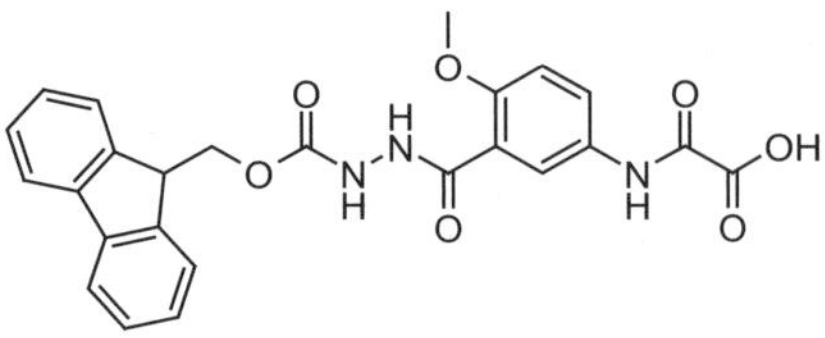

27

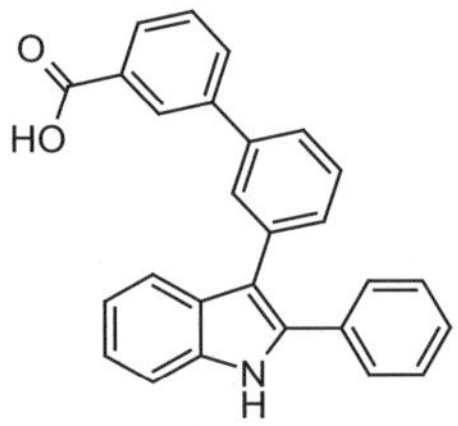

28

Figure 6.5

Another example of a β-strand mimetic is coming from the antimicrobial field. Z-interacting protein A (ZipA) is a 36.6-kDa membrane-anchored protein in *Escherichia coli* that interacts with FtsZ, a homolog of eukaryotic tubulins, forming a septal ring structure that mediates bacterial cell division. Thus, the ZipA/FtsZ PPI is a potential target for the development of new antibacterial agents. Structural studies revealed that FtsZ binds to a hydrophobic cleft on the solvent-exposed side of a β-sheet on ZipA [78]. Following a structure-based strategy, compound 28 (Figure 6.5) was identified as a weak inhibitor of the ZipA-FtsZ interaction ($IC_{50} = 192\ \mu M$) [79]. ^{1}H- and ^{15}N-NMR chemical shift perturbation analysis demonstrated the it interacts with binding site on ZipA.

Contrary to some of the α-helix and β-turn mimetics previously described, compounds 27 and 28 appear unsuitable as general β-strand scaffolds to inhibit PPIs that involved extended β-sheet conformations.

Parallel to the efforts to design general templates to mimic specific secondary conformations, molecular modeling and rational approaches have been able to identify novel pharmaceutical leads that deliberately incorporate precise functional groups in a complementary manner to that of the intended protein target. As illustrated with a few representative examples of this challenging strategy, computer-aided design efforts have successfully match the key protein recognition elements – spatial orientation, and electrostatic and hydrophobic complementarities – of a few therapeutically relevant PPIs.

Several groups have reported in the past few years the discovery of ligands for XIAP -a member of the so-called 'Inhibitors-of-apoptosis proteins' (IAPs) [80]. The pro-apoptotic protein Smac binds to the BIR3 domain of XIAPs, where caspase-9 binds, thereby relieving the inhibition of this and other caspases, and reactivating apoptosis. Inhibiting this PPI could help to restore apoptosis and thereby off-set a cancerous condition by inducing tumor cell death.

Structural studies have shown that the physical interaction between Smac and XIAP is mediated by the *N*-terminal four residues –Ala-Val-Pro-Ile- in Smac and a well-defined surface cleft in XIAP-BIR3. This binding site is ideally suited for the design of low molecular mass Smac mimetics that antagonize XIAP. Following structure based design principles, a series of scaffolds have been obtained (e.g. compounds 29–33, Figure 6.6) [80–83]. In

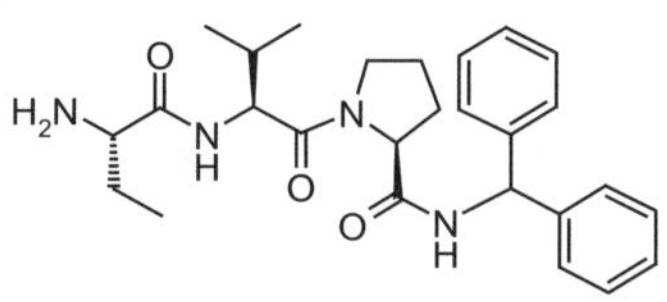

29

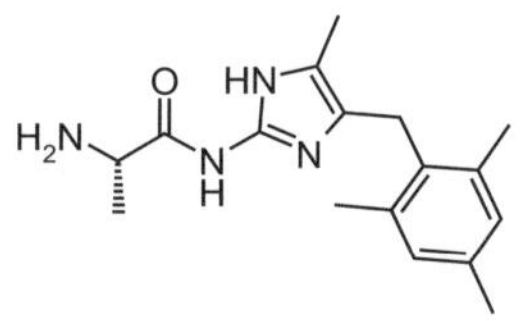

30

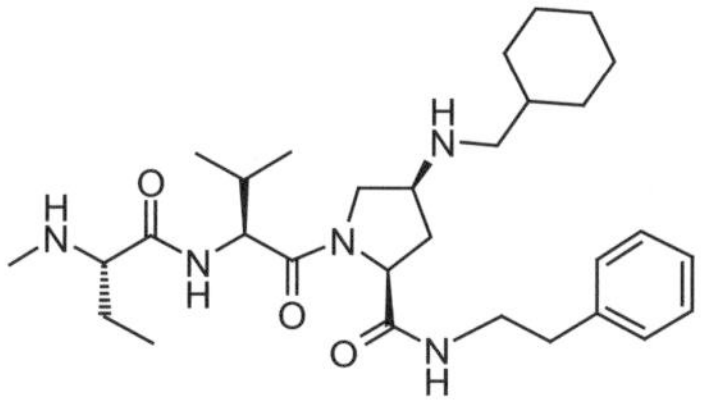

31

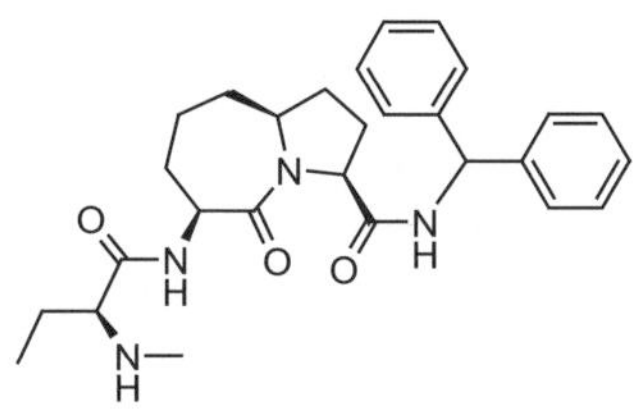

32
SM-131

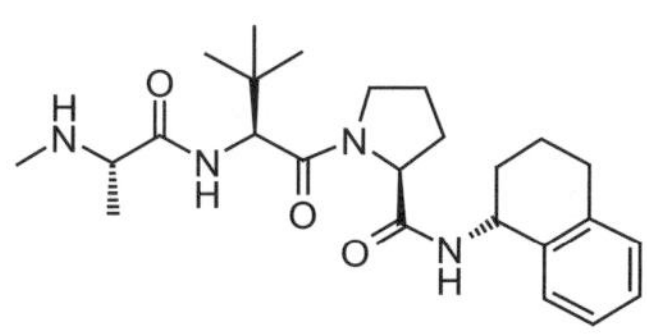

33

Figure 6.6

addition to be highly conformationally constrained, a common feature of all these compounds is the presence of an N-terminal amino group. This functional group forms an extensive hydrogen bond network with a number of negatively charged residues in XIAP, including Asp-390 and Glu-314. The critical importance of this hydrogen-bonding network was also demonstrating by SAR; thus, incorporation of an acetyl group made the resulting compounds inactive in binding to XIAP-BH3. As shown in Figure 6.6, the greatest diversity among the different Smac mimetic involved variation at the C-terminus. Substituents at this part of the molecule have the potential to establish hydrophobic interactions with different parts of the BIR3 domain and may also be solvent exposed, allowing for attachment of groups that can modulate the pharmaceutical properties of the molecule. One of the most active antagonist – compound 32; Ki = 61 nM, *in vitro* competitive binding assay- [81] significantly inhibits tumor cell growth (IC_{50} = 100 nM) and induces cell death and activation of caspase-9 and -3 in the MDA-MB-231 human breast cancer cell line.

As in the previous case, a number of structure-based design molecules that target PPIs mediated by members of the Bcl-2 family of antiapoptotic proteins have been reported over the past few years (compounds 34–36, Figure 6.7) [43, 64, 67]. These antagonists have been

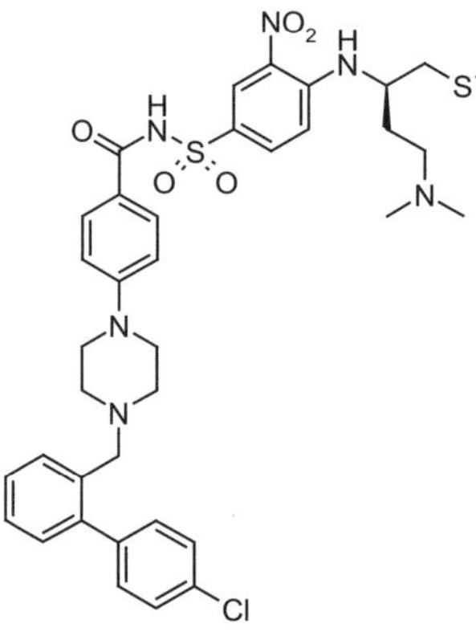

34
ABI-737

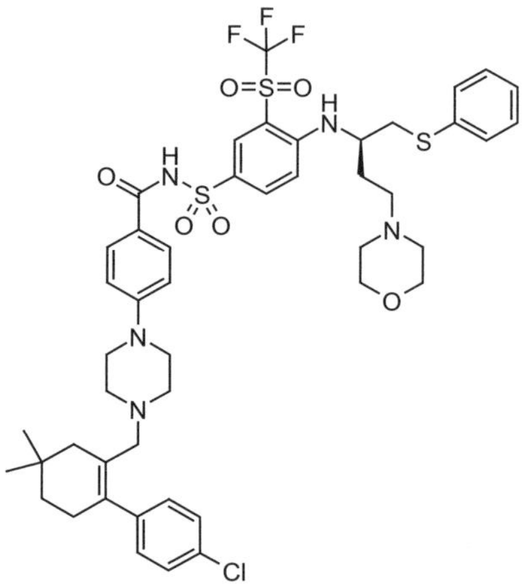

35
ABT-263

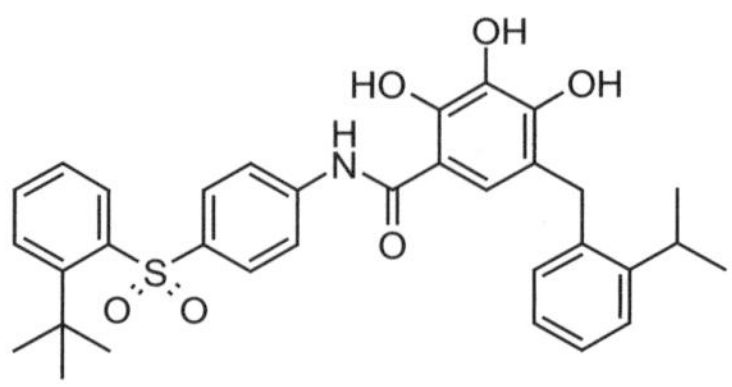

36
TW-37

Figure 6.7

obtained though the course of rounds of structure-guided chemistry efforts to optimize potency, selectivity and pharmaceutical properties of hits identified by different screening approaches (e.g. HTS, NMR or virtual screening). Although the Bak peptide binding to Bcl-x_L involves only approximately 500 Å of protein surface, it is obvious by looking at the size and chemical complexity of the Bcl-x_L antagonists that potency has been achieved by identifying and optimizing key interactions of the molecule with multiple hot spots on the surface of the target protein. The optimization of these interaction and the fine-tuning of the pharmaceutical properties have resulted in the identification of clinical candidates. Thus, compound 35 (ABT-263, Figure 6.7) is currently in Phase I/IIa clinical trials in patients with lymphomas. ABT-263 binds with high affinity to several Bcl-2 family proteins, including Bcl-xL and Bcl-2, with Ki values <1 nM. In human tumor cell lines derived from small-cell lung cancer and lymphoid malignancies, ABT-263 displayed potent cytotoxicity. In addition, the drug potently enhanced the cytotoxicity of chemotherapy and radiation, and induced tumor shrinkage in a variety of human xenograft models.

6.4 Conclusions and Outlook

The identification and development of antagonists of PPIs represents a growing field in medicinal chemistry. The results reported in this chapter illustrate in a succinct manner the progress made in this area of drug discovery. From the above examples, it is clear that as successes in discovering and optimizing protein–protein antagonists accumulate, principles and key factors in judging whether or not a particular pair of proteins is amenable for pharmaceutical intervention [84] and the most suitable strategy to identify hit compounds for a specific PPI target will be more effective. This should allow the establishment of platforms to integrate drug discovery activities for this family of targets, and better characterized the activity and selectivity profile of hits and drug candidates. In this context, it is remarkable that target selectivity is often not specifically address in the available scientific literature reflecting the lack of a strategy to identify, elucidate and address potential off-target effects of the currently published modulators of PPIs.

We can only hope that all the current research and drug discovery efforts dedicated to the identification of antagonists of protein–protein interactions will result in the near future in more effective modalities of therapeutic treatment.

References

1. D. C. Fry, Protein–protein interactions as targets for small molecule drug discovery, *Biopolymers* 84, 535–52 (2006).
2. Y. Che, B. R. Brooks and G. R. Marshall, Development of small molecules designed to modulate protein–protein interactions. *J. Comput. Aided Mol. Des.* 20, 109–30 (2006).
3. D. Gonzalez-Ruiz and H. Gohlke, Targeting protein–protein interactions with small molecules: challenges and perspectives for computational binding epitope detection and ligand finding, *Curr. Med. Chem.* 13, 2607–25 (2006).
4. M. Arkin, Protein–protein interactions and cancer: small molecules going in for the kill, *Curr. Opin. Chem. Biol.* 9, 317–24 (2005).
5. H. Yin and A. D. Hamilton, Strategies for targeting protein–protein interactions with synthetic agents, *Angew. Chem. Int. Ed.* 44, 4130–63 (2005).

6. M. R. Arkin and J. A. Wells, Small-molecule inhibitors of protein–protein interactions: progressing towards the dream, *Nature Rev.* 3, 301–17 (2004).
7. D. A. Ockey and T. R. Gadek, Inhibitors of protein–protein interactions, *Expert. Opin. Ther. Pat.* 12, 393–400 (2002).
8. M. Z. Lin, M. A. Teitell and G. J. Schiller, The evolution of antibodies into versatile tumor-targeting agents, *Clin. Cancer Res.* 11, 129–38 (2005).
9. T. Matthews, M. Salgo, M. Greenberg, J. Chung, R. DeMasi and D. Bolognesi, Enfuvirtide: the first therapy to inhibit the entry of HIV-1 into host CD4 lymphocytes, *Nat. Rev. Drug Discov.* 3, 215–25 (2004).
10. C. Vita, E. Drakopoulou, S. Vizavona, *et al.*, Rational engineering of a miniprotein that reproduces the core of the CD4 site interacting with HIV-1 envelope glycoprotein, *Proc. Nat. Acad. Sci. U S A* 96, 13091–6 (1999).
11. J. W. Chin and A. Schepartz, Design and evolution of a miniature Bcl-2 binding protein, *Angew. Chem. Int. Ed.* 40, 3806–9 (2001).
12. S. E. Rutleges, H. M. Volkman and A. Schepartz, Molecular recognition of protein surfaces: high affinity ligands for the CBP KIX domain, *J. Am. Chem. Soc.* 125, 14336–47 (2003).
13. E. Drakopoulou, S. Zinn-Justin, M. Guenneugues, B. Gilquin, A. Menez and C. Vita, Changing the structural context of a functional beta-hairpin. Synthesis and characterization of a chimera containing the curaremimetic loop of a snake toxin in the scorpion alpha/beta scaffold, *J. Biol. Chem.* 271, 11979–87 (1996).
14. P. M. Fischer and D. P. Lane, Small-molecule inhibitors of the p53 suppressor hdm2: have protein–protein interactions come of age as drug targets?. *Trends Pharmacol. Sci.* 25, 343–6 (2004).
15. A. Böttger, V. Böttger, C. Garcia-Echeverria, *et al.*, Molecular characterization of the mdm2-p53 interaction. *J. Mol. Biol.* 269, 744–56 (1997).
16. C. Garcia-Echeverria, P. Chene, M. J. J. Blommers and F. Pascal, Discovery of potent antagonists of the interaction between human double minute 2 and tumor suppresor p53, *J. Med. Chem.* 43, 3205–8 (2000).
17. K. Sakurai, C. Schubert and D. Kahne, Crystallographic analysis of an 8-mer p53 peptide analogue complexed with MDM2, *J. Am. Chem. Soc.* 128, 11000–1 (2006).
18. P. M. Fischer, Peptide, peptidomimetic, and small-molecule antagonists of the 53-hdm2 protein–protein interaction, *Int. J. Pept. Res. Ther.*, 12, 3–19 (2006).
19. P. Chene, Inhibition of the p53-mdm2 interaction: targeting a protein–protein interface, *Mol. Cancer Res.*, 2, 20–8 (2004).
20. P. Chene, J. Fuchs, I. Carena, P. Furet and C. Garcia-Echeverria, Study of the cytotoxic effect of a peptidic inhibitor of the p53-hdm2 interaction in tumor cells. *FEBS Lett.*, 529, 293–7 (2002).
21. P. Chene, J. Fuchs, J. Bohn, C. Garcia-Echeverria, P. Furet and D. Fabbro. A small synthetic peptide, which inhibits the p53-hdm2 interaction, stimulates the p53 pathway in tumour cell lines, *J. Mol. Biol.*, 299, 245–253 (2000).
22. X. Li and J. Liang, Computationsl design of combinatorial peptide library for modulating protein–protein interactions, *Pac. Symp. Biocomput.* 28–39 (2005).
23. J. Zeng, T. Nheu, A. Zordet, B. Catimel, E. Nice, H. Maruta, A. W. Burgess and H. R. Treutlein, Design and inhibitors of Ras-Raf interaction using a computational combinatorial algorithm, *Protein Eng.*, 14, 39–45 (2001).
24. L. O. Sillerud and R. S. Larson, Design and structure of peptide and peptidomimetic antagonists of protein–protein interaction, *Curr. Prot. Pept. Sci.*, 6, 151–69 (2005).
25. S. J. DeMarco, H. Henze, A. Lederer, *et al.*, Discovery of novel, highly potent and selective b-hairpin mimetic CXCR4 inhibitors with excellent anti-HIV activity and pharmacokinetic profiles, *Bioorg. Med. Chem.*, 14, 8396–8404 (2006).
26. R. Fasan, R. L. Dias, K. Moehle, *et al.*, Structure-activity studies in a family of b-hairpin protein epitope mimetic inhibitors of the p53-hdm2 protein–protein interaction, *Chembiochem.*, 7, 515–26 (2006).
27. R. Fasan, L. A. R. Dias, K. Moehle, O. Zerbe, J. W. Vrijbloed, D. Obrecht and J. A. Robinson, Use a β-hairpin to mimic an a-helix: cyclic peptidomimetic inhibitors of the p53-hdm2 protein–protein interaction, *Angew. Chem. Int. Ed.*, 43, 2109–12 (2004).

28. J. D. Sadowsky, W. D. Fairlie, E. B. Hadley, *et al.*, (a/b + a)-Peptide antagonists of BH3 domain/Bcl-x_L recognition: toward general strategies for foldamer-based inhibition of protein–protein interactions, *J. Am. Chem. Soc.*, 129, 139–54 (2007).
29. D. C. Fry and L. T. Vassilev, Targeting protein–protein interactions for cancer therapy, *J. Mol. Med.* 83, 955–63 (2005).
30. (a) L. T. Vassilev, B. T. Vu, B. Graves, *et al.*, In vivo activation of the p53 pathway by small-molecule antagonists of mdm2, *Science*, 303, 844–8 (2004). (b) D. C. Fry, S. D. Emerson, S. Palme, B. T. Vu, C.-M. Liu and F. Podlaski, NMR structure of a complex between MDM2 and a small molecule inhibitor,J. Biomol. *NMR*, 30, 163–73 (2004).
31. (a) B.L. Grasberger, T. Lu, C. Schubert, *et al.*, Discovery and cocrystal structure of benzodiazepinedione HDM2 antagonists that activate p53 in cells, *J. Med. Chem.*, 48, 909–12 (2005). (b) D.J. Parks, L.V. LaFrance, R.R. Calvo, *et al.*, 1,4-Benzodiazepine-2,5-diones as small molecule antagonists of the HDM2-p53 interaction: discovery and SAR, *Bioorg. Med. Chem. Lett.*, 15, 765–70 (2005). (c) D.J. Parks, L.V. LaFrance, R.R. Calvo, *et al.*, Enhanced pharmacokinetic properties of 1,4-benzodiazepine-2,5-dione antagonists of the HDM2-p53 protein–protein interaction through structure-based drug desig, *Bioorg. Med. Chem. Lett.*, 16, 3310–14 (2006).
32. M. Ono, Y. Wada, Y. M. Wu, *et al.*, FP-21399 blocks HIV envelope protein-mediated membrane fusion and concentrates in lymph nodes, *Nat. Biotechnol.*, 15, 343–8 (1997).
33. J. L. Zhang, H. Choe, B. J. Dezube, *et al.*, The bis-azo compound FP-21399 inhibits HIV-1 replication by preventing viral entry, *Virology* 244, 530–41 (1998).
34. D. L. Boger, J. Goldberg, S. Silletti, T. Kessler and D. A. Cheresh, Identification of a novel class of small-molecule antiangiogenic agents through the screening of combinatorial libraries which function by inhibiting the binding and localization of proteinase MMP2 to integrin $\alpha_v\beta_3$, *J. Am. Chem. Soc.*, 123, 1280–8 (2001).
35. T. Berg, S. B. Cohen, J. Desharnais, *et al.*, Small-molecule antagonists of myc/max dimerization inhibit myc-induced transforamtion of chicken embryo fibroblasts, *Proc. Nat. Acad. Sci. USA*, 99, 3830–5 (2002).
36. A. Jordan, J. A. Hadfield, N. J. Lawrence and A. T. McGown, Tubulin as a target for anticancer drugs: agents which interact with the mitotic spindle, *Med. Res. Rev.*, 18, 259–96 (1998).
37. R. B. Ravelli, B. Gigant, P. A. Curmi, *et al.*, Insight into tubulin regulation from a complex with colchicine and a stahmin-like domain, *Nature*, 428, 198–202 (2004).
38. S. Desbene and S. Giorgi-Reanult, Drugs that inhibit tubulin polymerization: the particular case of podophyllotoxin and analogues, *Curr. Med. Chem. Anticancer Agents*, 2, 71–90 (2002).
39. L. A. Amos and J. Lowe, How Taxol stabilises microtuble structure, *Chem. Biol.*, 6, R65–R69 (1999).
40. S. J. Duncan, S. Gruschow, D. H. Williams, *et al.*, Isolation and structure elucidation of chlorofusin, a novel p53-mdm2 antagonist from a Fusarium sp., *J. Am. Chem. Soc.*, 2, 554–60 (2001).
41. P. Desai, S. S. Pfeiffer and D. L. Boger, Synthesis of the chlorofusin cyclic peptide: assignment of the asparagine stereochemistry, *Organic Lett.*, 5, 5047–50 (2003).
42. M. Lepourcelet, Y. N. Chen, D. S. France, *et al.*, Small-molecule antagonist of the oncogenic Tcf/b-catenin protein complex, *Cancer Cell*, 5, 91–102 (2004).
43. G. Wang, Z. Nikolovska-Coleska, C.-Y. Yang, *et al.*, Structure-based design of potent small-molecule inhibitors of anti-apoptotic Bcl-2 proteins, *J. Med. Chem.*, 49, 6139–42 (2006).
44. J. B. Baell and D. C. S. Huang, Prospects for targeting the Bcl-2 family of proteins to develop novel cytotoxic drugs, *Biochem. Pharmacol.*, 64, 851–63 (2002).
45. K. M. Kim, C. D. Giedt, G. Basanez, *et al.*, Biophysical characterization of recombinant human Bcl-2 and its interaction with an inhibitory ligand, *Biochemistry*, 40, 4911–4922 (2001).
46. S. Ghosh, A. Nie, J. An and Z. Huang, Structure-based virtual screening of chemical libraries for drug discovery, *Curr. Opin. Chem. Biol.*, 10, 194–202 (2006).
47. P. S. Galatin and D. J. Abraham, A nonpeptidic sulfonamide inhibits the p53-mdm2 interaction and activates p53-dependent transcription in mdm2-overexpressing cells, *J. Med. Chem.*, 47, 4163–5 (2004).

48. G. E. Kellogg, G. S. Joshi and D. J. Abraham, New tools for modeling and understanding hydrophobicity and hydrophobic interactions, *Med. Res. Rev.*, 1, 453 (1992).
49. D. C. Chan, D. Fass, J. M. Berger and P. S. Kim, Core structure of gp41 from the HIV envelope glycoprotein, *Cell*, 89, 263–73 (1997).
50. D. C. Chan, C. T. Chutkowski and P. S. Kim, Evidence that a prominent cavity in the coiled coil of HIV type 1 gp41 is an attractive drug target, *Proc. Natl. Acad. Sci. USA*, 95, 15613–17 (1998).
51. A. K. Debnath, L. Radigan and S. Jiang, Structure-based identification of small molecule antiviral compounds targeted to the gp41 core structure of the human immunodeficiency virus type 1, *J. Med. Chem.*, 42, 3203–9 (1999).
52. A. E. Edling, S. Choksi, Z. Huang and R. Korngold, An organic CD4 inhibitor reduces the clinical and pathological symptoms of acute experimental allergic encephalomyelitis, *J. Autoimmun.*, 18, 169–79 (2002).
53. M. Sattler, H. Liang, D. Nettesheim, *et al.*, Structure of Bcl-x_L-Bak peptide complex: recognition between regulators of apoptosis, Science, 275, 983–6 (1997).
54. J. L. Wang, D. Liu, Z. J. Zhang, *et al.*, Structure-based discovery of an organic compound that binds Bcl-2 protein and induces apoptosis of tumor cells, *Proc. Nat. Acad. Sci. USA*, 97, 7124–9 (2000).
55. I. J. Enyedy, Y. Ling, K. Nacro, *et al.*, Discovery of small-molecule inhibitors of Bcl-2 through structure-based computer screening, *J. Med. Chem.*, 44, 4313–24 (2001).
56. Y. Lu, Z. Nikolovska-Coleska, X. Fang, *et al.*, Discovery of a nanomolar inhibitor of the human murine double minute 2 (mdm2)-p53 interaction through an integrated, virtual database screening strategy, *J. Med. Chem.*, 49, 3759–62 (2006).
57. T. S. Rush III, J. A. Grant, L. Mosyak and A. Nicholls, A shape-based 3-D-scaffold hoping method and its application to a bacterial protein–protein interaction. *J. Med. Chem.*, 48, 1489–95 (2005).
58. D. A. Erlanson, Fragment-based lead discovery: a chemical update, *Curr. Opin. Biotech.*, 17, 643–52 (2006).
59. N. Baurin, F. Aboul-Ela, X. Barril, *et al.*, Design and characterization of libraries of molecular fragments for use in NMR screening against protein targets, *J. Chem. Inf. Compt. Sci.*, 44, 2157–66 (2004).
60. R. J. Boerner, D. B. Kassel, S. C. Barker, B. Ellis, P. DeLacy and W. B. Knight. Correlation of the phosphorylation states of pp60c-src with tyrosine kinase activity: the intramolecular pY530-SH2 complex retains significant activity if Y419 is phosphorylated, *Biochemistry*, 35, 9519–25 (1996).
61. M. Pellecchia, B. Becattini, K. J. Crowell, *et al.*, NMR-based techniques in the hit identification and optimisation processes, *Expert. Opin. Ther. Targets*, 8, 597–611 (2004).
62. S. B. Shuker, P. J. Hajduk, R. P. Meadows and S. W. Fesik, Discovering high-affinity ligands for proteins: SAR by NMR, *Science*, 274, 1531–4 (1996).
63. D. A. Erlanson, J. A. Wells and A. C. Braisted, Tethering: fragment-based drug discovery, *Annu. Rev. Biophys. Biomol. Struct.*, 33, 199–223 (2004).
64. M. Bruncko, T. K. Oost, B. A. Belli, *et al.*, Studies leading to potent, dual inhibitors of Bcl-2 and Bcl-x_L, *J. Med. Chem.*, 50, 641–62 (2007).
65. A. M. Petros, J. Dinges, D. J. Augeri, *et al.*, Discovery of a potent inhibitor of the antiapoptotic protein Bcl-x_L from NMR and parallel synthesis, *J. Med. Chem.*, 49, 656–63 (2006).
66. M. D. Wendt, W. Shen, A. Kunzer, *et al.*, Discovery and structure-activity relationship of antagonists of B-cell lymphoma 2 family proteins with chemopotentiation activity *in vitro* and *in vivo*, *J. Med. Chem.*, 49, 1165–81 (2006).
67. T. Oltersdorf, S. W. Elmore, A. R. Shoemaker, *et al.*, An inhibitor of Bcl-2 family proteins induces regression of solid tumours, *Nature*, 435, 677–81 (2005).
68. A. C. Braisted, J. D. Oslob, W. L. Delano, *et al.*, Discovery of a potent small molecule IL-2 inhibitor through fragment assembly, *J. Am. Chem. Soc.*, 125, 3714–15 (2003).
69. B. C. Raimundo, J. D. Oslob, A. C. Braisted, *et al.*, Integrating fragment assembly and biophysical methods in the chemical advancement of small- molecule antagonists of IL-2: an approach for inhibiting protein–protein interactions. *J. Med. Chem.*, 47, 3111–30 (2004).
70. M. G. Bursavich and D. Rich, Designing non-peptide peptidomimetics in the 21st century: inhibitors targeting conformational ensembles, *J. Med. Chem.*, 45, 541–58 (2002).
71. A. S. Ripka and D. Rich, Peptidomimetic design, *Curr. Opin. Chem. Biol.*, 2, 441–52 (1998).

72. J. M. Davis, L. K. Tsou and A. D. Hamilton, Synthetic non-peptide mimetics of alpha-helices, *Chem. Soc. Rev.*, 36, 326–34 (2007).
73. O. Kutzki, H. S. Park, J. T. Ernst, B. P. Orner, H. Yin and A. D. Hamilton, Development of a potent Bcl-x_L antagonist based on a-helix mimicry, *J. Am. Chem. Soc.*, 124, 11838–9 (2002).
74. H. Yin, G. Lee, K. A. Sedey, *et al.*, Terephthalamide Derivatives as Mimetics of Helical Peptides: Disruption of the Bak interaction Bcl-x_L/Bak interaction, *J. Am. Chem. Soc.*, 127, 5463–8 (2005).
75. H. Bittermann and P. Gmeiner, Chirospecific synthesis of spirocyclic beta-lactams and their characterization as potent type II beta-turn inducing peptide mimetics, *J. Org. Chem.*, 71, 97–102 (2006).
76. A. B. Smith, L. D. Cantin, A. Pastermak, *et al.*, Design, synthesis, and biological evaluation of monopyrrolinone-based HIV-1 protease inhibitors, *J. Med. Chem.*, 46, 1832–44 (2003).
77. M.-C. Song, S. Rajesh, Y. Hayashi and Y. Kiso, Design and synthesis of new inhibitors of HIV-1 protease dimerization with conformationally constrained templates, *Bioorg. Med. Chem. Lett.*, 11, 2465–8 (2001).
78. L. Mosyak, Y. Zhang, E. Glasfeld, S. Haney, M. Stahl, J. Seehra and W. S. Somers, The bacterial cell-division protein ZipA and its interaction with an FtsZ fragment revealed by X-ray crystallography, *EMBO J.*, 19, 3179–91 (2000).
79. A. G. Sutherland, J. Alvarez, W. Ding, *et al.*, Structure-based design of carboxybiphenylindole inhibitors of the ZipA–FtsZ interaction, *Org. Biomol. Chem.*, 1, 4138–40 (2003).
80. S. K. Sharma, C. Straub and L. Zawel, Development of peptidomimetics targeting IAPs, *Int.-J.-Pept.-Protein.-Res. Ther.*, 12, 21–32 (2006).
81. H. Sun, Z. Nikolovska-Coleska, J. Lu, *et al.*, Design, synthesis, and evaluation of a potent, cell-permeable, conformationally constrained second mitochondria derived activator of caspase (smac) mimetic, *J. Med. Chem.*, 49, 7916–20 (2006).
82. H. Sun, Z. Nikolovska-Coleska, J. Chen, *et al.*, Structure-based design, synthesis and biochemical testing of novel and potent Smac peptido-mimetics, *Bioorg. Med. Chem. Lett.*, 15, 793–7 (2005).
83. C.-M. Park, C. Sun, E. T. Olejniczak, *et al.*, Non-peptidic small molecule inhibitors of XIAP, *Bioorg. Med. Chem. Lett.*, 15, 771–5 (2005).
84. P. J. Hajduk, J. R. Huth and S. W. Fesik, Druggability indices for protein targets derived from NMR-based screening data, *J. Med. Chem.*, 48, 2518–25 (2005).

Part III

Techniques

7

High-throughput Methods of Chemical Synthesis Applied to the Preparation of Inhibitors of Protein–Protein Interactions

Annaliese K. Franz, Jared T. Shaw and Yuchen Tang
Department of Chemistry, University of California at Davis, Davis, CA, USA

7.1 Introduction

The design of chemical libraries and application of high-throughput organic synthesis (HTOS) is an established strategy with emerging applications to the discovery of small molecules that disrupt protein–protein interactions (PPIs). The chemical intervention of protein–protein interactions represents the opportunity to design new therapeutic agents to control cellular processes associated with disease. While the synthesis of complex molecules remains a challenge in chemistry, the discovery of PPI inhibitors poses its own significant challenge in biology. Although the areas of organic synthesis and investigations of protein–protein interactions have largely undergone parallel development, recent examples highlight instances where synthesis and biological discovery can work together to establish guidelines for the chemical intervention of protein–protein interactions. The design of targeted libraries for enzyme inhibitors has been advanced by knowledge of substrate structures, X-ray crystal structures of the enzyme, and co-crystal structures of the enzyme bound to substrates and inhibitors. While there is not the same level of detailed information regarding the molecular recognition between most protein–protein interaction pairs, recent evidence indicates that protein surfaces frequently contain 'hot spots', which are small patches of the key surface

Protein Surface Recognition: Approaches for Drug Discovery Ernest Giralt, Mark W. Peczuh and Xavier Salvatella. © 2011 John Wiley & Sons, Ltd

residues that mediate binding affinity [1–4]. In many cases, this information does not allow the same targeted design for PPI inhibitors as it does for enzyme inhibitors. As such, libraries of small molecules play a crucial role to help establish the chemical features that are important for the molecular recognition of PPIs. Small molecules can modulate/disrupt protein–protein interactions despite their relative small size as long as interactions with protein surface 'hot spots' are maximized. Once molecular recognition features have been identified, follow-up libraries can be designed to optimize these interactions and obtain more potent or more selective inhibitors.

The development of combinatorial (split-pool) and parallel automation techniques has modernized the discovery of drug candidates. As these techniques have become more sophisticated, chemists have developed efficient synthetic methods capable of assembling larger libraries of chemical structures that resemble the diversity and complexity of natural products. As high-throughput screening techniques have also undergone a parallel path of development, a series of empirical observations has emerged to restrict the size and complexity of compounds made for drug discovery in order to maximize their future success as therapeutics. Since the potent and selective inhibitors of PPIs discovered to date often have higher molecular weights and greater molecular complexity than most drug candidates, it is likely that a new set of guidelines will emerge as more PPI inhibitors are revealed. As such, the high-throughput synthesis of compounds that feature diverse core structures and complex architectures has become important to establish these guidelines for the chemical intervention of protein–protein interactions. To identify important molecular recognition features, the design and synthesis of libraries with the potential to inhibit PPIs has derived inspiration from three basic strategies: (1) the synthesis of compounds that mimic the secondary structure features of proteins, (2) the synthesis of compounds that feature structural motifs common to natural products known to inhibit PPIs, and (3) the diversity-oriented synthesis (DOS) of compounds in pathways that maximize core-structure diversity in discovery-based libraries. Each of these strategies must balance both design and execution in order to derive the maximum benefit of HTOS in the search for new PPI inhibitors.

The challenge of synthesizing libraries for the discovery of compounds that modulate biological pathways is nearly equally divided between design and execution. Although library syntheses require an initial design to incorporate the desired structural features of the final products, the feasibility to execute the library synthesis is equally important. Library design can de divided into three basic strategies: (1) a targeted/biased library, (2) a discovery/unbiased library, or (3) an optimization library. A biased library is designed using existing biological or chemical structural data when a target is known, while a discovery library is designed to identify a lead structure using chemical diversity when structural data is unknown. An optimization library is designed to improve the activity or solubility of an initial lead compound. Based on the structural diversity, the molecular complexity, and the purpose of the library synthesis, the size of the library can vary from less than a hundred compounds up to tens of thousands. Although the synthesis of millions of compounds has been demonstrated [5], a recent trend toward milligram-scale synthesis generally limits the size of most libraries to 10 000 compounds or fewer.

This chapter will provide an overview of the design and high-throughput synthesis of libraries of structurally diverse small molecules relevant to the discovery of PPI inhibitors. Although the subject of PPI inhibition has been reviewed [6–15], this review library synthesis-based approaches to the discovery of PPI inhibitors [16]. This chapter will discuss the synthesis of potential PPI inhibitors in three broad sections:

1. *'Peptide-inspired' library synthesis.* This section will discuss the design and synthesis of nonpeptide small molecule structures based on protein secondary (α-helix and β-turn) motifs. Although there are examples of oligomeric motifs, such as peptide and β-peptide structures, that have been investigated to successfully inhibit PPIs, these examples are outside the scope of this review and will not be discussed here.
2. *'Natural product-inspired' library synthesis.* Building on the total synthesis efforts of natural product structures that exhibit biological activity relevant to PPI inhibition, this section will discuss the synthetic efforts to produce libraries of natural product substructures as candidates for the discovery of PPI inhibitors.
3. *'Chemistry-inspired' library synthesis.* This section will examine strategies for achieving molecular complexity in the synthesis of unbiased libraries and highlight recent examples where this approach has been successful in the discovery of PPI inhibitors. Although the synthetic inspiration may draw overlapping inspiration from peptide and natural product structures, the diversity of resulting scaffolds often provide new motifs that contribute to an understanding of the molecular recognition between protein–protein interaction pairs

7.2 Survey of High-throughput Organic Synthesis

The invention of split-pool synthesis in 1986 by Richard Houghten [17] was the beginning of a rich period for the development of new techniques for synthesizing organic molecules *en masse* [18]. Solid-phase synthesis of nonoligomeric small molecules was first demonstrated in 1991 [19], and this technique has subsequently evolved into several variants for different applications. The primary benefit of solid-phase synthesis is the ability to 'pool' and 'split' batches of synthetic intermediates, which results in an exponential increase in the number of potential products relative to the number of reactions (Figure 7.1a) [20]. The use of solid-phase synthesis techniques to achieve split-pool amplification allows the rapid synthesis of thousands, and in some cases millions, of compounds for screening [5, 21]. Although early solid-phase synthesis efforts using one-bead/one-compound techniques were successful at generating large numbers of compounds [22, 23], subsequent synthesis efforts have focused on overcoming some of the potential disadvantages that can arise in library synthesis: (1) the small amounts of product generated from a single bead, (2) tracking the identity of compounds during synthesis, and (3) variable purity of the final products. To address these issues, advances in automation technology [24], phase separation, and synthetic design have focused on preparing larger quantities of each compound and ensuring the purity of the final product. For example, technology that couples robotics with microfluidic reactors, or microwave chemical reactors, is steadily gaining ground as useful tools for performing multiple parallel and sequential reactions.

In order to produce larger quantities of each compound in a library synthesis, larger resin formats are now available, and methods have been developed to handle batches of beads. Compared to the microgram quantities obtained from traditional peptide synthesis resin, polystyrene macrobeads can provide up to 0.1 mg and polystyrene-coated lanterns can

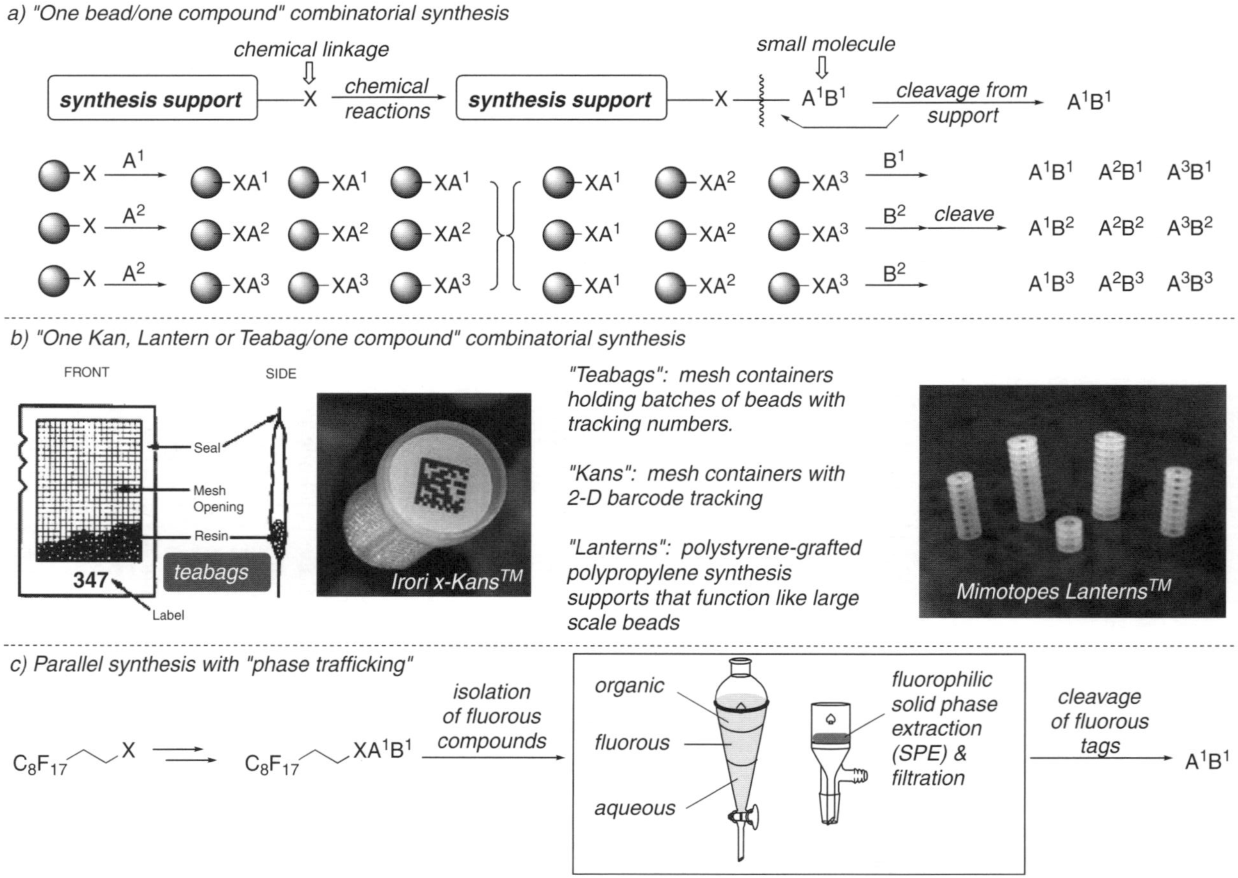

Figure 7.1 *Overview of HTOS techniques: (a) split-pool synthesis using solid support, (b) large scale split-pool solid phase synthesis using tea bags, Irori Kans (Nexus Biosystems), and polystyrene-grafted polypropylene lanterns (Mimotopes), (c) phase trafficking using fluorous tags (Fluorous technologies. Depiction of 'tea bag' reprinted with permission from [17]. Copyright (1985) National Academy of Sciences, U.S.A.)*

provide up to 10 mg of compound [25, 26]. The most popular method for executing larger scale library synthesis involves permeable membrane containers called 'Kans,' which allow batches of any beads to be pooled and split in reactions that yield milligram quantities of final products (Figure 7.1b). Kans are particularly attractive as a platform for split-pool synthesis because reactions can be tracked with either radiofrequency (RF) encoding [27, 28] or 2-D bar-coding [21]. Alternatively, the original 'tea-bag' batching technique reported by Houghten still enjoys a certain amount of popularity due to the ease of operation, which requires little specialized equipment [29]. The increase in synthetic scale offered by larger resins and permeable membrane containers affords sufficient quantity of compound so that the final products can be purified using preparative LCMS. As a result, solid-phase synthesis now offers the ability to prepare pure compounds on milligram-scale while still using the amplification offered by pooling and splitting.

Fluorous-tagged synthesis has emerged as an alternative to solid-phase synthesis that exploits the benefits of phase separation while allowing multistep syntheses to be conveniently conducted and monitored in solution (Figure 7.1c) [30–33]. The fluorous tag can be attached chemically to a broad range of functional groups in the same way that substrates are traditionally linked to solid phase. Because the fluorous substrates retain the properties of individual molecules rather than polymers, standard analysis and chromatographic purification techniques can still be employed. After a reaction is conducted and monitored in an organic solvent, the tagged compound can be temporarily captured onto a 'fluorophilic' resin, separated from the reaction byproducts, and then released for the next reaction. Because fluorous-tagged synthesis is conducted in solution, there is no difficulty to transfer a reaction from solution to the solid phase synthesis platform, which is one of the principle challenges for solid phase synthesis. Using phase-tag assisted purification/phase separation at each step of the synthetic sequence, reactions can still be driven to completion with excess reagents that are simply washed away from the tagged-substrate, similar to solid-phase synthesis. As with solid phase techniques, traceless cleavage of the fluorous tag is also possible. Although fluorous-mixture synthesis can capture some of the advantages of separating mixtures [34], split-pool synthesis is not possible with fluorous-tag library synthesis.

Recent advances in technology take advantage of liquid automation and microfluidics to accelerate the execution of multistep syntheses for both solution and solid-phase chemistry. Advances in automation have largely emanated from industrial labs focused on drug discovery, but many of these techniques have now been adopted by academic research labs for various synthetic applications. Reaction blocks and semi-automated synthesis stations allow chemists to conduct batches of reactions, for which the reagents can be added by a liquid-handling robot [35–38]. In addition, synthesis workstations featuring liquid handling, solid reagent addition, aqueous or solid-phase extraction (SPE), and purification are now available as single-user units [39–41]. These and other new automation technologies allow the execution of many chemical reactions with little or no modification of their original conditions. Furthermore, the products are easily transferred into plate formats that can be used for analysis of compound purity, high-throughput chromatographic purification, or directly in preliminary screening experiments. Very recently, the use of microreactors has been explored as a means to control and optimize reaction conditions on very small scale, allowing the rapid execution of sequential reactions using hydrodynamic pumping and capillary flow [42]. The parallel development of technology for flow-through reactors has also emerged in which a catalyst is deposited on a column similar to an HPLC column, and then solution phase and gaseous reagents (e.g. hydrogen) can be passed through the column [43–45]. These reactors open the door for reactions that were once not practical for high-throughput synthesis, while at the same time combining reaction execution with purification. As microfluidic reactors become increasingly available to academic researchers, their impact on high-throughput organic synthesis is likely to be significant [46].

The use of microwave chemical reactors has emerged as a standard technique in both industry and academe to execute chemical reactions rapidly and accelerate multistep synthesis [47, 48]. Microwave heating offers the ability to achieve higher temperatures and pressures that were previously only accessible using specially designed reaction vessels. Early studies, in which microwave ovens were often used, focused on an alleged 'nonthermal effect' [49, 50], although many observations are consistent with simple thermal acceleration [51]. In a microwave reactor, many reactions that typically require extensive heating in refluxing solvent for hours or days can be optimized to reduce reaction times to only minutes. To further increase reaction throughput, many microwave reactors now use simple robotics with liquid-handling capabilities to queue and conduct a series of reactions at programmed reactions temperatures. The combination of liquid handling capabilities, programmed automation, and microwave acceleration allows a single user to execute a large number of reactions sequentially to increase throughput for the synthesis of a small library of compounds resulting from related chemistry.

As outlined above, the choice of solution- or solid-phase chemistry and the implementation of technology are important factors for the effective design and synthesis of libraries of compounds to discover PPI inhibitors. Recent advances in automation technology and library synthesis now make it possible for many organic chemists to participate in high-throughput synthesis by adopting technology that suits their synthetic pathway development and/or library synthesis. In addition to considering the structural features of library compounds, the efficient synthesis of a library of molecules remains important to facilitate screening and follow-up chemistry. In each library described below, the choice of technology is a crucial component of the design process. Since there is no universal solution, each synthesis is designed and executed on a case-by-case basis utilizing variations of the options described above.

7.3 Synthesis of 'Peptide-Inspired' Compounds and Libraries

The analysis of peptide motifs that mimic protein secondary structure (α-helix, β-turn and β-strands) provides an excellent starting point to discover small molecules that can modulate protein–protein interactions [52–54]. This section will specifically focus on nonpeptide proteomimetics, where cyclic structures and small molecule proteomimetics often demonstrate improved activity due to increased bioavailability, reduced conformational entropy for binding, and resistance to proteolytic degradation. The synthesis of peptidomimetic compounds has been successful in the discovery of substrates and inhibitors of proteases, receptors (e.g. GPCRs), and many other drug targets. In the search for PPI inhibitors, the hypothesis is that by mimicking one of the two interacting protein surfaces, the interaction between proteins can be disrupted. Since several recent reviews have discussed the use of peptide [14], peptide-turn [52, 55], β-peptide [56], and other repeating oligomeric motifs to

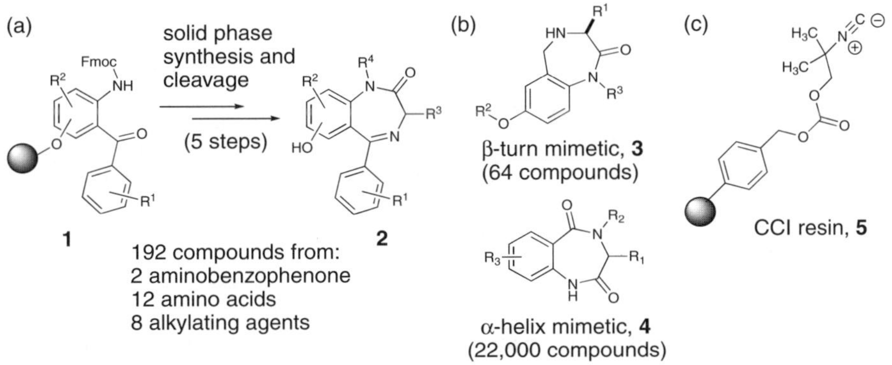

SCHEME 7.1 *(a) First benzodiazepine library synthesis, (b) benzodiazepine structure for the synthesis of β-turn and α-helix mimetics, and (c) a resin-bound carbonate convertible isonitrile*

inhibit PPIs, these topics will not be discussed here. This section will focus on the design and synthesis of nonpeptide small molecule structures amenable to library synthesis, whose mode of binding is believed to mimic the shape and chemical interactions of a peptide structural motif in order to disrupt the interactions of a protein–protein complex.

Benzodiazepines (**2**) were among the first classes of small molecules to be synthesized as libraries on solid supports and several examples of benzodiazepine libraries have generated small molecules that are effective at inhibiting protein–protein interactions (Scheme 7.1a) [19]. Efficient syntheses of benzodiazepines have made these structures widely available and heavily screened such that extensive structural and biological data is available. Benzodiazepines have been described as both α-helix and β-turn mimetic structures, depending on the substitution pattern and orientation of the functional groups on the scaffold (Scheme 7.1b). For example, Kim and coworkers used computational guidance to identify tetrahydro-1,4-benzodiazepine-2-one scaffold **3** with three points of diversity as a β-turn peptidomimetic for the synthesis of 64 derivatives on solid phase [57]. Due to the high level of investigation into library synthesis with benzodiazepines, these synthesis efforts also provide an excellent opportunity to develop new synthetic strategies, as well as solid-phase modification and linker chemistries. For example, Bräse and coworkers have demonstrated the use of triazene linkers, which can be exploited for post-cleavage aza-Wittig reactions and is a general strategy of post-cleavage modification reactions used to increase diversity [58]. The development of a new resin-bound carbonate convertible isonitrile (CCI resin, **4**) for the Ugi 4CR has also been an important advance for the synthesis of 2,5-diketopiperazines and 1,4-benzodiazepine-2,5-diones [59].

Using computational guidance from Directed Diversity®, Lu and coworkers designed a library of 1,4-benzodiazepine-2,5-diones (BDPs) as α-helix mimetic structures to screen for inhibitors of the p53-HDM2 interactions [60]. Using the highly efficient Ugi four-component condensation reaction, a library of 22 000 diverse 1,4-benzodiazepine-2,5-dione structures (**4**) was synthesized in two steps using 1-isocyanocyclohexene as a convertible isocyanide to facilitate a post-condensation lactamization reaction of the Ugi product **7** (Scheme 7.2) [61]. Lactamization proceeds by an acid-catalysed cleavage of the cyclohexenamide and simultaneous Boc deprotection to expose the anthranilic nitrogen. Initial screening of the library was performed using high-throughput Thermoflor screening

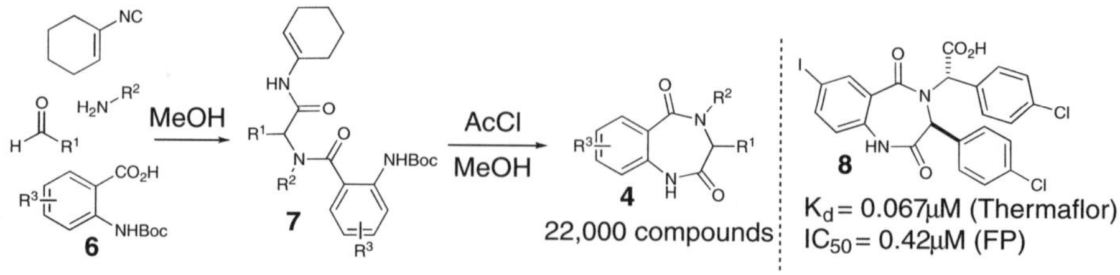

SCHEME 7.2 *Synthesis of a benzodiazepine library using an Ugi-lactamization strategy to discover an inhibitor of the p53-HDM2 interaction*

technology to measure the affinity of each compound toward the HDM2 protein target. The hits identified in the Thermoflor assay were confirmed and optimized using a secondary fluorescence polarization peptide displacement assay, providing an inhibitor (**8**) with an IC_{50} of 0.42 μM. A binding model was predicted on the basis of conformational analysis, and subsequently confirmed by obtaining a BDP-HDM2 co-crystal structure [62]. Subsequent studies have also demonstrated enantioselective syntheses and cell-based activity for benzodiazepine **8** [63, 64].

Elegant approaches have been described for the rational design of α-helix mimics based on rigid nonpeptide scaffolds, including terphenyl (**9**) [65, 66], biphenyl (**10**) [67], indane (**11**) [68–70], terpyridine [71], polycyclic ether (**13**) [72], terephthalamide [73], trispyridylamide (**12**) [74], and trisbenzamide (**14**) structures (Figure 7.2) [75, 76]. Hamilton and coworkers have used modeling to demonstrate that the rigid conformation of the staggered terphenyl scaffold adopts an angular orientation of functional groups, similar to the orientation of amino acid functional groups in an α-helix, and thus mimics the key elements of the protein surface that are involved in the recognition function of an α-helix. In several cases, these low molecular weight nonpeptido synthetic scaffolds have been successfully

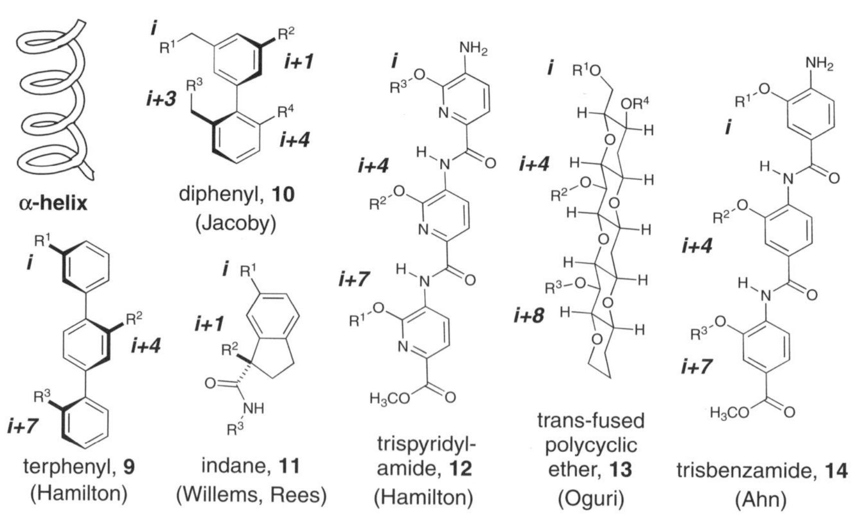

Figure 7.2 *A summary of synthetic α-helix mimics based on rigid scaffolds, where the relative location of amino acid sequence in the α-helix is indicated*

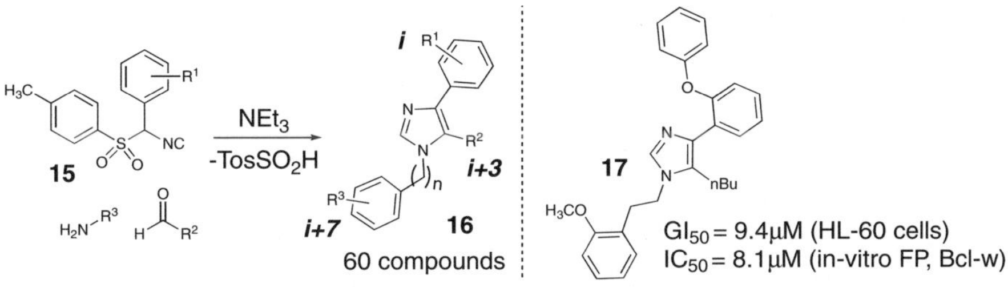

SCHEME 7.3 *Van Leusen MCR synthesis of trisubstituted imidazoles **16** leading to the discovery of a specific inhibitor of the Bcl-w/Bak interaction (**17**)*

identified to exhibit the full inhibitory activity of the protein domain. Early designs of terphenyl scaffolds examined in this approach demonstrated limited utility based on their low aqueous solubility; however, further design of terephthalamide and trispyridylamide compounds have been pursued to identify compounds that exhibit properties more suitable for cellular conditions.

Building on the success of these rigid nonpeptide scaffolds to mimic the key interactions of a protein surface, library strategies have also been designed for the efficient synthesis of α-helix mimetics that inhibit PPIs. Prompted by Hamilton's terphenyl α-helix mimetic concept, Dömling and coworkers designed a modular library strategy for the synthesis of an alternate diphenyl imidazole scaffold **16** that can be efficiently prepared from a multicomponent reaction (MCR) in one step (Scheme 7.3) [77].The trisubstituted imidazole scaffold **16** was selected based on molecular modeling and synthetic accessibility using a van Leusen MCR synthesis strategy with tosylmethyl isocyanides (**15**), for which numerous commercially-available substrates are available. From a parallel library synthesis, 60 trisubstituted imidazoles were synthesized and screened to identify compounds that disrupt the Bcl-w/Bak interaction. Compounds were first screened in a cell proliferation assay with HL-60 cells and the induction of apoptosis was assessed in a DNA fragmentation assay. To identify the mode of action for the induction of apoptosis, preliminary in vitro screening was performed with a fluorescence polarization binding assay for the disruption of the Bcl-w/Bak interaction. Three compounds such as imidazole **17** were identified as inhibitors that disrupt the Bcl-w/Bak interaction. The authors note that these inhibitors are specific to Bcl-w and all three compounds were inactive against other Bcl family members, such as Bcl-2 and Bcl-X_L.

Guy and coworkers utilized a multistep computational design process in an effort to establish a reproducible method for the design and synthesis of α-helix mimetic structures [78]. The p53-MDM2 interaction motif was selected as a model system and a series of structure-based computational analysis methods were used to identify structures suitable for library design (Scheme 7.4a). The top 40 structures resulting from computational searching and docking were assessed for synthetic accessibility, from which four structures (**18–21**) were considered for library synthesis. Structure **21** was selected for library synthesis based on the ability to prepare compounds in a modular fashion with simple chemistry. Based on this analysis, a library of 13 dimeric (**24**) and 117 trimeric (**21**) structures was synthesized to investigate as α-helix mimetic inhibitors of the p53-MDM2 interaction. The library strategy was implemented using parallel synthesis in 48-well Flex-Chem reaction blocks and using a collection of building blocks that was prepared by a common Suzuki coupling procedure (Scheme 7.4b). Instead of preparing the complete matrix of building block combinations,

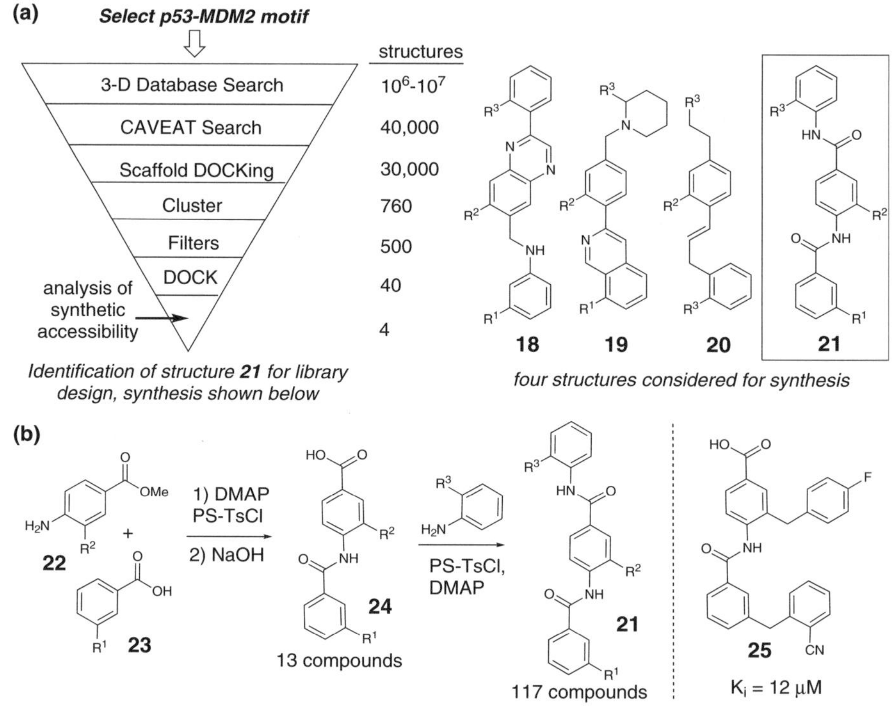

SCHEME 7.4 *(a) Computational guidance to design an inhibitor for the p53-MDM2 interaction, (b) synthesis of dimeric and trimeric structures to identify a 12 μM inhibitor (**25**) of p53-MDM2*

13 random combinations of the dimeric scaffold **24** were prepared and then coupled to nine R3 elements to prepare trimeric scaffold **21**. Members of both amide trimer (**21**) and amide dimer (**24**) scaffolds were tested as inhibitors of the p53-MDM3 interaction using a fluorescence polarization peptide displacement assay. Three of the 13 dimeric structures (**24**), such as inhibitor **25**, demonstrated activity with K_d's from 12–27 μM, while none of the trimeric structures (**21**) showed activity ($K_d > 30$ μM). The authors propose that the carboxylic acid present in the dimeric structures plays a key role to disturb the salt bridge between Lys51 and Glu25 of MDM2. Support for this hypothesis is provided based on NMR studies of the Glu25 chemical shift and the loss of activity observed for the corresponding methyl ester analog of compound **25**. Both the competitive fluorescence polarization experiments and preliminary NMR structural studies provide evidence that inhibitor **25** binds to MDM2 in the same pocket in which the p53 helix binds. Synthesis of a subsequent optimization library yielded only a large number of weakly binding compounds, all with K_d greater than 30 μM. As observed for previous a-helix mimetic compounds, the limited aqueous solubility of these compounds may not be suitable for cellular conditions. The authors comment that this method of computational design and synthesis should be applicable for other protein–protein interaction substructures with helical motifs. Furthermore, scaffolds **21** and **24** may also have activity as inhibitors for other protein–protein interactions involving an α-helix, such as Bak, NF-κB, and VP16.

While there are somewhat limited examples of library syntheses for α-helix mimetic structures, there has been significant development for the design and synthesis of diverse

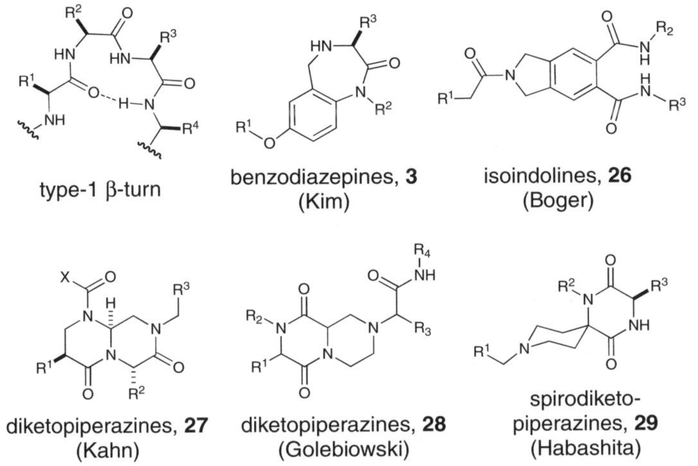

Figure 7.3 *Synthetic β-turn mimetic scaffolds used for library synthesis compared to the type-1 β-turn motif*

libraries of β-turns mimetics [79]. β-turns represent an important structural element of proteins and have been shown to play a key role in many of the molecular recognition events for protein–protein interactions. There are several types of diverse β-turn mimetic scaffolds that have been employed in library strategies to discover inhibitors of protein–protein interactions, including benzodiazepine, diketopiperazine, and isoindoline structures (Figure 7.3). These scaffolds often feature bicyclic or modified amide structures that mimic the β-turn motif while also exhibiting improved bioavailability, reduced conformational entropy for binding, and enhanced resistance to proteolytic degradation.

Boger and coworkers have prepared an Arg-Gly-Asp (RGD) β-turn mimetic library based on the bicyclic isoindoline-5,6-dicarboxylic acid template **26** (Scheme 7.5a) [9, 80]. This template has three positions that can be functionalized with nucleophiles and acylating groups using standard chemistry. The efficient design of the synthesis employed solution phase combinatorial library relying on acid-base extractions for removal of excess reagents in each step (Scheme 7.5c) [81]. A total of 240 compounds were prepared with 120 acid derivatives and 120 amide derivatives, and compound **30** was identified as a c-Myc-Max inhibitor in a high-throughput FRET screen. The binding activity was further confirmed with a gel shift assay (EMSA), and also with an in vitro assay ($IC_{50} = 20\ \mu M$) using Myc-induced oncogenic transformation of chicken embryo fibroblasts in cell culture [82].

Bicyclic diketopiperazines (DKPs) represent a class of molecules that has become a recent focus for the investigation of rigid nonpeptide cyclic β-turn mimetic scaffolds. Kahn and coworkers first reported the synthesis of bicyclic DKPs **27** on solid phase and performed extensive conformational analysis of these templates based on molecular mechanics, solution-phase NMR spectroscopy, and X-ray crystallography [83]. With four diversity points for library synthesis, a library of 5000 diketopiperazine structures was prepared using an efficient solid phase synthesis that exploits an acid-catalysed tandem cleavage and acyliminium cyclization of amide **33** to afford the bicyclic structure **27** (as a single diastereomer) in the final step (Scheme 7.6). Evaluation of these compounds led to the

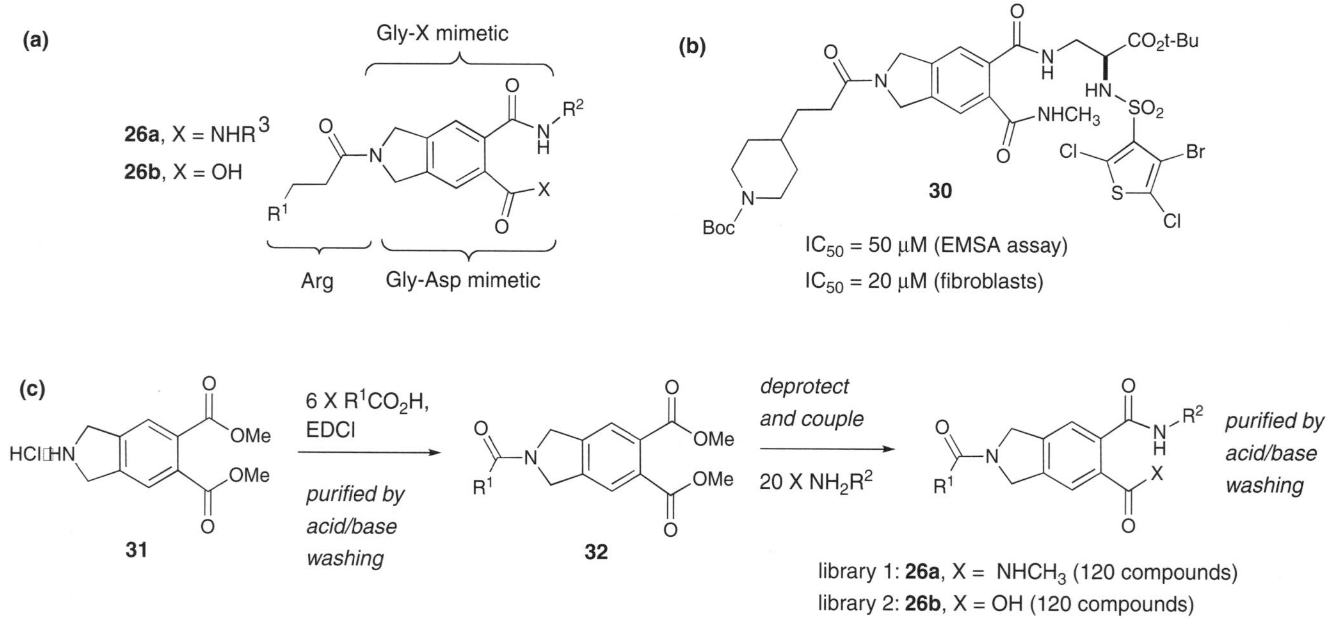

SCHEME 7.5 *(a) Isoindoline structure as an RGD β-turn mimic, (b) an isoindoline inhibitor of the c-Myc-Max interaction discovered from screening library in part c, and (c) synthesis of library of isoindoline structures*

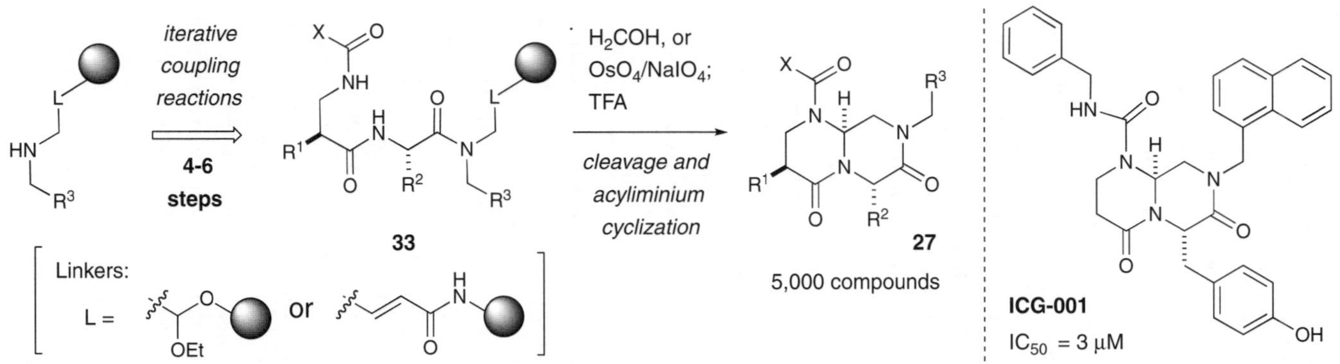

SCHEME 7.6 *Library synthesis of diketopiperazines* ***27*** *leading to the discovery of an inhibitor (ICG-001) that blocks the CBP/β-catenin binding interaction*

discovery of a small molecule (ICG-001) that selectively inhibits β-catenin/TCF-mediated transcription upon blocking the interactions between β-catenin and cyclic AMP response element-binding protein (CBP) [84, 85]. The authors have demonstrated that ICG-001 selectively blocks the CBP/β-catenin interaction, without interfering with the interaction of β-catenin with the CBP-related transcriptional coactivator p300, through a minimal region of interaction at the NH_2 terminus of CBP (amino acids 1-111). Further data for ICG-001 suggests that this molecule has significant therapeutic potential for the treatment of cancer due to its selective induction of apoptosis in colon carcinoma cells, but not in normal colonic epithelial cells.

One of the unique features of diketopiperazine structures is the diversity of bicyclic fused and spiro structures that are accessible by different synthetic sequences and cyclization strategies that can mimic β-turn motifs. Golebiowski has reported the solid phase synthesis of bicyclic diketopiperazine structures **28** with up to five points of diversity using an Ugi four component reaction (4CR, Scheme 7.7a) [86, 87]. Similar to the approach employed by Kahn, a linear amide structure (**34**) can be rapidly prepared for use in a subsequent cyclization reaction. Although the diastereoselectivity of the Ugi reaction is not high, this synthetic method allows control of the three remaining chiral centers. The synthetic scope for this strategy was examined for 23 final products with a diversity combination of seven α-bromo acids, six Boc-amino acids, four aldehydes, and four isocyanides. Although no biological data has yet to be reported, it is likely that these structures will provide rich activity to modulate protein–protein interactions. In a recent advance to facilitate synthesis and purification, Blackwell and Lin have reported the rapid synthesis of diketopiperazine macroarrays **38** using an Ugi four-component reactions on cellulose planar solid supports (Scheme 7.7b) [88].

Porco and Panek [89] have investigated an alternate approach to diketopiperazines using a convergent synthesis with diverse, enantioenriched pipecolic acid monomers. Pipecolic acid monomers such as cis-**41** were obtained by a scandium-triflate catalysed aza-annulation reaction from crotylsilane **39**, followed by liberation of the pipecolic acid in a resin 'catch and release' approach to streamline the synthesis and purification of these building blocks. Cross-coupling reactions were employed for further diversification of the pipecolic acid monomers to give coupled products such as **42**. Once prepared, a diverse series of derived pipecolic acid monomers were homo- or heterodimerized using a one-pot cyclodimerization strategy (in parallel fashion followed by essential LC-MS purification) to access 30 complex diketopiperazine structures such as **43** and **45** (Scheme 7.8). Preliminary studies indicate that the stereochemical variation of the diketopiperazine structures has a profound effect on the three-dimensional conformation of the structure, which contributes to the structural diversity of this collection. A matrix screening experiment was performed to profile the activity of the diketopiperazine compounds against a diverse panel of various GPCR targets. Several receptor-active compounds were identified in this preliminary assay for which secondary screens were performed to calculate the K_i value. For example, diketopiperazine **43** was identified as an 0.77 μM inhibitor of the 5-HT7 receptor, while diketopiperazine **45** was identified as an 0.33 μM inhibitor of the α-2B receptor. Of particular note, it was observed as a general trend that the stereochemistry of the DKP core appears to influence the selectivity for particular GPCR receptors.

Habashita and coworkers have designed a combinatorial library based on spirodiketopiperazine **29** in a specific effort to identify sub-type selective inhibitors/antagonists for

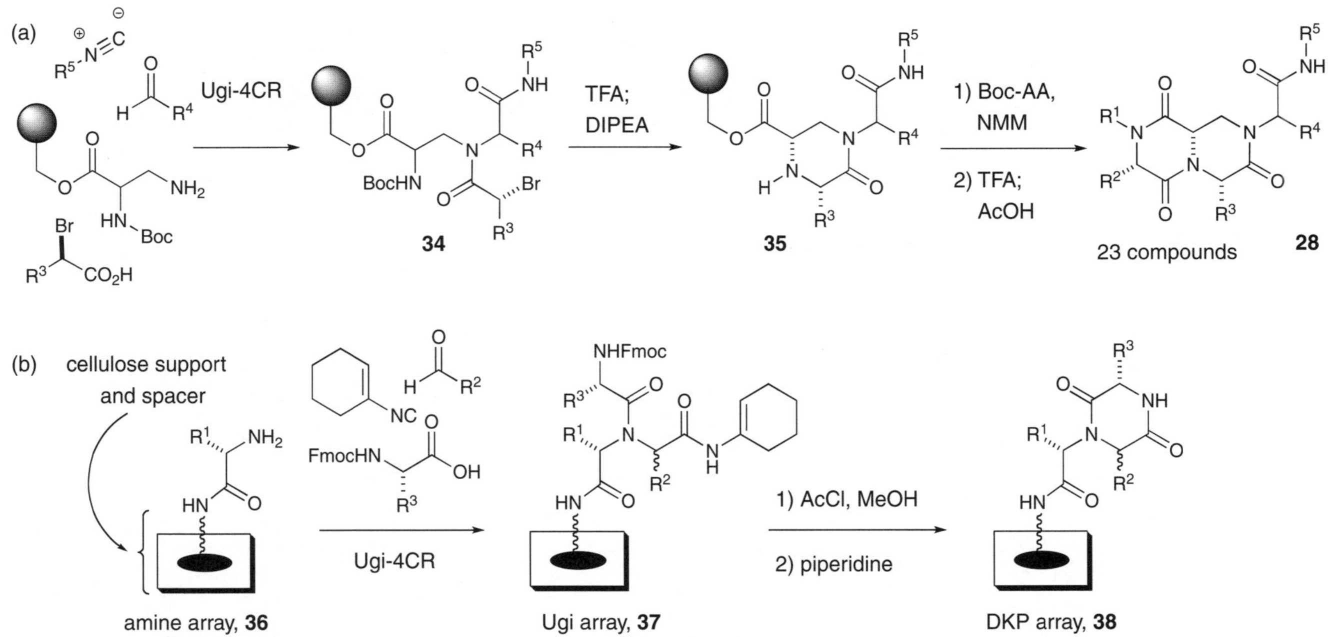

SCHEME 7.7 *(a) Library synthesis of diketopiperazine structures derived from an Ugi 4CR, and (b) synthesis of an diketopiperazine array using the Ugi 4CR on cellulose solid support*

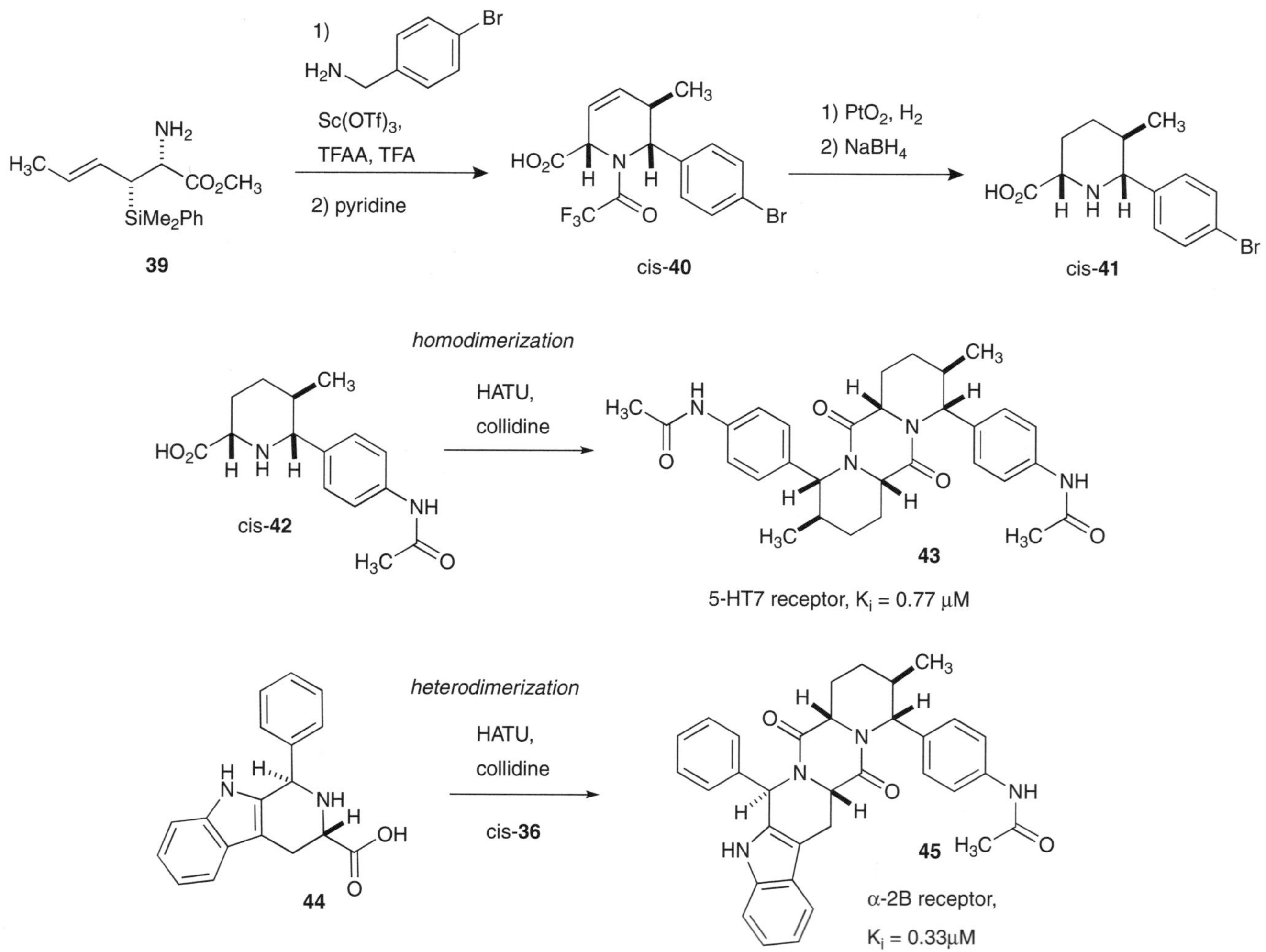

SCHEME 7.8 *A dimerization strategy with pipecolic acid monomers to obtain stereochemically-diverse diketopiperazine structures*

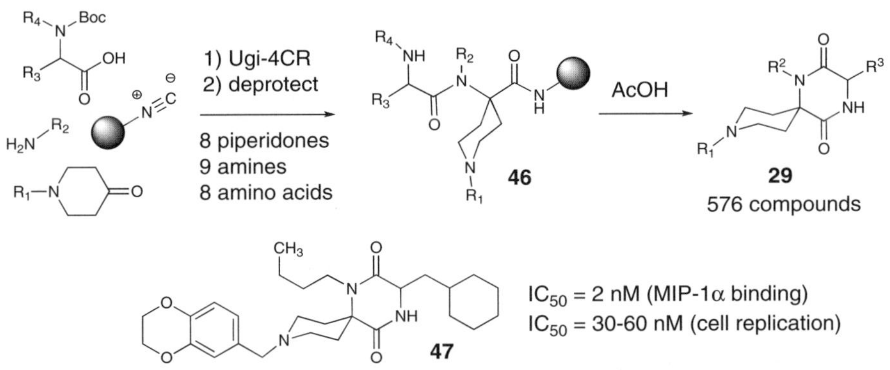

SCHEME 7.9 *Synthesis of a spirodiketopiperazine library based on the Ugi 4CR*

GPCRs (Scheme 7.9) [90, 91]. The synthesis and screening of a library of 576 spirodiketopiperazine identified several compounds as inhibitors of the CCR5-Mip-1α binding interaction. Molecular modeling demonstrates that spirodiketopiperazines mimic the orientation of the three side chains in a type-1 β-turn. An iterative library strategy was employed using an Ugi 4CR, first to prepare 576 compounds on solid phase for hit identification, and then to explore optimal structures by preparation of a focused library of 80 compounds using parallel solution phase chemistry. The strategy involves the use of a resin-bound isonitrile to facilitate high-throughput synthesis for the Ugi products **46**. Primary screening was performed with cellular assays where the chemokine has been over-expressed in CHO cells stimulated by the endogenous ligand. Subsequent secondary screening involved direct binding assays with MIP-1α for confirmation of activity and specificity. Spirodiketopiperazine **47** was identified to have nanomolar activity in both assays. Interference of binding of chemokines or HIV to the CC chemokine receptor 5 (CCR5), a member of the seven-transmembrane G-coupled family of receptors (GCPRs), is one of the most attractive drug targets to treat HIV infection, rheumatoid arthritis, and acute/chronic transplant rejections.

Perez-Paya, Messeguer, and coworkers have prepared a positional scanning library of acyclic N-alkylglycine trimers (peptoids) that led to the discovery of inhibitors of the Apaf-1 (apoptotic protease-activating factor) interaction with procaspase-9 (Scheme 7.10) [92, 93]. Building on the experience from their previous peptoid library synthesis [94], an optimized library of 5120 acyclic peptoid compounds **50** was prepared (consisting of 52 controlled mixtures) with diversity optimized for composition and concentration/yield of the products. The lateral side-chain substitution on the N-position allows an efficient synthesis strategy, a flexible display of functional groups, and contributes to enhanced proteolytic stability. Defined mixtures of compounds were screened for their ability to inhibit the apoptosome-mediated activation of procaspase-9, and four discrete compounds were identified as inhibitors, with peptoid **52** identified as the most potent inhibitor. Due to solubility problems resulting from the intrinsic hydrophobicity of these peptoids, more soluble analogs of peptoid **52** were examined. Iterative synthesis and structure optimization to improve membrane permeability led to the identification of several improved analogs, including cyclo-peptoid derivative **53**. The soluble analogs demonstrated efficacy in independent cell models and were shown to bind to both recombinant and human endogenous Apaf-1 in a

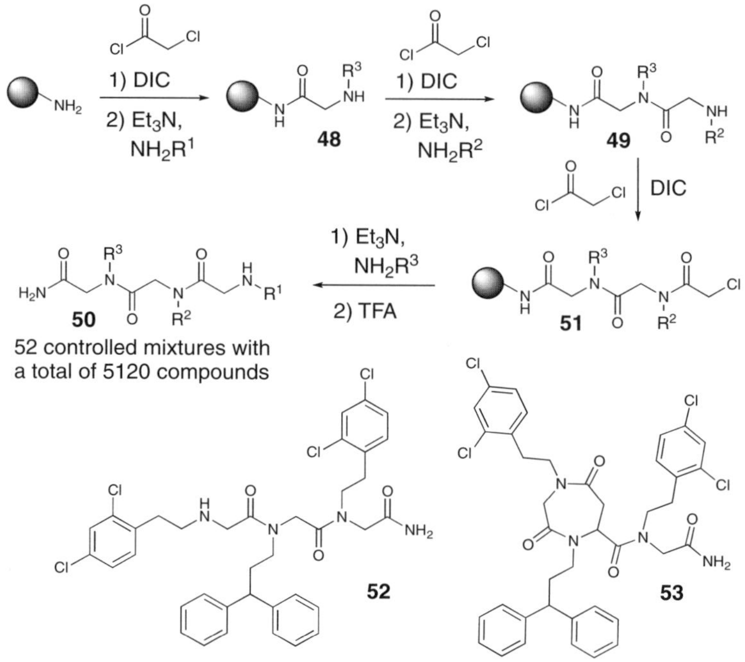

SCHEME 7.10 Synthesis of a peptoid library leading to the discovery of an inhibitor of the Apaf-1 interaction with procaspase 9

cytochrome *c*-noncompetitive mechanism, thus inhibiting the recruitment of procaspase-9 by the apoptosome.

7.4 Synthesis of 'Natural Product-Inspired' Compounds and Libraries

Natural products have traditionally provided the richest source of biologically active small molecules. Even with recent advances in high-throughput synthesis and screening, many current drugs have their structures derived from natural products [95]. The merits of natural products as starting points for biological activity may be attributed to many features, one of which is an evolutionary 'preselection' for properties that favor interactions with proteins [96]. To date, many natural products have been discovered that derive their activity from the inhibition of PPIs. The challenge is to understand the structural features responsible for PPI inhibition and incorporate these principles in to libraries of synthetic compounds. Since many synthetic efforts have been targeted toward the assembly of natural products, the synthesis of natural product-inspired libraries [97, 98] using stereoselective reactions is a logical starting point for the discovery of PPI inhibitors. Spirooxindoles, spiroketals, and bicyclo-guanidines are three classes of natural products that have been sources of PPI inhibitors as well as targets of library syntheses.

Using substructure comparison with natural product-derived scaffolds as well as molecular modeling, Wang and coworkers [99] designed a spirooxindole structure (**55**) as potent, specific, nonpeptide small-molecule inhibitors of the MDM2-p53 interaction.

The spirooxindole core structure (**54**) is found in several alkaloid natural products, including spirotryprostatin A, pteropodine, and alstonisine, all of which exhibit interesting biological activity [100]. From the crystal structure of MDM2-p53, the authors rationalized that the oxindole motif can mimic the interaction of the side chain of Trp23 with MDM2. The designed spirooxindole compound (**59**) was synthesized using an asymmetric Williams 1,3-dipolar cycloaddition reaction as a key-step (Scheme 7.11) [101–103]. Initial lead compound **60** exhibited modest ($K_d = 8.46\ \mu M$) disruption of the MDM2-p53 interaction, while optimized analog **61** was later found to bind with thousand-fold greater affinity ($K_d = 3\ nM$) [104].

Given the prevalence of naturally-occurring spirooxindoles, several high-throughput methods suitable for library synthesis have been developed. Schreiber [105] and coworkers utilized the Williams three-component strategy to construct a diverse library of natural product-like spirooxindoles scaffold using split-pool synthesis. Macrobead-supported aldehydes are treated with Williams' chiral auxiliary (**62**), isatin-derived dipolarophiles bearing an allyl ester (**63**), and a Lewis acid promoter to yield 16 spirooxindole cores (**64**, Scheme 7.12). These core structures were diversified to more than 3000 derivatives **67** using the following methods: (1) Sonagashira coupling with 8 alkynes results in concomitant allyl ester cleavage, (2) amidation between the carboxylic acid and 8 amines, (3) *N*-acylation of the resultant lactams with 3 electrophiles. Chemical encoding was used to track the identity of the compounds and HPLC analysis of a subset of the library indicated that the majority of the compounds were of >75% purity. Subsequent screening of this library using small molecule microarray technology has revealed several interesting new compounds that exhibit novel biological activity [106].

A second-generation library of spirooxindoles was constructed in a convergent approach by coupling products of two different sub-libraries (Scheme 7.13) [107]. A collection of 24 diketopiperazines (**69**) was prepared on solid phase using an asymmetric catalytic variant of the azomethine ylide dipolar cycloaddition [108]. The diketopiperazines were deprotected to reveal carboxylic acids that were pooled, split, and then coupled with the 16 spirooxindole alcohols **70** from scheme 12, resulting in a hybrid library of 384 diverse compounds **71** featuring two domains linked by an ester linkage. Shortly after this work was published, Porco disclosed a related strategy that he termed 'domain shuffling' in which libraries of compounds are combined in a matrix of variations to yield molecules featuring the structural elements of the two libraries from which the compounds were derived [109].

An impressive feat of synthesis automation was realized by Fokas [110] and coworkers who used the solution phase cycloaddition reaction of azomethine ylides and chalcones to construct a library of 25,600 spiropyrrolidines (Scheme 7.14). The intermediate ylide **77** was generated from isatin **72** and an α-amino acid, then captured by an electron-deficient chalcone to form spirooxindole core **79** in a regio- and stereocontrolled fashion [111–113]. Presumably, the stereochemistry was attributed to the *exo* addition of the chalcone to the W-conformer of *anti*-ylide **77**. A similar strategy was later reported by Bergman [114] in which various dipolarophiles were used in place of the chalcone to obtain diverse polycyclic core structures.

Recently, Franz and Schreiber [115] have employed functionalized crotylsilane **80** in a Lewis acid-mediated annulation strategy with a series of isatin reagents (**81**) for the solid-phase synthesis of structurally and stereochemically diverse spirooxindoles (Scheme 7.15). Although Lewis acids are not often used in the presence of the solid-phase silyl linkage, a

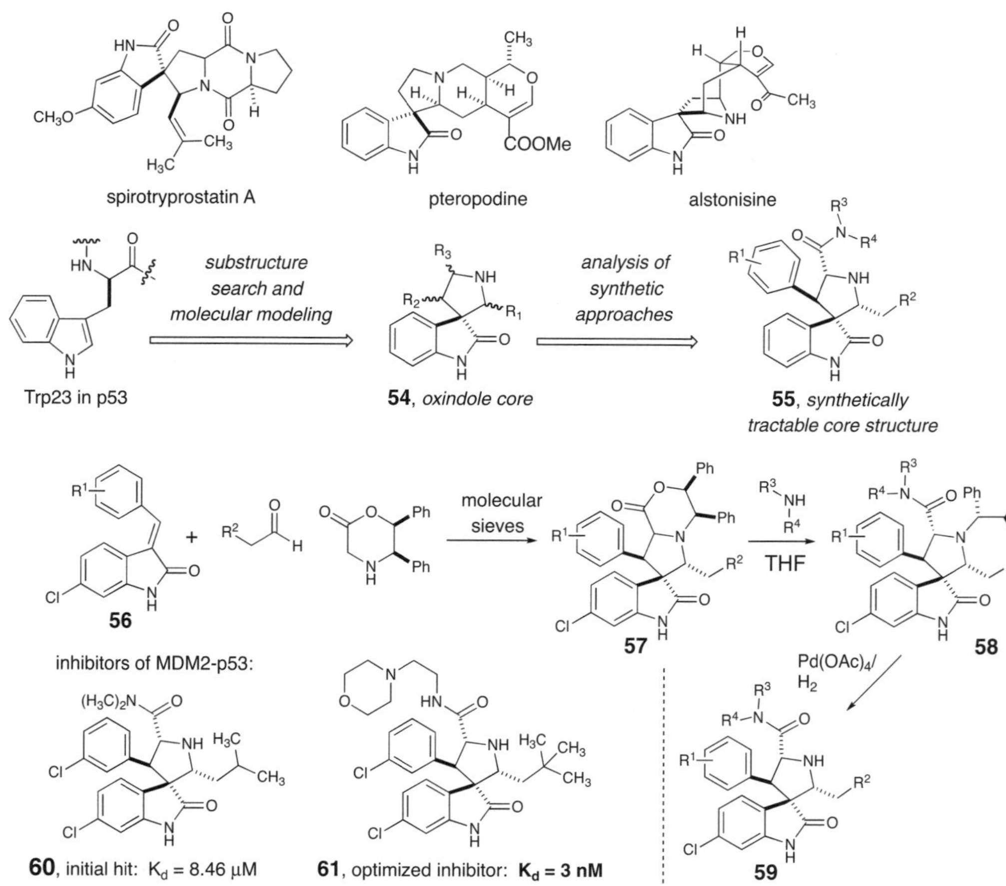

SCHEME 7.11 *(a) Spirooxindole natural products, (b) design of spirooxindole-derived MDM2-p53 inhibitors, (c) synthesis of designed MDM2 inhibitors based on the spirooxindole core*

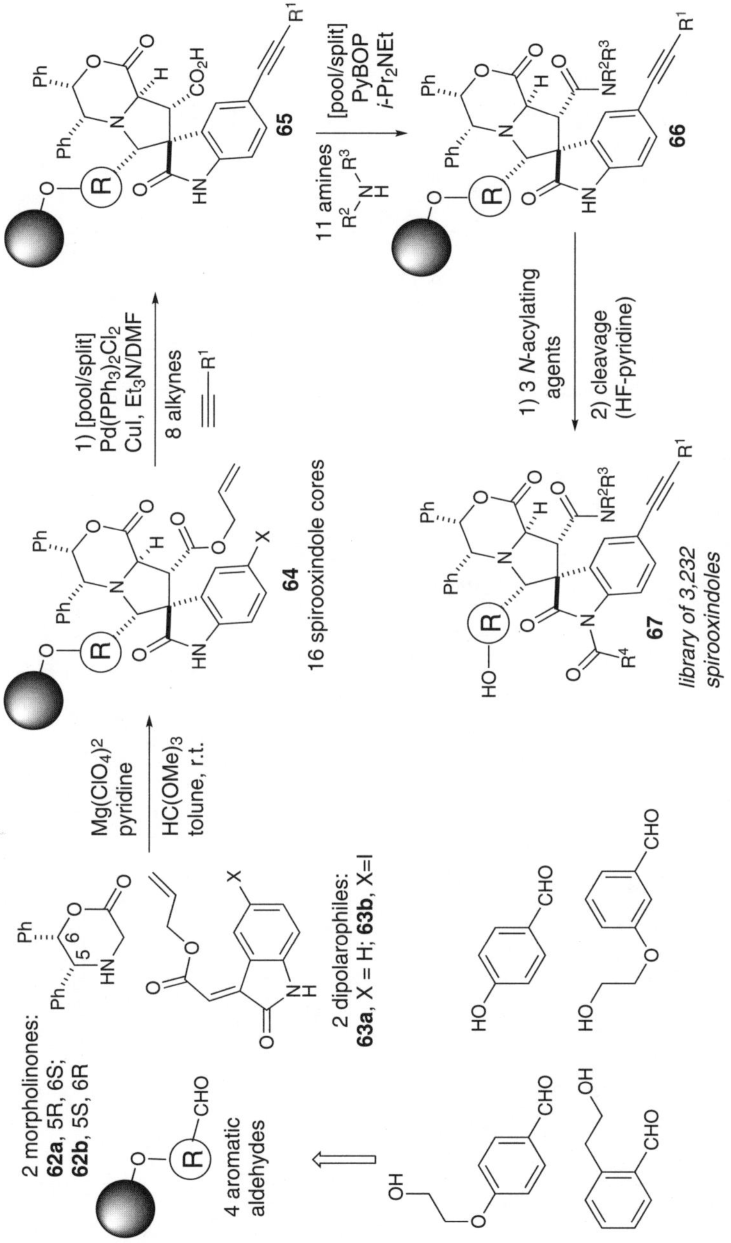

SCHEME 7.12 *Solid phase split-pool synthesis of diverse spirooxindoles*

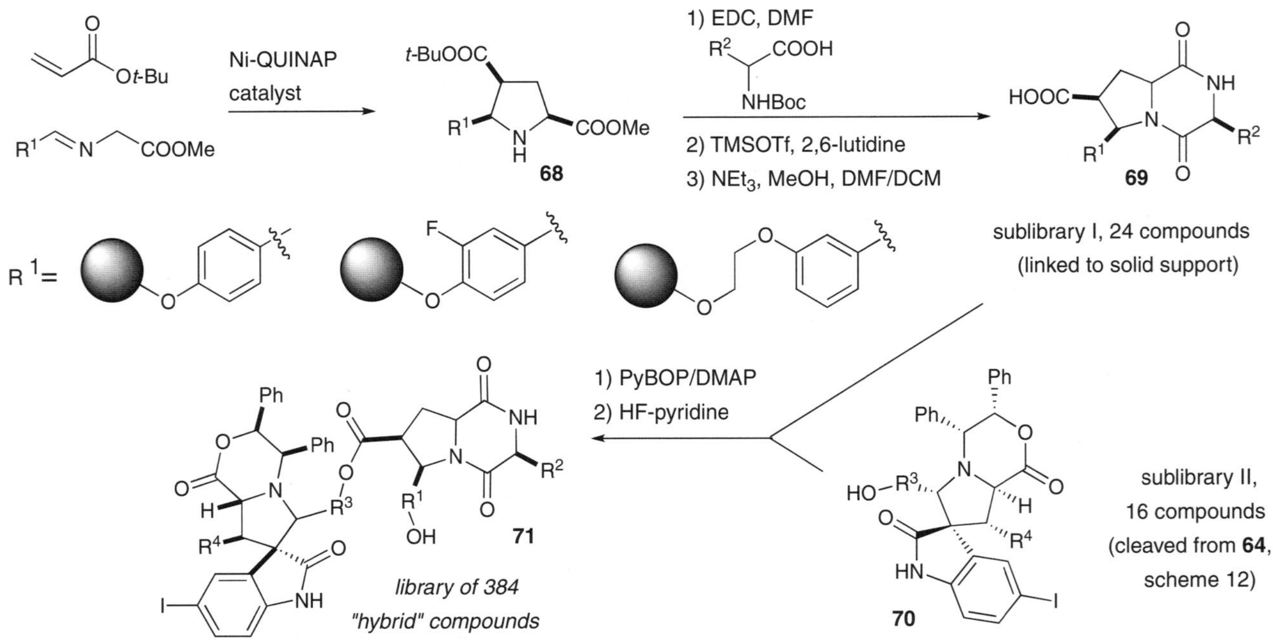

SCHEME 7.13 *Convergent synthesis of diverse 'hybrid' spirooxindole/diketopiperazines*

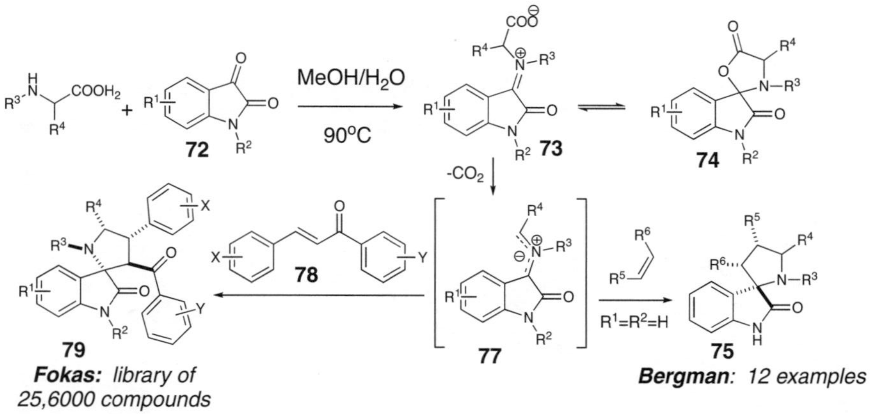

SCHEME 7.14 *Synthesis of spirooxindoles by a three-component reaction between isatins, α-amino acids, and dipolarophiles*

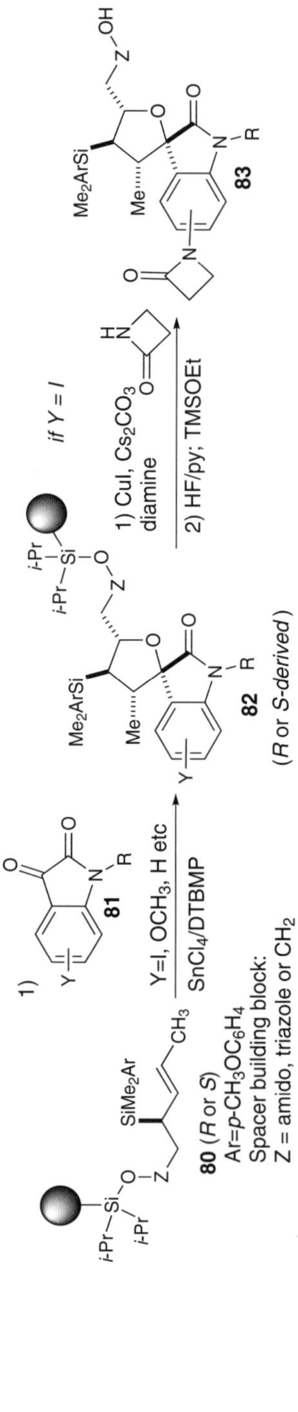

SCHEME 7.15 Split-pool solid phase synthesis of crotylsilane-derived spirooxindoles

protic acid scavenger (2,6-di-*tert*-butyl-4-methylpyridine, DTBMP) was sufficient to prevent Si-O cleavage. By employing isatin building blocks containing an aryl iodide, the resultant oxindoles (**82**) could be further diversified using a solid-phase copper-catalysed Buchwald amidation to give products with the general structure of **83**. The synthetic scope for this strategy was examined with a diversity combination of six enantiomerically-pure functionalized crotylsilanes (*R* or *S*), twelve isatins, and two amides. A preliminary collection of 90 compounds was evaluated in a series of 41 cellular profiling experiments to determine the signature of biological activity for each compound as dictated by the stereochemistry and substitution of the spirooxindole. By combining this approach with small molecule or protein microarray experiments, this approach may eventually facilitate the identification of compounds that affect common protein targets and/or pathways by comparison of their respective signatures.

A recent report from Scheidt and coworkers describes a novel and efficient synthesis of spirooxindoles **86** using a diastereoselective, Cu(I)-catalysed 3-component reaction (3CR) of an imine, diazoester and substituted olefin dipolarophile **85** (Scheme 7.16) [116]. Twelve spirooxindoles **86** with four contiguous stereogenic centers were synthesized from commercially available catalysts in a single operation with high diastereoselectivity resulting from the E-*exo* transition state. Given the demonstrated importance of spirooxindoles as PPI inhibitors and the recent advances in their synthesis using multicomponent and split-pool techniques, it is likely that more PPI inhibitors will emerge from this class of compounds.

The spiroketal motif is found in a variety of structurally complex and biologically important natural products and, as such, this class of natural products has been the subject of intense synthetic investigation. The altohyrtins (a.k.a. spongistatins, Figure 7.4) were synthesized by the Evans and Kishi groups, and subsequent biological investigations revealed that these cytotoxic compounds inhibit microtubule formation [117–119]. Bistramide A, a cell proliferation suppressor, has been shown to inhibit G-actin polymerization with a high binding affinity, K_d of 7 nM [120, 121]. Didemnaketals A and B inhibit HIV-1 protease with an IC_{50} of 2 μM and 10 μM, respectively, by disrupting the dimerization of HIV-1 protease monomer subunits [122]. The activity of spiroketals against a variety of protein targets, and specifically the didemnaketals' activity as PPI inhibitors, has singled out these unique structures for a variety of high-throughput synthesis approaches.

Porco and coworkers have devised an efficient synthesis of a spiroketal core structure (**90**) that forms the basis of a library synthesis [123]. Epoxide **87**, available in enantiomerically pure form using Jacobsen's kinetic resolution, is elaborated to methyl ketone **88** in three steps (Scheme 7.17). Aldehyde **89** is assembled in four steps, combined with **88**, and converted to pivotal spiroketal intermediate **90** in four more steps. A library of 90 spiroketal members was synthesized using a series of diversification steps. First, cyclic ketone **90** is diastereoselectively reduced and converted to carbamates **91** using three different amines. Next, the alkene is cleaved and oxidized to a carboxylic acid, which is then converted to three different amides represented by **92**. Finally, a second carbamate is installed using ten amines to give spiroketals **93**. The ease with which a high level of structural complexity can be assembled around the central spiroketal core makes this chemistry an ideal starting point for preparing potential PPI inhibitors.

The Ley group has employed an ynal-based approach to the synthesis of spiroketal scaffolds (Scheme 7.18) [124]. The ynal precursor **94** is easily prepared from benzyl glycidol, and subsequently converted to spiroketal core structure **95** in three steps.

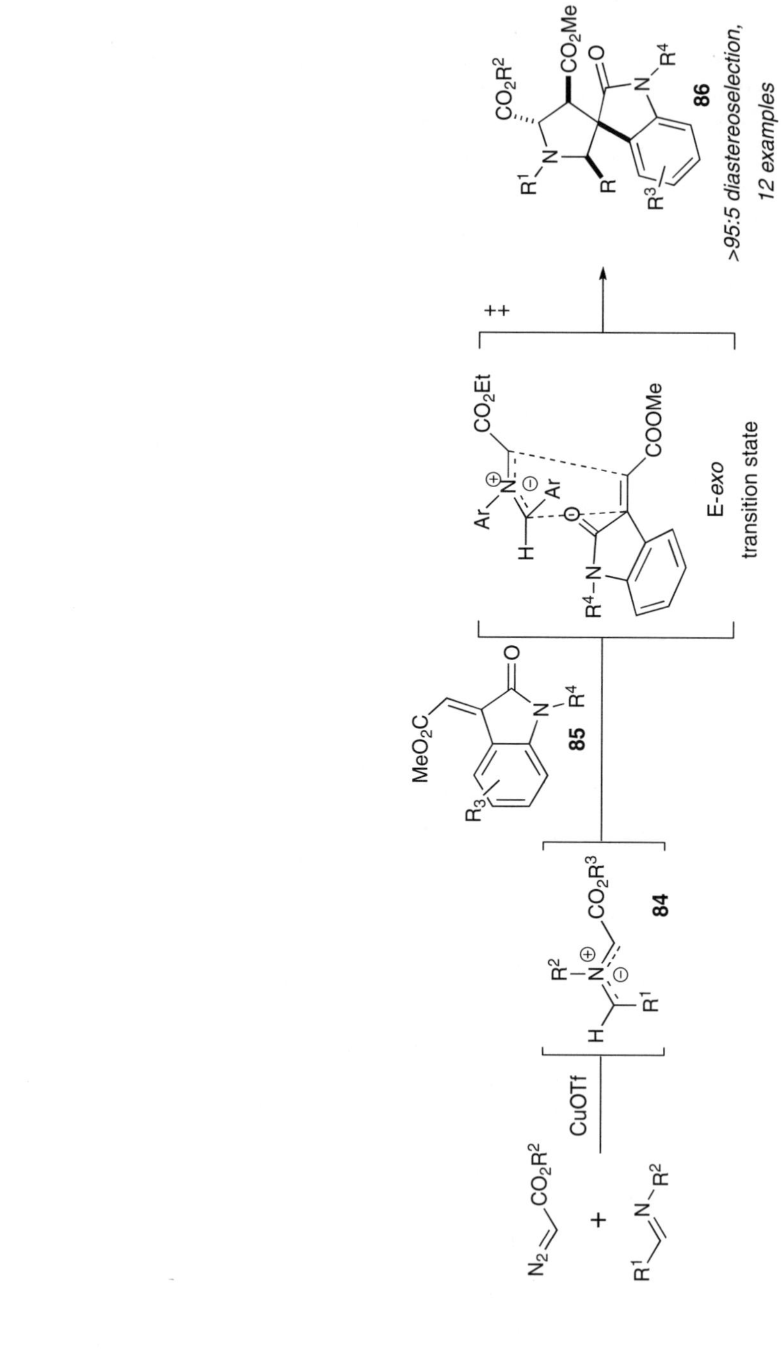

SCHEME 7.16 *CuOTf-catalysed three-component synthesis of spirooxindoles*

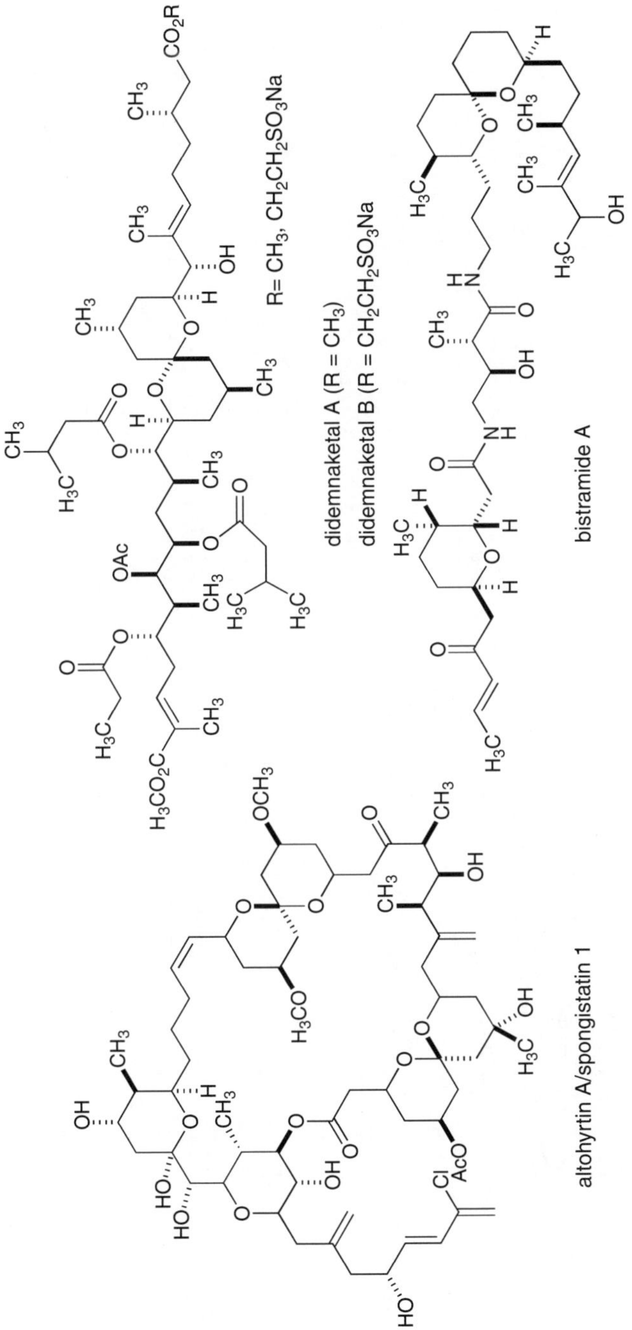

Figure 7.4 *Spiroketal natural products identified as PPI inhibitors*

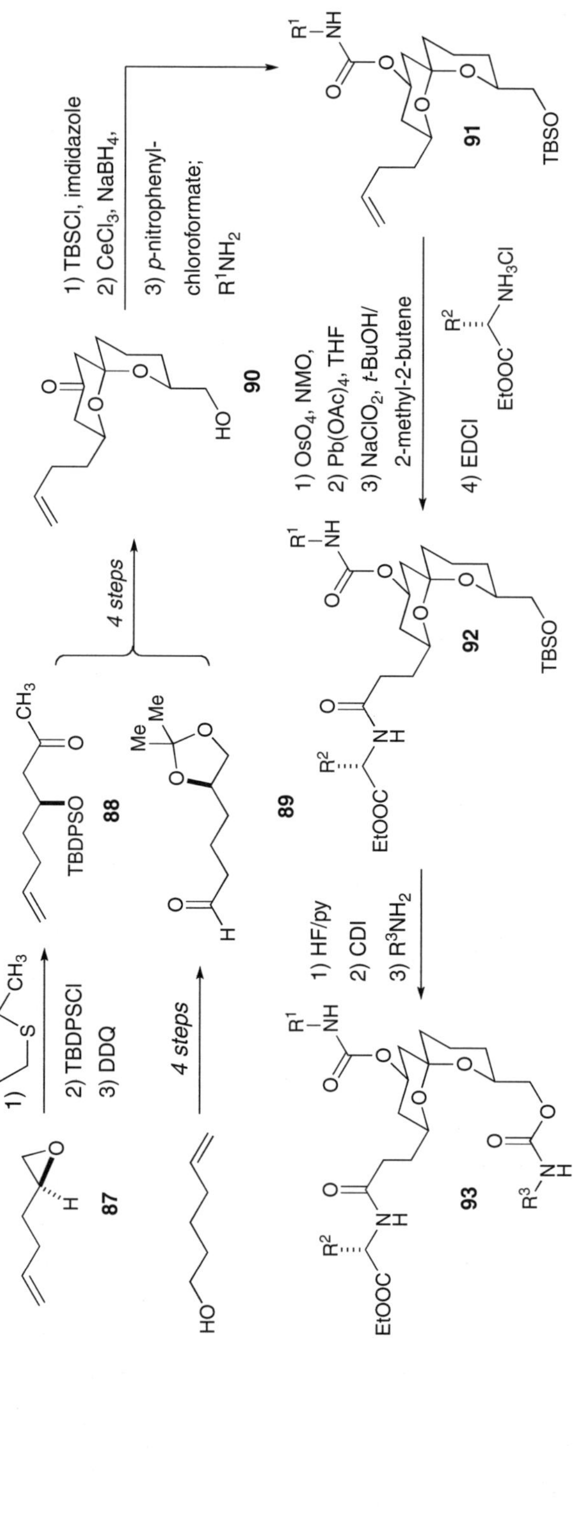

SCHEME 7.17 *Porco's library of diverse spiroketals*

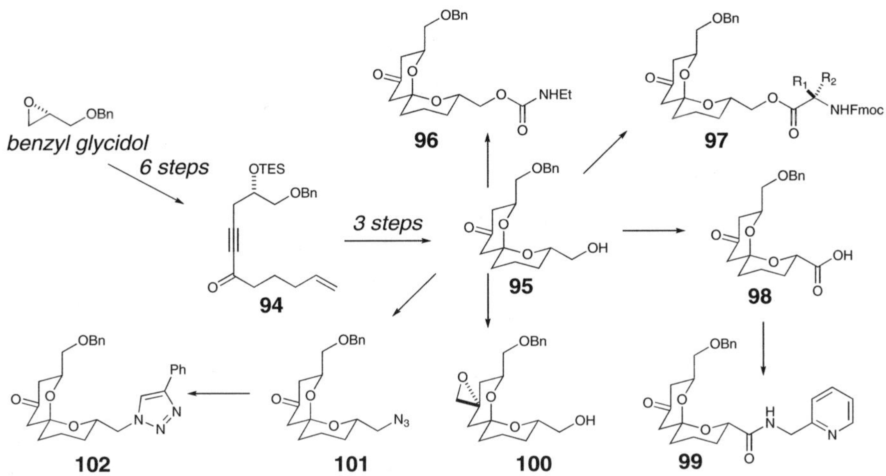

SCHEME 7.18 *Ley's synthesis of diverse spiroketal structures*

Spiroketal **95** is poised for a variety of functionalizations. The free alcohol **95** can be acylated with isocyanates to give **96** or with coupled to a protected amino ester capable of engaging in subsequent peptide coupling reactions. Oxidation of the free alcohol to acid **98** enables conversion to amide **99** whereas conversion of the alcohol to azide **101** serves as a starting point for CuI-catalysed 'click' reactions. Finally, the cyclic ketone can be efficiently and diastereoselectively converted to spirocyclic epoxide **100** which is likely capable of a variety of nucleophilic ring-opening reactions. Similar to Porco's library, Ley's synthesis allows for a rapid assemble of a spiroketal core while offering complementary opportunities for functionalization to a variety of diverse products.

Tan [125, 126] and coworkers have developed a method to systematically synthesize stereo-diversified spiroketals using kinetic, rather than thermodynamic, spiroketal formation (Scheme 7.19). In this approach, systematic stereodiversification is accomplished by stereoselective epoxidation reactions of glycals **106** and **107** followed by a kinetic spirocyclization to afford spiroketals **110–113**. Reactions proceed with either inversion or retention at the anomeric carbon under different conditions. As a complement to the methods of Ley and Porco (*vide supra*), this synthetic method is amenable to the preparation of a wide variety of spiroketal structures wherein stereochemical diversity is maximally exploited.

The solid phase synthesis of a spiroketal library was completed by Waldmann and coworkers employing a double intramolecular addition as a key step (Scheme 7.20) [127, 128]. The strategy utilizes homoallylic alcohols or β-hydroxyesters to form aldehyde functionalized polymers **114** in a short sequence. These solid phase-bound aldehydes are then converted to alkynone intermediates **115** after addition of alkynyl Grignard reagents, and IBX oxidation. Cleavage of the acid-labile THP protecting group in MeOH releases alcohol **116** from the solid phase and initiates the double intramolecular conjugate addition. The keto-spiroketals **117** were reduced to two spiroketal epimers (**118** and **119**), which could be separated by HPLC. In total, 146 spiroketals were synthesized in a 7-step sequence.

Tricyclic guanidine-containing alkaloids, such as the batzelladines and crambescidins, are marine natural products that have demonstrated the propensity to inhibit several different PPIs (Figure 7.5). Batzelladines A-E can inhibit gp120–CD4 binding [129, 130], while

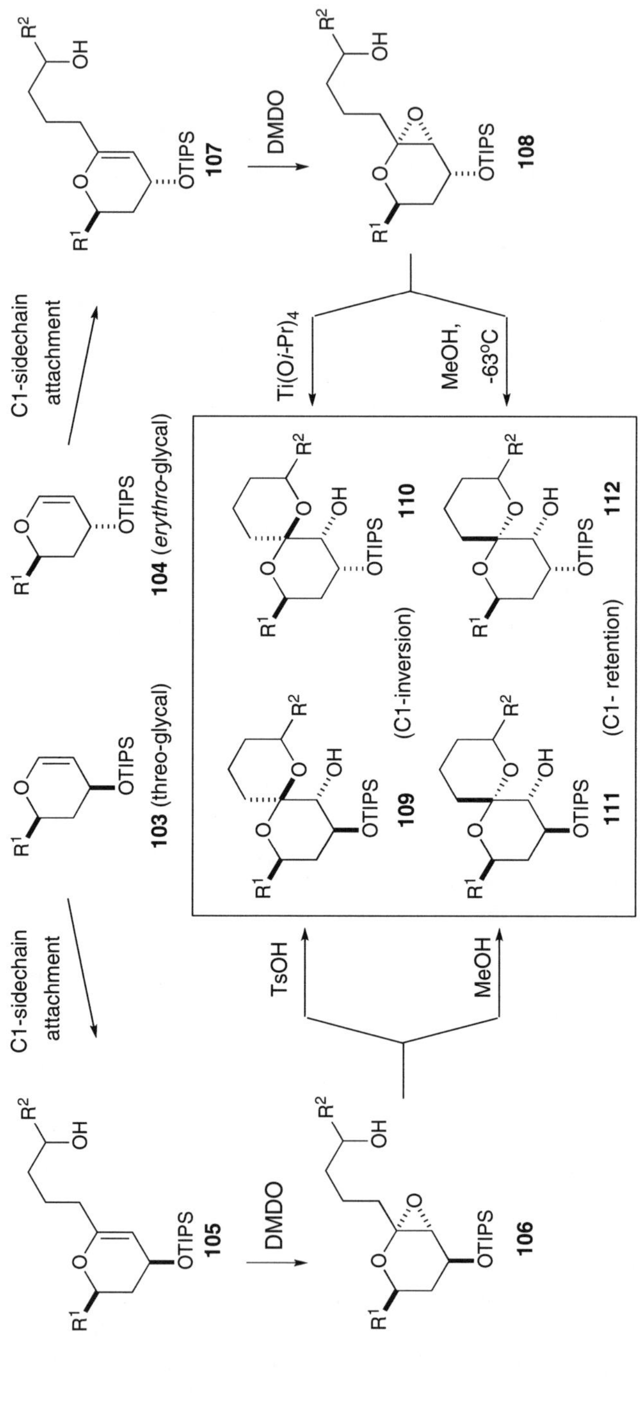

SCHEME 7.19 *Synthesis of stereochemically diverse spiroketal core structures*

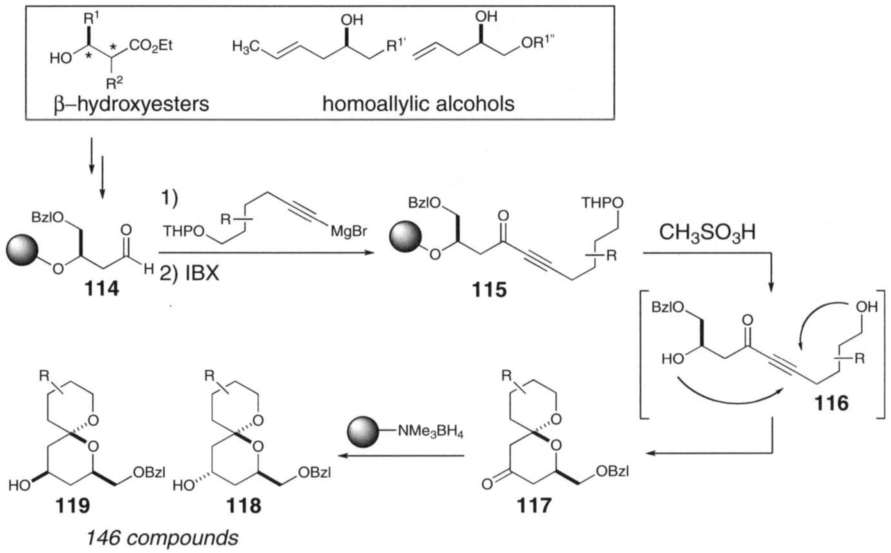

SCHEME 7.20 Waldmann's solid phase synthesis of spiroketal library

batzelladines F-I [131–133] and related synthetic analogs [134] are able to induce CD4–p56lck complex dissociation. Analogs of both batzelladine and crambescidin can disrupt Nef–p53, Nef– actin, and Nef–p56lck interactions [131, 135]. Although some biological activity of this family of compounds was noted at the time of their isolation, much of the PPI inhibition activity was revealed by screening experiments that were directly enabled by developing total syntheses of these complex natural products.

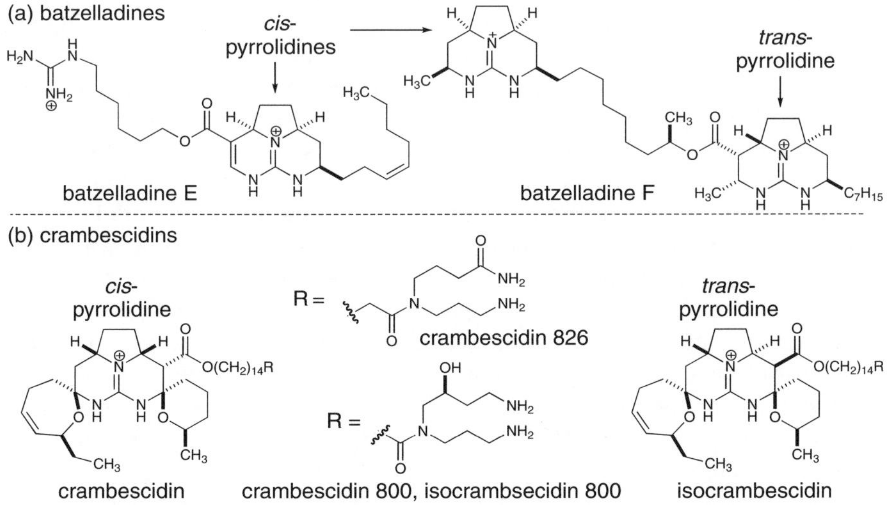

Figure 7.5 *Guanidine tri-cyclic ammonium natural products: (a) batzelladines and (b) crambescidins*

Overman and coworkers developed syntheses of crambescidin and batzelladine alkaloids using tethered variants of the Biginelli reaction [136]. Among the key advances that enabled the synthesis of these compounds were differential Biginelli conditions that permit the synthesis of either the *cis*- or *trans*-configured central pyrrolidines of the tricycloguanidine cores (Scheme 7.21) [137]. For the synthesis of crambescidin 800 and isocrambescidin 800, use of either the guanidinium substrate **120** or urea substrate **122** gave access to the *cis* and *trans* tethered Biginelli products, respectively [138–140]. For batzelladine F, which contains one tricyclic guanidine of each configuration, *cis*-selectivity was achieved using bicyclic guanidinium precursor **125** [136, 141].

Overman and coworkers have disclosed two efforts aimed at libraries of synthetic compounds related to the crambescidins and batzelladines. A small library of crambescidin alkaloid analogs that differ in their C14 side chain was synthesized using compound **130** (Scheme 7.22a), an intermediate from the total synthesis efforts [142]. Crambescidin 657 analogs were generated by Pd(0)-catalysed cleavage of the ester side chain, alkylation using iodides with various chain lengths, and de-allylation with palladium. Intermediate acids **132a–f** were coupled with protected hydroxyspermidine and the resulting amides were deprotected to provide crambescidin 800 analogs. A similar deprotection-alkylation-deprotection method was also applied to prepare crambescidin analogs containing simple hydrocarbon ester side chains. These compounds were evaluated for the ability to inhibit the growth of several murine and human cancer cell lines. In addition, a series of batzelladine analogs (**140–142**) has been assembled using tethered ketoesters **137–139** as substrates (Scheme 7.22b) [143].

7.5 Diversity Oriented Synthesis (DOS) in the Discovery of PPI Inhibitors

The synthesis and screening of collections of compounds with diverse structures has played an important role in the discovery of drug leads and biological compounds [144, 145]. Without prior knowledge of the structural requirements for a compound that will inhibit a biological pathway, one approach is to examine the activity of as many diverse structures as possible in the search for a lead. This approach has been refined significantly in the discovery of drug leads so that only compounds with a high probability of surviving the rigors of 'hit-to-lead' optimization are selected. This refinement often takes the form of enforcing structural requirements based on past successes, such as Lipinski's 'rule of 5' [146] or drug-likeness or Oprea's 'rule of 3' [147] for lead-likeness [148]. Although the discovery of PPI inhibitors is still in its infancy when compared to the discovery of enzyme inhibitors, it is apparent that PPI inhibitors are generally more lipophilic, have higher molecular weights, and exhibit more complex structures that put these compounds outside of most filters for drug-likeness. Although many enzyme-substrate and enzyme-inhibitor crystal structures are now available, the analogous information that would allow the rational design of PPI inhibitors is often lacking. These factors suggest that the synthesis of libraries of unbiased compound through diversity-oriented synthesis (DOS) [149–153], will play an increasingly important role in the discovery of PPI inhibitors.

Boger and coworkers initiated an effort to discover PPI inhibitors by building compound libraries from a minimal set of chemical reactions that would maximize structural diversity (Scheme 7.23). These compounds were of generally high molecular weight, based on the

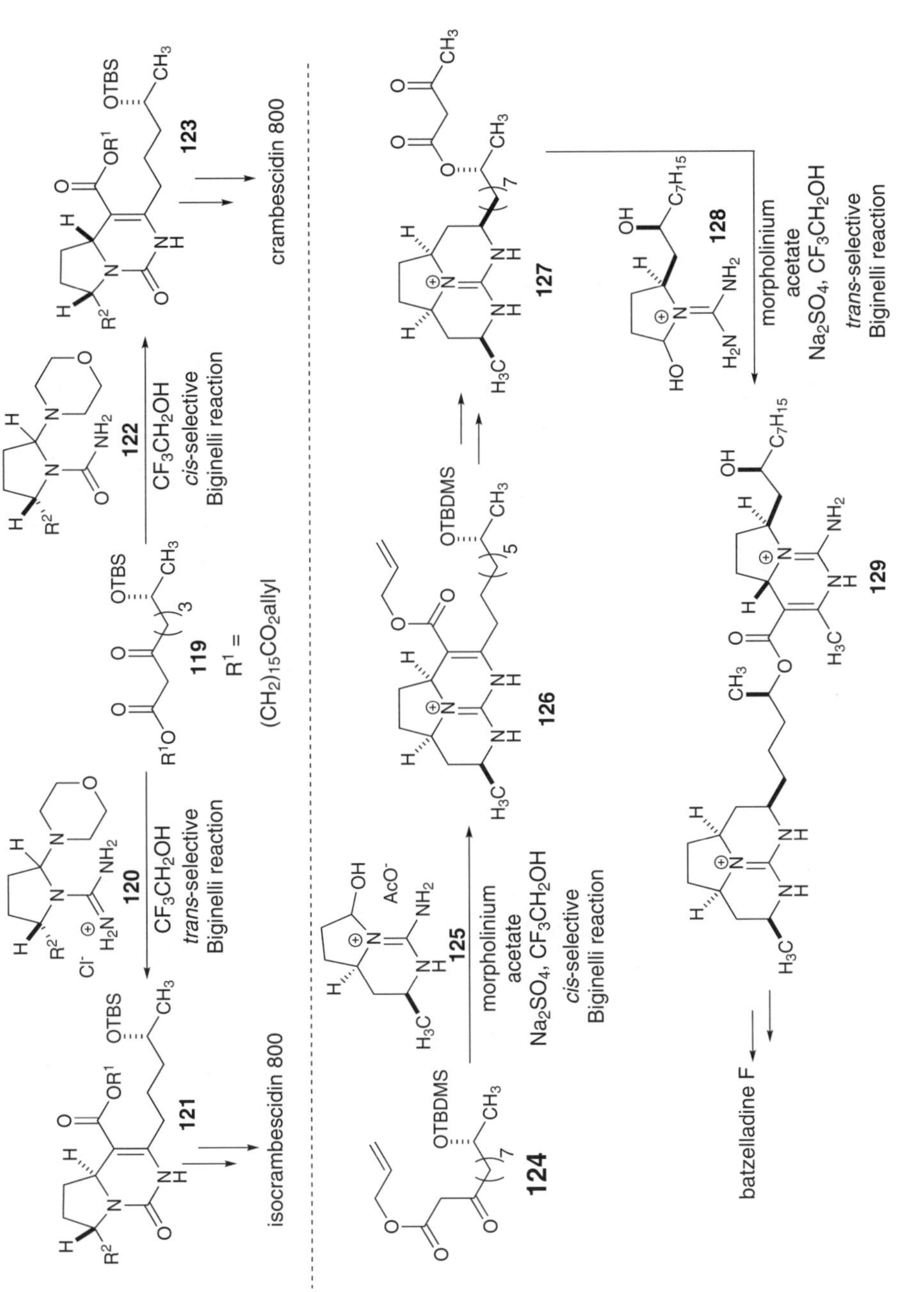

SCHEME 7.21 *Synthesis of batzelladine core structures using tethered Biginelli reactions*

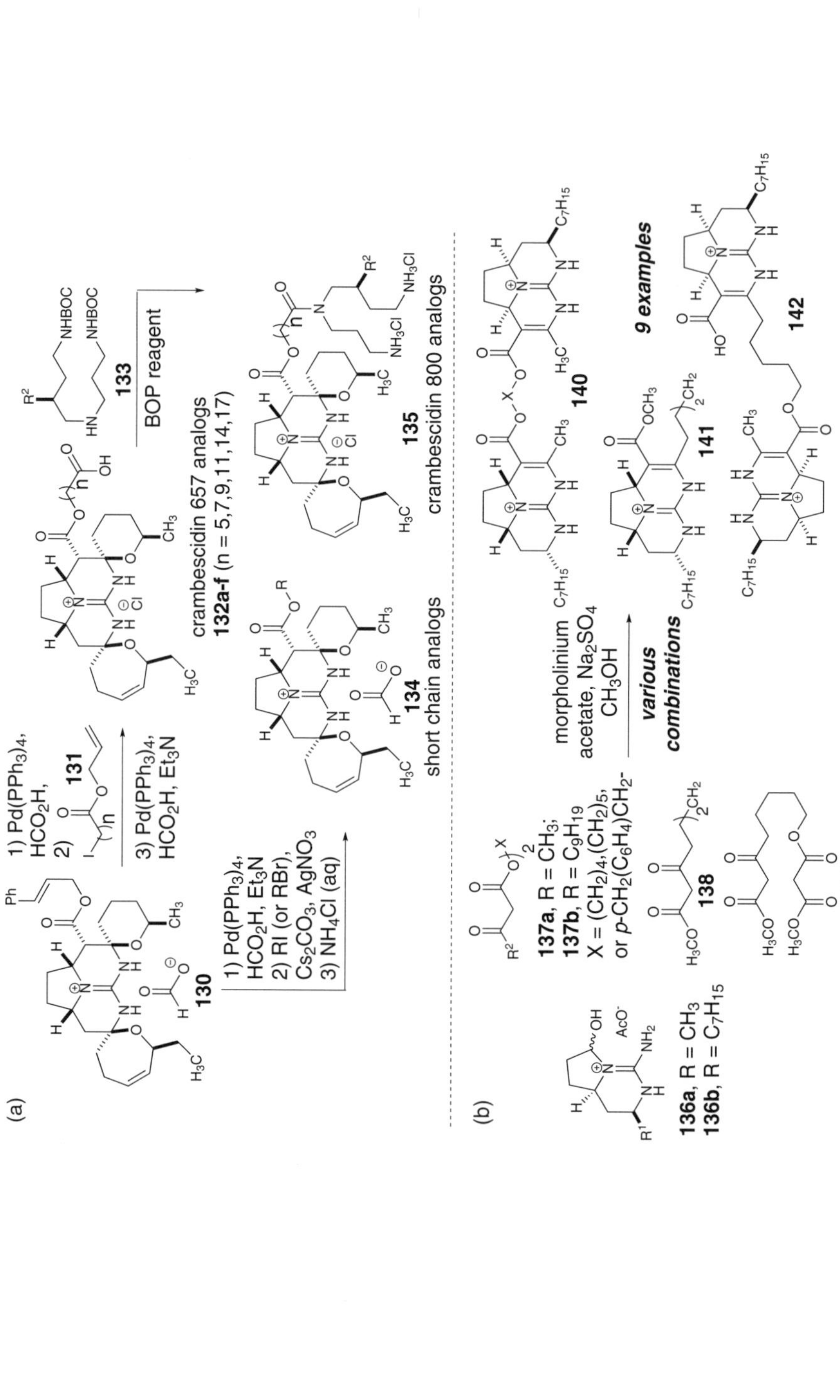

SCHEME 7.22 (a) Library of crambescidin analogs and (b) library bis-tricyclic guanidines resembling the batzelladines

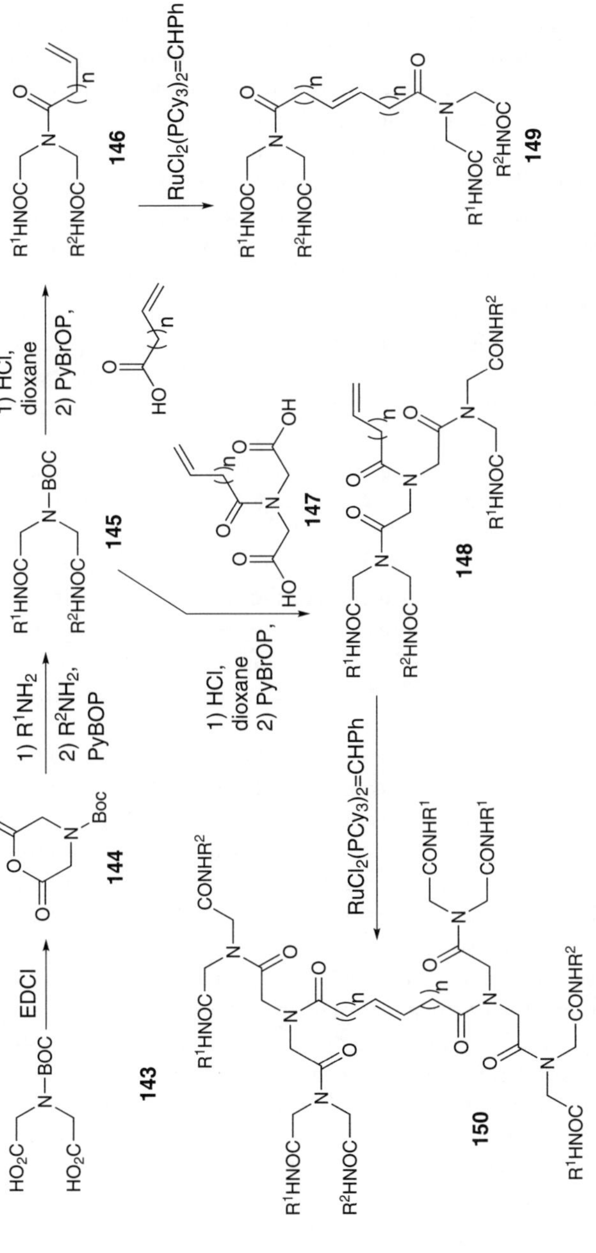

SCHEME 7.23 *Divergent synthesis of alkene dimer mixtures using cross metathesis*

hypothesis that a high level of protein surface contact would be necessary for effective binding and PPI disruption. The basic strategy hinges on rapid assembly of imino-diacetic acid-derived diamides **146** and **148**, which are then dimerized to yield structures such as **149** and **150**. The first efforts employed alkene metathesis as the dimerization step [154, 155], which was maximally efficient for the preparation of mixtures of compounds that could be screened and 'deconvoluted' using deletion, re-synthesis, or a combination of both [156]. The mixture consists of the pairwise homo- and heterodimeric compounds, each of which occurs as either the *cis* or *trans* alkene. The advantage of this technique is that a high level of molecular complexity can be accessed rapidly and active compounds can be identified in a minimal set of screening experiments. The disadvantage is that deconvolution, while minimized by keeping the number of compounds in any one mixture low, identifies potent compounds at the expense of providing preliminary SAR data based on the activity of less potent compounds.

A second library approach took advantage of a convergent dimerization strategy, replacing alkene metathesis with a diacyl linkage to install a more rigid linker and reduce the number of potential compounds in each mixture by eliminating the possibility for alkene isomers (Scheme 7.24) [157–159]. These compounds were prepared as 60 individual mixtures of ten compounds each that were screened for their ability to disrupt the binding of MMP2 to integrin $\alpha v\beta 3$ [160]. Three of these mixtures exhibited activity in this assay, all corresponding to the *m*-phthaloyl linkage of the symmetrical halves of the compounds. The 30 compounds from these mixtures were prepared and tested individually, and from this study, **153a** emerged as the most potent lead. A series of 77 analogs was prepared, which revealed that **153b**, featuring a more electron-poor aromatic ring on the lysine carbamate, exhibited improved activity. From this lead, the more water-soluble analog **153c**, which retained much of the original activity, was prepared by replacing the iminodiacetic acid residues with glycine linkages and converting the lysine methyl ester to the free acid. The ability of these compounds to inhibit angiogenesis and tumor growth *in vivo* was also demonstrated [161].

The synthesis of libraries focused on a particular structural feature that is hypothesized to aid in disrupting PPIs can lead to the discovery of new inhibitors. Janda and coworkers have synthesized collections of compounds that feature a flat aromatic core capable of occupying a cleft area between two proteins and have dubbed this strategy the 'credit-card library approach' (Scheme 7.25) [162]. In an effort to maximize the amount of diversity while minimizing the chemical effort, this approach relies on the assembly of the flat 'credit-card' region that is then elaborated on one end using the Ugi four-component reaction (4CR). The first library utilized naphthalene core **157**, whereas the second library employed quinoline core **155**. In three different screens, the naphthalene-based library has produced the most potent hits, yielding a 17 μM inhibitor of cMyc-Max interaction (**158**), a compound that inhibits gp41 six-helix bundle (6-B) formation (**159**)[163], and a molecule that binds acetylcholine esterase (AChE) and inhibits the deposition of β-amyloid (**160**) [164].

Solid-phase, split-pool synthesis has been employed in the discovery of compounds that inhibit inducible nitric oxide synthase (iNOS) dimerization (Scheme 7.26) [165]. In this case, the library was prepared on single-bead scale using chemical tags to track the synthesis, a technique known as ECLiPS, or 'Encoded Combinatorial Library on Polymeric Support'. The use of chloroaromatic tagging allows for the determination of the chemical history of the beads used for the synthesis on extremely small scale, and was later translated into the macrobead based technique of Schreiber,[25] which produces discrete solutions of compounds

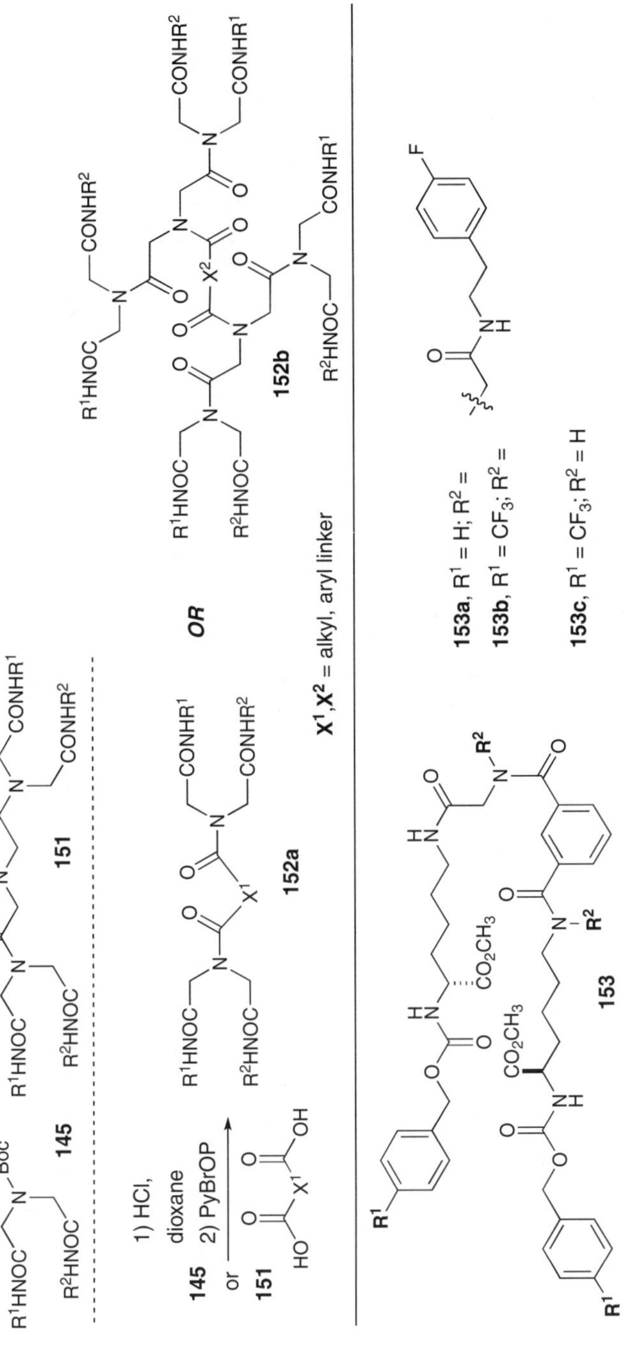

SCHEME 7.24 *Divergent synthesis of a diamide library yielding new inhibitors of the interaction of MMP2 with integrin* $\alpha v\beta_3$

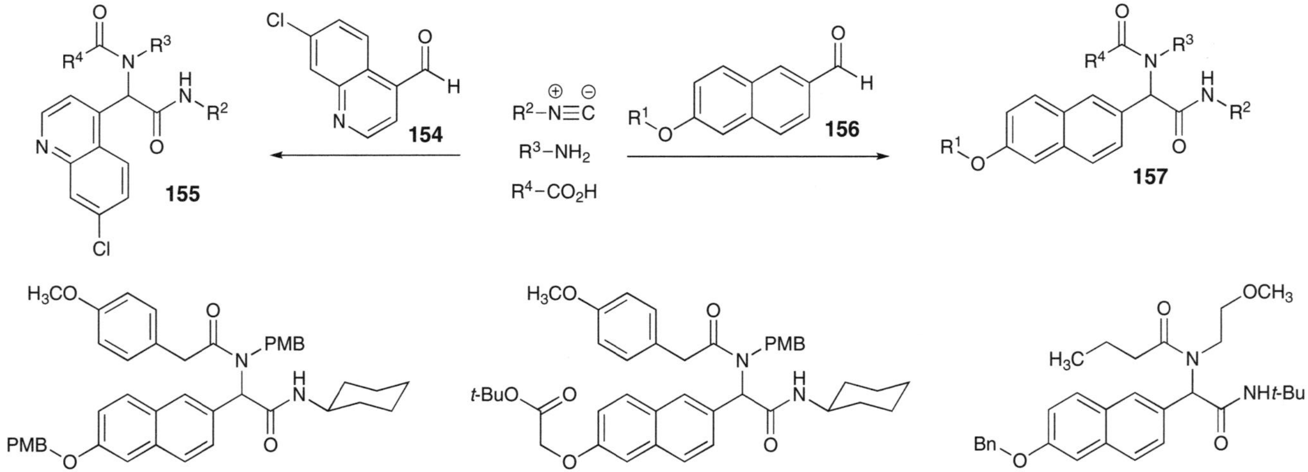

158: Inhibits c-Myc-Max Interaction
IC_{50} = 17 μM (EMSA)

159: Inhibits HIV-1 gp41 fusogenic 6-helix bundle core formation
IC_{50} = 38 μM (ELISA)

160: Inhibits AChE-induced aggregation of β-amyloid (Aβ)
K_i (AChE) = 1.9 μM

SCHEME 7.25 *Ugi four-component reaction (4CR) synthesis of 'credit card' libraries*

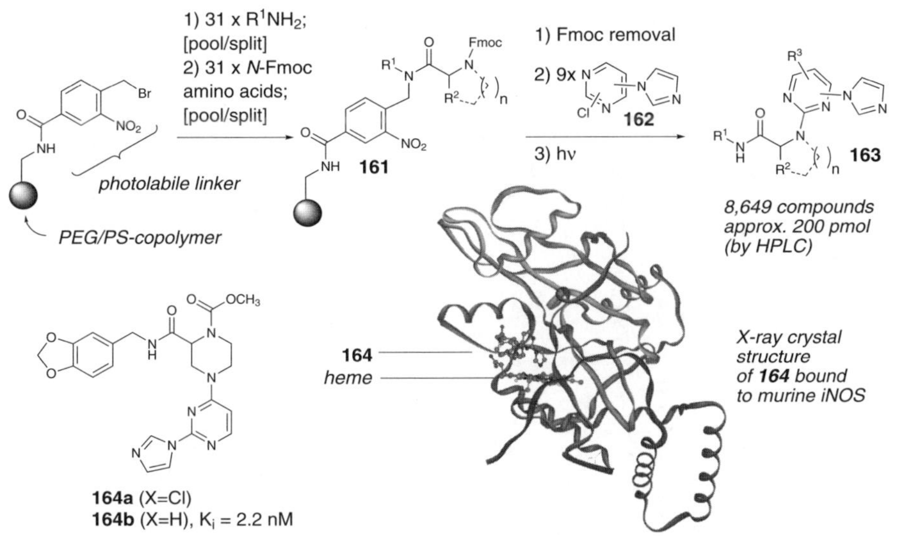

SCHEME 7.26 *Split-pool, single-bead encoded synthesis of N-heterocyclic amino amides. Discovery of* ***164****a and* ***164****b, which inhibit the dimerization of inducible nitric oxide synthase (iNOS)*

and was dubbed 'one bead, one stock solution'. For the discovery of iNOS dimerization inhibitors, a photolabile linkage was employed for ease of cleavage from the polyethyleneglycol-polystyrene copolymer (PEG/PS), in analogy to previously described methods. Primary amines were introduced by displacement of the bromides, and the resulting secondary amines were pooled and split into 31 acylations with *N*-Fmoc-protected amino acids to give amides **161**. These amide intermediates were pooled and split into nine groups for the final *N*-heteroarylation step using various imidazole-substituted chloropyrimidines **162**. A high-throughput cell-based screen for NO production was used as a preliminary evaluation, and subsequent studies established compound **164**a as the most active lead. Subsequent studies focused on **164**b, which lacks the potentially reactive aryl chloride of **164**a. X-ray crystallography established that **164**b bound to the heme iron center of iNOS and that the peripheral substituents prevented dimerization of the protein. Although this PPI inhibitor acts by first binding to the active site of an enzyme, this example demonstrates the use of diverse structures, screened as mixtures, in the discovery of PPI inhibitors.

The Schreiber group has successfully merged a split-pool combinatorial chemistry platform with small molecule microarray, a screening technique that is uniquely suited to the discovery of PPI inhibitors. The synthetic platform employed utilizes large (500–600 μm) polystyrene 'macrobeads' to obtain much higher quantities of compounds than the smaller beads typically used for single-bead synthesis (Scheme 7.27) [25]. Chemical encoding, by the method of Still [166], was employed to track the identity of compounds through the synthesis. A chemically inert silicon-oxygen linkage is employed so that maximally diverse chemistry can be employed and all of the compounds that are produced feature a primary alcohol [167]. This primary alcohol can be subsequently employed in a variety of experiments, including 'printing' of the small molecules on functionalized microscope slides [168] using several different chemical capture techniques [169]. The first library (Scheme 7.27a) used an epoxide opening reaction to generate a series of 1,3-diols

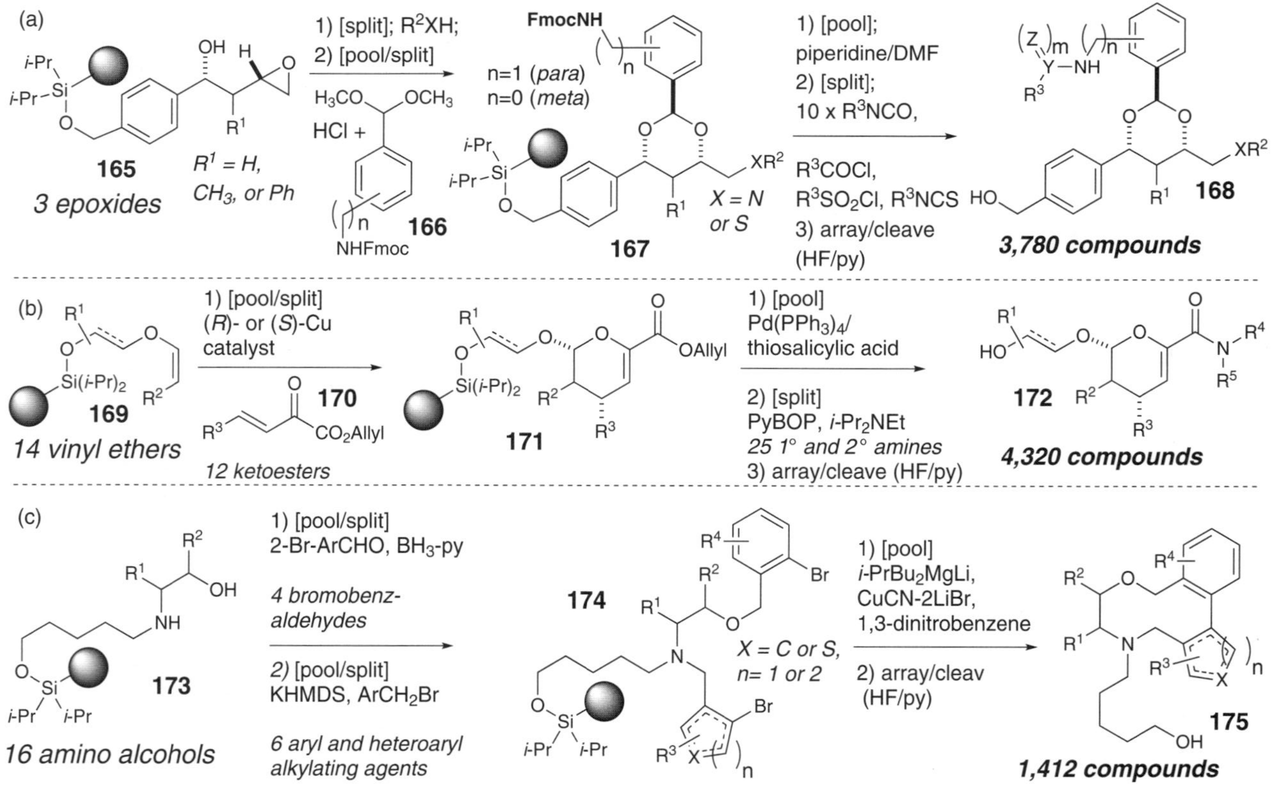

SCHEME 7.27 *Split-pool, single-bead encoded synthesis of diverse compounds featuring a common primary alcohol for subsequent attachment to functionalized slides. (a) 1,3-dioxanes, (b) dihydropyran carboxamides, (c) macrocyclic biaryl alkaloids*

which are used to form protected 3-amino or 4-aminomethyl benzylidene acetals **167** [170]. The amine is deprotected and then functionalized with a series of acyl chlorides, sulfonyl chlorides and other electrophiles to yield diverse dioxolanes **168**. A second library (Scheme 7.27b) used solid phase-bound vinyl ethers **169** in enantioselective catalytic hetero-Diels-alder reactions with unsaturated α-ketoesters **170** [171]. The resultant allyl esters **171** were cleaved to carboxylic acids that were functionalized with a wide variety of primary and secondary amines. In a third library (Scheme 7.27c), a reductive coupling reaction was used to prepare a series of biaryl macrocyclic ethers **175**[172]. In each case, representative library members were assayed with HPLC to assess compound purity.

In one of the first examples where chemistry and small-molecule microarray (SMM) were used in concert, compounds from three different library syntheses were employed in an effort to discover new inhibitors of transcription factors (Scheme 7.28). The activity of inhibitors can result indirectly from disruption of protein complexes that interact with DNA or directly by inhibiting the binding of the protein to the target DNA sequence [173]. The difficulty in targeting transcription factors with small molecules lies both in the requirement that a small molecule disrupt the interaction of two macromolecules and in the difficulty of assaying for such inhibitors in a high-throughput format. SMM addresses both of these challenges by providing an efficient way to merge the synthesis of potential inhibitors with diverse structures while simultaneously facilitating an assay whereby thousands of interactions are detected in a single experiment. In this case, small molecules that bind to Hap3p-GST fusion were detected in an effort to find compound that would disrupt the Hap2/3/4/5p transcription factor complex [174]. Haptamide A (**176**) emerged from the primary screen, and a series of derivatives was subsequently synthesized. Extensive modifications of the anilide portion and investigating the impact of changing the absolute configuration revealed that haptamide B (**178**) was twice as potent *in vivo*, as judged by a reporter gene assay, and exhibited a significantly lower K_d.

Multicomponent reactions (MCRs) offer an efficient alternative to multistep synthesis in the preparation of libraries for the discovery of PPI inhibitors. While Janda's work has focused on one particular structural feature in the 'credit card approach', the discovery of new MCRs allows the products of these reactions to contribute unique and diverse structural features to libraries for screening. Bienaymé assembled a collection of 8160 compounds derived from MCRs, and from this a novel indolizine inhibitor of the interaction of VEGF and NRP1 was discovered [175]. The MCR chemistry facilitated optimization of the potency by allowing each site to be modified (Scheme 7.29). A fluoride atom adjacent to the indolizine nitrogen can be displaced by a variety of heteroatom nucleophiles (**179** $\rightarrow$ **183**) and a dimethoxybenzyl (DMB) isonitrile that is incorporated from the isonitrile can be removed and the resulting amine acylated (**185** $\rightarrow$ **186**). In the case of the VEGF-NRP1 disruptor, neither of these modifications resulted increased potency, but the substituents R^1 and R^2 could be varied by employing a variety of aldehydes and isonitriles respectively, resulting in an overall tenfold increase in activity as judged by ELISA and cell migration. Although each MCR only allows access to one central scaffold, the ease with which derivatives can be made makes this approach particularly attractive for follow-up and optimization.

A recent report from Stadler, Shaw and coworkers has demonstrated that a library composed of diverse small molecules was useful for the discovery of a transcription factor inhibitor that acts directly on the protein-DNA interface (Scheme 7.30) [176]. Fluorescence

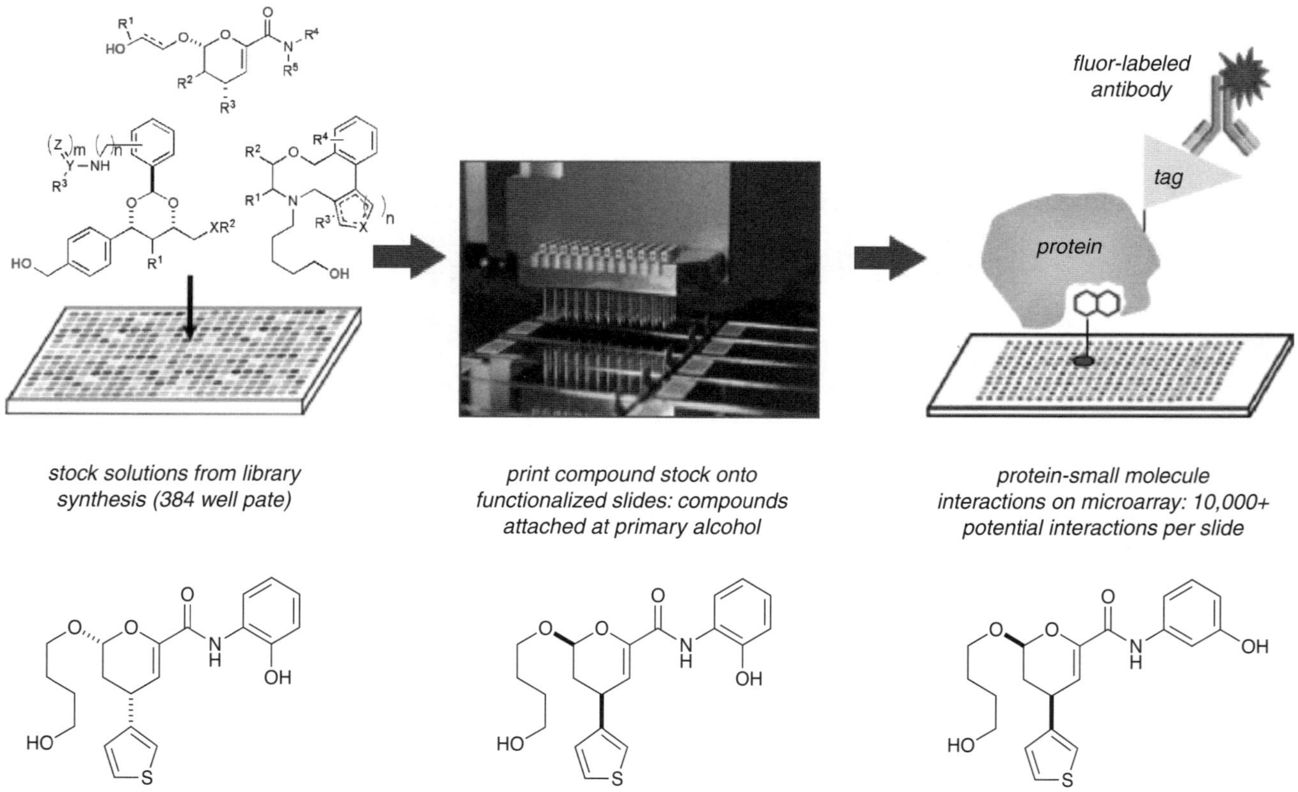

176, haptamide A
$IC_{50} = 42.1\ \mu M$; $K_D = 5.03\ \mu M$

177, *ent*- haptamide A
$IC_{50} = 30.6\ \mu M$; $K_D = 0.66\ \mu M$

178, haptamide B
$IC_{50} = 23.8\ \mu M$; $K_D = 0.33\ \mu M$

SCHEME 7.28 *Preparation of small molecule microarrays (SMMs) from stock solutions. Detection of protein-small molecule interactions using SMM; structures of compounds that disrupt the Hap2/3/4/5p transcription factor complex (haptamides) (See Plate 13.)*

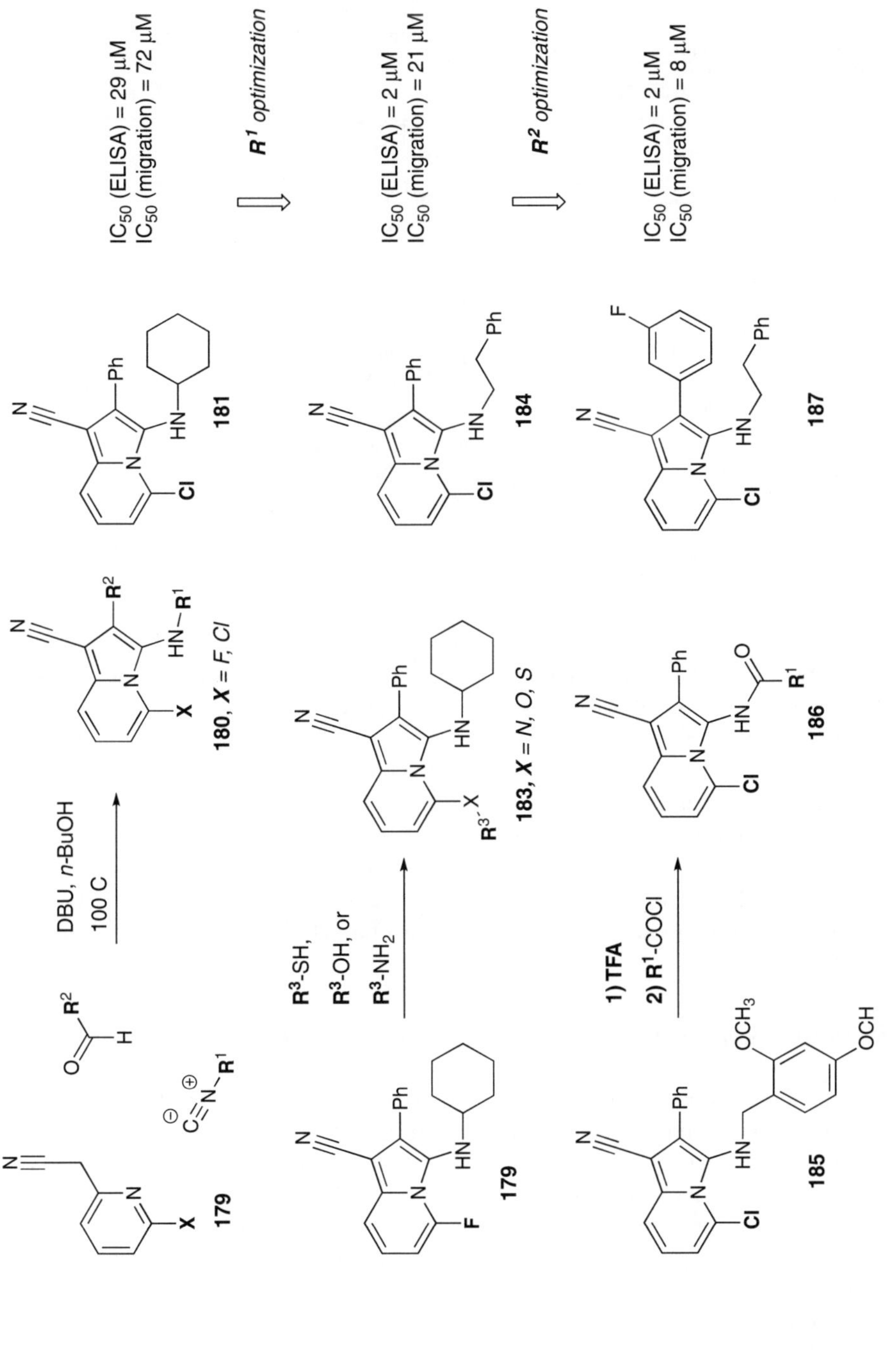

SCHEME 7.29 *Development of new MCRs for the discovery of VEGF-NRP1 inhibitors*

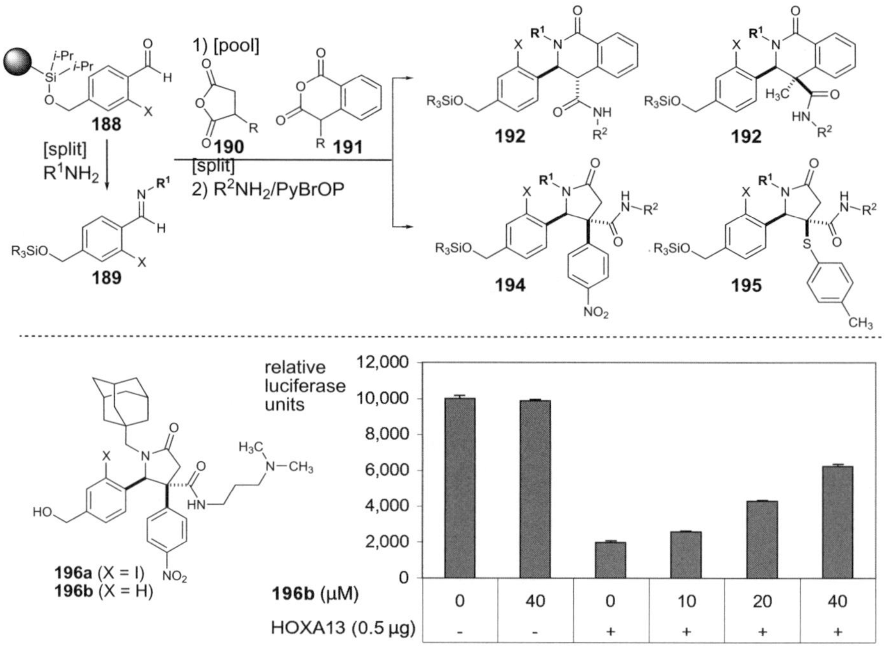

SCHEME 7.30 *Discovery of an inhibitor of HOXA13 transcription from a diverse library of small molecules*

polarization was used to detect the disruption of a complex between the HOXA13 transcription factor and an oligonucleotide containing its DNA target. From this initial screen, which included a mixture of commercially available library compounds, natural products, and diverse library molecules synthesized on solid phase, two hits were revealed with nearly identical structures (**196**a and **196**b). A sample of **196**b was re-synthesized and an IC_{50} for fluorescence polarization of 6.7 μM was determined. Subsequent examination of this compound in a reporter-gene assay indicated that it effectively inhibited the activity of HOXA13 in cells. Optimization of this compound is underway to find a derivative that is more potent, water soluble, and cell-permeable for future studies of mammalian development and cancer.

7.6 Summary and Outlook

The direct impact of library synthesis on the discovery of PPIs has been substantial and will only grow as the knowledge base of how to approach these difficult targets increases. Peptide-inspired libraries will continue to be important since the design and synthesis of peptidomimetics is relatively mature, and the full potential of these chemical approaches to discover PPIs is only now being realized. As the number of relevant natural product motifs increases, so will the efforts to synthesize and screen libraries of this type. Ongoing efforts aimed at natural product synthesis will ensure that new synthetic methods for assembling complex molecular architectures will continue to enable the synthesis of complex libraries. Synthesis and screening of unbiased libraries will continue to play an important role in defining the structural features necessary to inhibit PPIs, especially when structural knowledge of the proteins in question is unavailable.

References

1. T. Kortemme and D. Baker, A Simple Physical Model for Binding Energy Hot Spots in Protein–protein Complexes, *Proc. Natl. Acad. Sci. U. S. A.*, 99, 14116–21 (2002).
2. W. L. DeLano, Unraveling Hot Spots in Binding Interfaces: Progress and Challenges, *Curr. Opin. Struct. Biol.*, 12, 14–20 (2002).
3. B. Ma, T. Elkayam, H. Wolfson and R. Nussinov, Protein–protein Interactions: Structurally Conserved Residues Distinguish between Binding Sites and Exposed Protein Surfaces, *Proc. Natl. Acad. Sci. U. S. A.*, 100, 5772–7 (2003).
4. O. Keskin, B. Ma and R. Nussinov, Hot Regions in Protein–protein Interactions: The Organization and Contribution of Structurally Conserved Hot Spot Residues, *J. Mol. Biol.*, 345, 1281–94 (2005).
5. D. S. Tan, M. A. Foley, M. D. Shair and S. L. Schreiber, Stereoselective Synthesis of over Two Million Compounds Having Structural Features Both Reminiscent of Natural Products and Compatible with Miniaturized Cell-Based Assays, *J. Am. Chem. Soc.*, 120, 8565–6 (1998).
6. Z. Huang, The Chemical Biology of Apoptosis. Exploring Protein–protein Interactions and the Life and Death of Cells with Small Molecules, *Chem. Biol.*, 9, 1059–72 (2002).
7. P. L. Toogood, Inhibition of Protein–protein Association by Small Molecules: Approaches and Progress, *J. Med. Chem.*, 45, 1543–58 (2002).
8. T. Berg, Modulation of Protein–protein Interactions with Small Organic Molecules, *Angew. Chem. Int. Ed.*, 42, 2462–81 (2003).
9. D. L. Boger, J. Desharnais and K. Capps, Solution-Phase Combinatorial Libraries: Modulating Cellular Signaling by Targeting Protein–protein or Protein-DNA Interactions, *Angew. Chem. Int. Ed.*, 42, 4138–76 (2003).
10. M. R. Arkin and J. A. Wells, Small-Molecule Inhibitors of Protein–protein Interactions: Progressing Towards the Dream, *Nature Reviews Drug Discovery*, 3, 301–17 (2004).
11. M. Arkin, Protein–protein Interactions and Cancer: Small Molecules Going in for the Kill, *Curr. Opin. Chem. Biol.*, 9, 317–24 (2005).
12. A. Loregian and G. Palu, Disruption of Protein–protein Interactions: Towards New Targets for Chemotherapy, *J. Cell. Physiol.*, 204, 750–62 (2005).
13. H. Yin and A. D. Hamilton, Strategies for Targeting Protein–protein Interactions with Synthetic Agents, *Angew. Chem. Int. Ed.*, 44, 4130–63 (2005).
14. L. Zhao and J. Chmielewski, Inhibiting Protein–protein Interactions Using Designed Molecules, *Curr. Opin. Struct. Biol.*, 15, 31–4 (2005).
15. P. Chene, Drugs Targeting Protein–protein Interactions, *ChemMedChem*, 1, 400–11 (2006).
16. M. J. Vicent, E. Perez-Paya and M. Orzaez, Discovery of Inhibitors of Protein–Protein Interactions from Combinatorial Libraries, *Curr. Top. Med. Chem.*, 7, 83–95 (2007).
17. R. A. Houghten, General Method for the Rapid Solid-phase Synthesis of Large Numbers of Peptides: Specificity of Antigen-Antibody Interaction at the Level of Individual Amino Acids, *Proc. Natl. Acad. Sci. U. S. A.*, 82, 5131–5 (1985).
18. For personal perspectives on many of the early contributions during this period, see: M. Lebl, Parallel Personal Comments on 'Classical' Papers in Combinatorial Chemistry, *J. Comb. Chem.*, 1, 3–24 (1999).
19. B. A. Bunin and J. A. Ellman, A General and Expedient Method for the Solid-phase Synthesis of 1,4-Benzodiazepine Derivatives, *J. Am. Chem. Soc.*, 114, 10997–8 (1992).
20. K. C. Nicolaou, R. Hanko and W. Hartwig, *Handbook of Combinatorial Chemistry: Drugs, Catalysts, Materials*, Wiley-VCH, Weinheim, 2002.
21. K. C. Nicolaou, J. A. Pfefferkorn, H. J. Mitchell, *et al.*, Natural Product-Like Combinatorial Libraries Based on Privileged Structures. 2. Construction of a 10 000-Membered Benzopyran

Library by Directed Split-and-Pool Chemistry Using Nanokans and Optical Encoding, *J. Am. Chem. Soc.*, 122, 9954–67 (2000).
22. K. S. Lam, S. E. Salmon, E. M. Hersh, V. J. Hruby, W. M. Kazmierski and R. J. Knapp, A New Type of Synthetic Peptide Library for Identifying Ligand-Binding Activity, *Nature*, 354, 82–4 (1991).
23. R. A. Houghten, C. Pinilla, S. E. Blondelle, J. R. Appel, C. T. Dooley and J. H. Cuervo, Generation and Use of Synthetic Peptide Combinatorial Libraries for Basic Research and Drug Discovery, *Nature*, 354, 84–6 (1991).
24. N. Jung, A. Encinas and S. Bräse, Automated Synthesis of Heterocycles on Solid Supports, *Curr. Opin. Drug Discovery Dev.*, 9, 713–28 (2006).
25. H. E. Blackwell, L. Perez, R. A. Stavenger, *et al.*, A One-Bead, One-Stock Solution Approach to Chemical Genetics: Part 1, *Chem. Biol.*, 8, 1167–82 (2001).
26. J. G. Parsons, C. S. Sheehan, Z. Wu, I. W. James and A. M. Bray, A Review of Solid-Phase Organic Synthesis on Synphase Lanterns and Synphase Crowns, *Methods Enzymol.*, 369, 39–74 (2003).
27. K. C. Nicolaou, X.-Y. Xiao, Z. Parandoosh, A. Senyei and M. P. Nova, Radiofrequency Encoded Combinatorial Chemistry, *Angew. Chem. Int. Ed. Engl.*, 34, 2289–91 (1995).
28. E. J. Moran, S. Sarshar, J. F. Cargill, *et al.*, Radio Frequency Tag Encoded Combinatorial Library Method for the Discovery of Tripeptide-Substituted Cinnamic Acid Inhibitors of the Protein Tyrosine Phosphatase Ptp1b, *J. Am. Chem. Soc.*, 117, 10787–8 (1995).
29. J. I. Gavrilyuk, G. Evindar and R. A. Batey, Peptide-Heterocycle Hybrid Molecules: Solid-Phase Synthesis of a 400-Member Library of N-Terminal 2-Iminohydantoin Peptides, *J. Comb. Chem.*, 8, 237–46 (2006).
30. A. Studer, S. Hadida, R. Ferritto, *et al.*, Fluorous Synthesis: A Fluorous-Phase Strategy for Improving Separation Efficiency in Organic Synthesis, *Science*, 275, 823–6 (1997).
31. J. A. Gladysz and D. P. Curran, Fluorous Chemistry: From Biphasic Catalysis to a Parallel Chemical Universe and Beyond, *Tetrahedron*, 58, 3823–5 (2002).
32. W. Zhang and D. P. Curran, Synthetic Applications of Fluorous Solid-Phase Extraction (F-SPE), *Tetrahedron*, 62, 11837–65 (2006).
33. J. A. Gladysz, D. P. Curran and I. T. Horváth, *Handbook of Fluorous Chemistry*, Wiley-VCH, Weinheim, 2004.
34. Z. Luo, Q. Zhang, Y. Oderaotoshi and D. P. Curran, Fluorous Mixture Synthesis: A Fluorous-Tagging Strategy for the Synthesis and Separation of Mixtures of Organic Compounds, *Science*, 291, 1766–9 (2001).
35. H. N. Weller, A. E. Rubin, B. Moshiri, *et al.*, Development and Commercialization of the Miniblock Synthesizer Family: A Historical Case Study, *JALA*, 10, 59–71 (2005).
36. S. Manyem, M. P. Sibi, G. H. Lushington, B. Neuenswander, F. Schoenen and J. Aubé, Solution-Phase Parallel Synthesis of a Library of D2-Pyrazolines, *J. Comb. Chem.*, 9, 20–8 (2007).
37. J. A. Kowalski, S. F. Leonard and G. E. Lee, Jr., Diverse 2-Carboxamide-3-Amino-Substituted Quinoxalines: Synthesis and Reactivity Investigation for Library Generation, *J. Comb. Chem.*, 8, 774–9 (2006).
38. P. B. Madrid, A. P. Liou, J. L. DeRisi and R. K. Guy, Incorporation of an Intramolecular Hydrogen-Bonding Motif in the Side Chain of 4-Aminoquinolines Enhances Activity against Drug-Resistant *P. falciparum, J. Med. Chem.*, 49, 4535–43 (2006).
39. S. Bae, H.-G. Hahn, K. D. Nam and H. Mah, Solid-Phase Synthesis of Fungitoxic 2-Imino-1,3-Thiazolines, *J. Comb. Chem.*, 7, 7–9 (2005).
40. R. Touzani, S. Garbacia, O. Lavastre, V. K. Yadav and B. Carboni, Efficient Solution Phase Combinatorial Access to a Library of Pyrazole- and Triazole-Containing Compounds, *J. Comb. Chem.*, 5, 375–8 (2003).
41. R. Hoogenboom, M. A. R. Meier and U. S. Schubert, The Introduction of High-Throughput Experimentation Methods for Suzuki-Miyaura Coupling Reactions in University Education, *J. Chem. Educ.*, 82, 1693–6 (2005).
42. K. Geyer, J. D. C. Codee and P. H. Seeberger, Microreactors as Tools for Synthetic Chemists – the Chemists' Round-Bottomed Flask of the 21st Century?, *Chem. Eur. J.*, 12, 8434–42 (2006).

43. R. V. Jones, L. Godorhazy, N. Varga, D. Szalay, L. Urge and F. Darvas, Continuous-Flow High Pressure Hydrogenation Reactor for Optimization and High-Throughput Synthesis, *J. Comb. Chem.*, 8, 110–16 (2006).
44. V. Franckevicius, K. R. Knudsen, M. Ladlow, D. A. Longbottom and S. V. Ley, Practical Synthesis of (S)-Pyrrolidin-2-yl-1h-tetrazole, Incorporating Efficient Protecting Group Removal by Flow-Reactor Hydrogenolysis, *Synlett*, 889–92 (2006).
45. H. H. Horváth, G. Papp, C. Csajagi and F. Joo, Selective Catalytic Hydrogenations in a Microfluidics-Based High Throughput Flow Reactor on Ion-Exchange Supported Transition Metal Complexes: A Modular Approach to the Heterogenization of Soluble Complex Catalysts, *Catal. Commun.*, 8, 442–6 (2007).
46. G. M. Whitesides, The Origins and the Future of Microfluidics, *Nature*, 442, 368–73 (2006).
47. A. Loupy, *Microwaves in Organic Synthesis*, Wiley-VCH, Weinheim, 2006.
48. E. v. d. Eycken, C. O. Kappe and F. Almqvist. *Microwave-Assisted Synthesis of Heterocycles*, Springer, Berlin; New York, 2006.
49. A. de la Hoz, A. Diaz-Ortiz and A. Moreno, Microwaves in Organic Synthesis. Thermal and Non-Thermal Microwave Effects, *Chem. Soc. Rev.*, 34, 164–78 (2005).
50. N. E. Leadbeater, S. J. Pillsbury, E. Shanahan and V. A. Williams, An Assessment of the Technique of Simultaneous Cooling in Conjunction with Microwave Heating for Organic Synthesis, *Tetrahedron*, 61, 3565–85 (2005).
51. N. Kuhnert, Microwave-Assisted Reactions in Organic Synthesis – Are There Any Nonthermal Microwave Effects?, *Angew. Chem. Int. Ed.*, 41, 1863–6 (2002).
52. K. D. Stigers, M. J. Soth and J. S. Nowick, Designed Molecules that Fold to Mimic Protein Secondary Structures, *Curr. Opin. Chem. Biol.*, 3, 714–23 (1999).
53. M. J. Perez de Vega, M. Martin-Martinez and R. Genzalez-Muniz, Modulation of Protein–Protein Interactions by Stabilizing/Mimicking Protein Secondary Structure Elements, *Current Topics in Medicinal Chemistry*, 7, 33–62 (2007).
54. S. Fletcher and A. D. Hamilton, Protein Surface Recognition and Proteomimetics: Mimics of Protein Surface Structure and Function, *Curr. Opin. Chem. Biol.*, 9, 632–8 (2005).
55. K. Burgess, Solid-Phase Syntheses of β-Turn Analogues to Mimic or Disrupt Protein–protein Interactions, *Acc. Chem. Res.*, 34, 826–35 (2001).
56. J. A. Kritzer, O. M. Stephens, D. A. Guarracino, S. K. Reznik and A. Schepartz, β-Peptides as Inhibitors of Protein–protein Interactions, *Biorg. Med. Chem.*, 13, 11–16 (2004).
57. I. Im, T. R. Webb, Y.-D. Gong, J.-I. Kim and Y.-C. Kim, Solid-Phase Synthesis of Tetrahydro-1,4-Benzodiazepin-2-One Derivatives as a β-Turn Peptidomimetic Library, *J. Comb. Chem.*, 6, 207–13 (2004).
58. C. Gil and S. Bräse, Efficient Solid-Phase Synthesis of Highly Functionalized 1,4-Benzodiazepin-5-One Derivatives and Related Compounds by Intramolecular Aza-Wittig Reactions, *Chem. Eur. J.*, 11, 2680–8 (2005).
59. A. L. Kennedy, A. M. Fryer and J. A. Josey, A New Resin-Bound Universal Isonitrile for the Ugi 4CC Reaction: Preparation and Applications to the Synthesis of 2,5-Diketopiperazines and 1,4-Benzodiazepine-2,5-Diones, *Org. Lett.*, 4, 1167–70 (2002).
60. D. J. Parks, L. V. LaFrance, R. R. Calvo, *et al.*, 1, 4-Benzodiazepine-2,5-Diones as Small Molecule Antagonists of the Hdm2-p53 Interaction: Discovery and SAR, *Biorg. Med. Chem. Lett.*, 15, 765–70 (2005).
61. T. A. Keating and R. W. Armstrong, Postcondensation Modifications of Ugi Four-Component Condensation Products: 1-Isocyanocyclohexene as a Convertible Isocyanide. Mechanism of Conversion, Synthesis of Diverse Structures, and Demonstration of Resin Capture, *J. Am. Chem. Soc.*, 118, 2574–83 (1996).
62. B. L. Grasberger, T. Lu, C. Schubert, *et al.*, Discovery and Cocrystal Structure of Benzodiazepinedione HDM2 Antagonists That Activate p53 in Cells, *J. Med. Chem.*, 48, 909–12 (2005).
63. J. J. Marugan, K. Leonard, P. Raboisson, *et al.*, Enantiomerically Pure 1,4-Benzodiazepine-2,5-Diones as HDM2 Antagonists, *Biorg. Med. Chem. Lett.*, 16, 3115–20 (2006).

64. K. Leonard, J. J. Marugan, P. Raboisson, *et al.*, Novel 1,4-Benzodiazepine-2,5-Diones as Hdm2 Antagonists with Improved Cellular Activity, *Biorg. Med. Chem. Lett.*, 16, 3463–8 (2006).
65. B. P. Orner, J. T. Ernst and A. D. Hamilton, Toward Proteomimetics: Terphenyl Derivatives as Structural and Functional Mimics of Extended Regions of an a-Helix, *J. Am. Chem. Soc.*, 123, 5382–3 (2001).
66. H. Yin, G.-i. Lee, K. A. Sedey, *et al.*, Terphenyl-Based Bak BH3 α-Helical Proteomimetics as Low-Molecular-Weight Antagonists of Bcl-xL, *J. Am. Chem. Soc.*, 127, 10191–6 (2005).
67. E. Jacoby, Biphenyls as Potential Mimetics of Protein α-Helix, *Biorg. Med. Chem. Lett.*, 12, 891–3 (2002).
68. I. C. Kim and A. D. Hamilton, Diphenylindane-Based Proteomimetics Reproduce the Projection of the i, $i+3$, $i+4$, and $i+7$ Residues on an α-Helix, *Org. Lett.*, 8, 1751–4 (2006).
69. W. P. Nolan, G. S. Ratcliffe and D. C. Rees, The Synthesis of 1,6-Disubstituted Indanes Which Mimic the Orientation of Amino Acid Side-Chains in a Protein α-Helix Motif, *Tetrahedron Lett.*, 33, 6879–82 (1992).
70. I. R. Hardcastle, S. U. Ahmed, H. Atkins, *et al.*, Small-Molecule Inhibitors of the MDM2-p53 Protein–Protein Interaction Based on an Isoindolinone Scaffold, *J. Med. Chem.*, 49, 6209–21 (2006).
71. J. M. Davis, A. Truong and A. D. Hamilton, Synthesis of a 2,3';6',3''-Terpyridine Scaffold as an α-Helix Mimetic, *Org. Lett.*, 7, 5405–8 (2005).
72. H. Oguri, S. Tanabe, A. Oomura, M. Umetsu and M. Hirama, Synthesis and Evaluation of α-Helix Mimetics Based on a Trans-Fused Polycyclic Ether: Sequence-Selective Binding to Aspartate Pairs in α-Helical Peptides, *Tetrahedron Lett.*, 47, 5801–5 (2006).
73. H. Yin and A. D. Hamilton, Terephthalamide Derivatives as Mimetics of the Helical Region of Bak Peptide Target Bcl-xL Protein, *Biorg. Med. Chem. Lett.*, 14, 1375–9 (2004).
74. J. T. Ernst, J. Becerril, H. S. Park, H. Yin and A. D. Hamilton, Design and Application of an α-Helix-Mimetic Scaffold Based on an Oligoamide-Foldamer Strategy: Antagonism of the Bak BH3/Bcl-xL Complex, *Angew. Chem. Int. Ed.*, 42, 535–9 (2003).
75. J.-M. Ahn and S.-Y. Han, Facile Synthesis of Benzamides to Mimic an α-Helix, *Tetrahedron Lett.*, 48, 3543–7 (2007).
76. J. M. Davis, L. K. Tsou and A. D. Hamilton, Synthetic Non-Peptide Mimetics of α-Helices, *Chem. Soc. Rev.*, 36, 326–34 (2007).
77. W. Antuch, S. Menon, Q.-Z. Chen, *et al.*, Design and Modular Parallel Synthesis of a MCR Derived a–Helix Mimetic Protein–Protein Interaction Inhibitor Scaffold, *Biorg. Med. Chem. Lett.*, 16, 1740–3 (2006).
78. F. Lu, S.-W. Chi, D.-H. Kim, K.-H. Han, I. D. Kuntz and R. K. Guy, Proteomimetic Libraries: Design, Synthesis, and Evaluation of p53-MDM2 Interaction Inhibitors, *J. Comb. Chem.*, 8, 315–25 (2006).
79. A. J. Souers and J. A. Ellman, β-Turn Mimetic Library Synthesis: Scaffolds and Applications, *Tetrahedron*, 57, 7431–48 (2001).
80. S. Cheng, C. M. Tarby, D. D. Comer, *et al.*, A Solution-Phase Strategy for the Synthesis of Chemical Libraries Containing Small Organic Molecules: A Universal and Dipeptide Mimetic Template, *Biorg. Med. Chem.*, 4, 727–37 (1996).
81. D. L. Boger, J. K. Lee, J. Goldberg and Q. Jin, Two Comparisons of the Performance of Positional Scanning and Deletion Synthesis for the Identification of Active Constituents in Mixture Combinatorial Libraries, *J. Org. Chem.*, 65, 1467–74 (2000).
82. T. Berg, S. B. Cohen, J. Desharnais, *et al.*, Small-Molecule Antagonists of Myc/Max Dimerization Inhibit Myc-Induced Transformation of Chicken Embryo Fibroblasts, *Proc. Natl. Acad. Sci. U. S. A.*, 99, 3830–5 (2002).
83. M. Eguchi, M. S. Lee, H. Nakanishi, M. Stasiak, S. Lovell and M. Kahn, Solid-Phase Synthesis and Structural Analysis of Bicyclic β-Turn Mimetics Incorporating Functionality at the i to $i + 3$ Positions, *J. Am. Chem. Soc.*, 121, 12204–5 (1999).
84. K. H. Emami, C. Nguyen, H. Ma, *et al.*, A Small Molecule Inhibitor of β-Catenin/Cyclic AMP Response Element-Binding Protein Transcription, *Proc. Natl. Acad. Sci. U. S. A.*, 101, 12682–7 (2004).

85. M. Eguchi, C. Nguyen, S. C. Lee and M. Kahn, ICG-001, a Novel Small Molecule Regulator of TCF/β-Catenin Transcription, *Medicinal Chemistry*, 1, 467–72 (2005).
86. A. Golebiowski, S. R. Klopfenstein, X. Shao, *et al.*, Solid-Supported Synthesis of a Peptide β-Turn Mimetic, *Org. Lett.*, 2, 2615–17 (2000).
87. A. Golebiowski, J. Jozwik, S. R. Klopfenstein, *et al.*, Solid-Supported Synthesis of Putative Peptide β-Turn Mimetics via Ugi Reaction for Diketopiperazine Formation, *J. Comb. Chem.*, 4, 584–90 (2002).
88. Q. Lin and H. E. Blackwell, Rapid Synthesis of Diketopiperazine Macroarrays via Ugi Four-Component Reactions on Planar Solid Supports, *Chem. Comm.*, 2884–6 (2006).
89. S. Dandapani, P. Lan, A. B. Beeler, *et al.*, Convergent Synthesis of Complex Diketopiperazines Derived from Pipecolic Acid Scaffolds and Parallel Screening against GPCR Targets, *J. Org. Chem.*, 71, 8934–5 (2006).
90. H. Habashita, M. Kokubo, S.-I. Hamano, *et al.*, Design, Synthesis, and Biological Evaluation of the Combinatorial Library with a New Spirodiketopiperazine Scaffold. Discovery of Novel Potent and Selective Low-Molecular-Weight CCR5 Antagonists, *J. Med. Chem.*, 49, 4140–52 (2006).
91. R. Nishizawa, T. Nishiyama, K. Hisaichi, *et al.*, Spirodiketopiperazine-Based CCR5 Antagonists: Lead Optimization from Biologically Active Metabolite, *Biorg. Med. Chem. Lett.*, 17, 727–31 (2007).
92. I. Masip, N. Cortes, M.-J. Abad, *et al.*, Design and Synthesis of an Optimized Positional Scanning Library of Peptoids: Identification of Novel Multidrug Resistance Reversal Agents, *Biorg. Med. Chem.*, 13, 1923–9 (2005).
93. G. Malet, A. G. Martin, M. Orzaez, *et al.*, Small Molecule Inhibitors of Apaf-1-Related Caspase-3/-9 Activation That Control Mitochondrial-Dependent Apoptosis, *Cell Death and Differentiation*, 13, 1523–32 (2006).
94. M. Humet, T. Carbonell, I. Masip, *et al.*, A Positional Scanning Combinatorial Library of Peptoids as a Source of Biological Active Molecules: Identification of Antimicrobials, *J. Comb. Chem.*, 5, 597–605 (2003).
95. D. J. Newman, G. M. Cragg and K. M. Snader, Natural Products as Sources of New Drugs over the Period 1981–2002, *J. Nat. Prod.*, 66, 1022–37 (2003).
96. R. Breinbauer, I. R. Vetter and H. Waldmann, From Protein Domains to Drug Candidates: Natural Products as Guiding Principles in the Design and Synthesis of Compound Libraries, *Angew. Chem. Int. Ed.*, 41, 2878–90 (2002).
97. A. Reayi and P. Arya, Natural Product-Like Chemical Space: Search for Chemical Dissectors of Macromolecular Interactions, *Curr. Opin. Chem. Biol.*, 9, 240–7 (2005).
98. S. Shang and D. S. Tan, Advancing Chemistry and Biology through Diversity-Oriented Synthesis of Natural Product-Like Libraries, *Curr. Opin. Chem. Biol.*, 9, 248–58 (2005).
99. K. Ding, Y. Lu, Z. Nikolovska-Coleska, *et al.*, Structure-Based Design of Potent Non-Peptide MDM2 Inhibitors, *J. Am. Chem. Soc.*, 127, 10130–1 (2005).
100. C. V. Galliford, C. V. and K. A. Scheidt, Pyrrolidinyl-Spirooxindole Natural Products as Inspirations for the Development of Therapeutic Agents, *Angew. Chem. Int. Ed.* in press (2007).
101. P. R. Sebahar, H. Osada, T. Usui and R. M. Williams, Asymmetric, Stereocontrolled Total Synthesis of (+) and (−)-Spirotryprostatin B via a Diastereoselective Azomethine Ylide [1,3]-Dipolar Cycloaddition Reaction, *Tetrahedron*, 58, 6311–22 (2002).
102. P. R. Sebahar and R. M. Williams, The Synthesis of Spirooxindole Pyrrolidines via an Asymmetric Azomethine Ylide [1,3]-Dipolar Cycloaddition Reaction, *Heterocycles*, 58, 563–75 (2002).
103. P. R. Sebahar and R. M. Williams, The Asymmetric Total Synthesis of (+)- and (−)-Spirotryprostatin B, *J. Am. Chem. Soc.*, 122, 5666–7 (2000).
104. K. Ding, Y. Lu, Z. Nikolovska-Coleska, *et al.*, Structure-Based Design of Spiro-Oxindoles as Potent, Specific Small-Molecule Inhibitors of the MDM2-p53 Interaction, *J. Med. Chem.*, 49, 3432–5 (2006).

105. M. M. C. Lo, C. S. Neumann, S. Nagayama, E. O. Perlstein and S. L. Schreiber, A Library of Spirooxindoles Based on a Stereoselective Three-Component Coupling Reaction, *J. Am. Chem. Soc.*, 126, 16077–86 (2004).
106. Koehler, A. N., *et al.*, unpublished.
107. M. M. C. Lo, C. S. Neumann, S. Nagayama, E. O. Perlstein and S. L. Schreiber, A Library of Spirooxindoles Based on a Stereoselective Three-Component Coupling Reaction, *J. Am. Chem. Soc.*, 126, 16077–86 (2004).
108. C. Chen, X. Li and S. L. Schreiber, Catalytic Asymmetric [3 + 2] Cycloaddition of Azomethine Ylides. Development of a Versatile Stepwise, Three-Component Reaction for Diversity-Oriented Synthesis, *J. Am. Chem. Soc.*, 125, 10174–5 (2003).
109. S. Su, D. E. Acquilano, J. Arumugasamy, *et al.*, Convergent Synthesis of a Complex Oxime Library Using Chemical Domain Shuffling, *Org. Lett.*, 7, 2751–4 (2005).
110. D. Fokas, W. J. Ryan, D. S. Casebier and D. L. Coffen, Solution-Phase Synthesis of a Spiro [Pyrrolidine-2,3'-Oxindole] Library via a Three Component 1,3-Dipolar Cycloaddition Reaction, *Tetrahedron Lett.*, 39, 2235–8 (1998).
111. H. Ardill, R. Grigg, V. Sridharan, S. Surendrakumar, S. Thianpatanagul and S. Kanajun, Iminium Ion Route to Azomethine Ylides from Primary and Secondary Amines, *Chem. Commun.* 602–4 (1986).
112. R. Grigg, M. F. Aly, V. Sridharan and S. Thianpatanagul, Decarboxylative Transamination. A New Route to Spirocyclic and Bridgehead-Nitrogen Compounds. Relevance to α-Amino Acid Decarboxylases, Chem. Commun., 182–3 (1984).
113. R. Grigg and S. Thianpatanagul, Decarboxylative Transamination. Mechanism and Applications to the Synthesis of Heterocyclic Compounds, Chem. Commun., 180–1 (1984).
114. S. Rehn, J. Bergman and B. Stensland, The Three-Component Reaction between Isatin, α-Amino Acids, and Dipolarophiles, *Eur. J. Org. Chem.* 413–18 (2004).
115. A. K. Franz, P. D. Dreyfuss and S. L. Schreiber, Synthesis and Cellular Profiling of Diverse Organosilicon Small Molecules, *J. Am. Chem. Soc.*, 129, 1020–1 (2007).
116. C. V. Galliford, J. S. Martenson, C. Stern and K. A. Scheidt, A Highly Diastereoselective, Catalytic Three-Component Assembly Reaction for the Synthesis of Spiropyrrolidinyloxindoles, *Chem. Commun.* 631–3 (2007).
117. R. Bai, G. F. Taylor, Z. A. Cichacz, *et al.*, The Spongistatins, Potently Cytotoxic Inhibitors of Tubulin Polymerization, Bind in a Distinct Region of the Vinca Domain, *Biochemistry*, 34, 9714–21 (1995).
118. R. F. Luduena, M. C. Roach, V. Prasad, G. R. Pettit, Z. A. Cichacz and C. L. Herald, Interaction of Three Sponge-Derived Macrocyclic Lactone Polyethers (Spongistatin 3, Halistatins 1 and 2) with Tubulin, *Drug Dev. Res.*, 35, 40–8 (1995).
119. R. Bai, Z. A. Cichacz, C. L. Herald, G. R. Pettit and E. Hamel, Spongistatin 1, a Highly Cytotoxic, Sponge-Derived, Marine Natural Product That Inhibits Mitosis, Microtubule Assembly, and the Binding of Vinblastine to Tubulin, *Mol. Pharmacol.*, 44, 757–66 (1993).
120. A. V. Statsuk, R. Bai, J. L. Baryza, *et al.*, Actin Is the Primary Cellular Receptor of Bistramide A, *Nature Chemical Biology*, 1, 383–8 (2005).
121. A. V. Statsuk, D. Liu and S. A. Kozmin, Synthesis of Bistramide A, *J. Am. Chem. Soc.*, 126, 9546–7 (2004).
122. X. Fan, G. R. Flentke and D. H. Rich, Inhibition of HIV-1 Protease by a Subunit of Didemnaketal A, *J. Am. Chem. Soc.*, 120, 8893–4 (1998).
123. B. A. Kulkarni, G. P. Roth, E. Lobkovsky and J. A. Porco, Jr., Combinatorial Synthesis of Natural Product-Like Molecules Using a First-Generation Spiroketal Scaffold, *J. Comb. Chem.*, 4, 56–72 (2002).
124. G. Zinzalla, L.-G. Milroy and S. V. Ley, Chemical Variation of Natural Product-Like Scaffolds: Design and Synthesis of Spiroketal Derivatives, *Org. Biomol. Chem.*, 4, 1977–2002 (2006).
125. Moilanen, J. S. Potuzak and D. S. Tan, Stereocontrolled Synthesis of Spiroketals via Ti(O*i*-Pr)$_4$-Mediated Kinetic Spirocyclization of Glycal Epoxides with Retention of Configuration, *J. Am. Chem. Soc.*, 128, 1792–3 (2006).

126. J. S. Potuzak, S. B. Moilanen and D. S. Tan, Stereocontrolled Synthesis of Spiroketals via a Remarkable Methanol-Induced Kinetic Spirocyclization Reaction, *J. Am. Chem. Soc.*, 127, 13796–7 (2005).
127. S. Sommer and H. Waldmann, Solid Phase Synthesis of a Spiro[5.5]ketal Library, *Chem. Commun.* 5684–6 (2005).
128. O. Barun, S. Sommer and H. Waldmann, Asymmetric Solid-Phase Synthesis of 6,6-Spiroketals, *Angew. Chem. Int. Ed.*, 43, 3195–9 (2004).
129. A. D. Patil, A. J. Freyer, P. Offen, M. F. Bean and R. K. Johnson, Three New Tricyclic Guanidine Alkaloids from the Sponge Batzella Sp, *J. Nat. Prod.*, 60, 704–7 (1997).
130. A. D. Patil, N. V. Kumar, W. C. Kokke, *et al.*, Novel Alkaloids from the Sponge Batzella Sp.: Inhibitors of HIV Gp120-Human Cd4 Binding, *J. Org. Chem.*, 60, 1182–8 (1995).
131. A. Olszewski and G. A. Weiss, Library Versus Library Recognition and Inhibition of the HIV-1 Nef Allelome, *J. Am. Chem. Soc.*, 127, 12178–9 (2005).
132. B. B. Snider and M. V. Busuyek, Revision of the Stereochemistry of Batzelladine F. Approaches to the Tricyclic Hydroxyguanidine Moiety of Batzelladines G, H, and I, *J. Nat. Prod.*, 62, 1707–11 (1999).
133. A. D. Patil, A. J. Freyer, P. B. Taylor, *et al.*, Batzelladines F-I, Novel Alkaloids from the Sponge Batzella Sp.: Inducers of p56lck-CD4 Dissociation, *J. Org. Chem.*, 62, 1814–19 (1997).
134. C. A. Bewley, S. Ray, F. Cohen, S. K. Collins and L. E. Overman, Inhibition of HIV-1 Envelope-Mediated Fusion by Synthetic Batzelladine Analogues, *J. Nat. Prod.*, 67, 1319–24 (2004).
135. A. Olszewski, K. Sato, D. Aron Zachary, *et al.*, Guanidine Alkaloid Analogs as Inhibitors of HIV-1 Nef Interactions with p53, Actin, and p56lck, *Proc. Natl. Acad. Sci. U. S. A.*, 101, 14079–84 (2004).
136. F. Cohen and L. E. Overman, Enantioselective Total Synthesis of Batzelladine F and Definition of Its Structure, *J. Am. Chem. Soc.*, 128, 2604–8 (2006).
137. A. I. McDonald and L. E. Overman, Tuning Stereoselection in Tethered Biginelli Condensations. Synthesis of Cis- or Trans-1-Oxo- and 1-Iminohexahydropyrrolo[1,2-C]pyrimidines, *J. Org. Chem.*, 64, 1520–8 (1999).
138. D. S. Coffey, A. I. McDonald, L. E. Overman, M. H. Rabinowitz and P. A. Renhowe, A Practical Entry to the Crambescidin Family of Guanidine Alkaloids. Enantioselective Total Syntheses of Ptilomycalin A, Crambescidin 657 and its Methyl Ester (Neofolitispates 2), and Crambescidin 800, *J. Am. Chem. Soc.*, 122, 4893–4903 (2000).
139. Z. D. Aron and L. E. Overman, Total Synthesis and Properties of the Crambescidin Core Zwitterionic Acid and Crambescidin 359, *J. Am. Chem. Soc.*, 127, 3380–90 (2005).
140. D. S. Coffey, A. I. McDonald, L. E. Overman and F. Stappenbeck, Enantioselective Total Synthesis of 13,14,15-Isocrambescidin 800, *J. Am. Chem. Soc.*, 121, 6944–5 (1999).
141. F. Cohen and L. E. Overman, Evolution of a Strategy for the Synthesis of Structurally Complex Batzelladine Alkaloids. Enantioselective Total Synthesis of the Proposed Structure of Batzelladine F and Structural Revision, *J. Am. Chem. Soc.*, 128, 2594–2603 (2006).
142. Z. D. Aron, H. Pietraszkiewicz, L. E. Overman, F. Valeriote and C. Cuevas, Synthesis and Anticancer Activity of Side Chain Analogs of the Crambescidin Alkaloids, *Biorg. Med. Chem. Lett.*, 14, 3445–9 (2004).
143. F. Cohen, S. K. Collins and L. E. Overman, Assembling Polycyclic Bisguanidine Motifs Resembling Batzelladine Alkaloids by Double Tethered Biginelli Condensations, *Org. Lett.*, 5, 4485–8 (2003).
144. S. L. Schreiber, Small Molecules: The Missing Link in the Central Dogma, *Nature Chemical Biology*, 1, 64–6 (2005).
145. N. Tolliday, P. A. Clemons, P. Ferraiolo, *et al.*, Small Molecules, Big Players: The National Cancer Institute's Initiative for Chemical Genetics, *Cancer Research*, 66, 8935–42 (2006).
146. C. A. Lipinski, F. Lombardo, B. W. Dominy and P. J. Feeney, Experimental and Computational Approaches to Estimate Solubility and Permeability in Drug Discovery and Development Settings, *Advanced Drug Delivery Reviews*, 23, 3–25 (1997).
147. M. Congreve, R. Carr, C. Murray and H. Jhoti, A 'Rule of Three' for Fragment-Based Lead Discovery?, *Drug Discovery Today*, 8, 876–7 (2003).

148. M. M. Hann and T. I. Oprea, Pursuing the Leadlikeness Concept in Pharmaceutical Research, *Curr. Opin. Chem. Biol.*, 8, 255–63 (2004).
149. S. L. Schreiber, Target-Oriented and Diversity-Oriented Organic Synthesis in Drug Discovery, *Science (Washington, D. C.)*, 287, 1964–9 (2000).
150. M. D. Burke and S. L. Schreiber, A Planning Strategy for Diversity-Oriented Synthesis, *Angew. Chem. Int. Ed.*, 43, 46–58 (2004).
151. P. Arya, R. Joseph, Z. Gan and B. Rakic, Exploring New Chemical Space by Stereocontrolled Diversity-Oriented Synthesis, *Chem. Biol.*, 12, 163–80 (2005).
152. S. Tan Derek, Diversity-Oriented Synthesis: Exploring the Intersections between Chemistry and Biology, *Nature Chemical Biology*, 1, 74–84 (2005).
153. G. L. Thomas, E. E. Wyatt and D. R. Spring, Enriching Chemical Space with Diversity-Oriented Synthesis, *Current Opinion in Drug Discovery & Development*, 9, 700–12 (2006).
154. D. L. Boger, W. Chai, R. S. Ozer and C.-M. Anderson, Solution-Phase Combinatorial Synthesis via the Olefin Metathesis Reaction, *Biorg. Med. Chem. Lett.*, 7, 463–8 (1997).
155. D. L. Boger and W. Chai, Solution-Phase Combinatorial Synthesis: Convergent Multiplication of Diversity via the Olefin Metathesis Reaction, *Tetrahedron*, 54, 3955–70 (1998).
156. D. L. Boger, W. Chai and Q. Jin, Multistep Convergent Solution-Phase Combinatorial Synthesis and Deletion Synthesis Deconvolution, *J. Am. Chem. Soc.*, 120, 7220–5 (1998).
157. D. L. Boger, R. S. Ozer and C.-M. Andersson, Generation of Targeted C2-Symmetrical Compound Libraries by Solution-Phase Combinatorial Chemistry, *Biorg. Med. Chem. Lett.*, 7, 1903–8 (1997).
158. D. L. Boger, P. Ducray, W. Chai, W. Jiang and J. Goldberg, Higher Order Iminodiacetic Acid Libraries for Probing Protein–protein Interactions, *Biorg. Med. Chem. Lett.*, 8, 2339–44 (1998).
159. D. L. Boger, J. Goldberg, W. Jiang, *et al.*, Higher Order Iminodiacetic Acid Libraries for Probing Protein–protein Interactions, *Biorg. Med. Chem.*, 6, 1347–8 (1998).
160. D. L. Boger, J. Goldberg, S. Silletti, T. Kessler and D. A. Cheresh, Identification of a Novel Class of Small-Molecule Antiangiogenic Agents through the Screening of Combinatorial Libraries Which Function by Inhibiting the Binding and Localization of Proteinase MMP2 to Integrin $\alpha v\beta 3$, *J. Am. Chem. Soc.*, 123, 1280–8 (2001).
161. S. Silletti, T. Kessler, J. Goldberg, D. L. Boger and D. A. Cheresh, Disruption of Matrix Metalloproteinase 2 Binding to Integrin $\alpha v\beta 3$ by an Organic Molecule Inhibits Angiogenesis and Tumor Growth in Vivo, *Proc. Natl. Acad. Sci. U. S. A.*, 98, 119–24 (2001).
162. Y. Xu, J. Shi, N. Yamamoto, J. A. Moss, P. K. Vogt and K. D. Janda, A Credit-Card Library Approach for Disrupting Protein–protein Interactions, *Biorg. Med. Chem.*, 14, 2660–73 (2006).
163. Y. Xu, H. Lu, J. P. Kennedy, *et al.*, Evaluation of 'Credit Card' Libraries for Inhibition of HIV-1 gp41 Fusogenic Core Formation, *J. Comb. Chem.*, 8, 531–9 (2006).
164. T. J. Dickerson, A. E. I. V. Beuscher, C. J. Rogers, *et al.*, Discovery of Acetylcholinesterase Peripheral Anionic Site Ligands through Computational Refinement of a Directed Library, *Biochemistry*, 44, 14845–53 (2005).
165. K. McMillan, M. Adler, D. S. Auld, *et al.*, Allosteric Inhibitors of Inducible Nitric Oxide Synthase Dimerization Discovered via Combinatorial Chemistry, *Proc. Natl. Acad. Sci. U. S. A.*, 97, 1506–11 (2000).
166. M. H. J. Ohlmeyer, R. N. Swanson, L. Dillard, *et al.*, Complex Synthetic Chemical Libraries Indexed with Molecular Tags, *Proc. Natl. Acad. Sci. U. S. A.*, 90, 10922–6 (1993).
167. J. A. Tallarico, K. M. Depew, H. E. Pelish, *et al.*, An Alkylsilyl-Tethered, High-Capacity Solid Support Amenable to Diversity-Oriented Synthesis for One-Bead, One-Stock Solution Chemical Genetics, *J. Comb. Chem.*, 3, 312–18 (2001).
168. G. MacBeath, A. N. Koehler and S. L. Schreiber, Printing Small Molecules as Microarrays and Detecting Protein-Ligand Interactions En Masse, *J. Am. Chem. Soc.*, 121, 7967–8 (1999).
169. J. L. Duffner, P. A. Clemons and A. N. Koehler, A Pipeline for Ligand Discovery Using Small-Molecule Microarrays, *Curr. Opin. Chem. Biol.*, 11, 74–82 (2007).
170. S. M. Sternson, J. B. Louca, J. C. Wong and S. L. Schreiber, Split-Pool Synthesis of 1,3-Dioxanes Leading to Arrayed Stock Solutions of Single Compounds Sufficient for Multiple Phenotypic and Protein-Binding Assays, *J. Am. Chem. Soc.*, 123, 1740–7 (2001).

171. R. A. Stavenger and S. L. Schreiber, Asymmetric Catalysis in Diversity-Oriented Organic Synthesis: Enantioselective Synthesis of 4320 Encoded and Spatially Segregated Dihydropyrancarboxamides, *Angew. Chem. Int. Ed.*, 40, 3417–21 (2001).
172. D. R. Spring, S. Krishnan, H. E. Blackwell and S. L. Schreiber, Diversity-Oriented Synthesis of Biaryl-Containing Medium Rings Using a One Bead/One Stock Solution Platform, *J. Am. Chem. Soc.*, 124, 1354–63 (2002).
173. J. E. Darnell, Jr., Transcription Factors as Targets for Cancer Therapy, *Nature Reviews Cancer*, 2, 740–9 (2002).
174. A. N. Koehler, A. F. Shamji and S. L. Schreiber, Discovery of an Inhibitor of a Transcription Factor Using Small Molecule Microarrays and Diversity-Oriented Synthesis, *J. Am. Chem. Soc.*, 125, 8420–1 (2003).
175. K. Bedjeguelal, H. Bienayme, A. Dumoulin, S. Poigny, P. Schmitt and E. Tam, Discovery of Protein–protein Binding Disruptors Using Multi-Component Condensations Small Molecules, *Biorg. Med. Chem. Lett.*, 16, 3998–4001 (2006).
176. P. Y. Ng, Y. Tang, W. M. Knosp, H. S. Stadler and J. T. Shaw, Synthesis of Diverse Lactam Carboxamides Leading to the Discovery of a New Transcription Factor Inhibitor, *Angew. Chem. Int. Ed.*, 5352–5 (2007).

8

In Silico Screening

F. J. Luque[1,2] *and X. Barril*[1,2,3]

[1]*Department of Physical Chemistry, Faculty of Pharmacy, Barcelona, Spain*

[2]*Institut de Biomedicina de la Universitat de Barcelona (IBUB), Barcelona, Spain*

[3]*Institució Catalana de Recerca i Estudis Avançats (ICREA), Barcelona, Spain*

8.1 Introduction

Drug discovery can be described as a trial and error exercise where chemical compounds are tested in biological assays. As it is only feasible to explore a small proportion of the chemical and biological universes, the outcome of a drug discovery project crucially depends on making the right choice of compounds and assays. The decision on the latter is defined by the therapeutic area and the target product profile of the drug-to-be, which will help us design a screening cascade, ensuring that all relevant properties are considered while making the most efficient use of resources [1, 2]. Selecting the appropriate pool of compounds to test is also of fundamental importance because it will determine the hit rate (i.e. the proportion of active compounds) and the properties of those hits in terms of novelty, potency, pharmacokinetic profile, synthetic feasibility, selectivity, etc. As this is a difficult choice, a prevailing trend in pharmaceutical industry has been to use high-throughput screening (HTS) technologies to test as many compounds as possible, often in the order of 10^6. The underlying assumption was that such a systematic approach would provide active compounds against any target. Unfortunately, this potential has not been realized for the discovery of protein–protein inhibitors (PPIs) and other target families [3].

An alternative to this 'brute force' approach consists in testing a smaller subset of compounds, but with a greater probability of being active. This can be achieved either by preselecting active compounds, as in *in silico* screening, or by using more sensitive assays, as

Protein Surface Recognition: Approaches for Drug Discovery Ernest Giralt, Mark W. Peczuh and Xavier Salvatella. © 2011 John Wiley & Sons, Ltd

in fragment screening [4]. *In silico* screening (also known as virtual screening) can be applied at any stage of the drug discovery process, but is particularly useful in its early phases (hit identification and hit to lead), when the size and diversity of the chemical collections is at its peak and the biological properties to be modelled are relatively simple.

In this chapter we will first review the most common techniques used in virtual screening. Then the particularities of protein–protein interactions will be discussed, describing some computational methods that can be used to understand the molecular recognition capacities of protein surfaces and identify those features that can hinder or facilitate the discovery of new PPIs. Finally, we will examine two recent successful applications of virtual screening to the discovery of PPIs.

8.2 Methods for Virtual Ligand Screening

Traditionally, computational methods for drug discovery have been ascribed to two different categories according to the type of prior knowledge used to select compounds. If some active compounds are already known, ligand-based methods can be used to find similar molecules. If, on the other hand, what is known is the three-dimensional structure of the targeted site, structure-based methods can be used to identify potential binders. In the context of this book, the latter is much more relevant, but nowadays both approximations are often combined together.

8.2.1 Chemoinformatics and Ligand-based Methods

When dealing with large volumes of chemical data, it is absolutely necessary to have an information system able to store, interpret and classify compounds by their chemical structure. The new discipline of chemoinformatics (also known as chemical informatics) provides these capacities and deals, amongst many other issues, with the electronic processing of chemical information, file format conversion, databasing and data mining. Chemoinformatics is intimately linked to two-dimensional (2D) ligand-based methods that are universally used to predict molecular properties and profile chemical libraries. Three-dimensional (3D) ligand-based methods lay one step closer to structure-based methods and, thanks to synergies between them, their combination is often a good solution for virtual screening.

8.2.1.1 Chemoinformatics and 2D Methods

In order to do *in silico* screening, the first step is to have at one's disposal a virtual chemical library. Some popular choices are the National Cancer Institute (NCI) collection or the Available Chemical Directory (ACD). In other cases the databases will be generated from historical collections or from catalogues of fine chemicals providers. In any case, the experience gained in the HTS era has shown that a screening collection should fulfil certain conditions to ensure that the hits can be progressed into leads. Table 8.1 lists some of the items that should be monitored in chemical collections (both real and virtual) [5, 6]. All these tasks previous to virtual screening fall into the realm of chemoinformatics and 2D quantitative structure-activity relationships (2D-QSAR). Other important tasks include the enumeration of combinatorial libraries or the analysis of HTS results. Interested readers

Table 8.1 *Parameters related to the quality of screening collections that can be monitored using chemoinformatics*

Name	Description	References
Reactivity	Reactive functionalities result in false positives and have to be discarded.	[8, 9]
Solubility	Compounds must be reasonable soluble in water to be tested and to become suitable starting points.	[10]
Self-aggregation	A recently detected source of false positives.	[11, 12]
Drug-likeness	Physico-chemical properties that confer oral bioavailability to drugs.	[13, 14]
Lead-likeness	Leads tend to be smaller and more soluble that the drugs derived from them	[15, 16]
Diversity	A measure of (dis)similarity of chemical compounds, which is used to design libraries with high information contents.	[17]
Privileged scaffolds	Some chemical groups with good medicinal chemistry properties.	[18, 19]
Focussed libraries	Collections targeting specific protein families	[20]

are pointed to the book of Gasteiger and Engel [7] or other specialized publications on the subject.

8.2.1.2 3D Similarity

There are different metrics that one can use to compare the three-dimensional similarity of two molecules. The most popular ones are shape, molecular fields and pharmacophores [21]. Here only the latter will be discussed because it is the one most commonly used in combination with structure-based methods. According to the IUPAC, a pharmacophore is an ensemble of steric and electronic features that is necessary to ensure the optimal supramolecular interactions with a specific biological target and to trigger (or block) its biological response. Defining molecules as a set of pharmacophoric features (aromatic rings, H-bond donors, charged centres, etc.) provides a fuzzy description of the active compounds, making it possible to identify similarities between molecules that are structurally unrelated. The use of pharmacophores involves two sequential steps:

1. Hypothesis generation. Based on empirical information, an arrangement of pharmacophoric features is proposed as important for binding.
2. Pharmacophoric search. Compounds are compared to the hypothesis and a prediction about their activity is made. The prediction usually comes in the form of a score that gives an idea of how well the molecule fits the query.

Traditionally, both steps use only information about ligands, but as the availability of structural information is ever more common, the powerful concept of pharmacophore has been adapted and exploited in combination with structure-based methods. In the first step, the structure of ligand-receptor complexes can be used to bypass the alignment of the ligands, which is one of the bottlenecks in any three-dimensional ligand-based method [22, 23]. It is also possible to derive a pharmacophore from the structure of the target using hot spots analysis (*vide infra*) to identify those sites on the protein surface more avid for interactions [24]. The pharmacophoric search can also make use of the target structure, for instance filtering out those compounds that would clash with the protein. Reference [25] provides an exhaustive description of pharmacophore-related methods and applications.

8.2.2 Structure-based Methods

Structural information about biological macromolecules has been accumulating for over 30 years and is growing at a very fast rate thanks to the structural genomics initiatives and the introduction of high-throughput techniques in protein crystallography [26]. Using the principles of molecular recognition, it is possible to estimate the degree of complementarity between a target protein and candidate ligands, which is of great interest in drug design. The only requirements, besides structure, are an appropriate set of potentials to describe molecular interactions and a sampling algorithm capable of exploring the configurational space. Unfortunately, precise quantification of binding is very difficult because it involves many different terms, often of large magnitude and opposing signs and also because the configurational space of both ligand and receptor cannot be exhaustively sampled. As a consequence, different molecular potentials and search algorithms have been developed for specific applications. Here we will review some of the formalisms and methods more relevant for virtual screening and drug design.

8.2.2.1 Molecular Potentials

a. Force-fields. Within the Born-Oppenheimer approximation, the movement of electrons and nuclei can be considered independently; hence it is possible to study only the movement of the atomic centres assuming that the electron distribution is always in equilibrium. Molecular force-fields make use of this approximation to bypass the calculation of such electronic distribution and replace it with functions and parameter sets that describe its effects. As a result, molecules and atoms become classical particles, whose mutual interactions are governed by bonded and nonbonded terms that adopt simple forms (see Equation (8.1)); hence the name of molecular mechanics (MM). The parameters describing the force-field are derived from experimental data and/or high-level quantum mechanics results. The most common and well-validated force-fields for biological macromolecules are AMBER [27], CHARMM [28] and GROMOS [29], while force-fields such as MMFF94 [30] were designed to be as general as possible and are widely used to simulate small organic molecules. In spite of the roughness of the approximation, MM-methods have a long history of success and have become a fundamental tool not only for computational chemistry [31] but also for structural sciences [32–34]. DOCK, the first docking program [35], implemented a modified force-field; this strategy is still used by several docking programs. One of the limitations of force-fields is their lack of good quality parameters for organic molecules. The partial charges, for instance, are crudely approximated using electronegativity equalization principles [36] or are derived for functional groups and assumed to be transferable between molecules [37], which can be an important source of errors [38].

$$V = V_{\text{bonded}} + V_{\text{nonb}}$$

$$V_{\text{bonded}} = \sum\nolimits_{\text{bonds}} K_b(d-d_0) + \sum\nolimits_{\text{angles}} K_a(\theta-\theta_0) + \frac{1}{2}\sum_{\text{dihedrals}} K_d(1+\cos(m\phi-\gamma))$$

$$V_{\text{nonb}} = \sum_{\text{nonb}}\left(\frac{A}{r^{12}} - \frac{B}{r^6} + \frac{q_i q_j}{4\pi\varepsilon_0 r_{\text{ij}}}\right)$$

Equation 8.1. Form of the AMBER force-field equation. The bonded terms penalize the deviation of bond distances, bond angles and dihedral angles from their ideal values. The non-bonded part consists of Lennard-Jones and Coulombic terms that account for the dispersion-repulsion and electrostatic effects respectively.

b. Empirical Scoring Functions. Empirical scoring functions estimate the binding energy of a ligand-receptor complex in terms of the interactions formed in the interface. The contribution of hydrogen bonding, ionic and hydrophobic interactions towards the binding free energy are defined by functions that have been calibrated against complexes of known affinity. Most of the empirical scores in use (e.g. FlexX [39], ChemScore [40]) today derive ultimately from the pioneering work of Bohm [41], as incorporated in the LUDI program [42]. Empirical functions generally perform well in binding mode prediction and hit identification, but are less successful at accurately ranking active molecules by binding free energy. The hydrophobic terms of pure empirical scoring functions have been replaced by Lennard-Jones potentials from molecular force-fields to produce the so-called semiempirical scoring functions (e.g. GOLD [43], LigandFit [44]).

$$\Delta G_{\text{bind}} = \Delta G_0 + \Delta G_{\text{hb}} \sum\nolimits_{\text{h-bonds}} f(\Delta d) f(\Delta\alpha) + \Delta G_{\text{ion}} \sum\nolimits_{\text{ionic}} f(\Delta d) f(\Delta\alpha) + \Delta G_{\text{lipo}} A_{\text{lipo}} + \Delta G_{\text{aro}} \Delta N_{\text{aro}} + \Delta G_{\text{rot}} NR$$

Equation 8.2. Form of Bohm's empirical equation. The binding free energy of a ligand-receptor pair is approximated as a linear combination of a constant term (ΔG_0) which accounts for the entropic cost of losing rotational and translational degrees of freedom; four geometry dependent terms, which quantify the ligand-receptor interaction based on its hydrogen bond, ionic, lipophilic and aromatic complementarity; and a conformational entropic penalty related to the number of rotatable bonds of the ligand.

c. Statistical Potentials. A third class of parametric methods is that of knowledge-based statistical potentials, exemplified by potentials of mean force (PMFs). They are based on the principle that the observed distribution of distances between pairs of different atom types is a reflection of their energy of interaction. In practice, large training sets of protein-ligand structures are analysed to provide sets of distribution functions. These are then converted to atom-pair potentials using the inverse Boltzmann technique (Equation (8.3)), which provides an energy value for a given state based on observed probabilities;

no experimental binding affinities are, thus, needed. Examples of protein-ligand potentials include BLEEP [45], PMF Score [46] and Drug Score [47]. The various approaches differ in the sets of protein-ligand complexes used to obtain these potentials, the form of the energy function, the definition of protein and ligand atom types, the definition of reference states, distance cut-offs and several other parameters. In addition to scoring protein-ligand complexes, knowledge based potentials have also been derived to study protein–protein complexes [48] and for protein structure prediction [49]. For docking applications, PMFs are generally not used during the optimization phase but mostly to identify decoys or to use in combination with other scoring functions in virtual screening (VS) applications (consensus scoring [50]).

$$\text{score} = \sum_{\substack{kl \\ r < r^{ij}_{\text{cut-off}}}} A_{ij}(r) \quad \text{where} \quad A_{ij}(r) = -k_B T \ln \left[f^{i}_{\text{Vol\_corr}}(r) \, \frac{\rho^{ij}_{\text{seg}}(r)}{\rho^{ij}_{\text{bulk}}} \right]$$

Equation 8.3. In the knowledge-based approach the binding free energy (score in the leftmost formula) is described by interactions between pairs of groups (*i* and *j*) in the ligand and the protein, respectively. A_{ij} is a distance dependent function describing the interaction between two specific atom types, which is obtained from the relationship between de number density of atom pairs (ρ^{ij}_{seg}) at a certain atom pair distance (r) and in the reference state (ρ^{ij}_{seg}).

8.2.2.2 Solvation Effects

Water has a profound influence in all biochemical phenomena, including molecular recognition. It can alter the charge distribution, as noted by the changes in properties such as the dipole moment and the molecular electrostatic potential, affects the conformational, tautomeric and ionization preferences of both small molecules and their macromolecular targets and governs the hydrophobic effect, by which nonpolar groups tend to aggregate to reduce the solvent exposed hydrophobic surface, thus minimizing the entropy loss due to the ordering of water molecules [51, 52]. Since ligand-receptor non covalent association takes place in an aqueous environment, the role of water must be taken into account both to qualitatively understand this process and to obtain quantitative measures of the free energy.

If the system is described by a classical model based on a force-field, the simplest way of taking into account the effect of water is representing it by its discrete molecules. Using molecular dynamics or Monte Carlo simulations, it is possible to obtain an ensemble description of the molecular system. This allows to gain insight on the differential solvation of certain parts of the solute, to generate radial distribution functions, to determine whether water molecules bridge ligand and receptor, etc. When coupled to statistical mechanics, it can also be used to calculate differential free energies of solvation between different solutes with the help of thermodynamic cycles [53]. However, one of the most serious drawbacks of the explicit water treatment is its computational cost.

Less computationally demanding methods have been developed to account for solvent effect on a classical system. Among the most popular ones are the classical continuum electrostatic methods, where the solvent is treated as a continuum environment. Several formalisms have been developed to couple the charge distribution of the solute with the solvent reaction field. The most widely used models are Poisson, or Poisson-Boltzmann (PB) [54, 55] if the effect of counterions is also taken into account, and Generalized Born (GB) [56]. Besides the formalism used to treat the electrostatic interaction between solute and solvent, the quality of the results is largely affected by the definition of the dielectric permittivity assigned to the interior of the cavity [57], and by the definition of the solute/solvent boundary in PB computations [58] or the assignment of the coulomb radii in GB calculations.

Other implicit water treatment methods for a classical system are those derived empirically. Although less rigorous, they provide the benefit of a lower computational cost. A first group makes use of parameters for modelling screening of electrostatic interactions by water, replacing the macroscopic permittivity by a distance-dependent dielectric function. In the simplest models the latter can change linearly [59], but more complex models where it changes exponentially have also been developed [60]. A second group of empirical methods is based on the solvent accessible surface area (SASA). In them, it is assumed that solvation free energy can be calculated by the addition of the contribution of each atom or group of atoms. Each atom type is given a solvation parameter obtained by a fitting procedure, and the contribution of each atom is based on its solvent accessible surface area [61].

8.2.2.3 Sampling Algorithms

Once the chemical system under study has been defined and all necessary parameters have been obtained, one can proceed to run calculations on it. Usually flexible molecules have a complicated potential energy surface, with many minima and saddle points, which are a function of the nuclear coordinates. Especially interesting are the configurations that correspond to minima in the potential hypersurface as these are stable states of the system. The identification of these minima will generally consist of two steps: global exploration and local minimization. Although it is theoretically possible to systematically explore each degree of freedom, in practice this is only feasible for very small systems and stochastic methods are used instead. Evolutionary computational techniques such as genetic algorithms (GA) are particularly widespread. These methods start by generating a random collection of candidate solutions whose fitness is evaluated; the best individuals are then stochastically selected and mutated or recombined to obtain a new population. This process is repeated until a certain convergence criterion is met.

For local optimizations, there are several minimization algorithms that search the nearest minimum in the potential surface. These can be broadly classified into two groups: those that do not use the derivatives of the potential energy with respect to the coordinates, such as the simplex method, and those that do, such as the steepest descent and conjugate gradient methods. The latter operate in an iterative procedure: (1) calculate the potential energy of a given configuration; (2) determine the first (gradient) and second derivative of the energy with respect to the coordinates; (3) generate of a new set of coordinates in the direction of the minimum; and (4) evaluate again the energy of the system with the new coordinate set. The process is repeated until the energy is converged (gradient below a given threshold). Minimizations are often used in structure-based drug design in several contexts, such as the initial refinement of a protein structure obtained by experimental methods or to relax the geometry and eliminate unfavourable contacts of a ligand–receptor complex.

However, the biggest limitation of a minimized molecular system is its static character. Molecules vibrate and constantly change conformation, overcoming potential energy barriers and populating an ensemble of microstates, which are globally responsible for the properties of the system. To generate such an ensemble different algorithms can be used, molecular dynamics (MD) and Monte Carlo (MC) being the most widely used. MD is based on the application of Newton's equations of motion to describe the evolution of a classical system along time. When the system is defined with a force field, it is feasible to calculate the forces acting upon each particle (atom) by obtaining the gradient of the potential energy. Once the force on each particle is known, its acceleration can be derived which, in turn, determines the velocity and position after a time increment. The process is iteratively repeated, resulting in a set of structures that represent the evolution of the system along a time path. Since the initial velocities are randomly generated, the only prerequisite is a valid set of initial coordinates.

While in MD there is a time dependency, MC generates an ensemble of states in a stochastic fashion. Starting from a given conformation, a perturbation is introduced to the system by modifying a randomly selected degree of freedom by a random small quantity. Then a ratio of probabilities is computed for the trial and original configurations and, from this quantity, a decision is made to accept or reject the trial configuration. Usually, in molecular simulations, the metropolis criterion is used to decide whether the trial configuration is accepted or rejected. This is implemented as follows: first, the energy change (ΔU) due to the perturbation introduced to the system is computed. If the trial configuration has lower energy than the original, it is accepted; otherwise a function of probability (ω) is calculated:

$$\omega = \exp(-\beta \Delta \mathrm{U})$$

where β is a factor that depends on the temperature at which the simulation is carried out. Finally, a random number (r) between 0 and 1 is generated and the trial conformation is accepted if $r < \omega$. Thus, the probability of accepting a new configuration is greater if the increase in energy is small or if the temperature is high. The efficiency of MC greatly depends on how often a new configuration is accepted and on the capability to explore significantly different configurations; an acceptance ratio in the region of 50% is generally considered as a good trade-off between both parameters. MD and MC as sampling algorithms are fundamental tools for statistical mechanics, the most rigorous methods to determine changes in free energy of binding of ligand-receptor complexes [62, 63].

8.2.2.4 Docking

Molecular docking was first applied to drug design 25 years ago [35] as a computational tool that combines a search algorithm with a scoring function, both tailored to the specific problem of predicting the binding mode of a ligand into a receptor. Although the basic principles remain the same, many new algorithms and scoring functions have been (and continue to be) developed. A detailed survey of the progress in the field has been presented in recent reviews [64–66]. The three main factors defining a docking application are the scoring function, the search algorithm and to which extend the flexibility of the ligand and the receptor are considered. Most programs implement empirical scoring functions as they seem to provide the best quality of results. Regarding the sampling algorithm, Monte Carlo and

Genetic Algorithms are the main methods of choice, but in both cases, owing to the stochastic nature of the search, multiple independent docking runs per ligand are required to achieve convergence with a reasonable degree of confidence. At present, most docking applications consider the receptor as a rigid body and, on the ligand side, only the degrees of freedom corresponding to dihedral angles are explored, either during docking (flexible docking) or by means of pregenerated libraries of conformers (rigid docking). In contrast, protein flexibility is often not addressed at all, or only in a limited way, the main reason being the complexity and computational cost of dealing with a flexible receptor. However, docking calculations can be quite sensitive to small differences in the active-site of a protein [67]. Receptor flexibility can be partially accounted for using different approximations [68]:

- Soft docking. The protein is kept rigid, but a softer scoring function is used to allow a certain degree of overlap between ligand and protein.
- Side-chain flexibility. Some side-chains of residues in the active site are allowed to explore their torsional degrees of freedom.
- Water molecules. Protons and lone pairs can be reoriented, or interstitial water molecules may be displaced by the ligand [69].
- Ensembles. Multiple receptor structures can be used, either from experimental [70] or computational [71–73] sources. The structures can be used individually, combining the docking results *a posteriori*, or grid averages can be used to dock ligands to fuzzy description of the receptor [74, 75].

8.3 Binding Site Characterization

The tendency to engage in molecular association is unevenly distributed on the surface of proteins, with one or a few patches showing much greater interaction potential than the rest of the protein; these are the so-called 'hot spots' [76, 77]. The obvious implication for structure-based drug design is that one should know beforehand where those hot spots are located and target them to have a greater chance of finding ligands. Indeed, in a retrospective analysis of NMR-based fragment screening it was shown that all ligands were binding to a single site for 18 out of the 23 pharmacological targets under consideration, while the other 5 targets presented two binding sites [78]. Clefts or pockets on the protein surface are a special case of hot spots that have been extensively studied in pharmaceutical research and it is now possible to predict their ability to bind small molecules and even the likelihood of those ligands to become drugs. In this section we will review computational tools that can help us interpret the structure of a protein, predicting its binding potential, and providing clues about the best drug design strategy. Finally we will also address the issue of conformational flexibility and how this affects ligand binding.

8.3.1 Hot Spots Analysis

The notion of hot spots originates from alanine scanning experiments, where residues located at the interface between two proteins are systematically mutated to alanine, measuring the change in the association constant of the complex [76]. In the majority of cases studied, most of the binding free energy is contributed by a few residues only. The thermodynamic changes of the association process are related to the specific interaction that is disrupted when a residue is mutated to alanine, but also to changes in the unbound state; therefore the results should be interpret with caution. Solvent mapping by NMR [79] or X-ray crystallography [80, 81] is an alternative approach that identifies binding sites of solvent molecules, and the results show that binding tends to localize in a few preferential sites, which is consistent with the presence of hot spots on the surface of proteins. *In silico* screening techniques can make use of any existing knowledge about important binding epitopes, but if this is not available, there is a wide range of computational tools that can be used to predict their existence.

8.3.1.1 Knowledge-based Methods

The increasing availability of genetic and structural information means that often there is a great deal of latent knowledge about the binding preferences of our target. Inspection of crystallographic structures can provide insights into the flexibility of the ligand, the presence of structural water molecules and other useful information. This is illustrated here with β-catenin, an oncology target in the Wnt pathway [82]. β-catenin is a superhelix with a long and shallow groove where transcription factors and other proteins bind. The complexes between β-catenin and several partners can be found in the PDB, and a simple superimposition of those structures shows that there is a common binding site used not only by the transcription factors Tcf-3 and Tcf-4, but also by several members of the cadherin family, by APC and the inhibitory protein ICAT. In spite of the fact that these proteins are completely unrelated, and that their complexes with β-catenin play entirely different biological roles, the structural alignment reveals a consensus binding sequence or linear motif [83] of the form: Asp-X-X-Aliphatic-Aliphatic-X-Aromatic. Structurally conserved residues have been shown to correlate with the presence of hot spots [84], therefore the fact that at least 4 unrelated protein families have converged to a common binding motif should be taken as a strong indication that this is an important binding epitope (Figure 8.1). The aspartate binding site is particularly interesting because it binds at the base of a small pocket where a positive charge (K435) is found. Due to desolvation costs, salt bridges do not usually contribute to the free energy of binding. But if, as in this case, the charged group is already partially desolvated in the unbound state, ionic interactions can be exothermic [85–87]. The binding potential of this pocket is also supported by the observation that it is a binding site for urea, as revealed by an apo structure of β-catenin obtained in the presence of 2.4M urea [88]. An alanine scanning experiment on Tcf-4 confirms that mutation of the conserved Asp (D16A-Tcf-4) causes a 50-fold reduction in binding affinity [89].

8.3.1.2 Computational Methods

The most common approach to detect hot spots in protein-ligand interfaces is to calculate the interaction potential of the protein with several probe atoms representative of the most common chemical features (h-bond donors, h-bond acceptors, lipophilic groups). The interaction between the protein and the probe can be calculated using empirical scoring functions [41], statistical potentials [90, 91] or first-principle methods [92]. The two former methods have the advantage of implicitly accounting for desolvation and entropic effects, but parameterizations are heavily influenced by the composition of the training set [93], which compromises the ability of empirical and knowledge based potentials to identify hot spots in protein surfaces of arbitrary shape and chemical composition. Force-fields or even

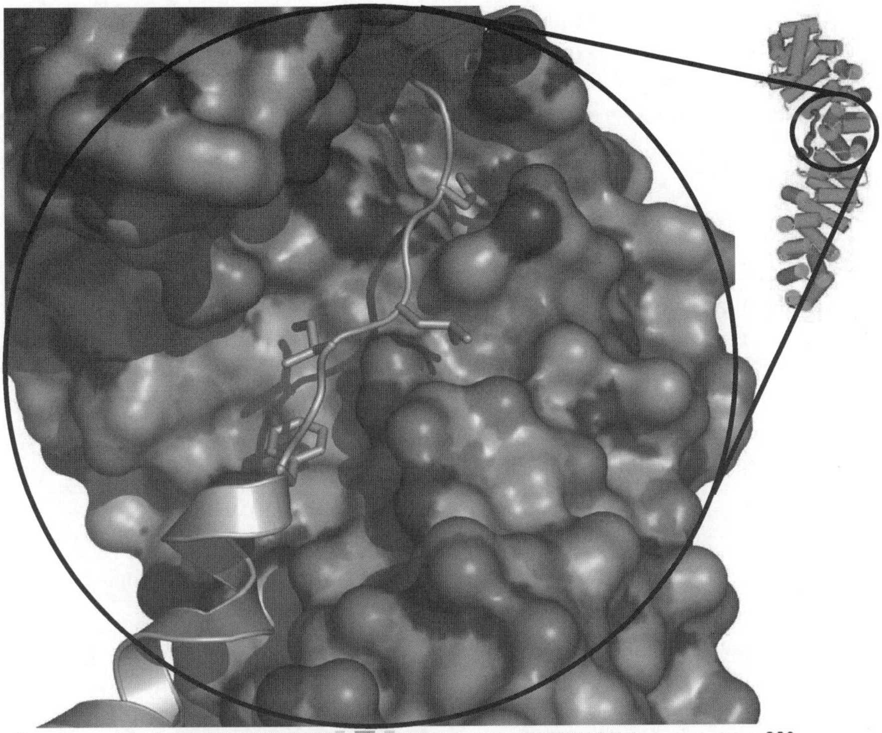

```
N-cad     841 FINEGLKAADNDPTAPPYDSLLVFDYEGS--------GSTAGSLSSLN------------ 880
DE-Cad   1322 FLGDKKENCDRDVGATTVDDVRHYAYEGD--------GNSDGSLSSLASCTDDGDLNFDY 1473
E-Cad     812 FIDENLKAADSDPTAPPYDSLLVFDYEGS--------GSEAASLSSLNSSESDQDQDYDY 863
ICAT       49 -NSQLSQLPPHSIDQGAEDVVMAFS---------------RSETEDRRQ----------- 81
tcf-3       1 -MPQLNSGGGDELGAN--DELIRFKDEGEQEEKSPGEGSAEGDLADVKSSLVNESENHSS 57
tcf-4       1 -MPQLNGGGGDDLGAN--DELISFKDEGEQEEKSSENSSAERDLADVKSSLVNESETNQN 57
apc      1004 LAHKIHSANHMDDNDGELDTPINYSLKYS-----------DEQLNSGRQS-PSQNERWA- 1050
```

Figure 8.1 *Structure of the complex between β-catenin and Tcf-4 (PDB code: 1JDH) and structural alignment of the sequences of other β-catenin-binding proteins. The residues forming the common binding motif are shown in the structure and are highlighted in the sequence alignment. All figures were prepared using PyMOL (http://www.pymol.org) (See Plate 14.)*

quantum mechanical methods can also be used to calculate molecular interaction potentials, but in this case solvation and entropic effects have to be accounted for explicitly [94]. Computational solvent mapping offers an efficient solution to the identification of preferential binding sites. In this approach, the energetically most favourable binding sites for up to eight prototypical solvents or small organic molecules are calculated using the CHARMM force-field and a continuum solvation method. Important binding sites can be distinguished because all probe molecules bind to them, while other locations bind only some of them [95].

Computational methods can also be used to predict the outcome of alanine scanning experiments, but in this case it is necessary to know the structure of the protein–protein complex. Again, methods based on first principles can be used; Massova and Kollman, for instance, applied the MM-PBSA approach to estimate the contribution of each residue to the formation of the MDM2-p53 complex [96]. Nevertheless this is a very expensive approach and the average unsigned error is in the region of 1 kcal/mol. Results of similar quality can be obtained at a much lower cost using FOLD-X and other scoring functions that use a linear combinations of different energetic terms, which have been optimized using large sets of experimental data [97, 98].

8.3.2 Cavity Druggability

Drugs are subjected to a double pressure. On the one hand they have to bind tightly to their target, but on the other they have to be able to cross membranes, be metabolically stable, safe, etc. Retrospective analysis of known drugs shows that there are certain physical and chemical properties of molecules that modulate their administration, distribution, metabolization, excretion and toxicity (ADMET) profile. Lipinki's rule of five [13] is without any doubt the best well know study of this class. It states that a molecule is more likely to have poor absorption or permeation when any of those conditions are met:

- It contains more than 5 hydrogen bond (H-bond) donors MW.
- The molecular weight (MW) is over 500.
- The calculated octanol/water partition coefficient (ClogP) is greater than 5.
- It contains more than 10 H-bond acceptors.

Amongst other more recent studies it is worth mentioning the work of Veber as it provides a predictor of oral absorption using only two descriptors: the number of rotatable bonds, which should be less than 10, and the polar surface area, which should be less than 140 Å^2 [14]. Regardless of the specific parameters, it is apparent that small molecules have a greater probability of being bioavailable than larger ones. Likewise, molecules with a balance of polar and apolar features are preferred. However, it should be emphasized that this is a game of probabilities; Lipinki's or Veber's rules are very useful guideposts, but there are many examples of drugs that do not adhere to these criteria and can cross membranes either by passive diffusion or acting as substrates of membrane transporters [99]. It should also be acknowledged that although oral bioavailability provides some key benefits, other administration routes are more permissive with respect to the nature of the drug, and even compounds that do not cross membranes can be very effective, as is the case of protein drugs, an increasingly important market [100]. It is also expected that progress on drug delivery systems may be useful in the future to facilitate oral absorption of compounds with poor pharmacokinetics [101]. All things considered, keeping a low molecular weight is highly recommended if ligands have to become drugs.

In order to achieve the required binding potency with a small molecule, each atom in the ligand has to make energetically favourable contacts with the receptor. The tightest noncovalent complexes attain up to 1.5 kcal/mol per nonhydrogen atom [102], and many drugs can achieve dissociation constants in the nM range using usually less that 40 nonhydrogen atoms, which approximately translates to a binding efficiency of 0.3 kcal/mol per atom [103]. These levels of efficiency can only be achieved when the ligand and the protein are closely intertwined, with the surface of the protein wrapped around the ligand. Enzymes, molecular receptors and other proteins whose biological role involves binding small substrates or ligands have adapted to realize this function, and their surfaces have an extraordinary capability to recognize small molecules. This explains why they are the targets of the immense majority of approved drugs [104]. For other protein classes finding active molecules and developing them into oral drugs is bound to be more difficult, but while in some cases it may be feasible, as demonstrated by the case studies provided in this book, in others it will simply lead to a waste of resources. Similarly, enzymes and receptors are considered to be druggable, but some cases can be completely intractable, as demonstrated by the failure of HTS to identify hits in entire families of GPCRs [3]. In consequence, there is a great interest in being able to predict which protein surfaces are druggable.

Two recent studies have obtained structure-based relationships to quantify the druggability of protein surfaces. In the first study, Hajduk and collaborators at Abbott used the hit rates obtained with NMR-based fragment screening as the measure of druggability for 28 binding sites on 23 protein targets, including protein–protein interaction sites such as the Bcl-xL/BAD interface (Figure 8.2). Then they tried to correlate those hit rates with several structural parameters (volume of the cavity, surface area, etc.). None of the descriptors alone

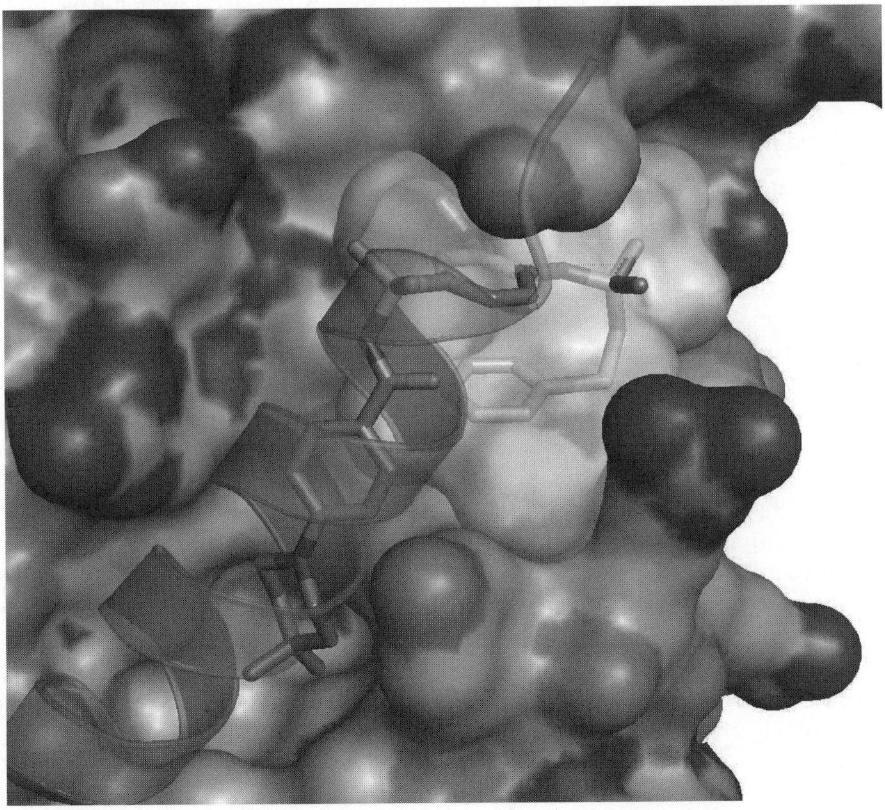

Figure 8.2 *The interaction between Bcl-xL and BAD (Bcl-2 antagonist of cell death) can be disrupted with drug-like compounds. The inhibitor (shown in sticks) binds to a cavity on the surface of Bcl-xL that can be predicted as druggable using computational methods. The PDB code of the ligand-protein complex is 2O22 (See Plate 15.)*

could predict druggability, but they obtained a mathematical model with eight variables that shows a good predicting capability (leave-one-out $q^2 = 0.59$) [78]. Scientists at Pfizer have followed a different approach that uses the maximum achievable potency as the measure of druggability. Starting from a physically based model and using very elegant approximations, they obtain a formula with only two variables, namely the curvature of the cavity and the fraction of apolar surface area, to predict the maximum binding free energy that any ligand could attain for this particular site. It is worth noting that the model is derived for typical drug-like molecules and does not account for strong ionic interactions or metal chelation. Therefore highly charged cavities are classified as undruggable, but in those cases the active molecules usually are constitutively charged, which renders them unable to cross membranes. As noted by the authors, successful drugs targeting these types of cavities either use an active transport mechanism or are administered as prodrugs [105].

In summary, one should distinguish binding that occurs in flat surfaces from that mediated by cavities. The latter is typical of protein-ligand interactions but also of protein–protein interactions, particularly those that occur between a globular protein and a linear motif [106]. Lipophilic and enclosed surfaces have intuitively considered better targets for drug design [107]; this have been confirmed by recent studies [78, 105], which have provided the means to predict small molecule binding sites and their druggability from the structure of the protein.

8.3.3 Binding Site Plasticity

Considering that biological macromolecules are made of very flexible elements and that most of the contacts are driven by weak forces, motion is consubstantial to their nature. Proteins are therefore free to explore a large number of conformational states, corresponding to minima on the potential energy hypersurface [108]; at certain intervals the protein will change conformation, escaping a local minimum and, eventually, reaching another one. As with structure, protein dynamics has been modelled by evolution to serve important biological roles and an understanding of protein function cannot be achieved without knowledge of both structure and dynamics. Nevertheless, protein dynamics has been fundamentally ignored in structure-based drug design and, although examples of incorrect models due to unexpected conformational changes of the protein are not unheard of (see for example the case of PU3, an inhibitor of the Hsp90 protein [109, 110]) it has been largely successful using essentially a rigid description of the receptor. This may be partially explained by the fact that small-molecule binding sites tend to be more rigid than the rest of the protein, and because most often ligand binding does not involve noticeable conformational changes. For instance, in the case of enzymes, backbone change is not significant and the functional atoms move generally less than 1 Å upon ligand binding [111]. But there are notorious exceptions to this rule and, as demonstrated by the nonpeptidic inhibitors of renin [112, 113] or the allosteric inhibitors of p38 [114], making use of protein flexibility can confer important advantages in drug design. Furthermore, some enzymes and receptors have naturally flexible binding sites, which enables them recognize a wide range of ligands. This is the case of P450 cytochromes or the liver X receptors [68].

In the particular case of PPIs, protein flexibility plays an even more important role for two reasons. In the first place, protein–protein association sites are more flexible than small-molecule binding sites [115], although one has to distinguish between core interface residues, which tend to be rigid, and the surrounding layer of residues, which are more flexible and can explore a number of conformations even at very short timescales [116]. In the second place, binding of small molecules to (often cryptic) allosteric sites can lead to noncompetitive inhibition of protein–protein interactions. This is a general mechanism, resulting from an alteration of the equilibrium between conformationally distinct states of the protein [117–119]. Although allostery has also been exploited in more traditional targets [120], it is particularly attractive as an alternative to direct inhibition of protein–protein interactions lacking druggable cavities in their interface. Allosteric PPIs include those disrupting the LFA-1/ICAM1 complex, the inducible nitric oxide synthase (iNOS) homodimer or the NGF/p75^{NTR} heterodimer [121]. It is expected that this mechanism will be increasingly important in the future [122].

A range of experimental techniques are being used to study the dynamics of proteins, including Mössbauer spectroscopy, neutron scattering, time resolved Laue X-ray diffraction and NMR spectroscopy [123–125], but all of them have in common that the observable is the result of the average state of the sample. In this regard, molecular simulations are an interesting alternative, because they can provide the finest details on each individual particle of the system (see Methods and references [31] and [126]). Ultimately, combinations of experimental and theoretical methods seem the best way to dissect the complexity of the protein conformational universe [33, 34, 127]. New computational methods to sample the conformational space of the receptor are also being developed in the context of docking applications [128, 129], but using experimentally determined conformations of ligand-receptor complexes seems to provide the best docking results [130]. One should also be reminded that sampling the conformational space of the receptor is only one side of the story: the internal energies of each conformer should also be known in order to distinguish rare (high energy) conformations from the most abundant ones (low energy). Otherwise flexible receptor docking can easily produce worst results than rigid docking [70, 72]. Unfortunately, the internal energy of a protein is a very difficult property to measure, and even very small differences can have important consequences for binding. In a recent example, it was shown that two experimentally determined conformations of acetylcholinesterase, differing only by a flip of the peptide bond between G117 and G118, are 2.4 kcal/mol apart. Molecules binding to the least stable conformation are, therefore, far less active than expected. In contrast, the conformational flexibility of F330 produces much more important effects on the shape and size of the active site, but all the conformations are essentially isoenergetic [131].

8.4 Case Studies

Most PPIs discovered to date have been identified by means of random screening, usually with HTS methods, but *in silico* screening has also been used to discover ligands that disrupt the formation of complexes. The earliest example dates back to 1997, with the discovery of small molecules that bind to CD4, blocking the formation of CD4/MHC class II [132]. A recent revision identifies 7 protein–protein complexes that have been successfully targeted with computer-aided techniques [133]. In this section we will review two additional examples, highlighting the aspects that contributed to the successful outcome and can be generally applicable.

8.4.1 β-Catenin Inhibitors

There is a strong rationale supporting the therapeutic potential of small molecules disrupting the formation of the complex between β-catenin and the Tcf/LEF family of transcription factors [82], but until recently there were no accounts of successful hit finding, which may be partially due to the anticipated difficulty of targeting a protein–protein interaction with an extremely large surface contact area. Nevertheless, two papers have been published in the last few years disclosing the discovery of antagonists of the Tcf/β-catenin complex. In the first one, an ELISA assay amenable to HTS was used to screen approximately 7000 purified natural compounds, identifying eight compounds able to displace the molecular probe used in their assay with an IC_{50} lower than 10 μM. Interestingly, they also screened 45 000 synthetic compounds from the Novartis collection, finding no additional hits. The biological activity of the hits was further characterized, providing results consistent with the expected mode of action [134]. Trosset *et al.* used a completely different approach to skip altogether the necessity to develop a sophisticated biochemical assay, replacing it with a combination of virtual screening and biophysical screening techniques [135]. The NMR technique of WaterLOGSY and isothermal titration calorimetry (ITC) are very general and sensitive methods to detect molecular association, which do not require labelling. Unfortunately, these advantages are largely offset by a limited throughput and high demands on the amount of protein [136, 137]. The authors used *in silico* screening techniques to select compounds from their corporate collection with a greater chance of being active, hoping that the final list would be sufficiently enriched with actives to compensate for the poor capacity of the screening techniques. Obviously the outcome of this strategy is uncertain, but this work clearly shows that biophysical and *in silico* techniques can make a powerful combination, providing a fast and inexpensive opportunity to head start a discovery project or to target a system for which no biochemical assays are available.

As we have already discussed in Section 8.3, it is important to review the literature and to inspect the structure of the target in order to identify putative hot spots. In the case of β-catenin, this was particularly important because the interaction with the transcription factor expands over an area of approximately 4800 Å^2. The program PASS [138] identified 6 binding pockets, only one of which was used for virtual screening. The choice was based on the size of the cavity, the nature of the aminoacids and the type of interactions formed between Tcf and β-catenin at each site. A cavity located between armadillo repeats 9 and 10 was identified as the most likely to provide hits and was chosen as the docking site. It should be noted that the selected area includes K435 which, as discussed in Section 8.3.1.1, appears to be a hot spot for binding.

Preparation of the virtual screening collection started with 90 000 compounds from the corporate collection and followed fairly standard criteria to ensure that the compounds were available and druglike, resulting in 17 700 compounds for further consideration. Docking was done using the FLO_QXP package [139], allowing conformational flexibility of the side-chains of D390, C429, C466 and K508. The program uses a modified force-field as the scoring function and the Monte Carlo algorithm to sample the positional and conformational degrees of freedom. Most of the docking poses were discarded using strict cut-offs for each energy term, and 3000 top ranking compounds were selected for visual inspection. This is a labour-intensive and tedious task, but it is very common practice and, although difficult to quantify, it is largely responsible for the hit rates obtained in many structure-based virtual screening experiments [140]. Visual inspection led to a final selection of 42 compounds, but only 22 were available in sufficient quantity to be tested. Of those 22 compounds, 7 were identified as β-catenin binders by NMR and 3 of them were also active in the calorimetric experiment. The most active compound (PNU-74654) had a dissociation constant of 450 nM. Attempts to determine the binding mode using X-ray crystallography were unsuccessful, but the authors propose a binding mode that is consistent with the very scarce SAR available.

As noted by the authors, it is surprising that they could find 3 drug-like compounds out of 22 tested when Lepourcelet and coworkers had failed to identify a single active compound from their collection of synthetic compounds [134]. Even in the unlikely case that the virtual screening protocol produced no false negatives, the initial pool of compounds for HTS (45 000) more than doubled that of VS (17700). The different composition of the chemical collections may be an explanation, but the difference can also stem from the fact that HTS assays are generally less sensitive and more likely to produce false negatives than biophysical techniques.

8.4.2 Small Molecule Modulators of G$\beta\gamma$activity

G proteins coupled receptors (GPCRs) are a major family of therapeutic targets. Transduction of the signals received by GPCRs is mediated by the trimeric G proteins, which dissociate into Gα and G$\beta\gamma$. Free G$\beta\gamma$ interacts with many downstream effectors, mediating different physiological responses. This system would be of great pharmacological interest if certain G$\beta\gamma$-effector protein complexes could be specifically blocked. Recently Smrcka and collaborators have used *in silico* screening tools to identify compounds that modulate the interactions between G$\beta\gamma$ and their effectors [141].

The Smrcka group had previously screened phage-display libraries to identify peptides that bind to G$\beta\gamma$. This gave them information about the presence of a hot spot on the surface of the protein, and they also observed that these peptides could differentially affect the interaction with various effectors [142]. An analogue of the most interesting peptide (SIGK), was successfully crystallized [143] and this structure was used in docking studies to identify small molecule binders (Figure 8.3). In this case the initial library for virtual screening is the NCI diversity library, which contains only 1990 diverse compounds selected from the larger 250251-compound NCI library. FlexX was used to dock these compounds to the previously identified hot spot. The authors used all scoring functions available within FlexX, which include force-field based, empirical and knowledge-based functions. The 1% best scoring compounds from each list were selected, resulting in 85 nonredundant compounds.

The activity of the compounds was tested in an ELISA assay measuring the capability of the compounds to compete the SIGK peptide. Nine of the 85 compounds showed an IC_{50} ranging from 0.1 to 60 μM. Twenty-one analogues of M119, one of the most potent compounds, were selected from the NCI library, providing some initial SAR and identification of those chemical moieties in M119 more important for binding. M119 and M201 (a noncompetitor of the SIGK peptide) were extensively tested in *in vitro*, cellular and *in vivo* assays to characterize the biological consequences of G$\beta\gamma$ binding. While the former attenuated G$\beta\gamma$-dependent activation of phospholipase C (PLC) β2, PLCβ3 and phosphoinositide 3 kinase γ(PI3Kγ), the latter did not affect PLCβ2 activation by G$\beta\gamma$ but potentiated G$\beta\gamma$-dependent activation of both PLCβ3 and PI3Kγ. While a satisfactory explanation for

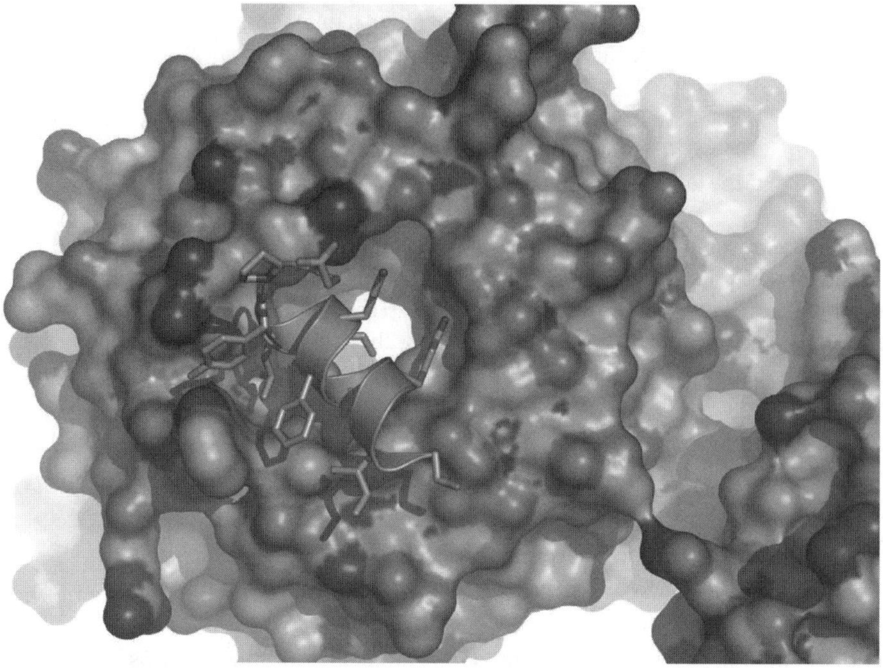

Figure 8.3 SIGK peptide bound to Gβ (PDB: 1XHM). The peptide binds to a hydrophobic patch located near the axis of this β-propeller structure. Helix 9 of Gα and the C-terminus of GPCR Kinase 2 also bind to the same site (PDB codes: 1GP2 & 2BCJ, respectively) (See Plate 16.)

this behaviour is still lacking, it suggests that it may be possible to develop PPI drugs capable of selectively interfering with specific complexes and pathways while preserving (or even increasing) other complexes mediated by the same protein surface. This work also highlights the importance of a good characterization of the target, which helped focussing the virtual screening search around the location of the hot spot.

8.5 Outlook and Conclusions

Virtual screening has repeatedly demonstrated its capacity to identify suitable starting points for drug discovery [140, 144, 145]. Nevertheless, the computational tools used in high-throughput VS rely on major approximations and are thus limited in their predicting capability, which is at best semi-quantitative. This advises against the use of *in silico* screening techniques in a mechanical, unsupervised way. Rather, its main strength lies in the capacity to integrate information from very disparate sources and to adapt the process to the specificities of the system of interest. Introduction of empirical information in nonautomated steps is, in fact, largely responsible for the success of structure-based virtual screening [140]. As the intrinsic difficulty of the target increases (e.g. poor druggability, high flexibility), this aspect of virtual screening should be particularly taken into consideration. In this regard, the combined, possibly iterative, use of virtual screening with robust but low-throughput experimental techniques (e.g. fragment screening, biophysical assays) offer an extremely promising prospect to deliver suitable leads faster and more cost-effectively. Further methodological and practical improvements will provide new opportunities to understand and tackle increasingly difficult targets.

References

1. P. Bussey, J. Pisani and Y. Bonduelle, Understanding the value of research, in: *Industrialization of Drug Discovery: From Target Selection through Lead Optimization*, J.S. Handen (ed.), CRC Press, Boca Raton, 2005.
2. L. Brown and T. Grundy, *Project Management for the Pharmaceutical Industry*, Gower Publishing, Ltd., 2004.
3. R. Macarron, Critical review of the role of HTS in drug discovery, *Drug Discov. Today*, **11**, 277–9 (2006).
4. P.J. Hajduk and J. Greer, A decade of fragment-based drug design: Strategic advances and lessons learned, *Nat. Rev. Drug Discov.*, **6**, 211–19 (2007).
5. J. Bajorath, Integration of virtual and high-throughput screening, *Nat. Rev. Drug Discov.*, **1**, 882–94 (2002).
6. A.K. Ghose, T. Herbertz, J.M. Salvino and J.P. Mallamo, Knowledge-based chemoinformatic approaches to drug discovery, *Drug Discov. Today*, **11**, 1107–14 (2006).
7. J.J. Gasteiger and T. Engel, *Chemoinformatics: A textbook*, Wiley-VCH, Weinheim, 2003.
8. Rishton, Reactive compounds and in vitro false positives in HTS, *Drug Discovery Today*, **2**, 382 (1997).
9. Hann, M Hudson, B Lewell, X Lifely, R Miller, L Ramsden, N., Strategic pooling of compounds for high-throughput screening. *Journal of Chemical Information and Computer Sciences*, **39**, 897 (1999).
10. C.A. Lipinski, Drug-like properties and the causes of poor solubility and poor permeability, *J. Pharmacol. Toxicol. Methods*, **44**, 235–49 (2000).
11. S.L. McGovern, B.T. Helfand, B. Feng and B.K. Shoichet, A specific mechanism of nonspecific inhibition, *J. Med. Chem.*, **46**, 4265–72 (2003).
12. J. Seidler, S.L. McGovern, T.N. Doman and B.K. Shoichet, Identification and prediction of promiscuous aggregating inhibitors among known drugs, *J. Med. Chem.*, **46**, 4477–86 (2003).
13. C.A. Lipinski, F. Lombardo, B.W. Dominy and P.J. Feeney, Experimental and computational approaches to estimate solubility and permeability in drug discovery and development settings, *Adv. Drug Deliv. Rev.*, **46**, 3–26 (2001).
14. D.F. Veber, S.R. Johnson, H.Y. Cheng, B.R. Smith, K.W. Ward and K.D. Kopple, Molecular properties that influence the oral bioavailability of drug candidates, *J. Med. Chem.*, **45**, 2615–23 (2002).
15. T.I. Oprea, A.M. Davis, S.J. Teague and P.D. Leeson, Is there a difference between leads and drugs? A historical perspective, *J. Chem. Inf. Comput. Sci.*, **41**, 1308–15 (2001).
16. T.I. Oprea, Current trends in lead discovery: Are we looking for the appropriate properties? *J. Comput. Aided Mol. Des.*, **16**, 325–34 (2002).
17. M.G. Bures and Y.C. Martin, Computational methods in molecular diversity and combinatorial chemistry, *Curr. Opin. Chem. Biol.*, **2**, 376–80 (1998).
18. G.W. Bemis and M.A. Murcko, The properties of known drugs. 1. molecular frameworks, *J. Med. Chem.*, **39**, 2887–93 (1996).
19. G.W. Bemis and M.A. Murcko, Properties of known drugs. 2. side chains, *J. Med. Chem.*, **42**, 5095–9 (1999).
20. J.L. Miller, Recent developments in focused library design: Targeting gene-families, *Curr. Top. Med. Chem.*, **6**, 19–29 (2006).
21. H. Kubinyi, *3D Qsar in Drug Design: Theory, Methods and Applications*, Kluwer Academic Pub, 1993.

22. P. Gratteri, C. Bonaccini and F. Melani, Searching for a reliable orientation of ligands in their binding site: Comparison between a structure-based (glide) and a ligand-based (FIGO) approach in the case study of PDE4 inhibitors, *J. Med. Chem.*, **48**, 1657–65 (2005).
23. W. Sippl, Development of biologically active compounds by combining 3D QSAR and structure-based design methods, *J. Comput. Aided Mol. Des.*, **16**, 825–30 (2002).
24. G. Klebe, Virtual ligand screening: Strategies, perspectives, and limitations, *Drug Discov Today*, **11**, 580–94 (2006).
25. T. Langer and R.D. Hoffmann, *Pharmacophores and Pharmacophore Searches*, Wiley-VCH, Weinheim, 2006.
26. Blundell TL and Patel S, High-throughput X-ray crystallography for drug discovery. *Curr. Opin. Pharmacol.*, **4**, 490–6 (2004).
27. J.W. Ponder and D.A. Case, Force fields for protein simulations, *Adv. Protein Chem.*, **66**, 27–85 (2003).
28. A. MacKerell Jr, D. Bashford, M. Bellott, *et al.*, All-atom empirical potential for molecular modeling and dynamics studies of proteins, *J Phys Chem B*, **102**, 3586–3616 (1998).
29. C. Oostenbrink, T.A. Soares, N.F. van der Vegt and W.F. van Gunsteren, Validation of the 53A6 GROMOS force field, *Eur. Biophys. J.*, **34**, 273–84 (2005).
30. T.A. Halgren, Merck molecular force field.I. basis, form, scope, parameterization, and performance of MMFF94, *Journal of Computational Chemistry*, **17**, 490–519 (1996).
31. W.F. van Gunsteren, D. Bakowies, R. Baron, *et al.*, Biomolecular modeling: Goals, problems, perspectives, *Angew. Chem. Int. Ed Engl.*, **45**, 4064–92 (2006).
32. C.M. Summa and M. Levitt, Near-native structure refinement using in vacuo energy minimization, *Proc. Natl. Acad. Sci. U. S. A.*, **104**, 3177–82 (2007).
33. Lindorff-Larsen K, Best RB, Depristo MA, Dobson CM and Vendruscolo M, Simultaneous determination of protein structure and dynamics. *Nature*, **433**, 128–32 (2005).
34. A. Cavalli, X. Salvatella, C.M. Dobson and M. Vendruscolo, Protein structure determination from NMR chemical shifts, *Proc. Natl. Acad. Sci. U. S. A.*, **104**, 9615–20 (2007).
35. Kuntz ID, Blaney JM, Oatley SJ, Langridge R and Ferrin TE, A geometric approach to macromolecule-ligand interactions. *J. Mol. Biol.*, **161**, 269–88 (1982).
36. W.J. Mortier, K. Van Genechten and J. Gasteiger, Electronegativity equalization: Application and parametrization, *J. Am. Chem. Soc.*, **107**, 829–35 (1985).
37. W.L. Jorgensen, D.S. Maxwell and J. Tirado-Rives, Development and testing of the OPLS all-atom force field on conformational energetics and properties of organic liquids, *Q. Rev. Biophys.*, **26**, 49 (1993).
38. A.E. Cho, V. Guallar, B.J. Berne and R. Friesner, Importance of accurate charges in molecular docking: Quantum mechanical/molecular mechanical (QM/MM) approach, *J. Comput. Chem.*, **26**, 915–31 (2005).
39. M. Rarey, S. Wefing and T. Lengauer, Placement of medium-sized molecular fragments into active sites of proteins, *J. Comput. Aided Mol. Des.*, **10**, 41–54 (1996).
40. M.D. Eldridge, C.W. Murray, T.R. Auton, G.V. Paolini and R.P. Mee, Empirical scoring functions: I. the development of a fast empirical scoring function to estimate the binding affinity of ligands in receptor complexes, *J. Comput. Aided Mol. Des.*, **11**, 425–45 (1997).
41. H.J. Bohm, The development of a simple empirical scoring function to estimate the binding constant for a protein-ligand complex of known three-dimensional structure, *J. Comput. Aided Mol. Des.*, **8**, 243–56 (1994).
42. H.J. Bohm, The computer program LUDI: A new method for the de novo design of enzyme inhibitors, *J. Comput. Aided Mol. Des.*, **6**, 61–78 (1992).
43. G. Jones, P. Willett and R.C. Glen, Molecular recognition of receptor sites using a genetic algorithm with a description of desolvation, *J. Mol. Biol.*, **245**, 43–53 (1995).
44. C. Venkatachalam, X. Jiang, T. Oldfield and M. Waldman, LigandFit: A novel method for the shape-directed rapid docking of ligands to protein active sites, *Journal of Molecular Graphics & Modelling*, **21**, 289–307 (2003).

45. J.B.O. Mitchell, R.A. Laskowski, A. Alex and J.M. Thornton, BLEEP – potential of mean force describing protein–ligand interactions: I. generating potential, *Journal of Computational Chemistry*, **20**, 1165–76 (1999).
46. I. Muegge and Y.C. Martin, A general and fast scoring function for protein-ligand interactions: A simplified potential approach, *J. Med. Chem.*, **42**, 791–804 (1999).
47. H. Gohlke, M. Hendlich and G. Klebe, Knowledge-based scoring function to predict protein-ligand interactions, *J. Mol. Biol.*, **295**, 337–56 (2000).
48. L. Jiang, Y. Gao, F. Mao, Z. Liu and L. Lai, Potential of mean force for protein–protein interaction studies, *Proteins*, **46**, 190–6 (2002).
49. M.J. Sippl, Knowledge-based potentials for proteins, *Curr. Opin. Struct. Biol.*, **5**, 229–35 (1995).
50. P.S. Charifson, J.J. Corkery, M.A. Murcko and W.P. Walters, Consensus scoring: A method for obtaining improved hit rates from docking databases of three-dimensional structures into proteins, *J. Med. Chem.*, **42**, 5100–9 (1999).
51. C.J. Cramer and D.G. Truhlar, Implicit solvation models: Equilibria, structure, spectra, and dynamics, *Chem. Rev.*, **99**, 2161–2200 (1999).
52. M. Orozco and F.J. Luque, Theoretical methods for the description of the solvent effect in biomolecular systems, *Chem. Rev.*, **100**, 4187–4226 (2000).
53. J.L. Miller and P.A. Kollman, Solvation free energies of the nucleic acid bases, *J. Phys. Chem.*, **100**, 8587–94 (1996).
54. W. Orttung, Direct solution of the poisson equation for biomolecules of arbitrary shape, polarizability density and charge distribution, *Ann. NY Acad. Sci*, **303**, 22–37 (1977).
55. M.K. Gilson and B.H. Honig, Energetics of charge-charge interactions in proteins, *Proteins*, **3**, 32–52 (1988).
56. W.C. Still, A. Tempczyk, R.C. Hawley and T. Hendrickson, Semianalytical treatment of solvation for molecular mechanics and dynamics, *J. Am. Chem. Soc.*, **112**, 6127–9 (1990).
57. C.N. Schutz and A. Warshel, What are the dielectric 'constants' of proteins and how to validate electrostatic models? *Proteins*, **44**, 400–17 (2001).
58. J.M. Swanson, J. Mongan and J.A. McCammon, Limitations of atom-centered dielectric functions in implicit solvent models, *J. Phys. Chem. B. Condens Matter Mater. Surf. Interfaces Biophys.*, **109**, 14769–14772 (2005).
59. J.M. Blaney, P.K. Weiner, A. Dearing, *et al.*, Molecular mechanics simulation of protein-ligand interactions: Binding of thyroid hormone analogs to prealbumin, *J. Am. Chem. Soc.*, **104**, 6424–34 (1982).
60. E.L. Mehler and T. Solmajer, Electrostatic effects in proteins: Comparison of dielectric and charge models, *Protein Eng.*, **4**, 903–10 (1991).
61. D. Eisenberg and A.D. McLachlan, Solvation energy in protein folding and binding, *Nature*, **319**, 199–203 (1986).
62. Ajay and M.A. Murcko, Computational methods to predict binding free energy in ligand-receptor complexes, *J. Med. Chem.*, **38**, 4953–67 (1995).
63. M.K. Gilson, J.A. Given, B.L. Bush and J.A. McCammon, The statistical-thermodynamic basis for computation of binding affinities: A critical review, *Biophys. J.*, **72**, 1047–69 (1997).
64. R.D. Taylor, P.J. Jewsbury and J.W. Essex, A review of protein-small molecule docking methods, *J. Comput. Aided Mol. Des.*, **16**, 151–66 (2002).
65. I. Halperin, B. Ma, H. Wolfson and R. Nussinov, Principles of docking: An overview of search algorithms and a guide to scoring functions, *Proteins*, **47**, 409–43 (2002).
66. N. Brooijmans and I.D. Kuntz, Molecular recognition and docking algorithms, *Annu. Rev. Biophys. Biomol. Struct.*, **32**, 335–73 (2003).
67. Murray CW, Baxter CA and Frenkel AD, The sensitivity of the results of molecular docking to induced fit effects: Application to thrombin, thermolysin and neuraminidase. *J. Comput. Aided Mol. Des.*, **13**, 547–62 (1999).
68. X. Barril and X. Fradera, Incorporating protein flexibility into docking and structure-based drug design, *Expert Opin. Drug Discov.*, **1**, 335–49 (2006).

69. Verdonk ML, Chessari G, Cole JC, *et al.*, Modeling water molecules in protein-ligand docking using GOLD. *J. Med. Chem.*, **48**, 6504–15 (2005).
70. Barril X and Morley SD, Unveiling the full potential of flexible receptor docking using multiple crystallographic structures. *J. Med. Chem.*, **48**, 4432–43 (2005).
71. Claussen H, Buning C, Rarey M and Lengauer T, FlexE: Efficient molecular docking considering protein structure variations. *J. Mol. Biol.*, **308**, 377–95 (2001).
72. Wei BQ, Weaver LH, Ferrari AM, Matthews BW and Shoichet BK, Testing a flexible-receptor docking algorithm in a model binding site. *J. Mol. Biol.*, **337**, 1161–2 (2004).
73. Lin JH, Perryman AL, Schames JR and McCammon JA, Computational drug design accommodating receptor flexibility: The relaxed complex scheme. *J. Am. Chem. Soc.*, **124**, 5632–3 (2002).
74. Knegtel RM, Kuntz ID and Oshiro CM, Molecular docking to ensembles of protein structures. *J. Mol. Biol.*, **266**, 424–40 (1997).
75. F. Osterberg, G.M. Morris, M.F. Sanner, A.J. Olson and D.S. Goodsell, Automated docking to multiple target structures: Incorporation of protein mobility and structural water heterogeneity in AutoDock, *Proteins*, **46**, 34–40 (2002).
76. T. Clackson and J.A. Wells, A hot spot of binding energy in a hormone-receptor interface, *Science*, **267**, 383–6 (1995).
77. W.L. DeLano, Unraveling hot spots in binding interfaces: Progress and challenges, *Curr. Opin. Struct. Biol.*, **12**, 14–20 (2002).
78. P.J. Hajduk, J.R. Huth and S.W. Fesik, Druggability indices for protein targets derived from NMR-based screening data, *J. Med. Chem.*, **48**, 2518–25 (2005).
79. E. Liepinsh and G. Otting, Organic solvents identify specific ligand binding sites on protein surfaces, *Nat. Biotechnol.*, **15**, 264–8 (1997).
80. K.N. Allen, C.R. Bellamacina, X. Ding, *et al.*, An experimental approach to mapping the binding surfaces of crystalline proteins, *J. Phys. Chem.*, **100**, 2605–11 (1996).
81. A.C. English, C.R. Groom and R.E. Hubbard, Experimental and computational mapping of the binding surface of a crystalline protein, *Protein Eng.*, **14**, 47–59 (2001).
82. N. Barker and H. Clevers, Mining the wnt pathway for cancer therapeutics, *Nat. Rev. Drug Discov.*, **5**, 997–1014 (2006).
83. V. Neduva and R.B. Russell, Linear motifs: Evolutionary interaction switches, *FEBS Lett.*, **579**, 3342–5 (2005).
84. B. Ma, T. Elkayam, H. Wolfson and R. Nussinov, Protein–protein interactions: Structurally conserved residues distinguish between binding sites and exposed protein surfaces, *Proc. Natl. Acad. Sci. U. S. A.*, **100**, 5772–7 (2003).
85. L. Serrano, A. Horovitz, B. Avron, M. Bycroft and A.R. Fersht, Estimating the contribution of engineered surface electrostatic interactions to protein stability by using double-mutant cycles, *Biochemistry*, **29**, 9343–52 (1990).
86. D.P. Sun, U. Sauer, H. Nicholson and B.W. Matthews, Contributions of engineered surface salt bridges to the stability of T4 lysozyme determined by directed mutagenesis, *Biochemistry*, **30**, 7142–53 (1991).
87. X. Barril, C. Aleman, M. Orozco and F.J. Luque, Salt bridge interactions: Stability of the ionic and neutral complexes in the gas phase, in solution, and in proteins, *Proteins*, **32**, 67–79 (1998).
88. A.H. Huber, W.J. Nelson and W.I. Weis, Three-dimensional structure of the armadillo repeat region of beta-catenin, *Cell*, **90**, 871–82 (1997).
89. M. Fasolini, X. Wu, M. Flocco, J.Y. Trosset, U. Oppermann and S. Knapp, Hot spots in Tcf4 for the interaction with beta-catenin, *J. Biol. Chem.*, **278**, 21092–8 (2003).
90. R. Brenk, L. Naerum, U. Gradler, *et al.*, Virtual screening for submicromolar leads of tRNA-guanine transglycosylase based on a new unexpected binding mode detected by crystal structure analysis, *J. Med. Chem.*, **46**, 1133–43 (2003).
91. H. Gohlke and G. Klebe, Approaches to the description and prediction of the binding affinity of small-molecule ligands to macromolecular receptors, *Angew. Chem. Int. Ed Engl.*, **41**, 2644–76 (2002).

92. M.R. Landon, D.R. Lancia Jr, J. Yu, S.C. Thiel and S. Vajda, Identification of hot spots within druggable binding regions by computational solvent mapping of proteins, *J. Med. Chem.*, **50**, 1231–40 (2007).
93. E. Furuichi and P. Koehl, Influence of protein structure databases on the predictive power of statistical pair potentials, *Proteins*, **31**, 139–49 (1998).
94. S. Vajda and F. Guarnieri, Characterization of protein-ligand interaction sites using experimental and computational methods, *Curr. Opin. Drug Discov. Devel.*, **9**, 354–62 (2006).
95. S. Dennis, T. Kortvelyesi and S. Vajda, Computational mapping identifies the binding sites of organic solvents on proteins, *Proc. Natl. Acad. Sci. U. S. A.*, **99**, 4290–5 (2002).
96. I. Massova and P.A. Kollman, Computational alanine scanning to probe protein–protein interactions: A novel approach to evaluate binding free energies, *J. Am. Chem. Soc.*, **121**, 8133–43 (1999).
97. R. Guerois, J.E. Nielsen and L. Serrano, Predicting changes in the stability of proteins and protein complexes: A study of more than 1000 mutations, *J. Mol. Biol.*, **320**, 369–87 (2002).
98. T. Kortemme and D. Baker, A simple physical model for binding energy hot spots in protein–protein complexes, *Proc. Natl. Acad. Sci. U. S. A.*, **99**, 14116–21 (2002).
99. P. Matsson, C.A. Bergstrom, N. Nagahara, S. Tavelin, U. Norinder and P. Artursson, Exploring the role of different drug transport routes in permeability screening, *J. Med. Chem.*, **48**, 604–13 (2005).
100. A.K. Pavlou and J.M. Reichert, Recombinant protein therapeutics–success rates, market trends and values to 2010, *Nat. Biotechnol.*, **22**, 1513–19 (2004).
101. M. Goldberg and I. Gomez-Orellana, Challenges for the oral delivery of macromolecules, *Nat. Rev. Drug Discov.*, **2**, 289–95 (2003).
102. I.D. Kuntz, K. Chen, K.A. Sharp and P.A. Kollman, The maximal affinity of ligands, *Proc. Natl. Acad. Sci. U. S. A.*, **96**, 9997–10002 (1999).
103. A.L. Hopkins, C.R. Groom and A. Alex, Ligand efficiency: A useful metric for lead selection, *Drug Discov. Today*, **9**, 430–1 (2004).
104. A.L. Hopkins and C.R. Groom, The druggable genome, *Nat. Rev. Drug Discov.*, **1**, 727–30 (2002).
105. A.C. Cheng, R.G. Coleman, K.T. Smyth, *et al.*, Structure-based maximal affinity model predicts small-molecule druggability, *Nat. Biotechnol.*, **25**, 71–5 (2007).
106. V. Neduva and R.B. Russell, Peptides mediating interaction networks: New leads at last, *Curr. Opin. Biotechnol.*, **17**, 465–71 (2006).
107. X. Barril and R. Soliva, Molecular modelling, in: *Structure-Based Drug Design; an Overview*, R.E. Hubbard (ed.), The Royal Society of Chemistry, 2006.
108. H. Frauenfelder, S.G. Sligar and P.G. Wolynes, The energy landscapes and motions of proteins, *Science*, **254**, 1598–1603 (1991).
109. G. Chiosis, M.N. Timaul, B. Lucas, *et al.*, A small molecule designed to bind to the adenine nucleotide pocket of Hsp90 causes Her2 degradation and the growth arrest and differentiation of breast cancer cells, *Chem. Biol.*, **8**, 289–99 (2001).
110. L. Wright, X. Barril, B. Dymock, *et al.*, Structure-activity relationships in purine-based inhibitor binding to HSP90 isoforms, *Chem. Biol.*, **11**, 775–85 (2004).
111. Gutteridge A and Thornton J, Conformational changes observed in enzyme crystal structures upon substrate binding. *J. Mol. Biol.*, **346**, 21–8 (2005).
112. C. Oefner, A. Binggeli, V. Breu, *et al.*, Renin inhibition by substituted piperidines: A novel paradigm for the inhibition of monomeric aspartic proteinases? *Chem. Biol.*, **6**, 127–31 (1999).
113. A.M. Davis, S.J. Teague and G.J. Kleywegt, Application and limitations of X-ray crystallographic data in structure-based ligand and drug design, *Angew. Chem. Int. Ed Engl.*, **42**, 2718–36 (2003).
114. C. Pargellis, L. Tong, L. Churchill, *et al.*, Inhibition of p38 MAP kinase by utilizing a novel allosteric binding site, *Nat. Struct. Biol.*, **9**, 268–72 (2002).
115. M.J. Betts and M.J. Sternberg, An analysis of conformational changes on protein–protein association: Implications for predictive docking, *Protein Eng.*, **12**, 271–83 (1999).

116. G.R. Smith, M.J. Sternberg and P.A. Bates, The relationship between the flexibility of proteins and their conformational states on forming protein–protein complexes with an application to protein–protein docking, *J. Mol. Biol.*, **347**, 1077–1101 (2005).
117. J.S. Marvin and H.W. Hellinga, Manipulation of ligand binding affinity by exploitation of conformational coupling, *Nat. Struct. Biol.*, **8**, 795–8 (2001).
118. Gunasekaran K, Ma B and Nussinov R, Is allostery an intrinsic property of all dynamic proteins? *Proteins*, **57**, 433–43 (2004).
119. J.F. Swain and L.M. Gierasch, The changing landscape of protein allostery, *Curr. Opin. Struct. Biol.*, **16**, 102–8 (2006).
120. S.J. Teague, Implications of protein flexibility for drug discovery, *Nat. Rev. Drug Discov.*, **2**, 527–41 (2003).
121. M.R. Arkin and J.A. Wells, Small-molecule inhibitors of protein–protein interactions: Progressing towards the dream, *Nat. Rev. Drug Discov.*, **3**, 301–17 (2004).
122. J.E. Lindsley and J. Rutter, Whence cometh the allosterome? *Proc. Natl. Acad. Sci. U. S. A.*, **103**, 10533–5 (2006).
123. R.M. Daniel, R.V. Dunn, J.L. Finney and J.C. Smith, The role of dynamics in enzyme activity, *Annu. Rev. Biophys. Biomol. Struct.*, **32**, 69–92 (2003).
124. F.G. Parak, Proteins in action: The physics of structural fluctuations and conformational changes, *Curr. Opin. Struct. Biol.*, **13**, 552–7 (2003).
125. L.E. Kay, NMR studies of protein structure and dynamics, *J. Magn. Reson.*, **173**, 193–207 (2005).
126. Karplus M and Kuriyan J, Molecular dynamics and protein function. *Proc. Natl. Acad. Sci. U. S. A.*, **102**, 6679–85 (2005).
127. G. Hummer, F. Schotte and P.A. Anfinrud, Unveiling functional protein motions with picosecond x-ray crystallography and molecular dynamics simulations, *Proc. Natl. Acad. Sci. U. S. A.*, **101**, 15330–4 (2004).
128. Sherman W, Day T, Jacobson MP, Friesner RA and Farid R, Novel procedure for modeling ligand/receptor induced fit effects, *J. Med. Chem.*, **49**, 534–53 (2006).
129. M.I. Zavodszky and L.A. Kuhn, Side-chain flexibility in protein-ligand binding: The minimal rotation hypothesis, *Protein Sci.*, **14**, 1104–14 (2005).
130. McGovern SL and Shoichet BK, Information decay in molecular docking screens against holo, apo, and modeled conformations of enzymes. *J. Med. Chem.*, **46**, 2895–2907 (2003).
131. P. Camps, E. Gomez, D. Munoz-Torrero, *et al.*, Binding of 13-amidohuprines to acetylcholinesterase: Exploring the ligand-induced conformational change of the gly117-gly118 peptide bond in the oxyanion hole, *J. Med. Chem.*, **49**, 6833–40 (2006).
132. S. Li, J. Gao, T. Satoh, T.M. Friedman, *et al.*, A computer screening approach to immunoglobulin superfamily structures and interactions: Discovery of small non-peptidic CD4 inhibitors as novel immunotherapeutics, *Proc. Natl. Acad. Sci. U. S. A.*, **94**, 73–8 (1997).
133. D. Gonzalez-Ruiz and H. Gohlke, Targeting protein–protein interactions with small molecules: Challenges and perspectives for computational binding epitope detection and ligand finding, *Curr. Med. Chem.*, **13**, 2607–25 (2006).
134. M. Lepourcelet, Y.N. Chen, D.S. France, *et al.*, Small-molecule antagonists of the oncogenic Tcf/beta-catenin protein complex, *Cancer. Cell.*, **5**, 91–102 (2004).
135. J.Y. Trosset, C. Dalvit, S. Knapp, *et al.*, Inhibition of protein–protein interactions: The discovery of druglike beta-catenin inhibitors by combining virtual and biophysical screening, *Proteins*, **64**, 60–7 (2006).
136. S.E. Harding and B.Z. Chowdhry, *Protein-ligand interactions: Structure and spectroscopy: A practical approach*, Oxford Univ Pr, Bath, 2001.
137. S.E. Harding and B.Z. Chowdhry, *Protein-ligand interactions: Hydrodynamics and calorimetry: A practical approach*, Oxford University Press, Bath, 2001.
138. G.P. Brady and P.F.W. Stouten, Fast prediction and visualization of protein binding pockets with PASS, *J. Comput. Aided Mol. Des.*, **14**, 383–401 (2000).
139. C. McMartin and R.S. Bohacek, QXP: Powerful, rapid computer algorithms for structure-based drug design, *J. Comput. Aided Mol. Des.*, **11**, 333–44 (1997).

140. Barril X, Hubbard RE and Morley SD, Virtual screening in structure-based drug discovery. *Mini Rev. Med. Chem.*, **4**, 779–91 (2004).
141. T.M. Bonacci, J.L. Mathews, C. Yuan, *et al.*, Differential targeting of gbetagamma-subunit signaling with small molecules, *Science*, **312**, 443–6 (2006).
142. J.K. Scott, S.F. Huang, B.P. Gangadhar, *et al.*, Evidence that a protein–protein interaction 'hot spot' on heterotrimeric G protein betagamma subunits is used for recognition of a subclass of effectors, *EMBO J.*, **20**, 767–76 (2001).
143. T.L. Davis, T.M. Bonacci, S.R. Sprang and A.V. Smrcka, Structural and molecular characterization of a preferred protein interaction surface on G protein beta gamma subunits, *Biochemistry*, **44**, 10593–10604 (2005).
144. B.K. Shoichet, Virtual screening of chemical libraries, *Nature*, **432**, 862–5 (2004).
145. H. Kubinyi, Success stories of computer-aided design, in: *Computer Applications in Pharmaceutical Research and Development*, S. Ekins (ed.), Wiley, Weinheim, 2006.

9.1

In Vitro Screening: Screening by Nuclear Magnetic Resonance

Ernest Giralt
Department of Organic Chemistry, University of Barcelona and Institute for Research in Biomedicine, Barcelona, Spain

Over the past decade, NMR has become paramount for studying molecular recognition at protein surfaces, and consequently, has had a tremendous impact on drug discovery [1]. Using either NMR or X-ray crystalography, research teams from around the world have deciphered the 3D structure of thousands of proteins. This knowledge has enabled the field of *structure-based drug design*, in which scientists harness computational power, together with information on molecular recognition events, to design compounds that interact with therapeutically relevant proteins at a specific area on the target protein, such as an active site or a surface-patch [2].

De novo design refers to designing a putative ligand from scratch, whereas *virtual screening* describes computational screening of a previously defined library of accessible compounds against a known target. Even in an ideal case of virtual screening, in which the 3D structure of the protein target is known, the binding between the ligand and the protein must still be confirmed experimentally. This is due to current limitations in both the theoretical study of protein-ligand interactions (chiefly the result of poor understanding of molecular recognition) and the technical study of these events (e.g. the difficulty encountered in trying to model protein flexibility or solvation using docking algorithms) [3].

The 3D structure of the therapeutic target is often unknown, making structure-based design impossible. In these cases drugs can be designed directly from a completely experimental approach, which generally entails screening of single compounds and/or libraries of compounds (whether natural, commercially available, or proprietary). NMR

Protein Surface Recognition: Approaches for Drug Discovery Ernest Giralt, Mark W. Peczuh and Xavier Salvatella. © 2011 John Wiley & Sons, Ltd

plays an important role in this work and is also used to validate ligands suggested from computation.

NMR offers *reliability*, *flexibility* and the *ability to detect weak binders*. *Reliability:* since the experiments are performed in solution, there is usually no risk of denaturing the protein target. *Flexibility*: since different nuclei can be studied (e.g. ^{1}H, ^{13}C, ^{15}N, etc.) and various experiments (e.g. 1D, 2D, NOESY, HSQC etc.) can be employed in function of the protein-ligand system. The *ability to detect weak binders*: this is invaluable in the initial stages of a drug discovery program, when even weak-binding hits are welcomed.

9.1.1 Saturation Transfer Difference (STD)

STD is now the most widely used NMR experiment for studying protein-ligand interactions. It was introduced in 1999 by Bernd Meyer in two seminal papers [4, 5], in which he described the experiment as well as its use in studying the interaction of wheat germ agglutinin with saccharides and its potential use in analyzing mixtures of putative ligands. Since then, several other STD experiments have been reported, the targets of which include various proteins [6, 7] and even whole cells [8].

What is an STD experiment? Imagine a sample consisting of a medium- or high-molecular mass protein and a small-molecular-mass ligand, in which the ligand is free for the majority of the time and is bound to the protein for a small fraction of the time [9, 10]. A conventional ^{1}H-1D-NMR spectrum of the sample would show a combination of broad peaks, corresponding to the protein's protons, and narrow peaks, corresponding to the ligand's protons. In a typical experiment, the concentration of protein is so low that its signals are difficult to see, but are distributed along the entire spectrum as follows: *ca.* δ10-δ6 (aromatic protons and NH), *ca.* δ6-δ4 (αH), and *ca.* δ4-δ(-1) (βH, γH, and δH).

The signals from the ligand vary strongly according to its structure, but usually appear at $\delta > 0.8$. If a sample is irradiated at δ(-1), then some of the protein protons will become saturated (normally those from methyl groups in proximity of aromatic side-chains). Due to the protein's high molecular mass, this saturation will be transferred to most of its protons, resulting in a drop in intensity in their signals. This saturation will also be transferred to the protons of ligand molecules that are bound to the protein, leading to a drop in the intensity of the ligand signals. A typical STD experiment consists of subtracting the ^{1}H-1D-NMR spectrum of the saturated (on the resonance spectrum) protein from that of the unsaturated protein. The ^{1}H-1D-NMR spectrum of the ligand will appear *only if* the ligand has spent some time bound to the protein. Therefore, if a mixture of ligands is used, then the only ligand signals obtained will be those corresponding to the ligands that actually bind to the protein (Figure 9.1.1). This is the underlying concept in screening compound mixtures via STD.

An example of such an STD experiment is shown in Figure 9.1.2. Our laboratory, in collaboration with the group of Javier de Mendoza at ICIQ (Tarragona, Spain), has a long-standing interest in the design of ligands that interact with the tetramerization domain of the protein p53 [11, 12]. p53 is a tetrameric, multidomain tumor suppressor protein that plays a central role in the cell cycle and in maintaining genomic integrity. p53 is a homotetramer of 4x393 residues. The tetramerization domain (p53TD), which comprises residues 323–359,

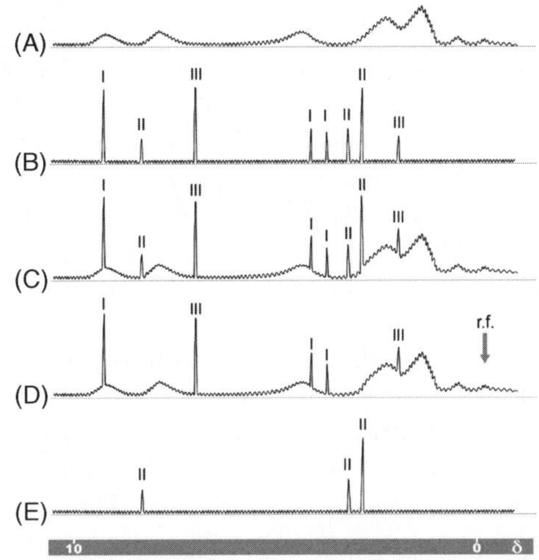

Figure 9.1.1 *Schematic representation of an STD experiment. (A) ^{1}H-1D-NMR spectrum of a protein; (B) ^{1}H-1D-NMR spectrum of a mixture of three ligand candidates (I, II and III); (C) off-resonance spectrum of the STD sample (protein + ligand candidates); (D) on-resonance spectrum of the STD sample (r.f. = radiofrequency irradiation); and (E) STD spectrum*

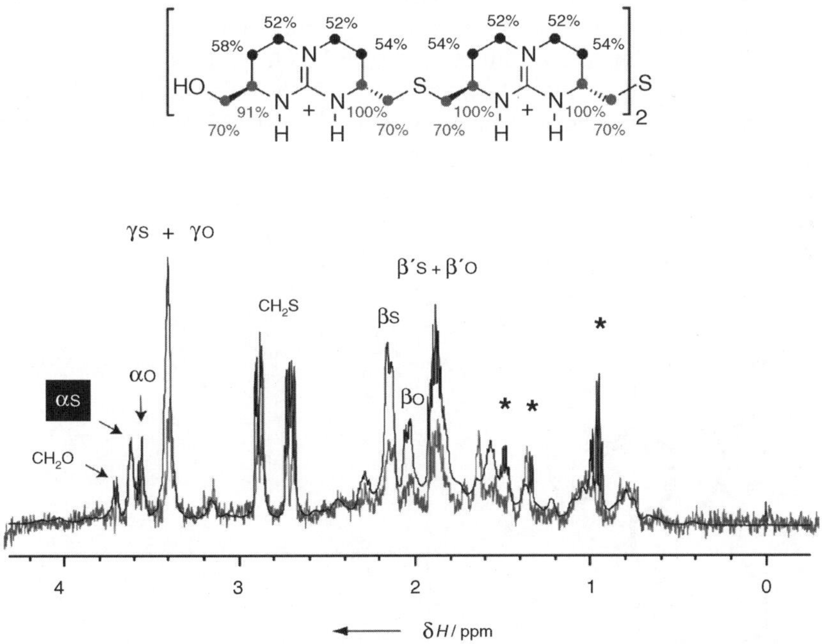

Figure 9.1.2 *STD experiment on a mixture of p53TD and α-tetraguanidinium ligand. A ^{1}H-1D-NMR spectrum of the sample is shown. The relative saturation transferred to each proton is shown on the chemical structure of the ligand.[11] (See Plate 17) Reprinted with permission from [11]. Copyright Wiley-VCH Verlag GmbH & Co. KGaA*

is crucial for establishing the tetrameric structure. The tetraguanidinium compound shown in the figure interacts with an anionic patch located in one of the p53 α-helices [11].

When we used STD to study a sample containing 25 nmoles of p53TD, we obtained the spectrum shown in Figure 9.1.2 (blue spectrum). For reference, the spectrum of the ligand alone has been provided in black. Most of the signals from the ligand appear in the STD spectrum. When we normalized the intensities of the signals, obtaining the percentage values shown in the figure, we deduced that the cationic face of the ligand is more involved in the binding event to the anionic face. This confirmed our proposed molecular recognition mechanism based on electrostatic interaction and hydrogen bound formation between the guanidinium groups of the ligand and the carboxylates at the p53 surface anionic patch [11].

A more accurate way to compose the intensity of the STD effect among the different protons of the same ligand is to compare build-up curves obtained using different saturation times [13]. This allows for correction of differences in relaxation times among the protons of the ligand. Figure 9.1.3 shows the results of a series of STD experiments on a mixture of p53TD and a new tetraguanidinium calixarene ligand in which the starting values for saturation time ranged from 0.3 to 3 s [14]. The relative STD values obtained by comparing the slopes of the STD build-up curves at a saturation time of 0 follow the order: blue > yellow > red > green. These results can be interpreted as demonstrating that interaction of this ligand with p53TD involves not only guanidinium-carboxylate contacts but also interactions between the hydrophobic moiety of the ligand with the hydrophobic areas of p53TD. In terms of methodology, the STD values observed using long relaxation times (1.5, 2 or 3 s) follow the order: red > yellow > blue > green. This discrepancy is due to the fact that in any STD experiment the saturation of the ligand protons in the bound state is counteracted by their longitudinal relaxation times (T1) the corresponding STD values will be biased by their different relaxation rates. Since the acquisition time in an STD experiment is chiefly defined by the saturation time, this bias is more pronounced when long saturation

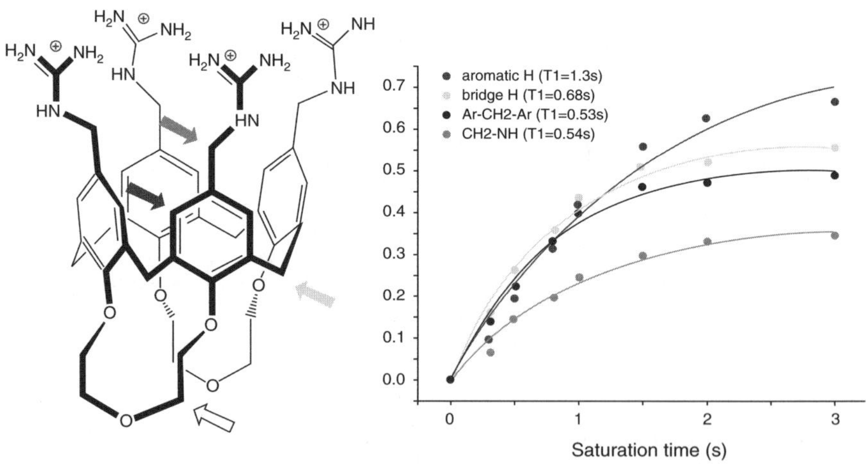

Figure 9.1.3 *STD build-up curves from the interaction between the tetramerization domain of protein p53 and a tetraguanidiniumcalixarene.*[14] *(See Plate 18.)* *Copyright (2008) National Academy of Sciences, U.S.A*

times are used. However, this problem can be circumvented by using build-up curves, or alternatively, by using very short saturation times, although this may lead to low signal-to-noise ratios [9].

As previously mentioned above, the position of the radio frequency is another strongly influential experimental parameter in STD. Its values tend to be *ca.* δ (-1), which is convenient since, first, no ligand resonances are normally found in this region, and secondly, the broad line width of the protein signals still enables a high degree of saturation. Nevertheless, if the ligand contains only nonaromatic protons, a saturation frequency in the aromatic region or even further downfield (δ 11–12) can be employed. For technical reasons, for the off-resonance spectrum, it is better to keep the radio frequency on and to irradiate far from protein and ligand resonances (*i.e.* δ (-40)). The best way to achieve selective irradiations is to use shaped pulses. A typical pulse sequence for performing STD experiments is shown in Figure 9.1.4 [9].

Ligand-protein stochiometry is a crucial parameter that must be optimized case by case; however, a good starting point is a 100-fold excess of ligand. It is advisable to perform experiments at a constant protein concentration and increasing amounts of ligand. *In order to demonstrate specific binding of the ligand to the protein, a saturation curve must be obtained; if this is not observed, then the ligand is a nonspecific binder.*

Another two properties that strongly influence the results of an STD experiment are the size of the protein and the strength of the binding. There is no limit on protein size; in fact, bigger is better. However, there must be a major difference between the molecular mass of the ligand and that of the protein; otherwise, the results will be compromised. As a general rule, for ligands of *ca.* 300 u special attention must be paid when using proteins that have a molecular mass less than 10 000 u.

Efficient saturation transfer from the protein to the bound ligands requires a high k_{off} value. This ensures a quick turnover of unsaturated ligands at the binding site. If the binding is very tight (Kd: subnanomolar), the k_{off} will normally be small and the build up of ligand saturation in solution will be inefficient (see below).

9.1.2 STD in Fragment-based Drug Design

Fragment-based drug design, or more precisely, fragment-based *lead* design, has recently solidified its place in medicinal chemistry as an alternative to the tandem use of

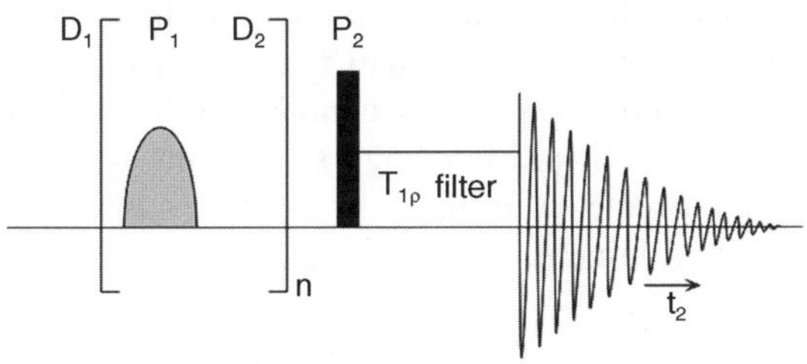

Figure 9.1.4 *NMR pulse sequence for an STD experiment.[9] P1 is a Gaussian pulse used for selective evolution either on-resonance or off-resonance; P2 is a hard pulse; and $T_{1\rho}$ is a spin-lock pulse that can be used optionally to eliminate the residual protein signals*

combinatorial chemistry and high-throughput screening (HTS). In a standard HTS campaign, numerous (>100 000) compounds of low molecular mass 250–600 u are screened *in vitro* using an activity-based bioassay. In contrast, in a fragment screening exercise, fewer compounds (500–2 000) of even lower molecular mass (150–300 u) are assayed using an affinity-based biophysical method (e.g. X-ray crystallography, NMR, or surface plasmon resonance). Any identified active fragments are then synthetically linked to assemble larger molecules that are subsequently screened in the hopes that they will exhibit stronger binding than any of their constituent fragments alone [15–18]. NMR is generally the method of choice in fragment screening, and STD, the preferred type of NMR experiment [9, 15, 16].

The main advantage of STD is its ability to detect weak binding interactions (Kd: millimolar range). Mixtures of five to ten compounds can be screened by STD [17, 18]. Success chiefly depends on the design of the fragment library [19–21], which is often laborious, and on the *drugability* of the target protein [22]. Moreover, since most STD experiments are performed in D_2O, it is essential that the fragments be water-soluble. However, the problem of limited water solubility can usually be overcome by adding a fixed amount (typically 5% v/v) of dimethylsulfoxide to each screening solution. Furthermore, the mixtures must be designed such that the STD results can be deconvoluted easily (i.e. the overlap of the signals corresponding to each compound in the mixture must be minimized). Lastly, the mixture must be chemically stable; the fragments must not react with each other in solution.

As explained above, active fragments are combined to construct a single hybrid compound that has higher binding than either component alone. A common technique for choosing which fragments to combine is to simultaneously test a fragment that is known to bind to the primary site of the protein with another fragment which is being screened for binding to a secondary site (Figure 9.1.5B) [23]. If the second fragment proves active, then the two fragments can be linked to form a ligand that, theoretically, should bind to both sites (Figure 9.1.5C).

There are other NMR experiments for detecting ligands that bind to secondary sites. These include the interligand NOE experiment (ILOE) [24], which is based on measuring the NOE build-up between protons on neighboring ligands (Figure 9.1.5D). Maurizio Pellecchia has pioneered the use of interligand NOEs in hit optimization via screening of libraries containing a few hundred fragments [25, 26]. An obvious advantage of this method is that the relative orientation of both ligands can be inferred from the interproton distances derived from interligand NOEs; this topological information can then be exploited to achieve optimal synthetic linkage of the two fragments. This strategy has been employed in the design of apoptosis inhibitors that function by binding to the protein Bid [26].

Another experiment is to attach a spin-label to the first ligand and then analyse the enhanced relaxation rates of protons on the secondary-site ligand (Figure 9.1.5E) [27].

Compared to STD, ILOEs and spin-labeling experiments are more prone to give false positives, but they are advantageous in that they provide invaluable information on the relative orientation of each fragment.

9.1.3 Chemical Shift Perturbation (CSP)

CSP is another popular NMR experiment for studying the interactions between ligands and proteins. The concept is very simple: since the chemical shifts of protein signals are affected

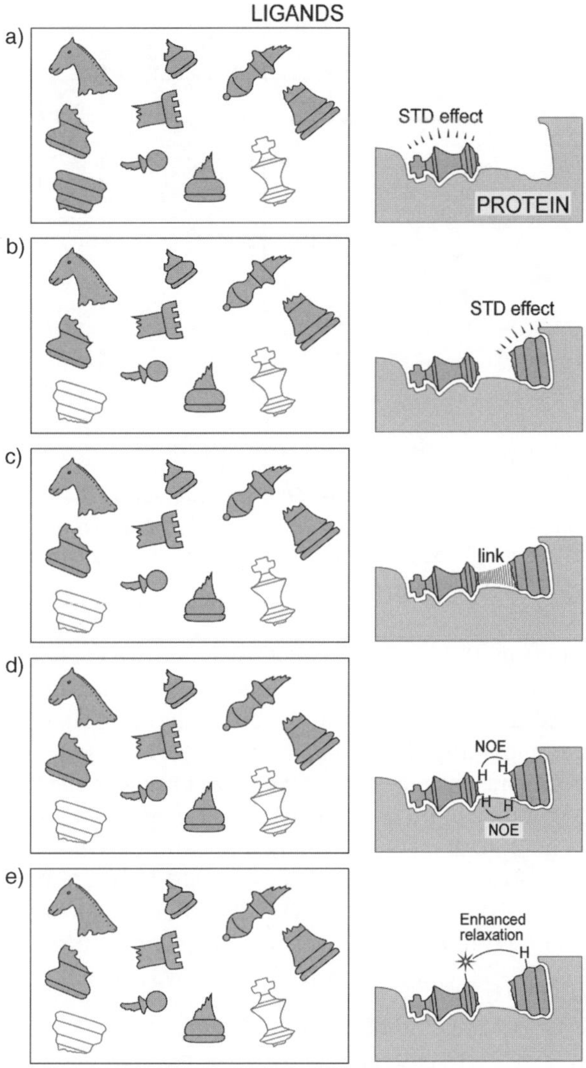

Figure 9.1.5 *Schematic representation of different strategies for linking active fragments to form larger, tighter-binding hybrid molecules (see text for details. The author acknowledges the assistance of Josep Ma. Agulló in the preparation of this figure)*

by ligand binding, monitoring of these chemical shifts enables detection of the binding event. Moreover, if the protein signals have already been assigned, then the exact location of the interaction on the protein surface can be mapped.

Although this approach was pioneered by several authors, including Gerhard Wagner [28], it is strongly associated with Stephen Fesik and his colleagues at Abbot, who coined the term *SAR by NMR* to describe the use of CSP for establishing structure-activity relationships in the context of drug discovery projects [29].

2D-[^{15}N,^{1}H]-HSQC is the mostly widely used experiment in CSP studies. Figure 9.1.6 shows CSP investigation of the binding of tetraguanidinium compound from Figure 9.1.2 to

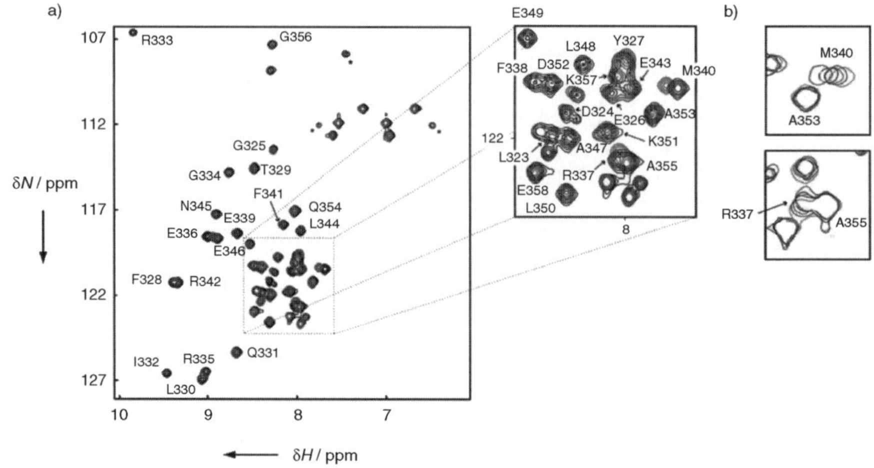

Figure 9.1.6 *(A) [^1H, ^{15}N]-HSQC spectrum of the tetramerization domain of protein p53 in the absence and in the presence of Ligand **1** (4 equiv.); and (B) titration results for residues Ala 353, Met 340, Arg 337 and Ala 355. (See Plate 19.) Reprinted with permission from [11]. Copyright Wiley-VCH Verlag GmbH & Co. KGaA*

p53TD. In Figure 9.1.6A the 2D-[^{15}N,^{1}H]-HSQC spectrum of p53TD alone (black contours) is compared with the spectrum of the protein in the presence of four equivalents of ligand (red contours). As observed, addition of the ligand caused significant changes in the ^{15}N and ^1H chemical shifts of residues Arg337, Met340, Leu344, Ala347 and Leu350 [11]. As seen in Figure 9.1.6B, for certain residues, such as Met340, the change in chemical shift is very easy to follow during the titration experiment, whereas for other cases, such as Arg337, signal overlap complicates interpretation of the results. The Kd values for the interaction between the protein and the ligand can be readily obtained by analyzing the plot of $\delta(^{15}\text{N})$ or $\delta(^1\text{H})$ *vs.* the concentration of the ligand, in which $\delta(^{15}\text{N})$ and $\delta(^1\text{H})$ represent the ^{15}N and ^{1}H chemical shifts, respectively, of a given residue at different points of the titration. Alternatively, an averaged weighted $\delta(^{15}\text{N}/^1\text{H})$ chemical shift – defined by Equation (9.1.1), below – can be used [30].

$$\delta(^{15}\text{N}/^1\text{H}) = (\delta(^1\text{H}) + \delta(^{15}\text{N}))/5 \qquad (9.1.1)$$

More importantly, mapping the perturbed residues on the 3D structure of the protein *enables delineation of the protein-surface patch to which the ligand binds* [11, 14]. It is this feature that makes CSP such a powerful method for studying protein surface recognition.

Comparison of STD and CSP reveals clear strengths and weaknesses for each type of experiment. Firstly, STD is a ligand-oriented approach: the only NMR signals tracked are those from the ligand. It provides evidence of binding and, in some cases, helps in identifying which of the ligand's atoms are most involved in the binding. Contrariwise, CSP is a protein-oriented approach: the only NMR signals followed are those from the protein. It can reveal evidence of binding, and provides data from which binding constants can be calculated as well as information on which protein residues are most affected by binding of the ligand. Whereas STD is only appropriate for proteins with molecular mass more than

10 000 u –indeed, protein size in STD is dictated by *the bigger the better* – CSP works well for small and medium size proteins but is difficult to apply to very big targets. Moreover, in terms of kinetic windows, CSP is more robust than STD (see below). However, unlike STD, CSP requires use of isotopically labeled proteins and preassignment of the protein signals, which is not trivial. Given that CSP experiments demand a higher quantity of protein, STD is more amenable to the screening of compound mixtures, especially large ones. Furthermore, unlike STD, CSP requires that the 3D structure of the protein is known. Lastly, contrary to STD, CSP cannot detect very weak binders; therefore, STD is better adapted to fragment screening.

The respective disadvantages of STD and CSP described above should only be taken as general considerations: depending on the case at hand, and the amount of effort dedicated to the experiment, these limitations can be overcome, or at least, minimized. For an example, affinity constants can actually be calculated via STD, by performing competition experiments in which a solution of the protein and a fixed amount of an established ligand, whose Kd has been predetermined by other means, is titrated with a new ligand.

There is one very clear practical challenge in CSP: the difficulty in *distinguishing between short and long distance effects*. Consequently, in CSP experiments, despite obtaining very clear results, it is impossible to know whether the observed changes in $\delta(^{15}N)$ and in $\delta(^{1}H)$ result from direct contact of the ligand with the protein or from some conformational change in the protein that is triggered by the binding event. Therefore, I suggest complementing NMR with other spectroscopic techniques that would be sensitive to such conformational changes (e.g. fluorescence or CD) and/or computational simulations (e.g. normal mode analysis) that could predict these changes.

Binding kinetics are highly influential in STD and CSP experiments. B. Meyer, the inventor of STD, has performed an exhaustive review of how kinetics affect this experiment [9]. To summarize, based on Meyer's findings, saturation between proteins and weak ligands (i.e. high k_{off}; Kd $\geq$ 100 nM) occurs rapidly, leading to efficient transfer of the saturation to the solution, and therefore, an intense STD effect. Contrariwise, saturation transfer between proteins and strong ligands (i.e. low k_{off} (0.1–0.01 $s^{-1}\,M^{-1}$); Kd 1–10 nM) is relatively slow, making saturation to the solution inefficient, which translates to weak or negligible STD signals. In the aforementioned cases k_{on} is assumed to be invariably fast: typically, $10^{-7}\,s^{-1}\,M^{-1}$ for a diffusion controlled process. However, k_{on} can be up to three orders of magnitude smaller if the binding entails major conformational rearrangements of the ligand and the protein.

The role of binding kinetics in CSP is more complex. Depending on the strength of the interaction, a standard titration experiment can occur on any of the three timescales. In the *slow exchange regime*, the ligand exchanges more slowly than the changes in chemical shifts. This generates two sets of NMR signals: one corresponding to free protein, and one, to bound protein. Upon titration, the free protein peaks begin to lose intensity. Once a high ligand to protein ratio is reached, the signals for the free protein disappear; only the signals for the bound protein remain visible. In the *intermediate exchange regime*, ligand exchange occurs at roughly the same rate as the changes in chemical shifts. In this case, chemical shift change is accompanied with peak broadening until the resonance ultimately disappears. Lastly, in the *fast exchange regime*, the ligand exchange rate is much higher than the rate at which the chemical shifts of the free and bound forms of the protein change. This leads to a gradual change in the chemical shift values of the affected NMR peaks [14].

Although CSP has been discussed here in the context of interactions between proteins and small ligands, there is increasing interest in its application to the study of protein–protein interactions [31]. The most common method for this type of experiment entails titrating a constant amount of an isotopically labeled protein with increasing amounts of an unlabeled putative protein partner. Protein–protein interactions are characterized by changes in the chemical shifts and/or broadening of the peaks that correspond to the labeled protein. Again, if the spectrum of the labeled protein has been previously assigned, then it can be used to extract information on the surface patch involved in binding [32].

NMR study of proteins in living cells (*in-cell NMR*) has recently become a reality [33]. Indeed, Sakakibara *et al.* have even reported the first 3D protein structure calculated exclusively on the basis of data obtained in living cells. As such, some examples of in-cell NMR studies of protein–protein interactions have already been reported [34, 35]. The method exploits the time-controlled, sequential expression of two (or more) proteins within a single bacterial cell. This enables overexpression of the target protein in a uniformly labeled [U-^{15}N] medium. The growth medium is then changed, and the unlabeled putative protein partner is then also overexpressed. By comparing HSQC spectra recorded before and after the overexpression of the second protein, the binding area on the target protein surface can be mapped.

9.1.4 ^{19}F-NMR in Molecular Recognition Studies

All the methods described so far, whether ligand- or target-based, are focused on the *affinity* of a ligand for a protein. Nevertheless, for studying enzymes, there are NMR experiments based on *catalytic activity*. Starting with the pioneering work of Claudio Dalvit [36, 37], ^{19}F has proven extremely useful for characterizing interactions between ligands and enzymes [38–45].

As a nucleus, ^{19}F is highly amenable to screening: it is highly sensitive to NMR and has maximum natural isotopic abundance (100%) [38, 39, 45]. Furthermore, ^{19}F chemical shifts are very sensitive to minor structural perturbations. This is highly advantageous in studying enzymatic reactions of fluorinated substrates, as it enables distinction of the substrate and the fluorinated product.

The protease prolyl oligopeptidase (POP; EC 3.4.21.26) is a therapeutic target for schizophrenia and bipolar disorder. Tarragó *et al.*, from our laboratory, used ^{19}F-NMR to test the POP inhibitory activity of traditional Chinese medicinal extracts [39]. We later identified the alkaloid berberine, a component of one of the original active extracts, as a human POP inhibitor [40]. In order to set up the ^{19}F-NMR screening methods for POP inhibitors, we synthesized a new POP substrate: ZGPF-4-CF3 (Figure 9.1.7). POP is a post-proline endoprotease; therefore, enzymatic cleavage of ZGPF-4-CF3 afforded 4-fluoromethylphenylalanine. To amplify the differences between the ^{19}F-NMR chemical shifts of the fluorinated substrate and those of the fluorinated product, the ^{19}F-NMR spectra were recorded at acidic pH (by quenching the reaction with HCl): protonation of the amino group provides an ammonium group, which is a stronger electron withdrawing group than the unprotonated carbonylamido group in the substrate. Despite the fact that the nitrogen and fluorine atoms are separated by eight bonds, the differences in $\delta(^{19}F)$ are clearly observed. Figure 9.1.8 shows the ^{19}F-NMR of the substrate ($t = 0$). The double peak is due to *cis-trans*

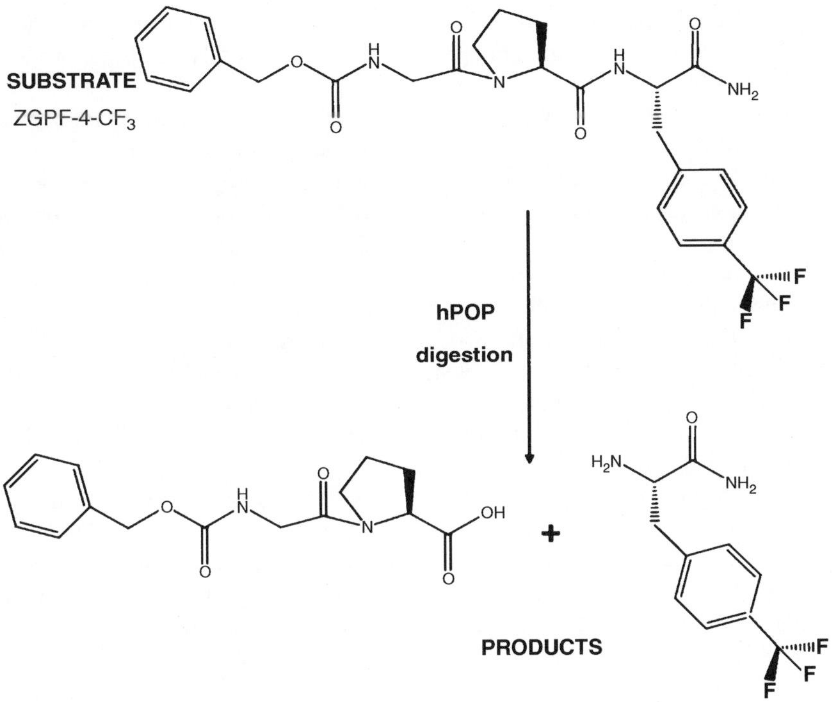

Figure 9.1.7 *Substrate ZGPF-4-CF_3 and its products from enzymatic digestion by prolyl oligopeptidase. Reprinted with permission from [39]. Copyright Wiley-VCH Verlag GmbH & Co. KGaA*

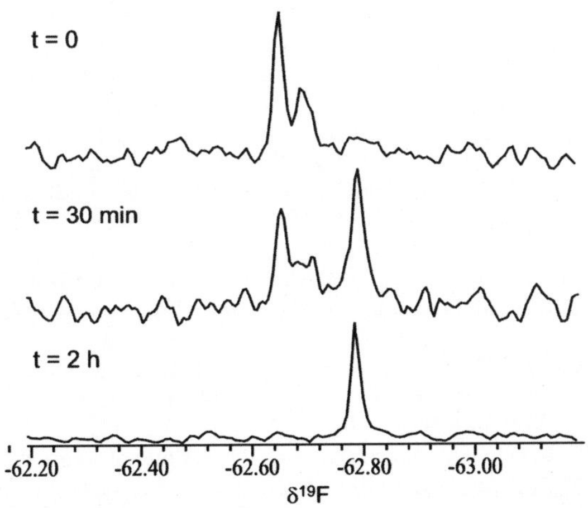

Figure 9.1.8 *Enzymatic hydrolysis of the substrate ZGPF-4-CF_3 by prolyl oligopeptidase releases Phe-4-CF_3, resulting in a change in the ^{19}F chemical shifts. As observed in the spectra, ^{19}F-NMR signals corresponding to the substrate* ($t = 0$), a mixture of the substrate and the product ($t = 30$ min), and the product alone ($t = 2$ h), can be distinguished at 400 MHz. Reprinted with permission from [39]. Copyright Wiley-VCH Verlag GmbH & Co. KGaA

conformational exchange around the Gly-Pro peptide bond. After 30 min, a new peak, corresponding to 4-trifluoromethylphenylalanine, arises. After 2 h, only the product spectrum remains visible.

This screening method is robust, fast and reliable. Its main advantage is its *bio-orthonogality* [46]. As can be imagined, it can be used to readily screen for inhibitors, determine IC50 values or perform various kinetic studies [39]. Unlike conventional fluorescence- or absorbance-based assays, ^{19}F-NMR is amenable to testing natural product extracts that contain high concentrations of colored or fluorescent compounds, provided that these compounds have a relatively low fluorine content. Furthermore, whereas conventional screening methods suffer from false positives ad false negatives – often the result of precipitation of the substrate or the product upon formation of aggregates inhibitor – ^{19}F-NMR enables tracking of both the substrate and the product signals, which ensures that only truly active compounds are identified as hits. In the short time since its inception, this approach has been applied to inhibitor discovery programs for various enzymes, including HIV-1 protease [38] and caspase [41, 42].

As illustrated throughout this chapter, NMR plays a central role in contemporary studies of molecular recognition at protein surfaces. In fact, STD, CSP, and ^{19}F-NMR screening represent only a small selection of the myriad experiments now used in academia and industry. For example, in addition to ^{1}H, ^{15}N and ^{19}F, other nuclei (e.g. ^{13}C and ^{31}P) have been used as reporter probes [47, 48]. For more detail on these and other approaches, the reader is referred to more specialized reviews [1, 16, 49, 50].

References

1. X. Salvatella and E. Giralt, NMR-based methods and strategies for drug discovery, *Chem. Soc. Rev.*, **32**, 365–72 (2003).
2. I. Belda, S. Madurga, X. Llorà, *et al.*, ENPDA: an evolutionary structure-based de novo peptide design algorithm. *J. Comput. Aided. Mol. Des.*, **3**, 1–17 (2005).
3. G.M. Withsides and V. M. Krishnamurty, Designing ligands to bind proteins, *Quaterly Rev. of Biophysics*, **38**, 385–95 (2005).
4. M. Mayer and B. Meyer, Characterization of ligand binding by saturation transfer difference NMR spectroscopy, *Angew. Chem. Int. Ed*, **38**, 1784–8 (1999).
5. J. Klein, R. Meinecke, M. Mayer and B. Meyer, Detecting binding affinity to immobilized receptor proteins and compound libraries by HR-MAS STD NMR, *J. Am. Chem. Soc.*, **121**, 5336–7 (1999).
6. A.T. Neffe, M. Bilang, I. Grüneberg and B. Meyer, Rational optimization of the binding affinity of CD4 targeting peptidomimetics with potential anti HIV activity, *J. Med. Chem.*, **50**, 3482–8 (2007).
7. M. Politi, M.I. Chávez, F.J. Cañada and J. Jiménez-Barbero, Screening by NMR: A new approach for the study of bioactive natural products? The example of *Pleurotus ostreatus* hot water extract, *Eur. J. Org. Chem.* 1392–6 (2005).
8. M. Pickhardt, G. Larbig, I. Khlistunova, *et al.*, Phenylthiazol-hydrazide and its derivative are potent inhibitors of τ aggregation and toxicity in vitro in cells, *Biochemistry*, **46**, 10016–23 (2007).
9. B. Meyer and T. Peters, NMR Spectroscopy techniques for screening and identifying ligand binding to protein receptors, *Angew. Chem. Int. Ed* **42**, 864–90 (2003).
10. J. Angulo, B. Langpap, A. Blume, *et al.*, Blood group B galactosyltransferase: insignts into substrate binding from NMR experiments, *J. Am. Chem. Soc.*, **128**, 13529–38 (2006).

11. X. Salvatella, M. Martinell, M. Gairí, *et al.*, A tetraguanidinium ligand binds to the surface of the tetramerization domain of proteína p53, *Angew. Chem. Int. Ed* **43**, 196–8 (2004).
12. M. Martinell, X. Salvatella, J. Fernandez-Carneado, *et al.*, Synthetic ligands able to interact with the p53 tetramerization domain. Towards understanding a protein surface recognition event, *ChemBioChem*, **7**, 1105–13 (2006).
13. M. Mayer and T.L. James, NMR-based characterization of phenothiazines as a RNA binding scaffold, *J. Am. Chem. Soc.*, **126**, 4453–60 (2004).
14. S. Gordo, V. Martos, E. Santos, *et al.*, Stability and structural recovery of the tetramerization domain of p53-R337H mutant induced by a designed templating ligand, *Proc. Natl. Acad. Sci. USA*, **105**, 16426–31 (2008).
15. D.C. Rees, M. Congreve, C.W. Murray and R. Carr, Fragment-based lead discovery, *Nat Rev Drug Discov.*, **8**, **3**, 660–72 (2004).
16. R.H. Jeffrey, S. Chaohong, D.R. Sauer and P.J. Hadjuk, Utilization of NMR-derived fragment leads in drug design, *Methods Enzymol.*, **34**, 349–571 (2005).
17. A.M. Mercier and R. Powers, Determining the optimal size of small molecule mixtures for high throughput NMR screening, *J. Biomol. NMR*, **31**, 243–58 (2005).
18. P.J., Hajduk and J. Greer, A decade of fragment-based drug design: strategic advances and lessons learned, *Nat. Rev. Drug Discov.*, **6**, 211–19 (2007).
19. M.M. Hann, A.R. Leach and G. Harper, Molecular complexity and its impact on the probability of finding leads for drug discovery, *J. Chem. Inf. Comput. Sci.*, **41**, 856–64 (2001).
20. N. Baurin, F. Aboul-Ela, X. Barril, *et al.*, Design and characterization of libraries of molecular fragments for use in NMR screening against protein targets, *J. Chem. Inf. Comput. Sci.* **44**, 2157–66 (2004).
21. P.J. Hadjuk, Fragment-based drug design: how big is too big? *J. Med. Chem.*, **49**, 6972–6 (2006).
22. J. Seco, F.J. Luque and X. Barril, Binding Site Detection and Druggability Index from First Principles, *J. Med. Chem.* **23**, 2363–71 (2009).
23. P. J. Hajduk, G. Sheppard, D. G. Nettesheim, *et al.*, Discovery of Potent Nonpeptide Inhibitors of Stromelysin Using SAR by NMR, *J. Am. Chem. Soc.*, **119**, 5818–27 (1997).
24. D. Li, E.F. DeRose and R.E. London, The inter-ligand Overhauser effect: a powerful new NMR approach for mapping structural relationships of macromolecular ligands, *J. Biomol. NMR*, **15**, 71–6 (1999).
25. M. Pellecchia, B. Becattini, K.J. Crowell, *et al.*, NMR-based techniques in the hit identification and optimisation processes, *Expert Opin. Ther. Targets*, **8**, 597–611 (2004).
26. B. Becattini, S. Sareth, D. Zhai, *et al.*, Targeting apoptosis via chemical design: inhibition of bid-induced cell death by small organic molecules, *Chem. Biol.*, **11**, 1107–17 (2004).
27. W. Jahnke, L.B. Perez, C. Gregory Paris, A. Strauss, G. Fendrich and C. M. Nalin, Second-site NMR screening with a spin-labeled first ligand, *J. Am. Chem. Soc.*, **122**, 7394–5 (2000).
28. M.A. Markus, T. Nakayama, P. Matsudaira and G. Wagner, Solution structure of villin 14T, a domain conserved among actin-severing proteins, *Protein Sci.*, **3**, 70–81 (1994).
29. S.B. Shuker, P.J. Hajduk, R.P. Meadows and S.W. Fesik, Discovering high-affinity ligands for proteins: SAR by NMR, *Science*, **274**, 1531–1534 (1996).
30. P.J. Hajduk, J. Dinges, G.F. Miknis, *et al.*, NMR-based discovery of lead inhibitors that block DNA binding of the human papillomavirus E2 protein, *J. Med. Chem.*, **40**, 3144–50 (1997).
31. S. Gordo and E. Giralt, Knitting and untying the protein network: Modulation of protein ensembles as a therapeutic strategy, *Protein Sci.*, **18**, 481–93 (2009).
32. E.R.P. Zuiderweg, Mapping protein–protein interactions in solution by NMR spectroscopy, *Biochemistry*, **41**, 1–7 (2002).
33. D. Sakakibara, A. Sasaki, T. Ikeya, *et al.*, Protein structure determination in living cells by in-cell NMR spectroscopy, *Nature*, **458**, 102–5 (2009).
34. D.S. Burz, K. Dutta, D. Cowburn, A. Shekhtman, In-cell NMR for protein–protein interactions (STINT-NMR), *Nat. Protoc.*, **1**, 146–52 (2006).
35. K. Inomata, A. Ohno, H. Tochio, *et al.*, High-resolution multi-dimensional NMR spectroscopy of proteins in human cells, *Nature*, **458**, 106–9 (2009).

36. C. Dalvit, P.E. Fagerness, D.T. Hadden, R.W. Sarver and B.J. Stockman, Fluorine-NMR experiments for high-throughput screening: theoretical aspects, practical considerations, and range of applicability, *J. Am. Chem. Soc.*, **125**, 7696–7703 (2003).
37. C. Dalvit, E. Ardini, M. Flocco, G.P. Fogliatto, N. Mongelli and M.A. Veronesi, A general NMR method for rapid, efficient, and reliable biochemical screening, *J. Am. Chem. Soc.* **125**, 14620–5 (2003).
38. S. Frutos, T. Tarragó and E. Giralt, A fast and robust 19F NMR-based method for finding new HIV-1 protease inhibitors, *Bioorg. Med. Chem. Lett.*, **16**, 2677–81 (2006).
39. T. Tarrago, S. Frutos, R.A. Rodriguez-Mias and E. Giralt. Identification by 19F NMR of traditional Chinese medicinal plants possessing prolyl oligopeptidase inhibitory activity, *ChemBioChem*, **7**, 827–33 (2006).
40. T. Tarrago, N. Kichik, J. Segui and E. Giralt, The natural product berberine is a human prolyl oligopeptidase inhibitor, *ChemMedChem*, **2**, 354–9 (2007).
41. R. Fattorusso, S. Frutos, X. Sun, N. Sucher and M. Pellecchia, Traditional Chinese medicines with caspase-inhibitory activity, *Phytomedicine*, **13**, 16–22 (2006).
42. R. Fattorusso, D. Jung, K.J. Crowell, M. Forino and M. Pellecchia. Discovery of a novel class of reversible non-peptide caspase inhibitors via a structure-based approach, *J. Med. Chem.*, **48**, 1649–56 (2005).
43. C. Dalvit, Ligand and substrate-based 19F NMR screening: Principles and applications to drug discovery, *Prog. Nuclear Magn. Reson. Spectrosc.*, **51**, 243–71 (2007).
44. C. Dalvit, E. Ardini, G.P. Fogliatto, N. Mongelli and M Veronesi, Reliable high-throughput functional screening with 3-FABS. *Drug Discov. Today*, **9**, 595–602 (2004).
45. D.B. Berkowitz, K.R. Karukurichi, R. de la Salud-Bea, D.L. Nelson and C.D. McCune, Use of fluorinated functionality in enzyme inhibitor development: Mechanistic and analytical advantages, *J. Fluchem.*, **129** 731–42 (2008).
46. C. Dong, F. Huang, H. Deng, *et al.*, Crystal structure and mechanism of a bacterial fluorinating enzyme. *Nature*, **427**, 561–5 (2004).
47. S. Frutos, R.A. Rodriguez-Mias, S. Madurga, *et al.*, Disruption of the HIV-1 protease dimer with interface peptides: structural studies using NMR spectroscopy combined with [2-(13)C]-Trp selective labelling, *Biopolymers*, **88**, 164–73 (2007).
48. F. Manzenrieder, A.O. Frank and H. Kessler, Phosphorus NMR spectroscopy as a versatile tool for compound library screening, *Angew. Chem. Int. Ed. Engl.*, **47**, 2608–11 (2008).
49. M. Pellecchia, I. Bertini, D. Cowburn, *et al.*, Perspectives on NMR in drug discovery: a technique comes of age, *Nat. Rev. Drug Discov.*, **7**, 738–45 (2008).
50. J. Klages, M. Coles and H. Kessler, NMR-based screening: a powerful tool in fragment-based drug discovery, *Analyst*, **132**, 692–705 (2007).

9.2

In Vitro Screening: Methods of High-throughput Screening

Wenjiao Song and Qing Lin
Department of Chemistry, State University of New York at Buffalo, Buffalo, NY, USA

9.2.1 Introduction

With rapid development in modern robotics, sensitive detectors, and data processing software, high-throughput screening (HTS) has become one of the most versatile tools in the development of potential drugs targeting protein–protein interactions. Historically, however, protein–protein interaction targets have shown lower probability of success in HTS campaigns [1]. One explanation is that commercially available libraries often do not contain compounds of sufficient size and complexity to match the large interfacial areas typically encountered in these targets. In this regard, natural products are considered to be a superior and more fruitful venue for identifying small-molecule inhibitors. The challenge, however, is to find natural products that exhibit inhibitory activities while possessing the desired drug-like properties, e.g. low molecular weight and favorable pharmacokinetic and pharmacodynamic properties. Alternatively, potent inhibitors can be obtained by a medium throughput screen of small-molecule fragments that inhibit protein–protein interactions with modest activities followed by a fragment tethering to generate moderate-size and potentially more potent compounds [2].

HTS is usually part of an integrated lead discovery process which typically consists of five consecutive steps: 1-target identification; 2-assay development and validation; 3-HTS implementation; 4-data capture, storage and analysis; and 5-lead identification (Figure 9.2.1). On the basis of assay format, HTS can be divided into two major categories: cell-free/biochemical assays, and cell-based assays. This chapter will discuss the pertinent

Protein Surface Recognition: Approaches for Drug Discovery Ernest Giralt, Mark W. Peczuh and Xavier Salvatella. © 2011 John Wiley & Sons, Ltd

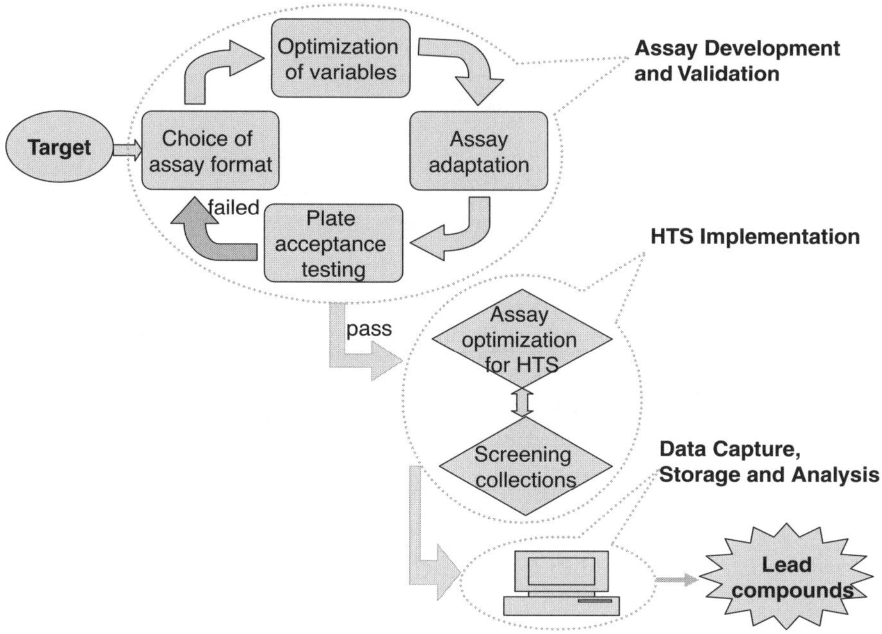

Figure 9.2.1 *Schematic representation of an integrated lead discovery process*

issues in HTS as it applies to the protein–protein interaction targets and discuss some of the more widely used HTS assay techniques in identifying protein–protein interaction inhibitors.

9.2.2 Statistical Evaluation of the HTS Assay Performance

Because HTS assays generate a large amount of data, a central concern in HTS is the ability to evaluate the assay performance to ensure the data is reliable and robust. To this end, statistical analysis is often performed on the acquired data sets in order to derive critical statistical parameters (Table 9.2.1). To attain an acceptable performance, an assay optimization step is often performed, during which individual steps such as reagent dispersal, plate transfer, washing, and plate reading are critically assessed. The assay optimization generally leads to significantly improved assay sensitivity, dynamic range, signal intensity, data stability and reproducibility against a particular target.

The statistical analysis of raw data involves the calculations of a number of parameters, including signal mean (μ), standard deviations (σ), and various combinations of these two parameters as listed in Table 9.2.1. For instance, signal-to-background (S/B) indicates the separation of signals between the positive and negative controls. It is usually dependent on the assay formats and, as a result, can be used early on to estimate the quality of a particular assay format. Signal window (SW) provides a measurement of the signal magnitude. The expression of signal-to-noise (S/N) in HTS is different from classic expression of S/N in that classic S/N does not include standard deviation of signals and therefore does not provide an

Table 9.2.1 *Statistical analysis of the performance of an HTS assay [7]*

Parameter	Equation	Measurement
signal to background	$S/B = \mu_{signal}/\mu_{background}$; μ = mean	separation between positive and negative controls
signal window or specific signal	$SW = \mu_{signal} - \mu_{background}$	signal magnitude
signal to noise	$S/N = (\mu_{signal} - \mu_{background})/[(\sigma_{signal})^2 + (\sigma_{background})^2]^{1/2}$; σ = standard deviation	combination of signal win dow and variability (>10 acceptable)
coefficient of variation of signal and background	$CV = 100 \times \sigma/\mu$	an indicator of relative signal variability
Z′-factor	$Z' = 1 - 3 \times (\sigma_{signal} - \sigma_{background})/\lvert\mu_{signal} - \mu_{background}\rvert$	combination of signal window and variability

accurate account of signal window and variability. The HTS S/N provides a complete picture of the statistic performance of a HTS assay; typically, values of S/N greater than 10 are considered acceptable. The coefficient-of-variation (CV) is a relative measurement of signal variability, reflecting the precision of liquid handling and detection instruments. The main parameter in assessing the statistical performance of an assay by the HTS community is the Z′-factor [3]. The value of the Z′-factor reveals relative separation of the signal and background populations. It is presumed that if the variability is due to random errors, there is a normal distribution for these populations. Z′-factor is a dimensionless parameter with the highest possible value of 1.0. When Z′ approaches 1.0, it indicates that the assay has a huge dynamic range with tiny standard deviations. When Z′ is between 0.5 and 1.0, the assay is considered excellent. When Z′ is between 0 and 0.5, the assay is considered marginal. When the Z′-factor value is less than 0, the signals from the positive and negative controls overlap, making the assay essentially useless for screening purposes.

Assays developed for HTS can be roughly divided into cell-free (biochemical) or cell-based in nature. The choice of either biochemical or cell-based assay is a balancing act between two considerations. On one side is the need to ensure that the measured signals provide relevant data to the desired biological process. On the other is the necessity of adaptability toward high throughput. The assays need to support reagents (or cells) to yield robust data on $10^5 \sim 10^6$ samples in microtiter plate formats.

9.2.3 Biochemical Assays

In biochemical assays, compounds are evaluated based on their interactions with the isolated targets in an artificial environment. Because biochemical assays can be easily adapted to microtiter plate formats, they have become the predominant approach for conducting HTS. Since the 1990s, when HTS became a central endeavor in drug discovery, numerous HTS assay methods have been successfully developed targeting protein–protein interactions, including: 1-prompt fluorescence intensity; 2-fluorescence polarization; 3-fluorescence resonance energy transfer (FRET); 4-homogeneous time-resolved fluorescence (HTRF) or luminescence resonance energy transfer (LRET); 5-scintillation proximity assay; 6-enzyme-linked immunosorbent assay (ELISA); and 7-surface plasmon resonance imaging. The advantages and disadvantages of each of these assay formats are summarized in Table 9.2.2. In the following, general features of each assay format and their applications in the HTS assay of protein–protein interaction inhibitors will be discussed.

9.2.3.1 Prompt Fluorescence Intensity (FLINT)

Given the industry-wide drive toward assay simplicity, miniaturization, speed, and sensitivity, fluorescence-based techniques have become the most widely used detection methods in HTS. The development of new fluorescent dyes and molecular reagents further fueled their popularity [4]. The simplest fluorescence-based method is prompt fluorescence intensity (FLINT) in which one of the interacting proteins is labeled with a fluorescent dye. When fluorescence intensity is used directly as assay readout, the most critical variable to consider during HTS is excitation wavelength; in general, short excitation wavelengths (<400 nm) should be avoided in order to minimize autofluorescence interference produced by test compounds.

To avoid separation and expedite data readout, Neubig and coworkers [6] developed an improved FLINT assay referred to as fluorescence-based flow-cytometric protein interaction assay (FCPIA) targeting the RGS-Gα_o interaction. Regulators of G-protein signaling (RGS) are important components of signal transduction pathways initiated through G-protein-coupled receptors (GPCRs). RGS proteins accelerate the intrinsic GTPase activity of G-protein α–subunits (Gα) and thus shorten the time course and reduce the magnitude of G-protein α- and β/γ-subunit signaling. Inhibiting RGS action has been proposed as a means to enhance the activity and specificity of GPCR agonist drugs. In their screen of small-molecule inhibitors of the RGS-Gα_o interaction, three key components were used: 1-avidin-coated microspheres; 2- a biotinylated target protein, e.g. RGS; and 3 – an interacting protein, e.g. Gα_o, labeled with Alexa-532 (Figure 9.2.2). In their experiment, the bead-immobilized RGS4 was coincubated in a 96-well plate with the AlF4 activated Gα_o in the presence of small-molecule compounds. Samples from each well were aspirated into the Luminex flow cytometer, and the bead-associated fluorescence was measured to provide a quantitative readout of the amount of Gα_o bound to RGS4. No wash steps were needed because flow cytometry only measures the bead-associated fluorescence. Using this assay, they identified the first small-molecule inhibitor of RGS protein, *N*-[(4-chlorophenyl)sulfonyl]-4-nitrobenzenesulfinimidoate (CCG-4986), from a 3,028-membered compound library. CCG-4986 inhibited the RGS4-Gαa_o interaction with IC_{50} values in the range of 3 to 5 μM.

Because many protein–protein interaction interfaces either remain uncharacterized or exhibit large, flat surfaces without discrete binding groves, e.g. Gα_o/RGS interaction, FCPIA is ideally suited to examine this type of protein–protein interactions. Whereas the throughput of flow cytometry has in the past not been very high, the availability of commercial 96-well plate-reading flow cytometry systems (e.g., Luminex) has improved this situation with the capability of reading a plate in less than 30 min. In addition, the Hypercyt system developed by Sklar and colleagues [7] was reported to collect flow cytometry data in as little as 2.5 min per 96-well plate, making this approach truly viable for HTS.

Table 9.2.2 *Advantages and disadvantages of various HTS methods* [5]

Technique	Advantages	Disadvantages
prompt fluorescence intensity (FLINT)	– simple – suitable for fluorigenic assays – readily miniaturized	– little information for quality control – suffer from inner-filter and autofluorescence interference
fluorescence polarization (FP)	– simple, reasonably predictive – insensitive to inner-filter effects – ratiometric technique – improved well-level quality control – suitable for small ligands (<15 KDa)	– local motion (propeller) effects – suffer from autofluorescence interference – limited by dye lifetime, ligand size and molecular weight change – limited dynamic range
fluorescence resonance energy transfer (FRET)	– simple, reasonably predictable – suitable for short inter/intramolecular distances (<5 nm) – a wide range of donors and acceptors are available	– suffer from inner-filter and autofluorescence interference – limited to short distances to obtain high signal changes – monitoring mostly on donor quenching
homogeneous time-resolved fluorescence (HTRF)/luminescence resonance energy transfer (LRET)	– reasonably predictable and robust – suitable for long distance (5–10 nm) – reduced autofluorescence interference – sensitive and miniaturizable	– labeling with donor/acceptor often problematic (e.g. steric hindrances) – limited choice of donor/acceptor – high background due to nonspecific energy transfer
scintillation proximity assay (SPA)	– homogeneous assay – well established technology	– hazardous due to the use of radioisotopes – limited reagent stability – relative long read-time – lower signal-to-background ratio
enzyme-linked immunosorbent assay (ELISA)	– flexible, simple – well-established technology – wide dynamic range	– very time-consuming, low throughput – nonspecific binding
surface plasmon resonance imaging (SPRi)	– label-free technology – high sensitivity	– low detection limit (1–10 nM) – relative low throughput screening

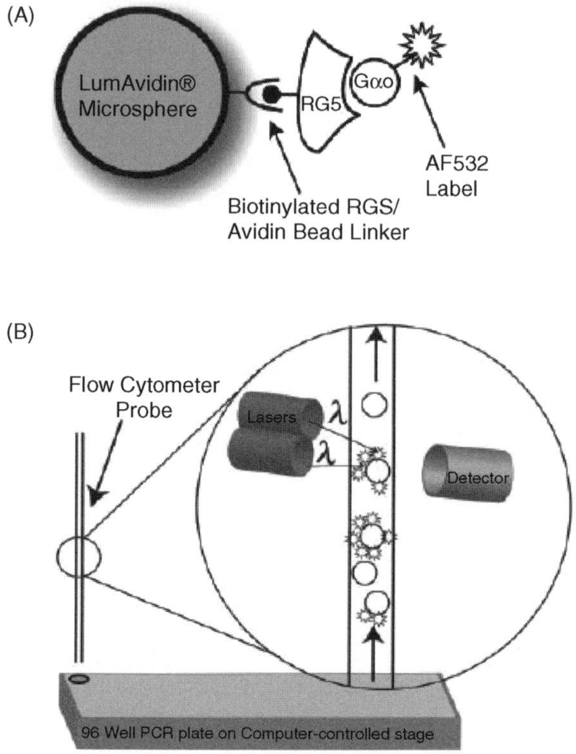

Figure 9.2.2 Schematic representation of a fluorescence-based flow cytometric protein interaction assay. Reprinted with permission from [6]. Copyright American Society for Pharmacology and Experimental Therapeutics

9.2.3.2 Fluorescence Polarization (FP)

Fluorescence polarization (FP; also known as fluorescence anisotropy, FA) is one of the most sensitive, robust, and widely used HTS methods for identifying inhibitors of protein–protein interactions in drug discovery. When a fluorophore is excited with plane-polarized light, the emitted light is polarized. The degree of polarization (mP) is a function of molecular properties. Specifically, Brownian molecular rotation serves as a sensitive molecular sensor. A ligand in its free state has a much lower mP compared to the identical ligand that is bound to a protein. FP assays are typically designed as competitive equilibrium binding assays in which potent inhibitors cause the release of the fluorescent ligand from the protein target, resulting in a decrease in mP (Figure 9.2.3). To obtain a robust ΔmP, it is advantageous to recapture a protein–protein interaction with a peptide-protein interaction. As a result, the design of a fluorescent peptide ligand that binds to the interface with high affinity becomes absolutely necessary before a competition based screening can be performed. A general framework for the FP-based HTS and strategies for assay development, sensitivity regimes, data quality control, analysis, and ranking has been recently established by Wagner and coworkers [8].

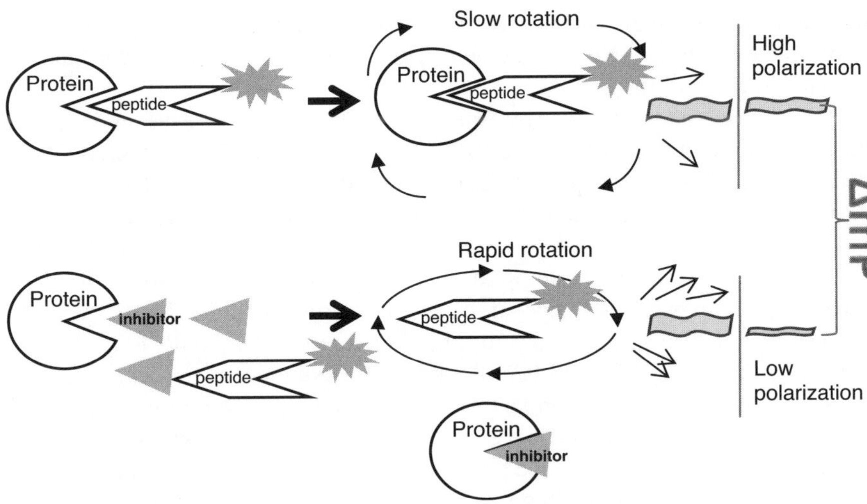

Figure 9.2.3 *Principle of an FP-based assay for identifying inhibitors of protein-protein interactions*

Because of its high sensitivity and ease of operation, FP has been widely employed in the HTS campaigns for the discovery of small-molecule inhibitors of protein–protein interactions in which structural information regarding the interface was available. For example, Yuan's group [9] reported the identification of a series of novel small molecules that inhibit the binding of Bak BH3 peptide to Bcl-xL using the FP-based HTS. The assay monitored the displacement of an Oregon Green labeled Bak BH3 peptide from a recombinant GST-Bcl-xL protein. A ChemBridge library consisting of 16 320 compounds was screened, and three compounds showing the highest potency in disrupting the BH3–Bcl-xL interaction were selected for further analysis. The follow-up NMR study confirmed that the compounds bound to the BH3-binding pocket of Bcl-xL. A similar screen targeting the BH3–Bcl-xL interaction was also reported by Zhang and coworkers [10]. They found that a fluorescein-labeled Bad BH3 peptide interacted strongly with Bcl-xL with K_d of 21.48 nM and gave a large dynamic change, $\Delta mP = 120.37 \pm 4.29$. When tested in the HTS format, the assay gave a signal-to-noise ratio of 15.37 and a Z′-factor of greater than 0.73.

Kenny and coworkers [11] applied an FP method to screen for inhibitors of the FtsZ–ZipA interaction, a critical mediator of cell division in *E. coli*. A phage display optimized peptide, FtsZ PD1, was developed for the FP assay. A screen of 250 000 compounds identified 29 hits with modest activities ($\geq$30 % inhibition at 50 mg/mL). A modeling study of a small-molecule lead ($K_i = 12\ \mu$M) in complex with $\mathrm{ZipA}_{185\text{-}328}$ revealed that the compound binds to the same hydrophobic pocket as the $\mathrm{FtsZ}_{367\text{-}383}$ peptide.

Wagner's group [12] reported the discovery of small-molecule inhibitors of the NFAT–calcineurin interaction, a protein–protein interaction critically important in mediating the activation of T cells and developmental genetic programs, through FP-HTS. They designed and optimized an NFAT-surrogate peptide ligand corresponding to the primary binding epitope for calcineurin. On the basis of the estimated rotational correlation times, they found that the experimentally observed anisotropy values were consistent with a specific binding model. The assay conditions were successfully tuned based on (1) FP sensitivity analysis and

(2) calculation of assay quality parameters, which led to the discovery of a number of novel small-molecule inhibitors of NFAT–calcineurin interaction with both complete and incomplete dissociation capabilities.

The FP method was also employed by Wang's group [13] to measure the binding affinities of small-molecule inhibitors toward the BIR3 domain of XIAP. The X-linked inhibitor of apoptosis protein (XIAP) is a potent cellular inhibitor of apoptosis. Designing small-molecule inhibitors that target the BIR3 domain of XIAP, where Smac/DIABLO (second mitochondria-derived activator of caspase/direct IAPbinding protein with low pI) and caspase-9 bind, is a promising strategy for inhibiting antiapoptotic activity of XIAP and for overcoming apoptosis resistance of cancer cells mediated by XIAP. Among four fluorescent probes tested, a mutated N-terminal Smac peptide, AbuRPFK(5-Fam)-NH_2, showed the highest affinity ($K_d = 17.92$ nM) and a large dynamic range (ΔmP $= 231 \pm 0.9$), and thus was selected as a suitable probe for the FP assay. Under the optimized conditions, a Z'-factor value of 0.88 was obtained in a 96-well plate format. Several known Smac peptides were analysed and the results indicated that the FP-based method can accurately determine the binding affinities of the Smac-based peptide inhibitors with a wide range of affinities.

Arnold and coworkers [14] reported the use of a FP-based HTS for inhibitors of the thyroid receptor (TR) interaction with its transcriptional coactivator. A fluorescently labeled coactivator peptide SRC2-2 (CLKEKHKILHRLLQDSSSPV, derived from the p160 family of nuclear receptor coactivators) was used to determine the binding constants of TR (hTRβ, $K_d = 0.44$ µM; hTRαLBD, $K_d = 0.17$ µM; hTRβLBD(C309A), $K_d = 0.17$ µM). A library comprising 138 000 compounds was screened for their ability to compete for coactivator binding to TR in the presence of the T3 ligand in a 384-well plate format. 27 hit compounds were identified that inhibited the interaction between TRβ LBD and the SRC2-2 coactivator peptide with IC_{50} values less than 30 µM. Hit compounds were further evaluated by several secondary assays, including dose-response analysis, glutathione-*S*-transferase pull-down assay, and hormone displacement assay.

Taylor and coworkers [15] developed a novel FP-based assay, dubbed ligand-regulated competition (LiReC), in their effort to find non-ATP competitive small-molecule regulators for type Iα cAMP-dependent protein kinase A (PKA-Iα). LiReC utilized a competitive fluorescent peptide probe to assess the integrity of the PKA-Iα complex composed of C- and R-subunits in the presence of a potential allosteric ligand. To identify compounds that operate through novel modes of action, the LiReC method shielded the ATP binding site and purposely excluded the ATP-competitive ligands by performing the assay at 2 mM ATP concentration. In a proof-of-principle screen, several cyclic nucleotide-derived agonists and antagonists were successfully identified from a test library with Z'-factor greater than 0.7, thereby validating this novel approach in identifying modulators of protein–protein interactions.

9.2.3.3 Fluorescence Resonance Energy Transfer (FRET)

Fluorescence resonance energy transfer (FRET) requires two fluorophores, a donor and an acceptor. Excitation of the donor by an energy source (e.g. flash lamp or fluorometer laser) generates its characteristic emission. This emission event then triggers an energy transfer to the acceptor provided that the two are in close proximity (<5 nm) and the emission wavelength of the donor fluorophore matches the excitation wavelength of the acceptor fluorophore. Therefore, protein–protein association can be studied by monitoring the energy transfer between a pair of fluorophores attached separately to the two interacting proteins. Importantly, the magnitude of acceptor emission, a measure of the degree of protein association, can be determined without a need to separate the unbound components from the bound complexes. This homogeneous assay format is extremely beneficial for HTS because it reduces both assay time and cost.

FRET was employed by Vogt's group [16] to screen a combinatorial peptidomimetics library encompassing approximately 7000 small organic molecules for their ability to inhibit Myc-Max dimerization. In their screen, the bHLHZip domains of Myc and of Max were fused to the N-termini of CFP (donor) and YFP (acceptor), respectively. Dimerization generated FRET signals characterized by a strong emission signal of YFP at 525 nm and a weaker emission signal of CFP at 475 nm. The Myc-CFP–Max-YFP heterodimers were incubated with the chemical library, and the ability of the compounds to dissociate the dimer was followed by monitoring the fluorescence intensities of CFP and YFP upon excitation of CFP. The presence of inhibitors resulted in a decrease of FRET signal characterized by the reduction in the ratio of emission intensities between 525 nm and 475 nm. The initial round of screening identified four compounds, which caused up to 38% of the dimers to dissociate at 25 μM. Inhibition of Myc–Max interaction was further validated by ELISA and electrophoretic mobility-shift assay. Two of the candidate inhibitors also interfered with the Myc-induced oncogenic transformation in the chicken embryo fibroblast cultures.

The main limitation of FRET is autofluorescence or background fluorescence from assay components such as buffers, proteins, chemical compounds and cell lysate. As a result, fluorescence intensities detected must be corrected for this autofluorescence, which greatly compromises assay sensitivity and complicates result interpretation. This type of background fluorescence is usually transient with a half-life in the nanosecond range and can be eliminated by using time-resolved methodologies.

9.2.3.4 Homogeneous Time-Resolved Fluorescence (HTRF)/Luminescence Resonance Energy Transfer (LRET)

HTRF® is a time-resolved, FRET-based method that uses the principles of both TRF (time-resolved fluorometry) and FRET. This combination brings together benefits of the low background of TRF with the homogeneous assay format of FRET. HTRF® has many unique features beyond simple TR-FRET, including the use of a lanthanide (Eu^{3+}) with an extremely long half-life, a large Stoke's shift for the Eu^{3+}-cryptate conjugate (an entity which also confers an increased assay stability), and the use of a patented ratiometric measurement that allows assay quench and sample interference correction. A more general term 'luminescence' as in LRET should be used instead of 'fluorescence' in HTRF because the lanthanide emission is technically not considered fluorescence (i.e., arising from a singlet-to-singlet transition). Given the long distance for effective energy transfer ($\sim$9 nm), HTRF/LRET is highly suitable for monitoring protein–protein interactions.

Burgess and coworkers [17] employed an LRET-based homogeneous assay to search for inhibitors of ρ70 binding to *E. coli* RNA polymerase (RNAP) from a crude natural product library composed of 100 extracts of marine sponges. In their assay, ρ70 was labeled with an europium-cryptate complex and the β'-subunit of $RNAP_{100\text{-}309}$ were labeled with an IC5 fluorophore (Figure 9.2.4). The LRET assay turned out to be very sensitive and reliable, showing a signal-to-noise ratio of greater than 10. Out of the 100 marine sponge extracts

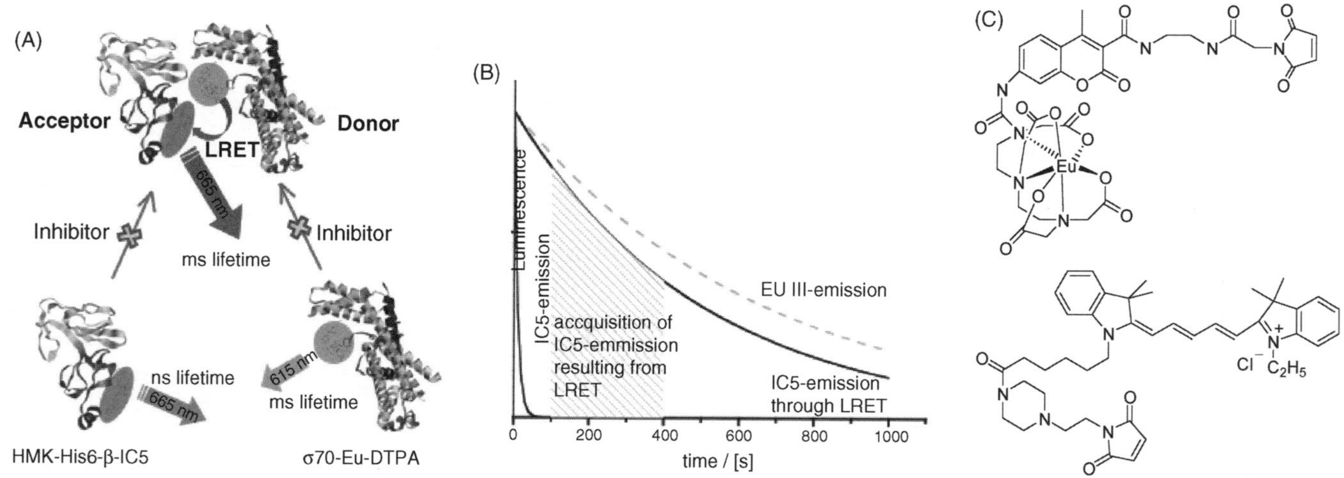

Figure 9.2.4 *Scheme for an LRET assay in the measurement of ρ70- β′(100–309) RNAP interactions. (A) Constructs of the labeled β′(100–309) RNAP and ρ70; (B) time-courses of IC5 emissions generated either intrinsically (time scale in ns) or by sensitization through LRET over 1 ms; (C) structures of the fluorophores used to label the proteins*

tested, sample D7 turned out to be the only one to reduce the LRET signal by roughly 90% in the assay. Furthermore, only D7 showed a dose-dependent inhibition of transcriptional activity, with an IC_{50} value around 1 μM. There was no significant difference between the IC_{50} values determined by LRET and those derived from the in vitro transcription assays.

9.2.3.5 Scintillation Proximity Assay (SPA)

SPA is a homogeneous assay technology for the rapid and sensitive measurement of a wide range of biological processes. It relies on beads that scintillate when they are in close proximity to radioisotopes. To apply SPA to protein–protein interactions targets, one protein needs to be bound to the beads while the other needs to be directly or indirectly labeled with radioisotopes. Because of its reliance on radioisotopes, SPA has several disadvantages compared to other nonradiometric assays, including safety, limited reagent stability, relative long read-time and little intrinsic information on the isotope environment. However, new technologies are now emerging to address the issue of read-time and assay miniaturization.

Staddon's group [18] reported the development of an SPA-based HTS method for detecting the heterodimerization of members of the Bcl-2 protein family which are critical regulators of apoptosis (Figure 9.2.5). They used Cu^{2+}-chelated scintillation beads to recruit the His_6-tagged Bcl-xL protein, and indirectly labeled the biotinylated BH3 peptide with ^{35}S-labeled streptavidin. When His6-Bcl-xL and BH3 peptide associates to form heterodimer, ^{35}S -radioisotope was brought in close proximity to the beads, resulting in robust scintillation signals. In a statistical analysis of a test screen targeting the Bim BH3–Bcl-xL interaction, a high Z′-factor value of 0.81 was obtained, indicating the assay was capable of detecting small-molecule inhibitors of Bcl-3 family dimerization in a HTS format.

9.2.3.6 Enzyme-Linked Immunosorbent Assay (ELISA)

ELISA, an immunochemical technique for the detection of an antibody or antigen in a sample, has been widely used in assaying protein–protein interaction due to its wide dynamic range. However, it can be very time-consuming because of many wash steps involved, which limits the throughput in the HTS campaigns. Using the ELISA-based HTS, Shivdasani's group [19] screened both a 7000-member natural product library and a 45 000 member synthetic small-molecule library for inhibitors of the Tcf4–β-catenin interaction. The interaction of β-catenin with Tcf4 plays a central role in the wnt-signaling pathway and has been discussed as a possible site of intervention for the development of anti-cancer drugs.

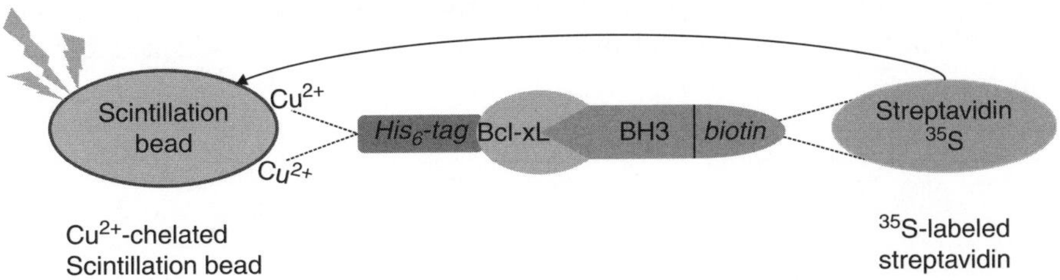

Figure 9.2.5 *Scheme for an SPA-based detection of the BH3–Bcl-xL dimerization*

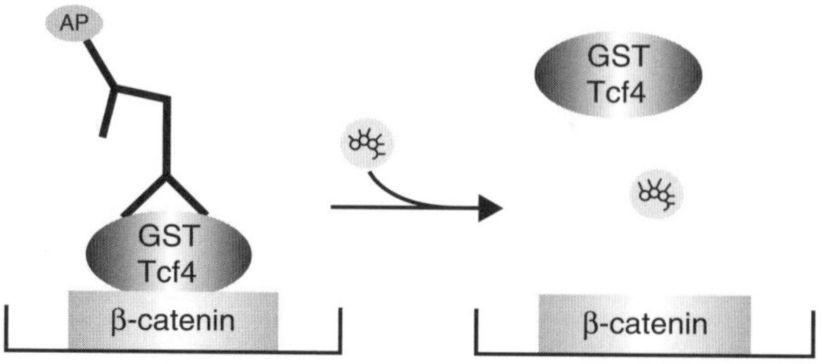

Figure 9.2.6 *An ELISA-based HTS of Tcf4–β-catenin interaction inhibitors*

To screen for small-molecule inhibitors, purified *β*-catenin (residues 134–668) was immobilized onto microtiter plates and incubated sequentially with GST-Tcf4 (residues 8–54), anti-GST antibody, and alkaline phosphatase (AP) conjugated secondary antibody (Figure 9.2.6). Compounds that disrupted the Tcf4–*β*-catenin complex were selected based on the reduction in the AP signals relative to the background. In the follow-up studies, these compounds were found to potently antagonize a number of *β*-catenin induced cellular activities, including reporter gene activation, *c*-myc and cyclin D1 expression, cell proliferation, and duplication of Xenopus embryonic dorsal axis. All the hit compounds were from the natural product library, whereas the synthetic compound library failed to produce any hits, underscoring the importance of screening the appropriate chemistry regardless of the assay format. Indeed, in an analysis of small-molecule protein–protein interaction inhibitors (SMPPIIs) and their cognate targets reported in the literature, only about 50% of SMPPIIs were covered by diversity space of the three commercial databases (Chemical Diversity database, Maybridge database, and Asinex database), and only four protein–protein interaction targets (Bak-BH3/Bcl-xL, MDM2/p53, NGF/p75, and LFA/ICAM-1) were well covered by these three commercial databases [20].

In another study, Kelly and coworkers [21] reported the discovery and characterization of BIRT 377, an orally available small molecule that interacts specifically with LFA-1 and prevents LFA-1 from binding to its ligand, ICAM-1. An ELISA-based HTS was established to measure the binding of purified LFA-1 to the plate-immobilized ICAM-1. Screening a proprietary compound collection produced an initial hit **1a** which inhibited the association of LFA-1 and ICAM-1 with K_d of 3.5 ± 1.0 μM. A structure-activity relationship study was performed around the hit structure **1a**, leading to a more potent compound, BIRT 377 (**1b**), with K_d of 25.8 ± 6.3 nM. BIRT 377 was a single stereoisomer and 35-fold more potent than its enantiomeric counterpart, indicating a specific interaction with LFA-1. Furthermore, the binding of LFA-1 to ICAM-1 was restored upon removal of BIRT 377, suggesting that the binding was reversible. BIRT 377 was highly selective for the LFA-1–ICAM-1 interaction as it did not inhibit the Mac-1–ICAM-1 interaction at concentrations as high as 225 μM.

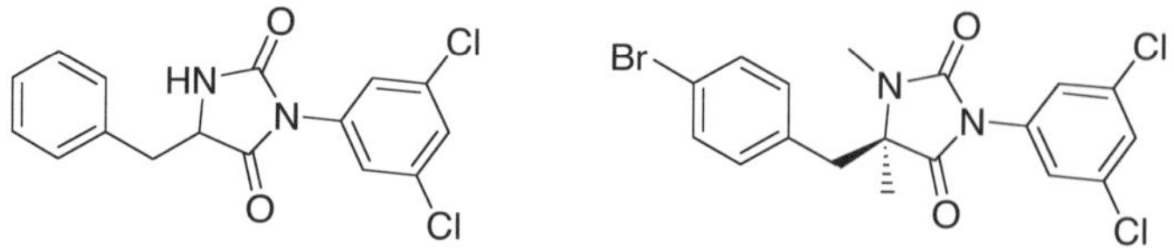

1a **1b**

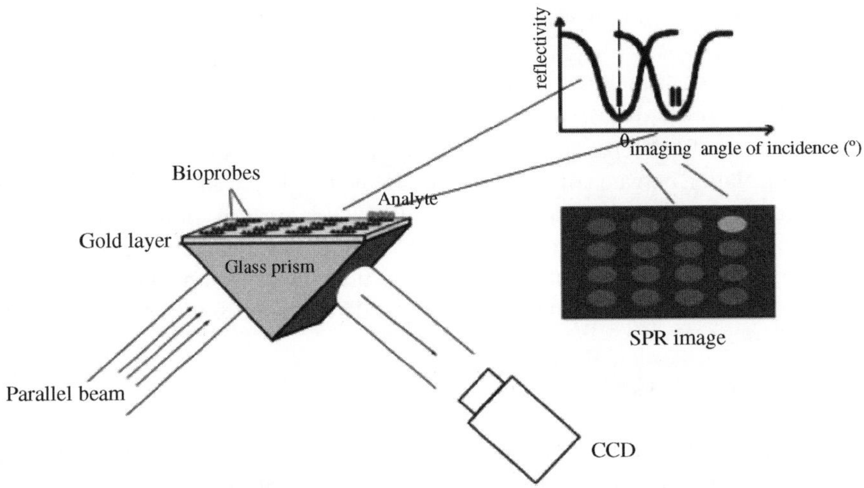

Figure 9.2.7 Principle of a label-free detection using surface plasmon resonance imaging method. A plane-polarized light is used to detect changes of refractive index by adsorption and/or desorption of analytes on the gold layer. Reprinted from [22]. Copyright WileyVCH Verlag GmbH & Co. KGaA

9.2.3.7 Surface Plasmon Resonance Imaging (SPRi)

SPRi is an optical surface technique based upon the generation of surface plasmons (SPs), which are oscillations of free electrons that propagate parallel to a metal/dielectric interface [22]. To excite SPs, plane-polarized light is reflected through a particular optical geometry typically involving a prism-gold film dielectric layer assembly. The SPs are evanescent waves that have maximum charge density at the interface and decay exponentially from the surface of the metal with a typical decay length of about 200 nm. As a result, SPR is sensitive to any changes in the refractive index of the dielectric layer adjacent to the metal film caused by adsorption or desorption of molecules on surface. In contrast to scanning angle SPR and scanning wavelength SPR, SPRi measures the reflectivity of the incident light at a fixed angle and the reflectivity is correlated to the changes on the surface. The unique feature of SPRi is that it can monitor hundreds of biomolecular interactions in real time simultaneously, as illustrated in Figure 9.2.7, and is well suited for the qualitative screening analysis of biomolecules as well as for quantitative kinetics studies.

Nearly all SPRi measurements were performed on a gold surface and anything absorbed to the gold surface will produce a detectable signal. As a result, adoption of proper surface chemistry is crucial for the success of an SPRi experiment. One surface modification strategy is to coat the gold surface with a polymeric layer, such as dextrin and polylysine. To test the feasibility of SPRi for small molecule screening, Chung and coworkers [23] fabricated a dextran-coated gold surface which was activated with glutathione. In a model study of the interaction between human papillomavirus E7 protein and retinoblastoma tumor suppressor RB protein, they immobilized the GST-tagged E7 protein to the fabricated gold surface, and examined whether SPRi can be used to screen potential inhibitors against the E7-RB interaction. Since RB interacts with the LxCxE motif of E7, they reasoned that peptides containing the LxCxE sequence should specifically disrupt the E7-RB interaction through competitive binding to the RB protein. Therefore, four peptides were synthesized: PepA is a control peptide derived from the p53 transactivation domain, which does not interact with RB; PepB is an E2F peptide that binds to a different site on RB; PepC and PepD contain the LxCxE motif, derived from the E7 protein. The peptides were mixed separately with the RB protein and then deposited to the E7-coated gold surface. It was found that the relative SPR imaging intensities (RSPRI) decreased gradually from peptide A to D, with PepC showing the highest correlation between the concentrations and RSPRI. In addition, they fabricated a 1500-spot protein array and demonstrated the utility of this array in conjunction with SPRi in the search of small-molecule inhibitors of E7-RB interaction. One drawback of the polymeric coating is that the thick layer limits analyte diffusion, which may lead to discrepancies between the SPRi-measured kinetic data and the solution-derived data.

9.2.4 Cell-based Assays

Traditional small-molecule drug discovery focuses primarily on the activity against purified targets, e.g. binding to cell-surface receptors or enzyme inhibition. Because of the complex intracellular environment, however, sometimes it is advantageous to screen compounds in a cellular context because the lead compounds identified have already demonstrated the ability to act within the biochemical complexity of the cell. The mechanism of action can then be addressed through a series of counterscreens, which progressively decrease the 'pathway space' in which the compounds might act. With the development of HTS-compatible confocal imaging tools and reporter-gene technology, cell-based assays have become one of the most valuable approaches in lead generation, including drug leads that act primarily by modulating the protein–protein interactions [20].

9.2.4.1 Green Fluorescent Protein-Assisted Readout for Interacting Proteins (GRIP)

GRIP is a universal, image-based protein interaction discovery system that treats protein translocation as activity readout in cellular signaling pathways. The technology uses a 'bait and prey' principle based on the distinct translocation behavior of one of the interacting protein partners, and offers several advantages over traditional HTS approaches. First, active compounds will be identified only when they possess adequate solubility, membrane permeability, and stability in a cellular environment. A degree of pharmacokinetic filtering is therefore inherent in the GRIP assay. Second, since the assay is image-based, cytotoxic compounds can be easily distinguished by examining the cellular morphological changes. Third, rapid kinetics of the GRIP assay is more desirable to HTS than the delayed kinetics of transcriptional activation in the reporter assays because potentially interesting, but mildly cytotoxic, compounds in the screen can be analysed during shorter incubation periods.

In a proof-of-concept study, Praestegaard and coworkers [24] performed a pilot screen of a diverse 3,165-membered small-molecule library for inhibitors of the p53-Hdm2 interaction. In their assay, they relied upon the unique translocation property of an inducible anchor protein, human cyclic AMP phosphodiesterase isoform 4A4 (PDE4A4), which localizes in the cytoplasm. The 'bait' protein Hdm2 was fused to PDE 4A4 while the 'prey' protein p53 was fused to green fluorescent protein (GFP) (Figure 9.2.8). Treatment with the PDE4A4 agonist RS253448 led to redistribution of PDE4A4 into compact foci, discernable dot-like structures within the cytoplasm. Because of the interaction between 'bait' and 'prey', the GFP tag was recruited to PDE4A4 focal sites, resulting in the fluorescent-labeled foci. On the

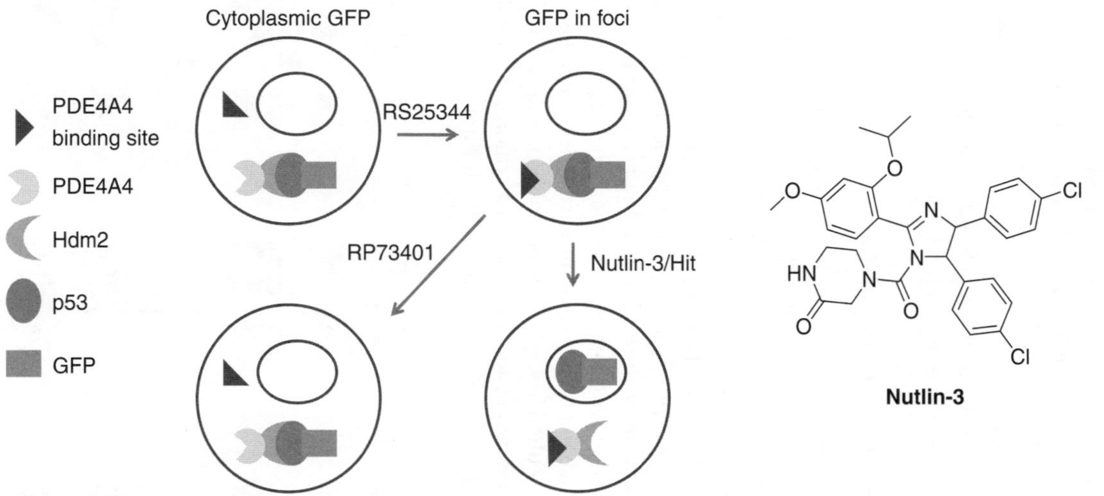

Figure 9.2.8 *Schematic representation of the GRIP technology for identifying small-molecule inhibitors of the p53-Hdm2 interaction*

other hand, treatment with PDE4A4 antagonist RP73401 caused foci dispersal. Using nutlin-3, a known small-molecule p53-Hdm2 interaction inhibitor as a positive control, they showed that treatment with nutlin-3 caused a loss of GFP fluorescence from the PDE4A4 foci by disrupting the interaction between PDE4A4-Hdm2 and GFP-p53, and a concomitant increase in nuclear fluorescence due to nuclear translocation of the dissociated GFP-p53. This phenomenon was not observed for the RP73401-mediated foci dispersal, indicating that only the unbound GFP-p53 was available for nuclear translocation. HTS implementation resulted in the identification of 6 potential inhibitors of the p53-Hdm2 interaction, including nutlin-3 that was spiked into a screening plate. Because of high content within an image, the assay allowed for a discrimination between compounds acting as general PDE4A4 inhibitors (RP73401-like) and compounds targeting the p53-Hdm2 interaction (nutlin-3-like).

9.2.4.2 β-Galactosidase Complementation Assay

Reporter gene assays have been widely used in the cell-based HTS, particularly the pathway screenings, due to their inherent high sensitivity [25]. Few, however, utilize β-galactosidase (β-gal) as a reporter gene, presumably due to its large size. Interestingly, the complementarity among the subunits of β-gal was found to be perfectly suitable for probing the protein–protein interactions in cellular systems [26–28]. To exploit this complementation, two β-gal mutants are utilized, $\Delta\alpha$ (N-terminal deletion mutant) and $\Delta\omega$ (C-terminal truncation mutant), which exhibit low affinity for each other and are incapable of forming a stable, functional β-gal complex. An active complex is restored when the two mutants are expressed as fusions with target protein pairs A and B that interact with one another (Figure 9.2.9). As a result, the enzyme activity resulted from β-gal reassembly provides an indirect measurement of protein A-protein B interaction inside the cell. Furthermore, the β-gal complementation assay is not restricted to a particular organism such as bacterium or yeast; nor is it restricted to

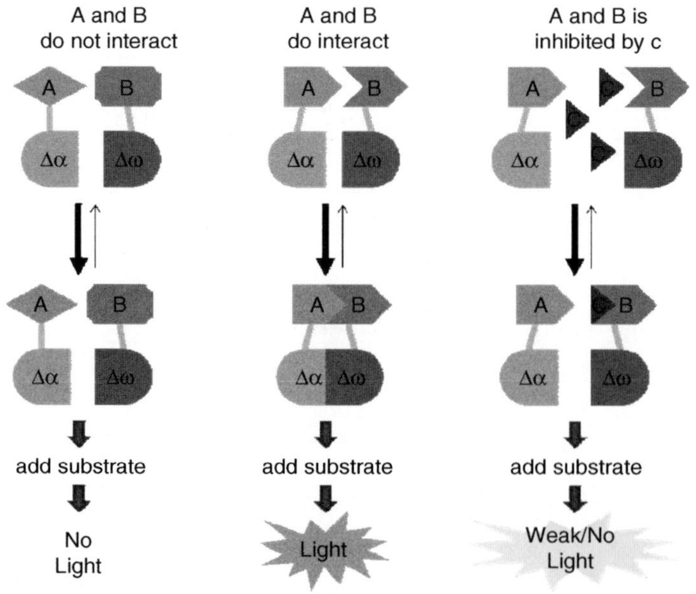

Figure 9.2.9 *Probing protein-protein interaction through β-galactosidase complementation*

any particular cellular compartment. Therefore, it offers a flexibility of monitoring protein interactions at their native physiological environments, e.g. cell surface, cytoplasm, nucleus, and other cellular compartments.

Using the β-gal complementation assay, Buttner and coworkers [26] performed a cell-based HTS in an effort to identify inhibitors of EGFR homodimerization. The cells were transfected with plasmids encoding two chimeric β-gal mutants, each fused to the extracellular and transmembrane portion of EGFR, respectively. Upon stimulation with EGF, EGFR dimerized which in turn led to the reassembly of an active β-galactosidase. In a homogeneous 384-well screen of about 20 000 diverse compounds, 31 compounds were identified as either potential EGFR dimerization inhibitors or EGF stimulation inhibitor.

9.2.4.3 Mammalian Reverse Two-Hybrid System

Genetic screens are uniquely capable of identifying individual molecules with desired properties from large libraries by using whole cells as reporters and correlating host growth to a desired property. Unlike affinity-based selections, an intracellular genetic screen can directly assay for effects of disrupting a protein–protein complex by a small-molecule inhibitor, thus bypassing the inherent limitations of in vitro approaches. Additionally, small-molecule compounds must function within the context of an entire host proteome, requiring positive candidates to have an enhanced level of selectivity for their target. This feature represents an important advantage over traditional biochemical methods by allowing both target affinity and selectivity to be simultaneously optimized.

A mammalian reverse two-hybrid system was developed by Shen and coworkers [29] for high-throughput screening of compounds that disrupt specific protein–protein interactions. To construct a tightly regulated system, they fused the inhibitory KRAB domain to the tetracycline repressor TetR to generate a potent TetR-KRAB suppressor. When TetR-KRAB binds to the tet operator OP in the absence of tetracycline, it completely suppresses the expressions of the two interacting proteins X and Y, which are respectively fused with Gal4 DNA binding domain (BD) and VP16 transcriptional activation domain (AD) of a transcription factor in an IRES (Internal Ribosome Entry Sequence) construct (Figure 9.2.10B). Three reporters, namely, GFP, β-lactamase, and luciferase, were employed to complement each other for the detection. Upon addition of Dox, the binding of TetR-KRAB to OP is abolished, resulting in an elevated expression of AD-X and BD-Y (Figure 9.2.10C), which in turn activates the reporter gene expression (Figure 9.2.10D). When small molecules and Dox are added together, any potential drug that disrupts the interaction between proteins X and Y can be selected based on their silencing effect on the reporter activity (Figure 9.2.10E).

In control experiments, they demonstrated that interaction of epidermal growth factor receptor (EGFR) with p85 can be effectively disrupted by EGFR kinase inhibitor AG1478. It is known that the cytoplasmic domain of EGFR is autophosphorylated on tyrosine residues when overexpressed, and the phosphorylated EGFR is recognized by p85, one of the signaling molecules. AG1478 is a potent inhibitor of the EGFR kinase, hence it was expected that treatment with AG1478 should abolish receptor tyrosine autophosphorylation, leading to the inhibition of the EGFR-p85 interaction. To demonstrate the power of this mammalian system for HTS, a library consisting of $\sim$20 000 drug-like compounds was first screened for compounds capable of disrupting the EGFR-p85 interaction in 293A cells stably transfected with the constructs expressing the interacting proteins and the GFP reporter. Compounds that exhibited signals significantly lower than the mean were further analysed in cells stably

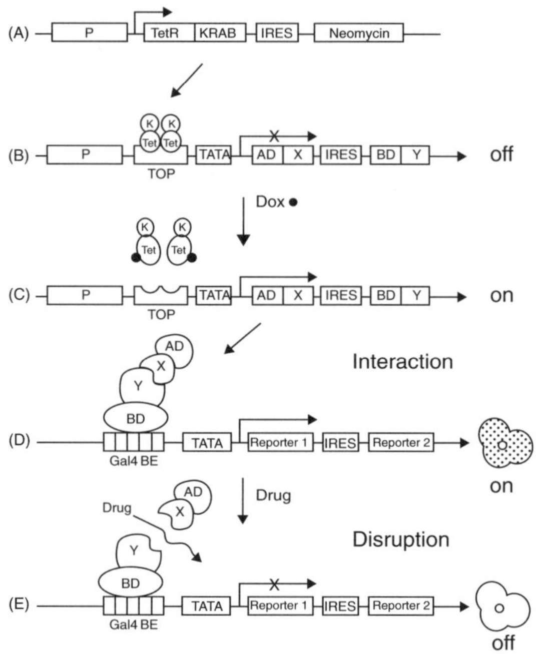

Figure 9.2.10 *Schematic representation of the mammalian reverse two-hybrid system for screening small molecules capable of disrupting a protein-protein interaction. Reprinted with permission from [29]. Copyright 2004 American Chemical Society*

transfected with the luciferase reporter, and the activity of lead compound BR20T2-B1 inhibited the EGFR-p85 interaction with an IC_{50} value of $\sim 5\ \mu M$. BR20T2-B1 was selective toward the EGFP-p85 interaction as it had no effect on the PDGFR-p85 interaction at a concentration of $10\ \mu M$.

9.2.4.4 Counterselection Yeast Two-Hybrid Screen

The yeast two-hybrid system has provided a rapid genetic platform to identify interacting proteins in biological systems. Franco and coworkers [30] described an HTS based on a counterselection yeast two-hybrid assay in their efforts to identify small molecules that modulate the N-type calcium channel $\beta 3$ subunit interaction with the $\alpha 1B$ subunit. In this assay, the $\beta 3$ protein coding sequence was fused to Gal4 DNA binding domain (BD), the $\alpha 1B$ intracellular loop was fused to Gal4 activation domain (AD), and a CYH2 was used as a counterselection reporter gene (Figure 9.2.11). Constitutive expression of fusion proteins was achieved by constructing stable plasmids containing chromosomal centromeres to provide very low copy number, and thus, enhanced sensitivity. It was expected that the presence of antagonists of $\beta 3$-$\alpha 1B$ interaction will downregulate the expression of CYH2, thereby promote yeast growth.

In the N-type calcium channel assay development, independent transformants expressing the AD and BD fusion proteins were evaluated for their reporter gene activities. Consistent histidine prototrophy (positive selection; HIS3 reporter gene) and cycloheximide sensitivity

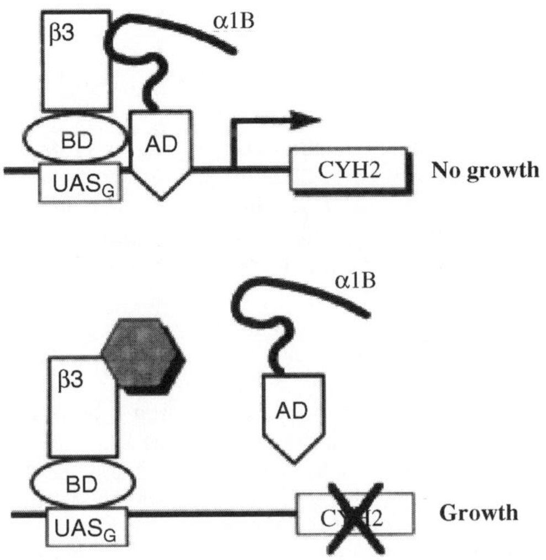

Figure 9.2.11 *A counterselection assay based on the yeast two-hybrid system. Reprinted by permission from Macmillan Publishers Ltd [30], copyright 1998*

(counterselection; CYH2 reporter gene) was observed. The counterselection yeast two-hybrid screen was conducted as a two-plate agar diffusion assay: One plate contained the test strain expressing the N-type fusions, while the other plate contained a negative control strain expressing an unrelated interacting protein pair. Over 156 000 diverse and random compounds were individually analysed. Ten compounds were found to rescue yeast growth in the N-type strain but not in the negative control strain, which resulted in a hit rate of 0.0064%.

9.2.5 Conclusion

The HTS approach to discovery of small-molecule inhibitors of protein–protein interaction has become well entrenched in the modern drug discovery program. In this chapter, we have provided an overview of various current protein–protein interaction assay methods. In particular, we have shown that both biochemical assays and cell-based assays can be successfully employed in the HST campaigns. Importantly, there is no single best method for performing HTS in the search for small-molecule inhibitors of protein–protein interactions: each method has its own advantages and disadvantages. In the practice of HTS, we should always be mindful about statistical performance of assay data that are generated during the screen in order to have a good grasp of data quality. And in most cases, secondary follow-up assays are needed in validating the hits originated from the initial screens. These methods, together with the continuing efforts by scientists in developing and further refining innovative HTS techniques – for example, relentless automation and miniaturization efforts to increase throughput, new reagents to improve assay sensitivity, new computer algorithm for data analysis, and biological innovation in the cell-based assay design, to name just a few – will, in the near future, bring the probability of success targeting protein–protein interactions to the same level as those targeting enzymes and receptors as we know today.

References

1. *High Throughput Screening: Methods and Protocols*Janzen, W. P., Ed.;Humana press, 2002; Vol.190.
2. Arkin, M. R.; Wells, J. A. Small-molecule inhibitors of protein–protein interactions: progressing towards the dream. *Nat Rev Drug Discov* **2004**, *3*, 301–17.
3. Zhang, J. H.; Chung, T. D.; Oldenburg, K. R. A Simple Statistical Parameter for Use in Evaluation and Validation of High Throughput Screening Assays. *J Biomol Screen* **1999**, *4*, 67–73.
4. Hertzberg, R. P.; Pope, A. J. High-throughput screening: new technology for the 21st century. *Curr Opin Chem Biol* **2000**, *4*, 445–51.
5. Pope, A. J.; Haupts, U. M.; Moore, K. J. Homogeneous fluorescence readouts for miniaturized high-throughput screening: theory and practice. *Drug Discov. Today* **1999**, *4*, 350–62.
6. Roman, D. L.; Talbot, J. N.; Roof, R. A.; Sunahara, R. K.; Traynor, J. R.; Neubig, R. R. Identification of small-molecule inhibitors of RGS4 using a high-throughput flow cytometry protein interaction assay. *Mol Pharmacol* **2007**, *71*, 169–75.
7. Young, S. M.; Bologa, C.; Prossnitz, E. R.; Oprea, T. I.; Sklar, L. A.; Edwards, B. S. High-throughput screening with HyperCyt flow cytometry to detect small molecule formylpeptide receptor ligands. *J Biomol Screen* **2005**, *10*, 374–82.
8. Roehrl, M. A.; Wang, J. Y.; Wagner, G. A General Framework for Development and Data Analysis of Competitive High-Throughput Screens for Small-Molecule Inhibitors of Protein–protein Interactions by Fluorescence Polarization. *Biochemistry* **2004**, *43*, 16056–66.
9. Degterev, A.; Lugovskoy, A.; Cardone, M.; Mulley, B.; Wagner, G.; Mitchison, T.; Yuan, J. Identification of small-molecule inhibitors of interaction between the BH3 domain and Bcl-xL. *Nat Cell Biol* **2001**, *3*, 173–82.
10. Zhang, H.; Nimmer, P.; Rosenberg, S. H.; Ng, S. C.; Joseph, M. Development of a high-throughput fluorescence polarization assay for Bcl-xL. *Anal. Biochem.* **2002**, *307*, 70–5.
11. Kenny, C. H.; Ding, W.; Kelleher, K.; *et al.* Development of a fluorescence polarization assay to screen for inhibitors of the FtsZ/ZipA interaction. *Anal. Biochem.* **2003**, *323*, 224–33.
12. Roehrl, M. H.; Wang, J. Y.; Wagner, G. Discovery of small-molecule inhibitors of the NFAT–calcineurin interaction by competitive high-throughput fluorescence polarization screening. *Biochemistry* **2004**, *43*, 16067–75.
13. Nikolovska-Coleska, Z.; Wang, R.; Fang, X.; *et al.* Development and optimization of a binding assay for the XIAP BIR3 domain using fluorescence polarization. *Anal. Biochem.* **2004**, *332*, 261–73.
14. Arnold, L. A.; Estebanez-Perpina, E.; Togashi, M.; *et al.* Discovery of small molecule inhibitors of the interaction of the thyroid hormone receptor with transcriptional coregulators. *J Biol Chem* **2005**, *280*, 43048–55.
15. Saldanha, S. A.; Kaler, G.; Cottam, H. B.; Abagyan, R.; Taylor, S. S. Assay Principle for Modulators of Protein–protein Interactions and Its Application to Non-ATP-Competitive Ligands Targeting Protein Kinase A. *Anal. Chem.* **2006**, *78*, 8265–72.
16. Berg, T.; Cohen, S. B.; Desharnais, J.; *et al.* Small-molecule antagonists of Myc/Max dimerization inhibit Myc-induced transformation of chicken embryo fibroblasts. *Proc Natl Acad Sci U S A* **2002**, *99*, 3830–5.
17. Bergendahl, V.; Heyduk, T.; Burgess, R. R. Luminescence resonance energy transfer-based high-throughput screening assay for inhibitors of essential protein–protein interactions in bacterial RNA polymerase. *Appl Environ Microbiol* **2003**, *69*, 1492–8.
18. Whitfield, J.; Harada, K.; Bardelle, C.; Staddon, J. M. High-throughput methods to detect dimerization of Bcl-2 family proteins. *Anal. Biochem.* **2003**, *322*, 170–8.
19. Lepourcelet, M.; Chen, Y. P.; France, D. S.; *et al.* Small-molecule antagonists of the oncogenic Tcf/β-catenin protein complex. *Cancer Cell* **2004**, *5*, 91–102.
20. Pagliaro, L.; Felding, J.; Audouze, K.; Nielsen, S. J.; Terry, R. B.; Krog-Jensen, C.; Butcher, S. Emerging classes of protein–protein interaction inhibitors and new tools for their development. *Curr. Opin. Chem. Biol.* **2004**, *8*, 442–9.

21. Kelly, T. A.; Jeanfavre, D. D.; McNeil, D. W.; *et al.* Cutting edge: a small molecule antagonist of LFA-1-mediated cell adhesion. *J Immunol* **1999**, *163*, 5173–7.
22. Yu, X.; Xu, D.; Cheng, Q. Label-free detection methods for protein microarrays. *Proteomics* **2006**, *6*, 5493–5503.
23. Jung, S. O.; Ro, H. S.; Kho, B. H.; Shin, Y. B.; Kim, M. G.; Chung, B. H. Surface plasmon resonance imaging-based protein arrays for high-throughput screening of protein–protein interaction inhibitors. *Proteomics* **2005**, *5*, 4427–31.
24. Lundholt, B. K.; Heydorn, A.; Bjorn, S. P.; Praestegaard, M. A Simple Cell-Based HTS Assay System to Screen for Inhibitors of p53-Hdm2 Protein–protein Interactions. *Assay Drug Develop. Tech.* **2006**, *4*, 679–88.
25. Fan, F.; Wood, K. V. Bioluminescent assays for high-throughput screening. *Assay Drug Dev Technol* **2007**, *5*, 127–36.
26. Buttner, F. H.; Kumpf, R.; Menzel, S.; Reulle, D.; Valler, M. J. Evaluation of the InteraX system technology in a high-throughput screening environment. *J Biomol Screen* **2005**, *10*, 485–94.
27. Graham, D. L.; Bevan, N.; Lowe, P. N.; Palmer, M.; Rees, S. Application of beta-galactosidase enzyme complementation technology as a high throughput screening format for antagonists of the epidermal growth factor receptor. *J Biomol Screen* **2001**, *6*, 401–11.
28. Yan, Y. X.; Boldt-Houle, D. M.; Tillotson, B. P.; *et al.* Cell-based high-throughput screening assay system for monitoring G protein-coupled receptor activation using beta-galactosidase enzyme complementation technology. *J Biomol Screen* **2002**, *7*, 451–9.
29. Zhao, H. F.; Kiyota, T.; Chowdhury, S.; *et al.* A mammalian genetic system to screen for small molecules capable of disrupting protein–protein interactions. *Anal Chem* **2004**, *76*, 2922–7.
30. Young, K.; Lin, S.; Sun, L.; *et al.* Identification of a calcium channel modulator using a high throughput yeast two-hybrid screen. *Nature Biotech.* **1998**, *16*, 946–50.

Part IV

Case Studies

10

Case Study: Inhibitors of the MDM2-p53 Protein–Protein Interaction

Sanjeev Shangary, Denzil Bernard and Shaomeng Wang[†]
Comprehensive Cancer Center and Departments of Internal Medicine, University of Michigan, Ann Arbor, MI, USA

10.1 MDM2-p53 Protein–Protein Interaction: A Case Study

Targeting biologically important protein–protein interactions is a general strategy pursued to design new drugs [1, 2]. One such biologically important and therapeutically relevant protein–protein interaction is the binding between tumor suppressor p53, discovered in 1979 [3–5], and murine double minute 2 oncoprotein (MDM2), discovered in 1991 (6). p53 is a powerful transcription factor of many genes and regulates a number of cellular processes, such as cell cycle, senescence, apoptosis, DNA repair and angiogenesis [7–10]. MDM2 directly binds to p53 regulating both its activity and levels in normal cells through an autoregulatory feedback mechanism (Figure 10.1). p53 suppresses oncogenesis and tumor formation but overexpression of MDM2 inhibits p53 function. In fact MDM2 was

[†]Corresponding author: Tel. 734-615-0362, Fax 734-647-9647; email: shaomeng@umich.edu

Protein Surface Recognition: Approaches for Drug Discovery Ernest Giralt, Mark W. Peczuh and Xavier Salvatella. © 2011 John Wiley & Sons, Ltd

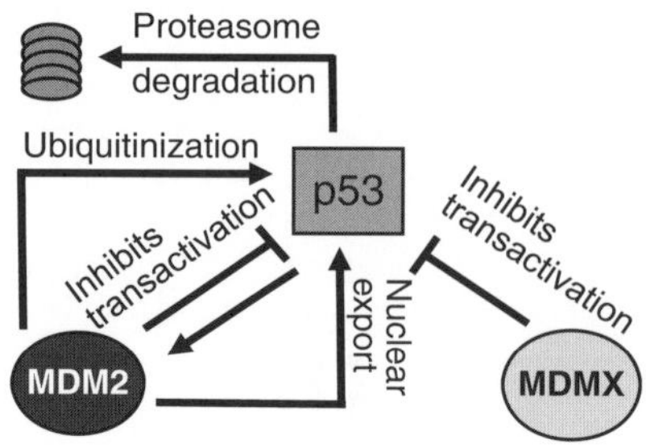

Figure 10.1 *Regulation of p53 by MDM2 and MDMX. Upon activation, p53 induces transcriptional upregulation of MDM2 and in turn, MDM2 inhibits p53 through an autoregulatory feedback loop. MDM2 directly binds to the transactivation domain of p53 and inhibits p53 transcriptional activity, facilitating nuclear export of p53. Through its E3 ligase activity, MDM2 causes the ubiquitinization and proteasomal degradation of p53. MDMX, a homologue of MDM2 also binds and conceals the transactivation domain of p53, inhibiting p53 activity. (See Plate 20.)*

discovered by its overexpression in a mouse tumor cell line [6] and overexpression of MDM2 is correlated with poor clinical prognosis and response to therapy in multiple types of human cancers [11–16].

Thus, disruption of the MDM2-p53 protein–protein interaction is an attractive approach for the reactivation of p53 function in cancer cells and has been pursued as a strategy for anticancer drug discovery [17–19]. The determination of a crystal structure of human MDM2 complexed with a p53 peptide in 1996 [20] (Figures 10.2 and 10.3a) revealed the precise structural requirements for the MDM2-p53 protein–protein interaction and suggested the possibility of designing small-molecule inhibitors to disrupt the MDM2-p53 interaction. Since then, intense efforts have been made in both academia and the pharmaceutical industry to design small-molecule inhibitors to target the MDM2-p53 protein–protein interaction (here, MDM2 inhibitors) [17–19]. In this chapter, we will review the different approaches employed, progress to date and the challenges in the design and development of small-molecule MDM2 inhibitors as a new class of anticancer drugs.

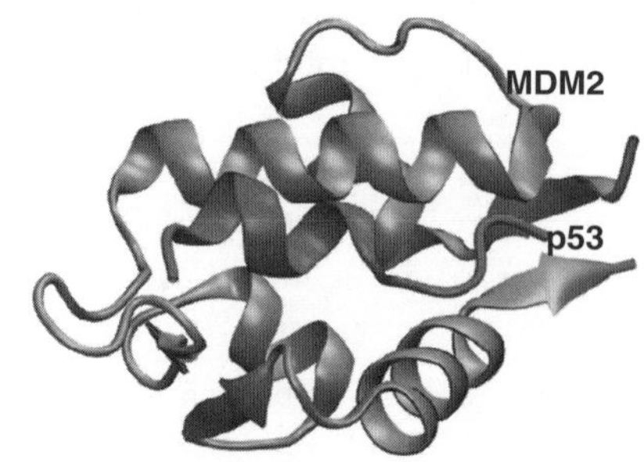

Figure 10.2 *Binding mode of p53 peptide (residues 15–29) to MDM2 (residues 25–109) [PDB ID:1YCR]. This figure was generated by the program VMD. (See Plate 21.)*

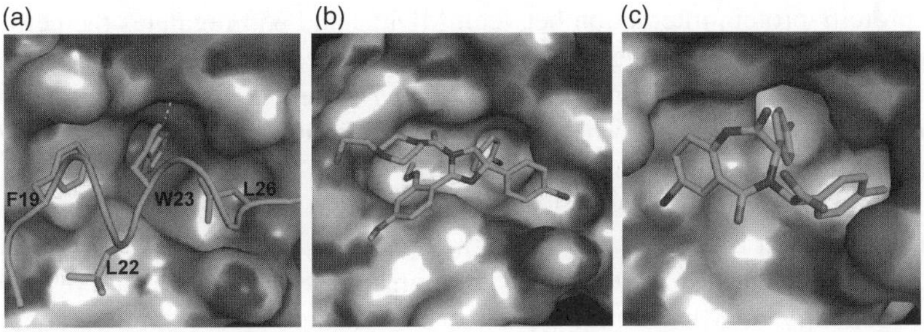

Figure 10.3 Binding mode of (a) p53 peptide, (b) Nutlin-2 and (c) benzodiazepinedione compound to MDM2 [PDB ID: (a) 1YCR, (b) 1RV1 and (c) 1T4E]. Ligands are shown with carbons. Key residues in the p53 peptide are shown in stick representation. The surface representation of MDM2 is shown with carbons. Hydrogen atoms are excluded for clarity with hydrogen bonds. This figure was generated by the program Pymol. (See Plate 22.)

10.2 Regulation of p53 by the MDM2-p53 Protein–Protein Interaction

p53 is a tumor suppressor which regulates diverse cellular processes and suppresses the development of tumors [7–10]. Due to its vital role in preventing carcinogenesis, it is not surprising that *p53* is one of the most frequently mutated genes in human cancers. Nearly half of human cancers have alterations in the *p53* gene, resulting in inactivation or loss of the p53 protein and its tumor suppressor function [8, 21]. Even in cancers retaining wild-type p53, the function of p53 can be effectively inhibited. This inhibitory role is performed primarily by MDM2 through direct interaction with the N-terminus transactivation domain of p53 [22].

The MDM2 protein is ubiquitously expressed and plays an important role in tissue development; however, overexpression of MDM2 promotes tumorigenesis [11–16, 23]. A variety of mechanisms, such as amplification of the *MDM2* gene and single nucleotide polymorphism at nucleotide 309 (SNP309) in the *MDM2* gene promoter can account for MDM2 overexpression [11, 23–25]. Mouse models have shown that overexpression of MDM2 at an early stage of differentiation predisposes animals to tumorigenesis [26]. Similar to the inherited Li-Fraumeni cancer predisposition syndrome in humans, mice lacking p53 develop normally but are predisposed to develop a variety of tumors [27]. MDM2 binds and inhibits p53 function and it could therefore be predicted that MDM2 overexpression and p53 mutations would represent alternative mechanisms of inactivation of p53 in tumors. A negative correlation between the occurrence of p53 mutations and MDM2 amplification in 28 different types of cancers, comprising more than 3000 tumors, largely supports this notion [16]. The MDM2-p53 protein–protein interaction is thus an important cancer therapeutic target.

Genetic studies in mouse models have revealed the antitumor efficacy of restoration of p53 function. In particular, it was shown that loss of p53 induces tumor formation and upon its restoration, a rapid and impressive regression of established tumors occurs, providing strong rationale to design drugs which can restore p53 function [28–30].

Direct protein–protein interaction between MDM2 and p53 regulates the cellular basal levels and activity of p53 through an autoregulatory feedback loop (Figure 10.1). Upon activation, p53 binds to the P2 promoter of the *MDM2* gene and transcriptionally induces MDM2 protein expression. In turn, the MDM2 protein directly binds to the p53 protein and inhibits it through multiple mechanisms: MDM2 (i) directly inhibits the transactivation function of p53, (ii) exports p53 out of the nucleus and (iii) promotes proteasome-mediated degradation of p53 by E3 ubiquitin ligase activity [31–33]. The physiological relevance of the MDM2-p53 loop was demonstrated by the genetic evidence that embryonic lethality of MDM2-null mice can be rescued by the simultaneous deletion of the *p53* gene [34, 35]. These mouse genetic studies demonstrate that MDM2 is a critical regulator of p53 function.

10.3 Structural Basis of the MDM2-p53 Interaction

An important breakthrough in MDM2-p53 studies was the report of the X-ray structure of the complex of the N-terminal domain of MDM2 with a 15-residue transactivation domain peptide of p53 (Figure 10.2 and 10.3a) [20]. This crystal structure provided valuable insights into the nature of the MDM2-p53 interaction and a structural basis for the design of small-molecule inhibitors to block the MDM2-p53 interaction.

In the crystal structure, the p53 peptide forms an amphipathic α-helix with most of the residues on the hydrophobic face buried into a cleft along the MDM2 surface. The binding cleft for the p53 peptide in MDM2 is formed by two pairs of α-helices along two sides and the bottom of the cleft, while a pair of β-strands line the remaining two sides. The residues forming this region are mostly hydrophobic and include Met^{50}, Leu^{54}, Leu^{57}, Gly^{58}, Ile^{61}, Met^{62}, Tyr^{67}, His^{73}, Val^{75}, Phe^{91}, Val^{93}, His^{96}, Ile^{99} and Tyr^{100}, which complement the hydophobic face of the p53 peptide. The key p53 residues involved in the interaction are Phe^{19}, Trp^{23} and Leu^{26} residues. While the primary interaction between MDM2 and p53 is hydrophobic in nature, two hydrogen bonds are observed between the amide backbone of Phe^{19} in p53 and the side chain of Gln^{72} in MDM2, and between the indole of Trp^{23} in p53 and the backbone carbonyl of Leu^{54} in MDM2. These structural insights into the MDM2-p53 interaction have had a significant bearing on the subsequent efforts in the design of small molecule inhibitors of MDM2-p53 inteaction.

While the crystallographic studies provide generally a static image of the complex, a more dynamic view can be obtained from NMR studies. In the case of the MDM2-p53 complex, NMR studies revealed that binding also involves some structural rearrangement of MDM2 in solution [36]. The NMR-based solution structures confirmed the secondary and tertiary structures of MDM2 (residues 25–109) obtained from X-ray crystallography, and also indicated that a portion of the N-terminal not resolved in the crystal structure (residues 19–24) is ordered in the absence of the ligand and that residues 16–24 are in close contact with the p53-binding site. It has been suggested that this N-terminal forms a flexible ‘lid’ that occupies the p53 binding site and serves to stabilize the MDM2 structure [36]. NMR studies reveal a more compact hydrophobic cleft in MDM2 which opens up during binding with the displacement of the lid region and movement of the sub-domains [37]. In its open form however, the crystal structure may be more suitable for the design of small-molecule inhibitors to target the MDM2-p53 interaction.

10.4 Design of p53-based Peptides

Co-crystallization of the MDM2 protein with a p53 peptide provided a major impetus to the drug design efforts targeting the MDM2-p53 interaction (Figure 10.2 and 10.3a) (20). However, it should be noted that even before this significant breakthrough, the p53 binding site in MDM2 had been mapped using biochemical approaches [38, 39]. These studies mapped the MDM2-p53 interaction site to the 106 residue N-terminal domain of MDM2 and the N-terminus of the transactivation domain of p53. A consensus hexapeptide sequence of ^{18}TFSDLW^{23} in human p53 was shown to be necessary for the interaction [39]. This peptide has an IC_{50} value of 700 μM for MDM2, indicating a weak interaction. To obtain high affinity p53 peptides that disrupt the MDM2-p53 interaction, phage display libraries were screened for the MDM2 binding phage [40]. A series of 12-mer and 15-mer p53 peptides were found to interact strongly with MDM2 and have high homology with the previously determined p53 consensus sequence. These peptides further revealed that molecular contact between MDM2 and p53 proteins extends to Leu 26 in p53. The most active 12-mer p53 peptide (Ac-MPRF^{19}MDYW^{23}EGL^{26}N-NH2) was 28-fold more potent than the 12-mer wild-type p53 peptide (^{16}Ac-QETFSDLWKLLP^{27}-NH2) in disrupting the MDM2-p53 interaction [1, 40, 41]. Based on the MDM2-p53 co-crystal structure, this phage-display peptide was further optimized using unnatural amino acids and modifications at the Trp^{23} residue of the p53 peptide. Optimization yielded a highly potent 8-mer (Ac-^{19}FMAibPmp(6-Cl-W) EAc$_3$cL^{26}-NH$_2$), containing four unnatural amino acids, and having an IC_{50} value of 5 nM, which is approximately 1700-fold better than that of the 12-mer wild-type p53 peptide [1, 40–42]. Despite its weak cellular activity, this octapeptide was, nevertheless, cell permeable without a carrier tag, and capable of disrupting the cellular MDM2-p53 interaction, inducing accumulation and activation of p53 [42]. However, the very weak cellular activity also suggested that this octamer peptide has poor cell permeability and/or stability.

Although p53-based peptides have been used to provide the important proof-of-principle for reactivation of p53 by disruption of the MDM2-p53 interaction, they have inherent limitations, such as poor cell permeability and stability, and are therefore not suitable as potential drugs. Consequently, there has been enormous interest in the design of nonpeptidic small-molecule inhibitors of the MDM2-p53 interaction in the last decade, which will be discussed below.

10.5 Design of Nonpeptidic Small-Molecule Inhibitors of the MDM2-p53 Interaction

A number of approaches have been employed for the design of nonpeptidic small-molecule inhibitors of the MDM2-p53 interaction. These include experimental approaches, such as screening of small focused chemical libraries, high-throughput screening of a diverse library of synthetic compounds, and computational approaches, such as three dimensional (3D) database screening of large libraries and structure based design.

10.5.1 Screening Chemical Databases

Screening of chemical libraries for the identification of lead compounds is an approach commonly employed in drug discovery. Its advantage is that it does not require detailed structural information of the target protein. The approach has been utilized with some success for screening of nonpeptidic small-molecules capable of disrupting the MDM2-p53 interaction.

10.5.1.1 High-throughput Screening of Large Chemical Libraries

cis-Imidazolines were discovered as inhibitors of the MDM2-p53 interaction through a high-throughput screen [43] of a diverse library of synthetic compounds. Extensive optimization of the initial lead compounds yielded a set of potent, specific, nonpeptidic MDM2-p53 interaction inhibitors called Nutlins [43]. Nutlin-1 (**1**) and Nutlin-2 (**2**), which are racemic compounds, and Nutlin-3a (**3**), which is the active enantiomer of racemic Nutlin-3, disrupt the MDM2-p53 interaction with IC_{50} values of 260 nM, 140 nM and 90 nM, respectively (Table 10.1). Determination of a high resolution crystal structure of Nutlin-2 in a complex with MDM2 shows that Nutlin-2 binds to the p53 binding site in MDM2 (Figure 10.3b) and mimics the key interactions of the p53 peptide with MDM2. While one bromophenyl moiety in Nutlin-2 sits deep in the Trp^{23} pocket, the other bromophenyl group occupies the Leu 26 pocket, and the ethyl ether side chain interacts with the Phe^{19} pocket. Nutlin-3a and its analogs are the first examples of potent, nonpeptidic small molecules that target the MDM2-p53 interaction site. Nutlin-3a has been extensively evaluated for its therapeutic potential in cell-based assays and in animal models of human cancer and has served as a powerful pharmacological tool since its initial publication in 2004 [43].

Another class of MDM2-p53 interaction inhibitors discovered through the use of high-throughput screens, in two independent studies, was the benzodiazepindiones (Table 10.1). Different libraries of compounds, designed employing Directed Diversity software, were synthesized using combinatorial chemistry. A library of more than 338 000 compounds of different classes [44] in one study, and another focused library of 22 000 benzodiazepinedione compounds [45] in the other, were prescreened for inhibition of MDM2-p53 binding with ThermoFluor microcalorimetry technology. Compounds identified from this initial screen were then confirmed for inhibition of the MDM2-p53 interaction in an FP-based peptide displacement binding assay. Following optimization, several compounds such as **4** and **5** with IC_{50} values of 420 nM and 490 nM respectively, were obtained (45). S,S enantiomer of **6**, the most potent compound from this class, has a binding affinity of 80 nM (Table 10.1) [44]. The co-crystal structure of human MDM2 and **6** verified that the compound occupies the same pockets as the side chains of Phe^{19}, Trp^{23} and Leu^{26} residues in p53 peptide (Figure 10.2c) [44]. Other benzodiazepinediones identified by ThermoFluor microcalorimetry technology and optimized by structure-based design are TDP521252 (**7**) and TDP665759 (**8**), which bind to MDM2 with IC_{50} values of 708 nM and 704 nM, respectively (Table 10.1) [46].

10.5.1.2 Screening of Focused Libraries

Using available structural knowledge of the target protein or its ligand, database screening focusing on a small library of compounds can be employed in a rational manner.

A small library of terphenyl compounds was screened for inhibitors of the MDM2-p53 interaction based on the rationale that the terphenyl compounds could mimic the α-helical nature of the p53 peptide [47]. Utilizing a fluorescence polarization-based binding assay, terphenyls capable of disrupting the MDM2-p53 complex were identified. The most potent inhibitor found, **9**, has a K_i value of 182 nM (Table 10.1) [47]. Computational docking studies

Table 10.1 *Non-peptidic small molecule inhibitors of the MDM2-p53 interaction*

Class of Compounds	Name	Chemical Structure	Affinity[a]	Reference
cis-Imidazoline	Nutlin-1 (**1**)	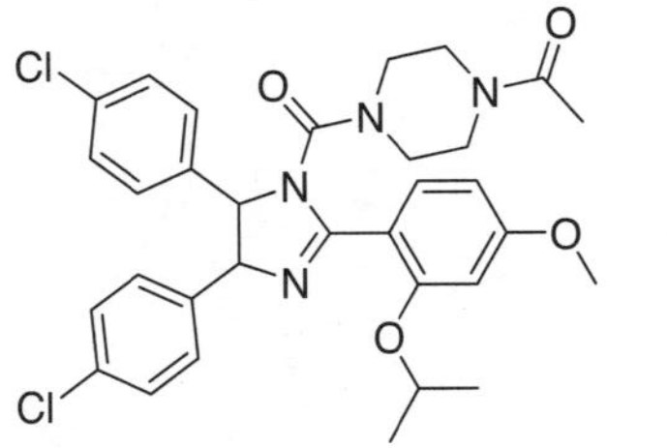	260 nM	Ref# 43 Vassilev et al. (2004) *Science* 303:844-848
	Nutlin-2 (**2**)	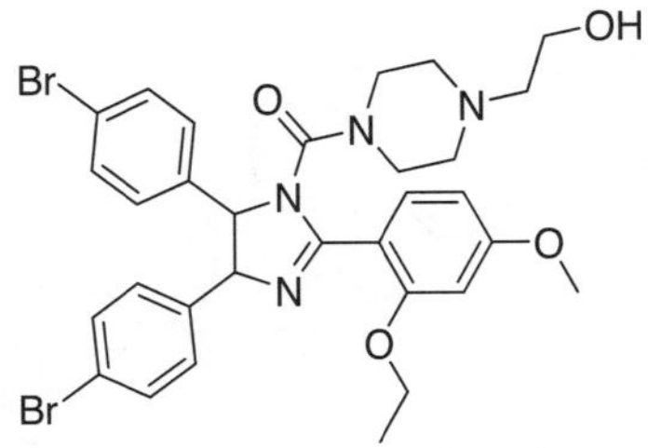	140 nM	
	Nutlin-3 (**3**)	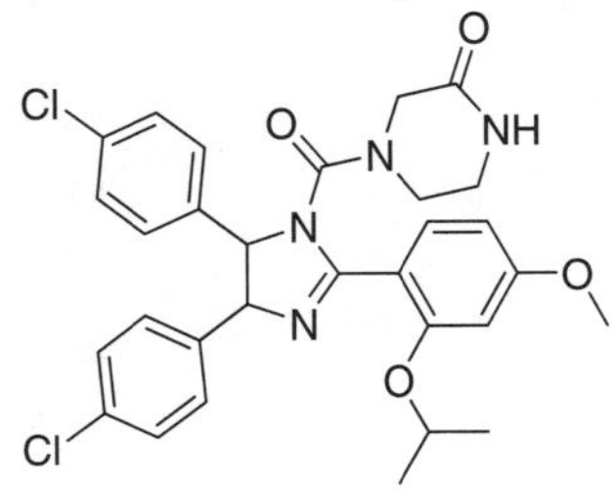	90 nM	

(continued)

Table 10.1 *(Continued)*

Class of Compounds	Name	Chemical Structure	Affinity[a]	Referencec
Benzodiazepinediones	(**4**)	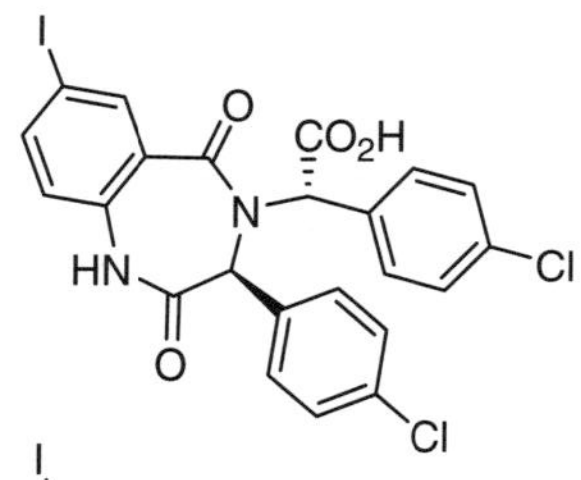	420 nM	Ref# 45 Parks et al. (2005) *Bioorg Med Chem Lett* 15:765-770.
	(**5**)	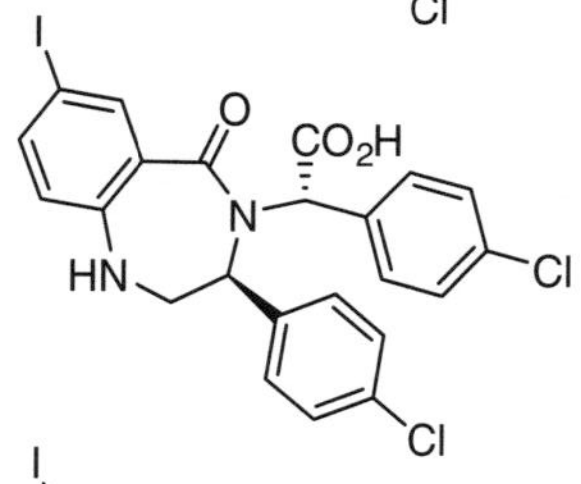	490 nM	
	(**6**)	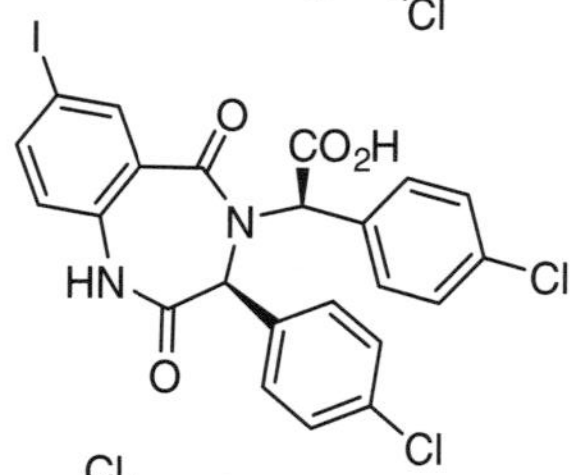	80 nM	Ref# 44 Grasberger et al. (2005) *J Med Chem* 48: 909-912
	TDP521252 (**7**)	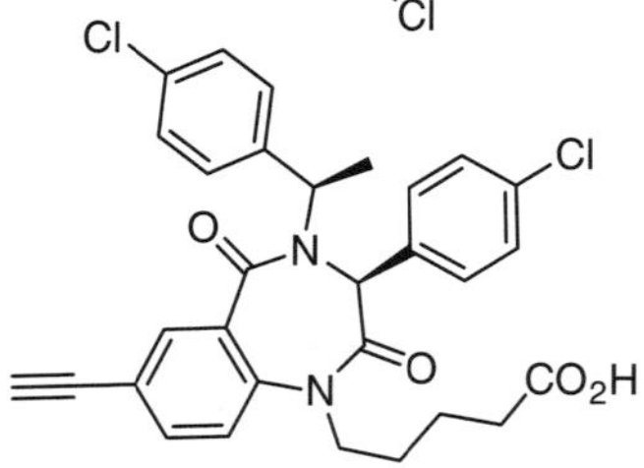	708 nM	Ref# 46 Koblish et al. (2006) *Mol Cancer Ther* 5:160-169

	TDP665759 (**8**)	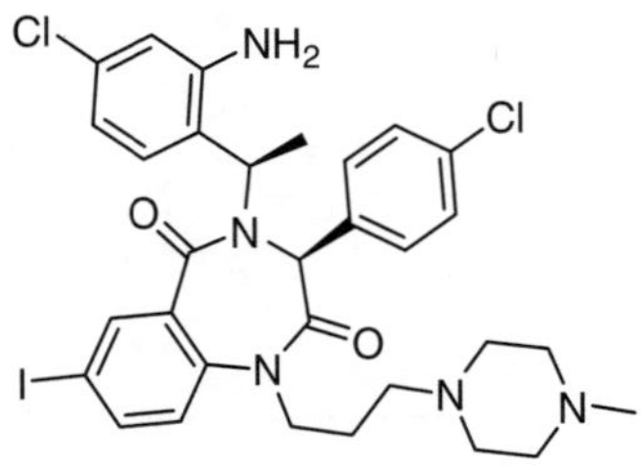	704 nM	
	TDP222669 (**21**)	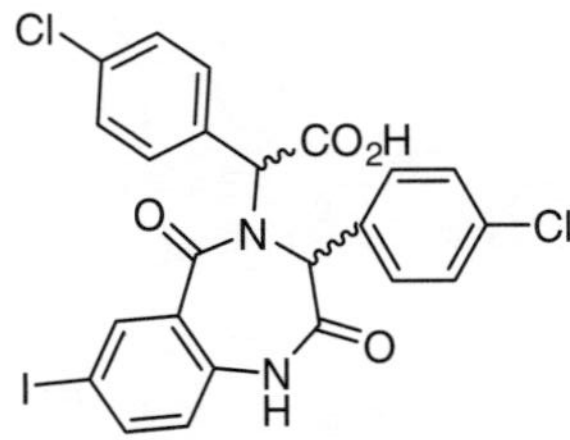	567 nM	
Terphenyl	(**9**)	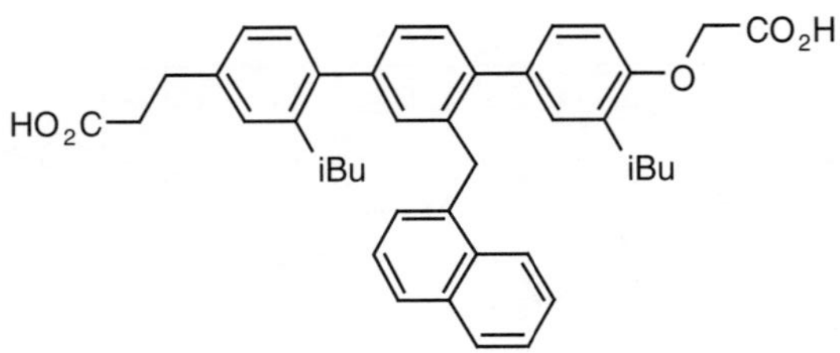	182 nM (K_i)	Ref# 47 Yin et al. (2005) *Angew Chem Int Ed Engl*. 44:2704-2707
	(**10**)	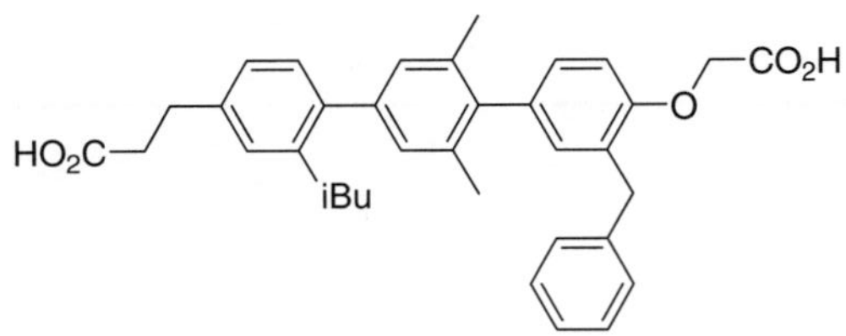	10 μM	Ref# 48 Chen et al. (2005) *Mol Cancer Ther*. 4:1019-1025.

(continued)

Table 10.1 *(Continued)*

Class of Compounds	Name	Chemical Structure	Affinity[a]	Referencec
	(**11**)	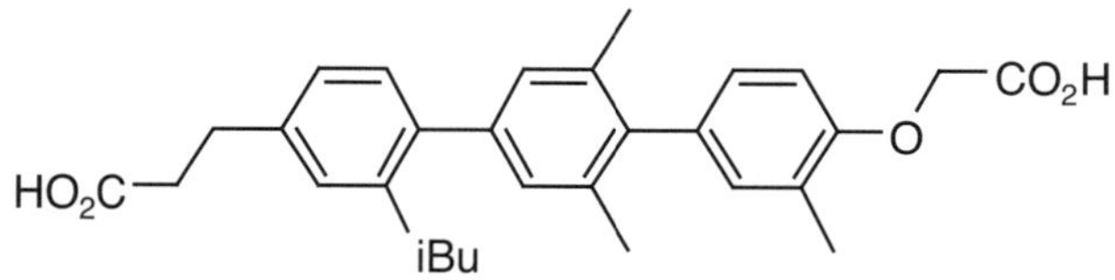	15 μM	
Chalcone	(**12**)	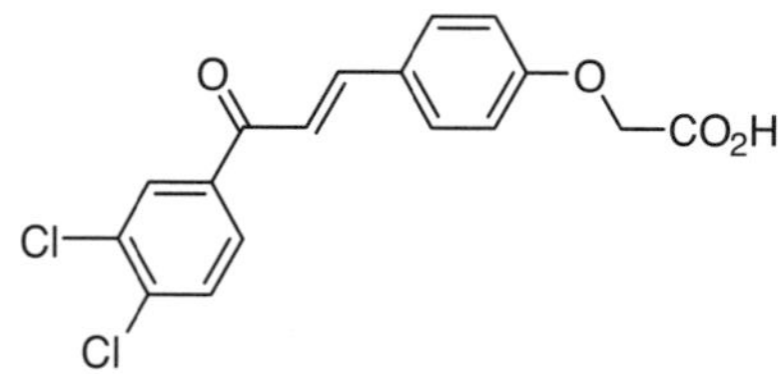	49 μM	Ref# 50 Stoll et al. (2001) *Biochemistry* 40:336-344
	(**13**)	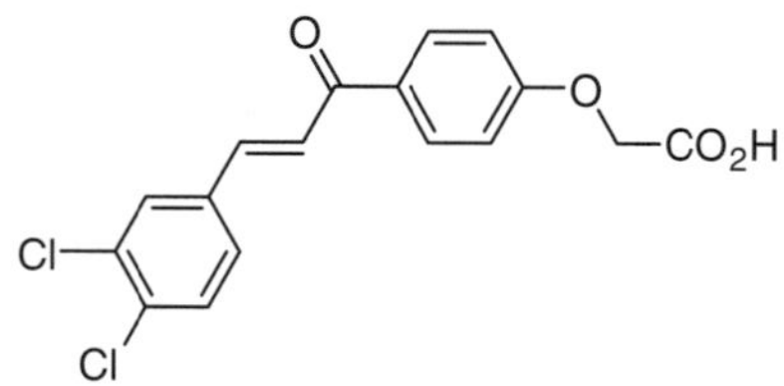	117 μM	
Sulfonamide	NSC279287 (**14**)	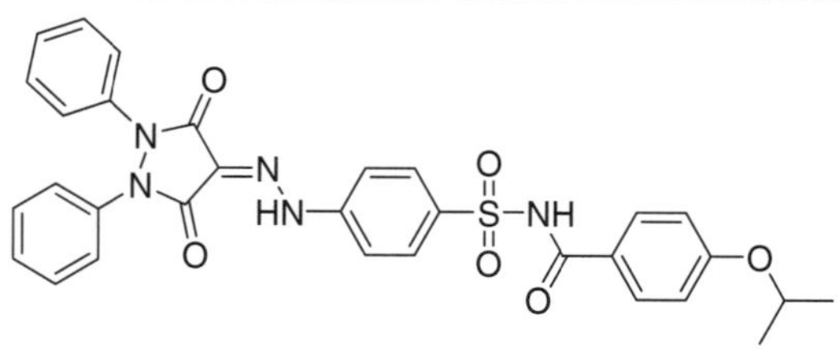	31.8 μM	Ref# 51 Galatin and Abraham (2004) *J Med Chem* 47:4163-4165

Quinolinol	NSC66811 (**15**)	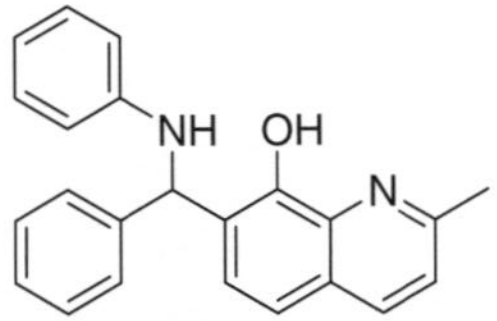	120 nM (K_i)	Ref# 52 Lu et al *J Med Chem* 49:3759-3762
	(**16**)	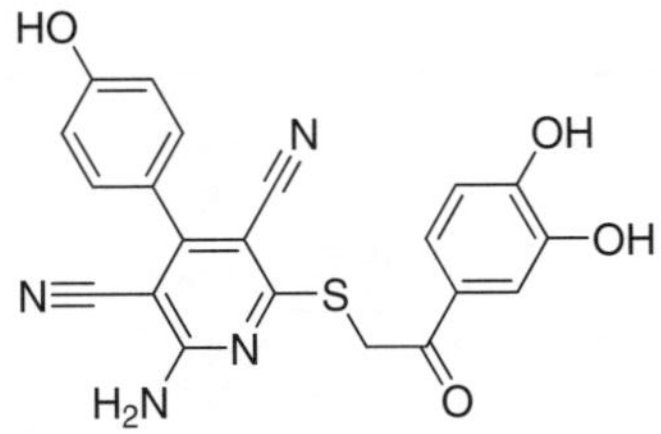	110 nM (K_i)	Ref# 59 Bowman et al. (2007) *J Am Chem Soc* 129:12809-12814.
Spiro-oxindole	(**17**)	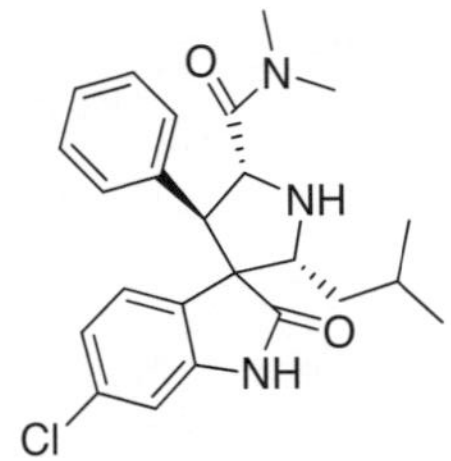	8.5 μM (K_i)	Ref# 60 Ding et al. (2005) *J Am Chem Soc* 127, 10130-10131
	(**18**)	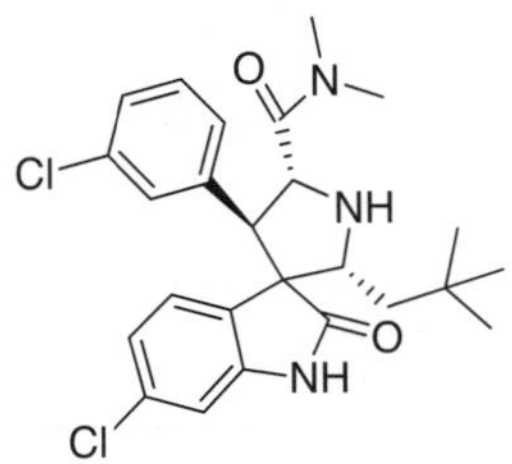	86 nM (K_i)	

(continued)

Table 10.1 *(Continued)*

Class of Compounds	Name	Chemical Structure	Affinity[a]	Referencec
	MI-63 (**19**)	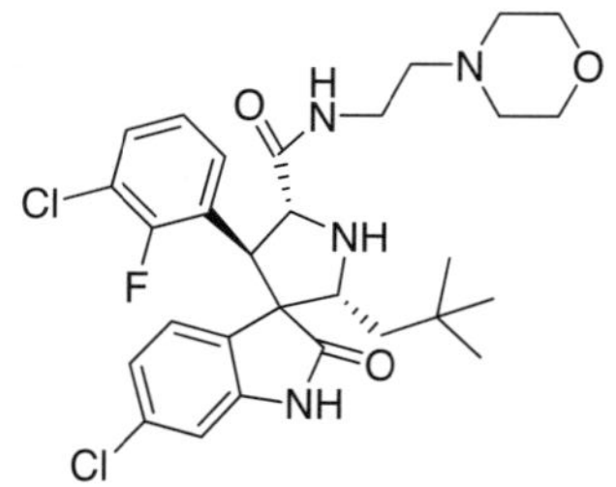	3 nM (K_i)	Ref# 61 Ding et al. (2006) *J Med Chem* 49:3432-3435
	MI-219 (**20**)	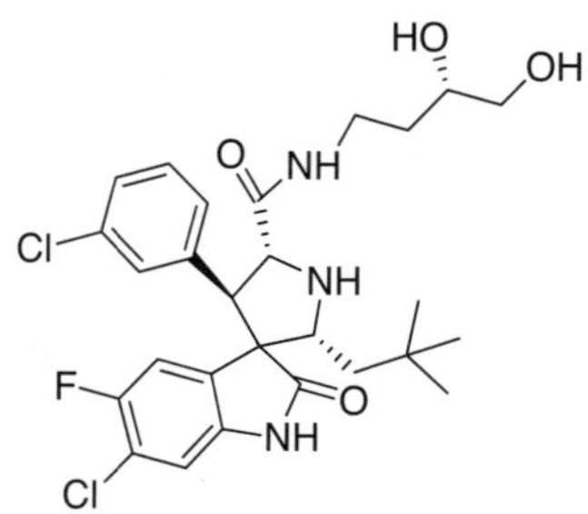	5 nM (K_i)	Ref# 63 Shangary et al. (2008) *Proc Natl Acad Sci USA* 105:3933-3938

[a]: IC_{50} values are reported except where noted.

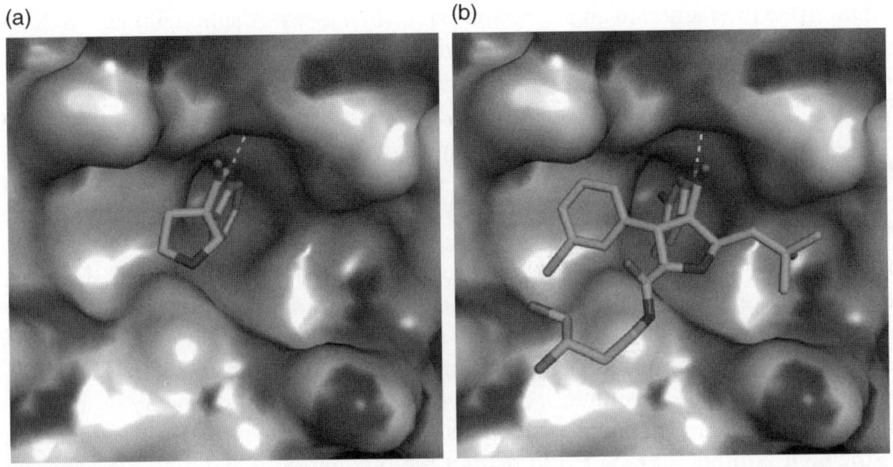

Figure 10.4 *Predicted binding mode of (a) spiro(oxindole-3,3′-pyrrolidine) and (b) MI-219 to MDM2 by GOLD. The surface representation of MDM2 is shown with carbons. Hydrogen atoms are excluded for clarity with hydrogen bonds. This figure was generated by the program Pymol. (See Plate 23.)*

suggested that the terphenyls target the MDM2-p53 interaction site, and this was further validated through ^{15}N heteronuclear single quantum coherence (HSQC) NMR spectroscopy. Additional terphenyl compounds were screened with an *in vitro* quantitative ELISA-based assay, and compounds **10** and **11** were shown to disrupt the MDM2-p53 interaction with IC_{50} values of 10 μM and 15 μM, respectively (Table 10.1) [48].

Chalcones are known anticancer agents [49] and a small library of 16 chalcones was also screened using an ELISA-based assay and NMR titration experiments [50]. Compounds **12** and **13** were found to be the two most potent chalcone inhibitors of MDM2, having IC_{50} values of 49 μM and 117 μM, respectively (Table 10.1).

10.5.2 Computational Database Screening

10.5.2.1 Ligand Pharmacophore-based Screening

A ligand pharmacophore defines critical binding elements (chemical groups) and their spatial relationships that are important for interaction with the target protein. Molecules that contain the pharmacophore can be found by pharmacophore searching and this technique has been used for the discovery of small-molecule inhibitors of the MDM2-p53 interaction.

Galatin and Abraham were the first to employ pharmacophore searching in this context [51]. They constructed a pharmacophore model based upon p53 mutagenesis data, peptide-based inhibitors from phage display and high affinity p53-based peptides incorporating unnatural amino acids [41]. A set of compounds identified by pharmacophore searching of the National Cancer Institute (NCI) database was tested for inhibition of the MDM2-p53 interaction in an ELISA-based *in vitro* assay. A sulfonamide, NSC279287 (**14**), was discovered to have an IC_{50} value of 31.8 μM (Table 10.1) [51].

10.5.2.2 Protein Structure-based Database Screening

Structure-based database screening is another computational approach to the discovery of lead compounds. Lu and colleagues have employed this method, in combination with pharmacophore searching, for the discovery of novel small-molecule inhibitors of the MDM2-p53 interaction [52].

In their study, a database of drug-like compounds from the NCI database [52–54] was first screened using a pharmacophore model based on the crystal structures of MDM2 complexed with the p53 peptide and various known small molecule MDM2 inhibitors. Compounds that were identified by pharmacophore searching were further subjected to screening for properties thought to be desirable in drugs, for example, hydrophobicity and a number of rotatable bonds. The remaining compounds were docked into the MDM2 binding pocket using the GOLD program [55, 56] to predict the binding pose and rank the compounds. Compounds were then rescored and the top 200 compounds, as ranked by Chemscore [57], and Xscore [58], were visually inspected to ensure that they mimicked the key binding residues in p53. A total of 67 compounds were obtained and tested in a competitive fluorescence polarization-based (FP-based) MDM2 binding assay, yielding 10 hits. The quinolinol NSC66811 (**15**), with a K_i value of 120 nM, was the most potent MDM2 inhibitor found (Table 10.1) [52].

A dynamic receptor-based pharmacophore model utilizing multiple protein structures has also been utilized to screen for inhibitors of the MDM2-p53 interaction [59]. Multiple conformations of the protein were obtained by molecular dynamics simulations of the MDM2-p53 complex and using small-molecule probes and consensus clusters of probes across all the protein conformations were identified. These pharmacophore elements were used to screen ~35 000 compounds leading to the identification of 27 compounds, of which 23 were tested in the FP-based competitive assay giving 4 hits, one of which was **16** with a K_i of 110 nM (Table 10.1).

10.5.3 Structure-based *de Novo* Design

The availability of the 3D structure of the protein complexed to a ligand provides structural information that can be utilized in rational '*de novo*' design of new ligands. Ding and colleagues have employed such a *de novo* design strategy for the discovery of a class of nonpeptide MDM2 inhibitors with a spiro-oxindole core (Table 10.1) [60, 61].

Analysis of the crystal structure of the MDM2-p53 interaction showed that four key hydrophobic residues of p53 (Phe^{19}, Leu^{22}, Trp^{23} and Leu^{26}) mediate the interaction with MDM2 [20] and led to the selection of the spiro(oxindole-3,3'-pyrrolidine) structure as the core for the design of new MDM2 inhibitors. The oxindole moiety mimics the key hydrogen-bond and hydrophobic interactions between Trp^{23} in p53 and MDM2, and the spiropyrrolidine ring provides a rigid scaffold from which two hydrophobic groups can be projected to mimic the side chain of Phe^{19} and Leu^{26} (Figure 10.4a). Using a structure-based approach, compounds designed using different functional groups to mimic these residues were docked into the MDM2 binding site. Compound **17** (Table 10.1), predicted to bind to MDM2 with good affinity, was synthesized. An FP-based binding assay showed that **17** was able to bind to

MDM2 with a K_i value of 8.5 μM, and proved to be an initial lead compound for subsequent optimization [60]. Modification of compound **17** yielded the spiro-oxindole **18** which binds to MDM2 with a K_i value of 86 nM. Computational docking suggested that **18** binds to MDM2 by mimicking Phe^{19}, Trp^{23}, and Leu^{26} residues in p53.

The X-ray structure of the MDM2-p53 complex suggests that Leu^{22} may also play an important role in the overall interaction between p53 and MDM2 [20]. This is supported by results from mutation analysis [62] and alanine scanning of p53 peptides [39]. Additional modifications made to capture this interaction between Leu^{22} in p53 and MDM2 yielded **19**, termed MDM2 inhibitor-63 or MI-63 [61], which binds to MDM2 with a K_i value of 3 nM (Table 10.1). Despite its high affinity, MI-63 was unsuitable for *in vivo* evaluation as a result of its poor *in vivo* pharmacokinetic properties. Subsequent optimization of MI-63 led to MI-219 (**20**), which not only binds to MDM2 with a K_i value of 5 nM but also is orally bioavailable (Table 10.1) [63]. Modeling showed that MI-219 mimics all four key residues in p53 (Figure 10.4b).

10.6 Challenges in the Design of Small Molecule Inhibitors of the MDM2-p53 Interaction

10.6.1 Binding Affinity and Specificity

For the purposes of employing MDM2 inhibitors as pharmacological tools and developing them as potential anticancer drugs, it is critical that they not only achieve high affinities in targeting the MDM2-p53 interaction but also show excellent specificity over other protein–protein interactions. To date, only a few MDM2 inhibitors have been evaluated for their binding specificity for MDM2 over other proteins.

MI-63 [61] and MI-219 [63] were evaluated for their binding specificity to other proteins such as Bcl-2 and Bcl-xL proteins. Like MDM2 protein, Bcl-2 and Bcl-xL proteins contain a large hydrophobic pocket and interact with Bid, Bim and Bad proteins/peptides via an amphipathic α-helix. However, MI-63 and MI-219 were shown to achieve more than 10 000-fold selectivity for MDM2 over Bcl-2 and Bcl-xL. Furthermore, MI-219 also shows greater than 10 000-fold selectivity over its closest homologous protein, MDMX. In comparison, Nutlin-3 has >250-fold selectivity for the MDM2-p53 interaction over the MDMX-p53 interaction [63].

10.6.2 Solubility and Cell Permeability

In addition to potency and selectivity, good solubility and cell permeability are highly desirable properties for MDM2 inhibitors. Achieving good aqueous solubility is particularly challenging in the case of MDM2 inhibitors. The hydrophobic nature of the p53 binding pocket in MDM2 has led to the compounds identified as inhibitors being hydrophobic as well and having poor aqueous solubility. MI-219 and Nutlin-3 have good solubility and MI-63 has an excellent aqueous solubility.

An MDM2 inhibitor may achieve high binding affinity to MDM2 but lack cellular permeability. For example, the benzodiazepinediones TDP222669 (**21**), TDP521252 and

TDP665759 bind to MDM2 with IC_{50} values of 567 nM, 708 nM and 704 nM, respectively, but only the latter two compounds are cell permeable [46].

10.6.3 *In Vivo* Pharmacological Properties

An MDM2 inhibitor with a potent binding affinity to MDM2, and excellent cellular activity and solubility may still be unsuitable for *in vivo* evaluations and for development as a new anticancer agent due to the lack of good *in vivo* pharmacological properties. A very short half-life and minimal exposure of the drug in tissues will both vitiate *in vivo* activity. For example, MI-219 achieves a good pharmacokinetic profile and is orally bioavailable [63], but its analogue MI-63 has poor oral bioavailability.

10.7 Reactivation of p53 by Inhibitors of the MDM2-p53 Interaction

Potent, specific and cell-permeable MDM2 inhibitors such as Nutlin-3, MI-219 and its analogues are pharmacological tools that can reactivate p53 in cells by blocking the MDM2-p53 interaction without causing DNA damage [17–19, 43, 61, 64]. Such potent and specific MDM2 inhibitors have been used to probe the p53 pathway.

A number of studies have clearly demonstrated that MDM2 inhibitors accumulate p53 without inducing DNA damage and do not require p53 phosphorylation [61, 63–71]. Accumulation of p53 as a result of the interaction between MDM2 and these small molecules leads to the transcriptional modulation of p53 target gene expression. An example of this is the upregulation of p53 target genes including $p21^{Waf1/Cip1}$ and *MDM2* [43, 63]. Of note, the mechanism of accumulation of p53 by MDM2 inhibitors is different from that induced by radiation and traditional chemotherapeutic agents, both of which induce DNA damage and lead to p53 accumulation through post-translational modifications, such as phosphorylation of p53.

Studies to date have shown that potent and specific MDM2 inhibitors suppress the proliferation of a variety of cancer cell lines retaining wild-type p53 and show a high degree of specificity over cells harboring mutated/deleted p53 [43, 61, 63]. In tumor cell lines with wild-type p53, activation of p53 by MDM2 inhibitors induces p53- and p21-dependent cell cycle arrest and p53-dependent cell death [43, 63, 64, 72]. In addition, overexpression of MDM2 in cancer cells inhibits p53 function and renders them particularly sensitive to p53 reactivation by MDM2 inhibitors (72). However in normal cells, p53 activation by MDM2 inhibitors leads to cell cycle arrest, but not cell death [43, 63, 64]. Thus, MDM2 inhibitors are selectively toxic to cancer cells with wild-type p53.

10.8 Development of MDM2 Inhibitors as New Anticancer Drugs

The therapeutic potential of MDM2 inhibitors has been evaluated in mouse xenograft models of human cancer and with specimens from patients with a variety of blood malignancies.

Since the activity of MDM2 inhibitors depends upon p53 activation in cells expressing wild-type p53, blood malignancies, such as acute myeloid leukemia (AML), B-chronic lymphocytic leukemia (B-CLL) and multiple myeloma, in which p53 mutation and deletion are rare, are attractive tumor types for the MDM2 inhibitor-based therapy. Indeed, *ex vivo* experiments on AML [73, 74], B-CLL [75–78] and specimens from multiple myeloma patients [79] show that Nutlin-3 and MI-63, as single agents, trigger apoptosis. Moreover, Nutlin-3 is synergistic with other chemotherapeutic agents in killing AML blasts and B-CLL patient specimens [75–77]. Importantly, both single agent and combination effects of Nutlin-3 are selective for cancer versus normal cells, as is revealed by lack of toxicity to peripheral blood mononuclear cells or hematopoietic progenitors and stromal epithelium cells from bone marrow [73, 75, 79].

Oral administration of Nutlin-3 [43, 72, 80] and MI-219 [63] strongly inhibits tumor growth in several xenograft models of human cancer with wild-type p53. MI-219 does not have a significant effect on the growth of xenograft tumors harboring mutant p53, indicating that the antitumor activity of MDM2 inhibitors correlates with p53 status [63].

It is significant that both Nutlin-3 [43] and MI-219 [63] achieve their antitumor activity without causing signs of toxicity in animals, as assessed by necropsy studies and body weight loss. Repeated administration of MI-219 to animals does not cause toxicity in normal mouse tissues, such as the small-intestine and thymus, which are sensitive to p53-induced apoptosis by radiation and chemotherapy [63, 81, 82].

10.9 Concluding Remarks

Targeting the MDM2-p53 protein–protein interaction with potent and specific small molecule inhibitors represents an attractive strategy for cancer treatment. Several distinct approaches have been employed to the discovery and design of nonpeptidic small small-molecule inhibitors of the MDM2-p53 interaction. These efforts have led to discovery of multiple classes of potent, specific, nonpeptidic small-molecule inhibitors of the MDM2-p53 interaction. Nutlin-3 and MI-219 are among the best characterized small molecule inhibitors. They are not only potent and specific in cell-based assays but also are highly effective in inhibition of tumor growth in xenograft models of human cancer in mice. The discovery of these compounds clearly demonstrates that it is possible to design potent, specific, cell-permeable and *in vivo* active small-molecule inhibitors of the MDM2-p53 interaction. The availability of Nutlin-3 and MI-219 provides a set of powerful pharmacological tools to probe the p53 pathway. A number of these MDM2 inhibitors have been advanced into the preclinical development stage and are expected to enter human clinical trials in the near future.

Acknowledgements

Funding from the National Cancer Institute/National Institutes of Health, the Prostate Cancer Foundation, the Leukemia and Lymphoma Society, and Ascenta Therapeutics is greatly appreciated. Page limitations have prevented exhaustive coverage and we apologize in advance to any investigators whose studies are not cited in this chapter.

Disclosure Statement

S.W. and the University of Michigan own equity in Ascenta Therapeutics, Inc., which has licensed the technologies related to the MDM2 inhibitors of the spiro-oxindole class. S.W. serves as a consultant for Ascenta and is the principal investigator on a research contract from Ascenta to the University of Michigan.

References

1. Murray, J. K. & Gellman, S. H. (2007) *Biopolymers* **88**, 657–86.
2. Fry, D. C. & Vassilev, L. T. (2005) *J Mol Med* **83**, 955–63.
3. Lane, D. P. & Crawford, L. V. (1979) *Nature* **278**, 261–3.
4. Linzer, D. I. & Levine, A. J. (1979) *Cell* **17**, 43–52.
5. DeLeo, A. B., Jay, G., Appella, E., Dubois, G. C., Law, L. W. & Old, L. J. (1979) *Proc Natl Acad Sci U S A* **76**, 2420–4.
6. Fakharzadeh, S. S., Trusko, S. P. & George, D. L. (1991) *Embo J* **10**, 1565–9.
7. Fridman, J. S. & Lowe, S. W. (2003) *Oncogene* **22**, 9030–40.
8. Hainaut, P. & Hollstein, M. (2000) *Adv Cancer Res* **77**, 81–137.
9. Vogelstein, B., Lane, D. & Levine, A. J. (2000) *Nature* **408**, 307–10.
10. Vousden, K. H. & Lu, X. (2002) *Nat Rev Cancer* **2**, 594–604.
11. Bond, G. L., Hu, W., Bond, E. E., *et al.* (2004) *Cell* **119**, 591–602.
12. Oliner, J. D., Kinzler, K. W., Meltzer, P. S., George, D. L. & Vogelstein, B. (1992) *Nature* **358**, 80–3.
13. Zhou, M., Gu, L., Abshire, T. C., *et al.* (2000) *Leukemia* **14**, 61–7.
14. Rayburn, E., Zhang, R., He, J. & Wang, H. (2005) *Curr Cancer Drug Targets* **5**, 27–41.
15. Gunther, T., Schneider-Stock, R., Hackel, C., *et al.* (2000) *Mod Pathol* **13**, 621–6.
16. Momand, J., Jung, D., Wilczynski, S. & Niland, J. (1998) *Nucleic Acids Res* **26**, 3453–9.
17. Vassilev, L. T. (2007) *Trends Mol Med* **13**, 23–31.
18. Chene, P. (2003) *Nat Rev Cancer* **3**, 102–9.
19. Shangary, S. & Wang, S. (2008) *Clin Cancer Res* **14**, 5318–24.
20. Kussie, P. H., Gorina, S., Marechal, V., *et al.* (1996) *Science* **274**, 948–53.
21. Feki, A. & Irminger-Finger, I. (2004) *Crit Rev Oncol Hematol* **52**, 103–16.
22. Momand, J., Zambetti, G. P., Olson, D. C., George, D. & Levine, A. J. (1992) *Cell* **69**, 1237–45.
23. Bond, G. L., Hu, W. & Levine, A. J. (2005) *Curr Cancer Drug Targets* **5**, 3–8.
24. Capoulade, C., Bressac-de Paillerets, B., Lefrere, I., *et al.* (1998) *Oncogene* **16**, 1603–10.
25. Momand, J., Wu, H. H. & Dasgupta, G. (2000) *Gene* **242**, 15–29.
26. Ganguli, G., Abecassis, J. & Wasylyk, B. (2000) *Embo J* **19**, 5135–47.
27. Kemp, C. J., Donehower, L. A., Bradley, A. & Balmain, A. (1993) *Cell* **74**, 813–22.
28. Ventura, A., Kirsch, D. G., McLaughlin, M. E., *et al.* (2007) *Nature* **445**, 661–5.
29. Martins, C. P., Brown-Swigart, L. & Evan, G. I. (2006) *Cell* **127**, 1323–34.
30. Xue, W., Zender, L., Miething, C., *et al.* (2007) *Nature* **445**, 656–60.
31. Freedman, D. A., Wu, L. & Levine, A. J. (1999) *Cell Mol Life Sci* **55**, 96–107.
32. Juven-Gershon, T. & Oren, M. (1999) *Mol Med* **5**, 71–83.
33. Wu, X., Bayle, J. H., Olson, D. & Levine, A. J. (1993) *Genes Dev* **7**, 1126–32.
34. Jones, S. N., Roe, A. E., Donehower, L. A. & Bradley, A. (1995) *Nature* **378**, 206–8.
35. Montes de Oca Luna, R., Wagner, D. S. & Lozano, G. (1995) *Nature* **378**, 203–6.
36. McCoy, M. A., Gesell, J. J., Senior, M. M. & Wyss, D. F. (2003) *Proc Natl Acad Sci USA* **100**, 1645–8.
37. Uhrinova, S., Uhrin, D., Powers, H., *et al.* (2005) *J Mol Biol* **350**, 587–98.
38. Chen, J., Marechal, V. & Levine, A. J. (1993) *Mol Cell Biol* **13**, 4107–14.
39. Picksley, S. M., Vojtesek, B., Sparks, A. & Lane, D. P. (1994) *Oncogene* **9**, 2523–9.

40. Bottger, V., Bottger, A., Howard, S. F., *et al.* (1996) *Oncogene* **13**, 2141–7.
41. Garcia-Echeverria, C., Chene, P., Blommers, M. J. & Furet, P. (2000) *J Med Chem* **43**, 3205–8.
42. Chene, P., Fuchs, J., Bohn, J., Garcia-Echeverria, C., Furet, P. & Fabbro, D. (2000) *J Mol Biol* **299**, 245–53.
43. Vassilev, L. T., Vu, B. T., Graves, B., *et al.* (2004) *Science* **303**, 844–8.
44. Grasberger, B. L., Lu, T., Schubert, C., *et al.* (2005) *J Med Chem* **48**, 909–12.
45. Parks, D. J., Lafrance, L. V., Calvo, R. R., *et al.* (2005) *Bioorg Med Chem Lett* **15**, 765–70.
46. Koblish, H. K., Zhao, S., Franks, C. F., *et al.* (2006) *Mol Cancer Ther* **5**, 160–9.
47. Yin, H., Lee, G. I., Park, H. S., *et al.* (2005) *Angew Chem Int Ed Engl* **44**, 2704–7.
48. Chen, L., Yin, H., Farooqi, B., Sebti, S., Hamilton, A. D. & Chen, J. (2005) *Mol Cancer Ther* **4**, 1019–25.
49. Aleskog, A., Larsson, R., Hoglund, M., Kristensen, J., Nygren, P. & Lindhagen, E. (2005) *Anticancer Drugs* **16**, 277–83.
50. Stoll, R., Renner, C., Hansen, S., *et al.* (2001) *Biochemistry* **40**, 336–44.
51. Galatin, P. S. & Abraham, D. J. (2004) *J Med Chem* **47**, 4163–5.
52. Lu, Y., Nikolovska-Coleska, Z., Fang, X., *et al.* (2006) *J Med Chem* **49**, 3759–62.
53. Milne, G. W., Nicklaus, M. C., Driscoll, J. S., Wang, S. & Zaharevitz, D. (1994) *J Chem Inf Comput Sci* **34**, 1219–24.
54. Voigt, J. H., Bienfait, B., Wang, S. & Nicklaus, M. C. (2001) *J Chem Inf Comput Sci* **41**, 702–12.
55. Jones, G., Willett, P., Glen, R. C., Leach, A. R. & Taylor, R. (1997) *J Mol Biol* **267**, 727–48.
56. Verdonk, M. L., Cole, J. C., Hartshorn, M. J., Murray, C. W. & Taylor, R. D. (2003) *Proteins* **52**, 609–23.
57. Eldridge, M. D., Murray, C. W., Auton, T. R., Paolini, G. V. & Mee, R. P. (1997) *J Comput-Aided Mol Des* **11**, 425–45.
58. Wang, R., Lai, L. & Wang, W. (2002) *J Comput-Aided Mol. Des* **16**, 11–26.
59. Bowman, A. L., Nikolovska-Coleska, Z., Zhong, H., Wang, S. & Carlson, H. A. (2007) *J Am Chem Soc* **129**, 12809–14.
60. Ding, K., Lu, Y., Nikolovska-Coleska, Z., *et al.* (2005) *J Am Chem Soc* **127**, 10130–1.
61. Ding, K., Lu, Y., Nikolovska-Coleska, Z., *et al.* (2006) *J Med Chem* **49**, 3432–5.
62. Lin, J., Chen, J., Elenbaas, B. & Levine, A. J. (1994) *Genes Dev* **8**, 1235–46.
63. Shangary, S., Qin, D., McEachern, D., *et al.* (2008) *Proc Natl Acad Sci U S A* **105**, 3933–8.
64. Shangary, S., Ding, K., Qiu, S., *et al.* (2008) *Mol Cancer Ther* **7**, 1533–42.
65. Hu, B., Gilkes, D. M., Farooqi, B., Sebti, S. M. & Chen, J. (2006) *J Biol Chem* **281**, 33030–5.
66. Laurie, N. A., Donovan, S. L., Shih, C. S., *et al.* (2006) *Nature* **444**, 61–6.
67. Patton, J. T., Mayo, L. D., Singhi, A. D., Gudkov, A. V., Stark, G. R. & Jackson, M. W. (2006) *Cancer Res* **66**, 3169–76.
68. Wade, M., Wong, E. T., Tang, M., Stommel, J. M. & Wahl, G. M. (2006) *J Biol Chem* **281**, 33036–44.
69. Thompson, T., Tovar, C., Yang, H., *et al.* (2004) *J Biol Chem* **279**, 53015–22.
70. Cheok, C. F., Dey, A. & Lane, D. P. (2007) *Mol Cancer Res* **5**, 1133–45.
71. Jones, R. J., Chen, Q., Voorhees, P. M., *et al.* (2008) *Clin Cancer Res* **14**, 5416–25.
72. Tovar, C., Rosinski, J., Filipovic, Z., *et al.* (2006) *Proc Natl Acad Sci U S A* **103**, 1888–93.
73. Kojima, K., Konopleva, M., Samudio, I. J., *et al.* (2005) *Blood* **106**, 3150–9.
74. Secchiero, P., Corallini, F., Gonelli, A., *et al.* (2007) *Circ Res* **100**, 61–9.
75. Secchiero, P., Barbarotto, E., Tiribelli, M., *et al.* (2006) *Blood* **107**, 4122–9.
76. Kojima, K., Konopleva, M., McQueen, T., O'Brien, S., Plunkett, W. & Andreeff, M. (2006) *Blood* **108**, 993–1000.
77. Coll-Mulet, L., Iglesias-Serret, D., Santidrian, A. F., *et al.* (2006) *Blood* **107**, 4109–14.
78. Saddler, C., Ouillette, P., Kujawski, L., *et al.* (2007) *Blood* **111**, 1584–93.
79. Stuhmer, T., Chatterjee, M., Hildebrandt, M., *et al.* (2005) *Blood* **106**, 3609–17.
80. Sarek, G., Kurki, S., Enback, J., *et al.* (2007) *J Clin Invest* **117**, 1019–28.
81. Lowe, S. W., Schmitt, E. M., Smith, S. W., Osborne, B. A. & Jacks, T. (1993) *Nature* **362**, 847–9.
82. Potten, C. S., Wilson, J. W. & Booth, C. (1997) *Stem Cells* **15**, 82–93.

11

Case Study: The Discovery of Potent LFA-1 Antagonists

Tom Gadek
SARcode Corporation, San Francisco, CA, USA

11.1 Introduction

Lymphocyte Function-Associated Antigen-1 (LFA-1) is a member of the integrin cell adhesion receptor family whose biological function in the immune system is dependent on the protein–protein interaction with its cognate ligands, the family of Intercellular Adhesion Molecules (e.g. ICAM-1) [1, 2]. The expression of LFA-1 in humans is almost exclusively limited to cells of the immune system, particularly the extracellular surface of leukocytes including lymphocytes. As such, LFA-1 regulates the adhesion, migration, proliferation and inflammatory/immune response of lymphocytes, particularly T-lymphocytes (T-cells), in normal immune function, as well as, in a number of inflammatory and autoimmune disease settings [3, 4]. Inflammatory signals including cytokines can upregulate the expression of ICAM in tissue in order to attract and captivate lymphocytes and engage them in the inflammatory process. Recent clinical trials have established that the humanized anti-LFA-1 monoclonal antibody, Raptiva™, can control T-cell mediated inflammatory disease via the inhibition of the LFA-1/ICAM-1 interaction. This combined with the regulatory approval of Raptiva™ for the treatment of moderate to severe psoriasis has validated LFA-1 as a target of interest to the pharmaceutical industry [5].

It is worth noting that in clinical trials, very high levels (e.g. >95 %) of LFA-1 occupancy by Raptiva™ were required in order to demonstrate a significant clinical benefit in patients [6]. Furthermore, in the absence of a predictive animal disease model and because

Protein Surface Recognition: Approaches for Drug Discovery Ernest Giralt, Mark W. Peczuh and Xavier Salvatella. © 2011 John Wiley & Sons, Ltd

of the human specificity of the monoclonal antibody, Raptiva™, the pharmacodynamic linkage of LFA-1 to psoriasis as a T-cell mediated disease was established in patients during Phase 1 and Phase 2 clinical trials [7, 8]. Four dose dependent effects were noted in these trials: circulating serum levels of the antibody, occupancy levels of LFA-1 on circulating leukocytes, T-cell counts in the dermis and epidermis of plaques, and patient response by PASI (psoriasis area severity index) score. This pharmacokinetic/pharmacodynamic linkage, between the pharmacokinetic circulating levels of antibody to its pharmacodynamic occupancy and inhibition of LFA-1 and the subsequent clinical benefit arising from the inhibition and depletion of T-cells in psoriatic tissue, has enabled the treatment of a large number of human diseases beyond psoriasis with a common origin in T-cell mediated inflammation. Additional clinical studies in renal transplantation [9, 10], asthma [11], eczema [12], as well as rheumatoid and psoriatic arthritis have explored the importance of LFA-1 and T-cells in the pathophysiology of these diseases. Ongoing clinical trials with Raptiva™ in Lupus Erthematosus, Sjogren's Syndrome, Uveitis, Eczema, Islet Cell Transplantion, Renal Transplantion, Atopic Dermatitis, and other indications (www.clinicaltrials.gov) seek to expand its use as an immunomodulating LFA-1 antagonist in a broad array of T-cell mediated inflammatory and immune diseases [13].

The success of monoclonal antibodies in patients and in a number of animal models of human diseases has 'validated' the clinical concept of LFA-1 antagonists and sparked interest in a search for small molecule 'equivalents' of Raptiva™ capable of antagonizing the LFA-1/ICAM protein–protein interaction. From a strategic perspective, the clinical and regulatory success of Raptiva™ in addition to the large number of emerging clinical indications have dramatically reduced the risk of failure for a small molecule antagonist research program. These lead discovery efforts offer the promise of better serving a large emerging targeted therapeutic market with the enhanced stability, bioavailability and lower cost of goods typically found in a small molecule. However, significant research and commercial challenges must be overcome by these small molecules. These include the perceived technical challenges in the inhibition of large protein–protein interactions with a small molecule, the high levels of receptor occupancy required for clinical benefit which require the engineering of extraordinary affinities into a small molecule framework and the commercial challenge of capturing the aggregate value presented by a drug with a mechanism of action common to a diffuse set of clinical indications (e.g. dermatology, transplant, pulmonary, gastro-intestinal...). This chapter will begin with a discussion of the structural and molecular biologies of LFA-1 and then describe the discovery of potent, high affinity, small molecule LFA-1 antagonists and their pharmacological activities in blocking the LFA-1/ICAM interaction in vitro and in vivo.

11.2 Structural, Molecular and Cellular Biologies of LFA-1

LFA-1, also known as CD11a/CD18 or the integrin $\alpha L\beta 2$, is a relatively large (e.g. >1800 amino acid, 210 kD) heterodimeric protein consisting of alpha (αL, CD11a) and beta subunits (β2, CD18) [1]. It is a member of the subclass of integrin family of adhesion receptors which exhibit an inserted or I-domain between the second and third of the seven beta sheets comprising beta-propeller of the αL subunit (see Figure 11.1) [14]. When isolated from the surface of circulating peripheral blood leukocytes, LFA-1 exhibits post

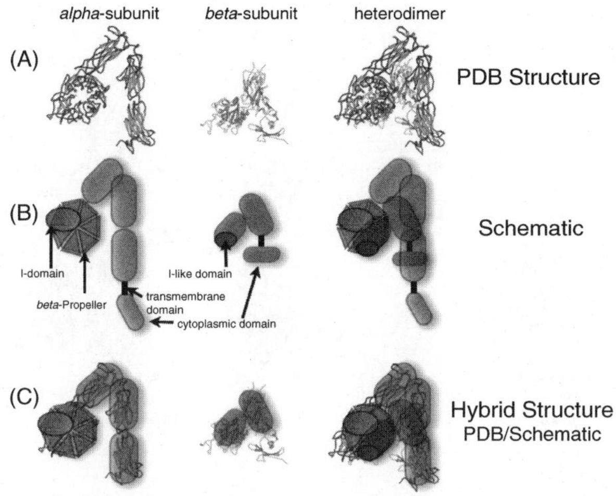

Figure 11.1 *(A) Ribbon diagrams showing the backbone folds of the extracellular domains of the integrin $\alpha V\beta 3$ [18]. File downloaded from the Protein Data Bank (PDB 1l5g). (B and C). Schematic model of LFA-1 extracellular, transmembrane and cytoplasmic domains with I-domain inserted between the second and third beta subunits of the beta-propeller. (See Plate 24.)*

translational modifications from both proteolysis and glycosylation [15]. Each of the alpha and beta subunits of LFA-1 contain: relatively short cytoplasmic domains which are capable of attaching to cytoskeleton; single transmembrane domains; and large extracellular domains which associate as a heterodimer and are involved in ligand binding. The tertiary structure of LFA-1 is stabilized by a number of intrachain disulfide bonds within the subunits and its heterodimeric structure is stabilized by a number of divalent cation (e.g. Ca + 2, Mg + 2) binding sites [16]. An additional metal ion dependent adhesion site (MIDAS) in the inserted domain (I-domain) of the alpha subunit has been demonstrated to be an integral component of the ICAM ligand binding site [17]. An X-ray crystallographic study [18] of the extracellular domain of the related integrin, alpha-V beta-3, has been used as the basis of a schematic structural model of the folding of an integrin heterodimer with the I-domain appended (Figure 11.1). However, it is not clear how the insertion of the I-domain sequence between units two and three of the beta-propeller perturbs the structural fold of this integrin substructural motif, the association of the alpha and beta subunits of LFA-1 or the I-domain's relation to a similar metal ion binding domain designated the I-like domain in the beta subunit of LFA-1 [19].

Additional X-ray crystallographic studies of the isolated I-domain of LFA-1 have demonstrated that the divalent metal cation in this MIDAS site is capable of bridging between LFA-1 and ICAM-1 binding domains via metal coordination of amino acid sidechains from both the I-domain of LFA-1's alpha subunit and the first immunoglobulin domain of ICAM-1 (Figure 11.2) [20]. Interestingly, the presence of different divalent metal cations (e.g. Ca + 2, Mg + 2 and Mn + 2) in the MIDAS site have been shown to effect the conformation of the I-domain and its affinity/off rate for ICAM and small molecule MIDAS ligands [21–24]. This has led to mechanistic proposals for 'activated' and 'resting'

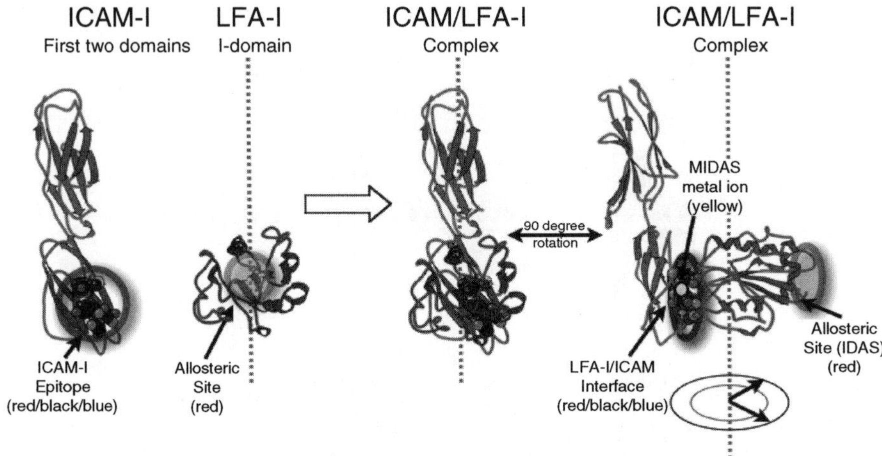

Figure 11.2 *Ribbon diagrams tracing the backbone folds in the protein-protein interaction between the first domain of ICAM-1 and the I-domain of LFA-1 [20]. File downloaded from the Protein Data Bank (PDB 1MQ8). Note that the sidechains of amino acid residues E-34, M-64, Y-66, N68 and Q73, comprising the epitope of ICAM-1 responsible for its binding to LFA-1, have been displayed. The I-domain allosteric site (IDAS) binding site for allosteric antagonists of LFA-1 is proximal to the ICAM-1 binding site on the I-domain and is denoted by a circle. (See Plate 25.)*

or 'open and 'closed' states of the I-domain with differential affinities for ICAM which is in line with similar proposals for other integrin family members [25, 26].

It should be noted that the crystallographic studies defining the 'open' and 'closed' affinity states involve only examination of the 200 amino acid I-domain fragment of the alpha subunit of LFA-1 crystallized under different buffer and salt conditions. Furthermore, only Ca + 2 and Mg + 2 ions are present at physiologically relevant levels in vivo and the octahedral coordination observed in the Mn + 2 structures of the I-domain may not be pertinent mechanistically. These isolated I-domain structures may not faithfully reflect its structure and function in the larger context of the LFA-1 heterodimer on the surface of a leukocyte circulating in a living, breathing patient [27]. A complete discussion of the structure and function of LFA-1 is beyond the scope of this chapter, but the reader is cautioned that the biophysical connection between the biological functions of LFA-1 at the macroscopic and microscopic levels on activated leukocytes remains controversial and unproven.

At sites of inflammation, in response to cytokines and other inflammatory signals, T-cells are activated and this stimulates an apparent increase in the affinity of LFA-1 for ICAM (Figure 11.3). This effect may result from a change in the activation state of the LFA-1/ICAM binding site induced by a conformational shift in LFA-1 transmitted from LFA-1's intracellular domain to the extracellular ICAM binding site (the inside out mechanism of integrin signaling) [28, 29]. Alternatively, a local clustering of LFA-1 and/or ICAM on the surface of interacting cells increasing the cooperativity or avidity of the LFA-1/ICAM interaction may appear to be an increase in affinity [30–33]. The local clustering of LFA-1 on the cell surface results from attachment of the relatively short cytoplasmic domains of the

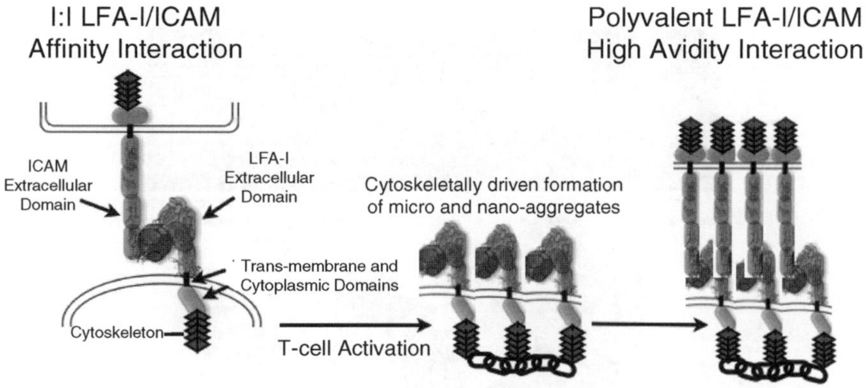

Figure 11.3 *Schematic representation of monovalent and polyvalent cell-cell interaction. Cytoskeletal engagement by the short cytoplasmic domains of LFA-1 can form nano/micro-aggregates of LFA-1 on the cell surface with enhanced avidity for ICAM. Similar surface aggregation of ICAM on the opposing cell surface will augment this avidity effect. (See Plate 26.)*

alpha and beta subunits to intracellular signaling proteins and/or the cell's actin cytoskeleton. Some portion of this cytoskeletal attachment has recently been shown to involve the binding of an NPxY sequence within the beta subunit's cytoplasmic domain to actin via Talin and/or Tensin1 [34].

In light of the previous discussion of LFA-1 structure and function, it would be reasonable to expect that antagonists of LFA-1 could be identified which inhibit the binding of ICAM to the extracellular domain of LFA-1 or the association of the intracellular domains of LFA-1 with either intracellular signaling or cytoskeletal proteins. Regardless of the mechanistic details or site of action, such antagonists of LFA-1 could be expected to show anti-inflammatory and immunosuppressive effects on lymphocytes in vitro and in T-cell mediated diseases in vivo. To date two major classes of protein, antibody and small molecule LFA-1 antagonists have been identified: allosteric inhibitors which bind the I-domain allosteric site (IDAS) of LFA-1's alpha subunit and modulate the I-domain's affinity for ICAM; and direct competitive inhibitors which bind to LFA-1's ICAM binding site directly and exclude ICAM binding.

Functional studies of the binding of LFA-1 and ICAM-1 have shown that this interaction is crucial in processes which contribute to disease mechanism including: leukocyte and lymphocyte adhesion to vascular endothelial cells; their extravasation from the vasculature at a site of inflammation; homotypic interactions between lymphocytes, as well as interactions between T-lymphocytes (T-cells) and dendritic cells in inflamed tissue; the formation of the immunologic synapse and in the transmission of costimulatiory signals in concert with MHC/T-cell Receptor crucial for lymphocyte proliferation and cytokine release [3, 4, 35, 36]. Consequently, antagonists of LFA-1/ICAM binding offer the promise of blocking the adhesive, migratory, proliferative and inflammatory signaling components of lymphocyte mediated inflammation (Figure 11.4). As such, they may provide more comprehensive inhibition of T-cell mediated diseases than is currently offered by steroid or calcineurin antagonist (e.g. cyclosporine) immunomodulators.

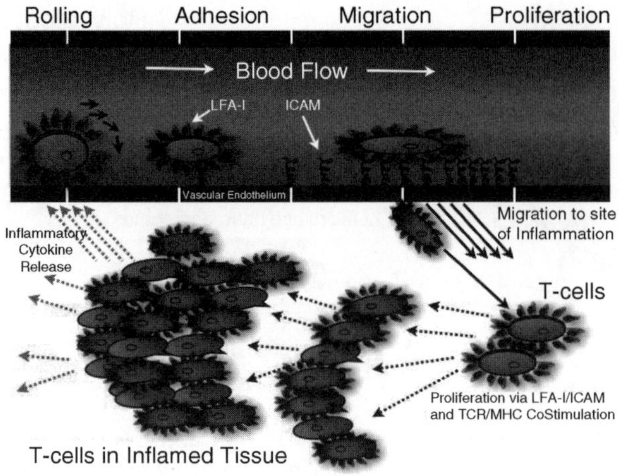

Figure 11.4 *T-cells involved in an inflammatory process. Cells rolling along the wall of a blood vessel encounter a locally high level of ICAM expression on the surface of vascular endothelial cells which has been induced by a local inflammatory signal. The adhesion of these cells to ICAM allows them to change their morphology from round to a more flattened adherent shape. Cells migrate out of the vessel and follow a cytokine concentration gradient to a site of inflammation. T-cells then self associate and proliferate in tissue via costimulation of LFA-1/ICAM and T-cell receptor/MHC expressed on their surface. These clonally expanding T-cells upregulate the expression of cytokines and cytokine receptors which continues a self propagating cycle of inflammation, attracting more cells from the local vasculature and stimulating their proliferation in the inflamed tissue. (See Plate 27.)*

11.3 The Search for Small Molecule LFA-1 Antagonists

The preceding discussion of lymphocyte biology and the structural biology of LFA-1 in its interaction with both the intracellular cytoskeleton and extracellular ICAM binding domain(s) in the process of lymphocyte activation suggests that small molecule antagonists could be identified which bind to and disrupt the function of either LFA-1's short cytoplasmic domains or the extracellular ICAM binding domain (i.e. the I-domain) of the alpha/CD11a subunit of LFA-1. A number of large pharmaceutical research organizations have initiated programs leveraging their organizational strengths in high-throughput screening (HTS), fragment screening, combinatorial chemistry, rational design and other lead and drug discovery methodologies. Key to these efforts was the establishment of a panel of in vitro assays with the sensitivity, dynamic range, sample throughput and pharmacodynamic relevance to structural, cell and disease biology to identify an inhibitor of the protein–protein interaction between LFA-1 and ICAM. In addition, in order to optimize initial lead molecules into clinical candidates, the sensitivity of these assays would be required to discriminate twofold differences in the structure activity relationships (SAR) of individual molecules even though these molecules might only differ by a single atom (e.g. sulfur vs. oxygen in a heterocyclic subtructural unit). Ideally these assays should also reflect the pharmacodynamic and pharmacokinetic challenges presented by increasingly complex biological matrices encountered as the molecule progresses through animal models of the human disease state to a reasoned and rational nomination of a candidate(s) for human clinical evaluation.

Toward this end, recombinant forms of ICAM-1, the I-domain of LFA-1 and full length LFA-1 were cloned, expressed in cell culture and purified [37, 38]. High-throughput assays were established which measured the binding of LFA-1 expressing cells (e.g. the HuT-78 human T-cell line) to recombinant ICAM-1 coated on the surface of a microtiter plate [37, 39], as well as the binding of LFA-1 [24] or its I-domain [37] to ICAM-1 in protein–protein binding assays. Emerging technologies were engaged to enable SAR by NMR studies to identify and optimize I-domain antagonists. In addition, rational design methods were utilized to identify LFA-1 antagonists which mimic ICAM-1's LFA-1 binding epitope [40]. Several of the small molecule antagonists identified to date have been optimized for binding potency and pharmaceutical properties enabling their entry into human clinical trials. Critical to the optimization of these molecules for human testing was the establishment of a panel of in vitro assays quantitating ICAM/I-domain, ICAM/LFA-1 binding as well as LFA-1/ICAM mediated T-cell adhesion and LFA-1/ICAM costimulation in T-cell proliferation. These in vitro assays studied increasingly complex aspects of the LFA-1/ICAM interactions as coupled to cellular function and proliferation which are reflective of and predictive of biological activities in vivo.

As noted above, a large number of large pharmaceutical research organizations initiated programs tailored to identify potent antagonists of LFA-1. Intellectual property concerns driving the commercial focus of these organizations leads to a high degree of secrecy in publication and public disclosure of the progress and current status of many of these efforts. However, it is clear from both publications and patent applications that at least five organizations have been successful in the identification of potent and selective antagonists of the protein–protein interaction between LFA-1 and ICAM-1 suitable for testing in humans. The following sections will compare and contrast the well documented efforts of Novartis, Boehringer Ingelheim, ICOS/Abbott, Bristol-Myers Squibb/Cerep and Genentech in this regard. The author apologizes to the other academic and commercial organizations working in the field for the incomplete discussion of their contributions.

11.4 Screening Assays

The fundamental challenge in establishing the assays which will enable the discovery and optimization of antagonists of LFA-1 or any other cellular target is to balance the desire to mimic the human disease within the confines of a microtiter plate and the need to rapidly acquire high fidelity data for a large number of compounds. In the real world, accommodations must be made which approximate the human disease and introduce artifact into the data at a level in proportion to the approximation of the assay to the human disease dynamics. This quandary is a translation of Heisenberg's quantum uncertainties to the macromolecular and biological scale. Every biological assay is an imprecise approximation of the reality it seeks to capture. Artifactual biases are embedded in every aspect of the assay protocol whether by intent or naivte. The assay cannot help but find what it is designed to find and the prudent researcher must manage the risk of artifact by constant correlation between in vitro and in vivo datasets in order to guide the discovery of functionally relevant lead molecules.

In an attempt to mimic aspects of the adhesion of lymphocytes circulating in the vasculature and their extravasation into tissue at sites of inflammation (Figure 11.4), researchers at Bristol-Myers Squibb (BMS) in a collaboration with Cerep established an assay which measured the adhesion of a human peripheral blood leukocytes (PBL's) to human umbilical vein endothelial cells (HUVEC) which had been stimulated with phorbol esters to upregulate their surface expression of ICAM-1 in culture. This assay was adapted to a high-throughput format and used to analyse the activities of a focused combinatorial library of molecules based on a hydantoin scaffold [41–43]. Major advantages of this assay approach include its use of human target tissues (e.g. both human vascular endothelial cells expressing human ICAM and human leukocytes expressing human LFA-1). The use of functioning cells 'casts a wide net' and allows for the identification of inhibitors of leukocyte attachment to vascular endothelium which do not involve the direct inhibition of LFA-1/ICAM (e.g. inhibitors of cell activation or the engagemenot of LFA-1 or ICAM cytoplasmic domains by cytoskeleton) in addition to antagonists of either LFA-1 or ICAM. Challenges present in this assay format include the day to day, donor to donor and lot to lot variabilities of the human derived HUVEC's and PBL's, the volume of each cell type necessary to execute a screening program and the plate to plate variability in the assay cell culture protocols. In this assay system, the adhesion occurs under static conditions in contrast to the dynamic and high shear stress conditions encountered by a PBL flowing over the static surface of blood vessel. The dynamics of the interaction of leukocytes with endothelium and LFA-1 with ICAM in vivo under shear flow conditions may effect the activation state of the leukocytes and the affinity/avidity of the LFA-1/ICAM interaction [44, 45].

In an effort to remove the variability inherent in primary cell culture with PBL's and HUVEC's while retaining the 'wide net' approach to lead identification, researchers at Novartis used a human T-cell line (HuT78) derived from a human lymphoma patient and available from the ATCC. They also isolated and purified recombinant ICAM from cell culture and studied HuT78 adhesion to ICAM-1 coated on a microtiter plate. In this system, it has been found that ICAM-1 'sticks' to the polystyrene surface of microtiter plates and at least some of this ICAM is oriented away from the polystyrene surface such that HuT78 cells can recognize and bind to the ICAM epitope in its first immunoglobulin domain. A variation of this assay format can use nonfunction blocking antibodies coated onto the polystyrene surface to 'capture' ICAM in a manner which orients its LFA-1 binding epitope away from the plate surface in an approximation of its presentation on the surface of an activated vascular endothelial cell. Novartis utilized this assay system in a high-throughput screen of more than 100 000 compounds from its corporate compound collection to identify LFA-1 antagonist activity in members of the statin class of HMG-CoA reductase inhibitors which have been approved for the control of human cholesterol levels [37, 46, 47]. Challenges presented by this assay include the lack of control in ICAM presentation, the relatively high density of ICAM on the plate surface relative to that found in endothelial cells and the static nature of the cell adhesion.

Workers at Abbott/ICOS, Boehringer Ingelheim (BI) and Genentech developed a protein/protein binding assay between ICAM coated on a polystyrene microtiter plate and LFA-1 isolated from cells in an Elisa format. This has the advantage that it is a 1:1 binding event enabling the detection of inhibition by low affinity antagonists which are not capable of inhibiting the more avid polyvalent cell attachment. In general, reported IC_{50} values for compounds tested in both an LFA-1/ICAM-1 Elisa and cell attachment assay formats may show 10-100 fold improvement in IC_{50} for the inhibition of LFA-1/ICAM-1 binding in the monovalent Elisa format relative to the polyvalent cell adhesion format. The Elisa assay also measures LFA-1 occupancy by antagonist, which was the pharmacodynamic marker correlated to clinical improvement with Raptiva™ in psoriatic patients [6–8]. This assay format is designed to capture direct antagonists of the LFA-1/ICAM protein interaction and will miss potential anti-inflammatory leads which inhibit leukocyte function by other mechanisms (e.g. cyclosoprine). Challenges with this assay format included the isolation and purification of the proteins from low expressing ATCC cell lines or the challenges of recombinant expression of the functional LFA-1 heterodimer [38]. The low inherent affinity of LFA-1 for ICAM-1 and the uncontrolled display of protein from the polystyrene surface can reduce the intensity of the fluorescent detection in the Elisa. This could be enhanced to robust levels by the use of divalent manganese salts in the assay buffer in place of the more physiologically relevant calcium. The presence of manganese has been shown to enhance the affinity of LFA-1 for ICAM-1, presumably by exchange of the metal cation in the MIDAS motif within the I-domain of LFA-1 [24]. The changes in this assay system from polyvalent cell attachment to monovalent protein binding and with the introduction of nonphysiological levels of manganese clearly enhance the possible introduction of artifact in the data [48].

In addition to using an Elisa binding assay, researchers at Abbott and ICOS studied the binding of ICAM-1 to the I-domain of LFA-1 by NMR [17]. They solved the structure of the I-domain by 2D NMR techniques and characterized the residues in and around the MIDAS and IDAS motifs of the I-domain whose resonances were perturbed in the presence of ICAM-1. In a variation of the SAR by NMR techniques developed at Abbott, they then surveyed mixtures of small molecule fragments to identify those fragments which perturbed the I-domain residues involved and generated models of the fragments bound to the I-domain as starting points for the elaboration and design of larger high affinity inhibitors of the I-domain/ICAM protein–protein interaction [49, 50]. In contrast to traditional HTS methods, this information rich data driven approach is not intended to identify a clinical candidate directly from a library or natural product screening campaign, but to gather the pieces of information necessary to solve the puzzle and assemble the candidate in a more iterative and interactive design/optimization process.

In addition to screening compound libraries with an LFA-1/ICAM-1 Elisa, researchers at Genentech initiated a rational discovery and design approach to lead discovery. This program sought to use ICAM itself as a lead, develop an SAR for this lead and then 'scaffold hop' from this protein to a small molecule. Toward this end, they developed a model of the first domain of ICAM-1 based on its structural homology to known structures of immunoglobulin domains. This homology model was used to guide alanine point mutagenesis studies where individual residues were mutated to alanine and the mutant protein was expressed, purified and its inhibition of wild type ICAM-1 binding tested [51]. The combined activity and structural data defined the SAR of ICAM-1. Molecular modeling studies guided the testing of a limited set of peptides and proteins from the company's library in order to identify leads [40]. Additional modeling studies were the basis of a rational transfer of the ICAM-1 epitope to a small molecule framework and guided the optimization of those small molecule leads.

There is no single assay system which is best suited for the discovery or optimization of an antagonist of a protein–protein interaction. As noted above, pharmaceutical research organizations balance the traditional strengths of their research and researchers to determine the method best suited to the particular target within the resources of their organization. The following section will detail the success of these various organizations and assay formats in the identification of small molecule antagonists of the LFA-1/ICAM-1 interaction.

11.5 Lead Identification and Optimization

11.5.1 Novartis

In a screen of over 100 000 compounds from its corporate library, several members of the statin family of HMG-CoA reductase inhibitors were identified as inhibitors of HuT-78 adhesion to ICAM [37, 46]. Several marketed members of the statin family including Lovastatin™, Simvastatin™ and Mevastatin™ along with a number of synthetic analogs showed moderate inhibiton of cell attachment in the 10–100 micromolar range. The inactivity of Pravastatin™, a close analog and potent HMG-CoA inhibitor, demonstrated differential statin SAR's for LFA-1 and their well understood HMG-CoA target. This prompted a wider evaluation of the SAR and the structural basis of statin's as LFA-1 inhibitors. X-ray crystallographic studies demonstrated that Lovastatin™ bound to the I-domain of LFA-1 to a site initially designated as the L-site. Subsequent studies have shown that this site regulates the binding of LFA-1 and ICAM in an allosteric manner and designated it as the I-domain allosteric site (IDAS) [17]. Comparative structural studies demonstrate the IDAS is on the opposite face of the I-domain from the MIDAS/ICAM binding site [20].

Semi-synthetic analogs of Lovastatin's™ lactone sidechain improved the cellular potency of these LFA-1 antagonists 100 fold (Figure 11.5) [46, 52]. Additional medicinal chemistry efforts have expanded beyond statins into other molecular scaffolds [53]. The ready availability of the statins from biological supply houses and their widespread human use has enabled a number of mechanistic investigations in vitro and in vivo [54]. In particular, off label uses of Lovastatin™ in various inflammatory disease settings is being explored [39, 55–57]. However, the relatively modest inhibition of LFA-1/ICAM-1 by Lovastatin™ makes interpretation of these studies difficult in light of the varied cellular effects downstream of its potent HMG-CoA inhibiton. Overall, researchers at Novartis identified potent inhibitors of LFA-1 and have published results of studies indicating an interest in transplant rejection as a clinical target, but the current status of the program is unknown.

11.5.2 Boehringer Ingelheim

Screening of a proprietary compound library through an LFA-1/ICAM-1 binding assay detected a concentration dependent inhibition by a compound containing a hydantoin core (Figure 11.5) [58]. Biophysical and biochemical studies demonstrated the compound binding to the IDAS site of the I-domain as the basis of allosteric inhibition of LFA-1/ICAM-1binding [59]. Medicinal chemistry rapidly generated a series of analogs and demonstrated an SAR linking enhanced potency in the LFA-1 binding assay to enhanced

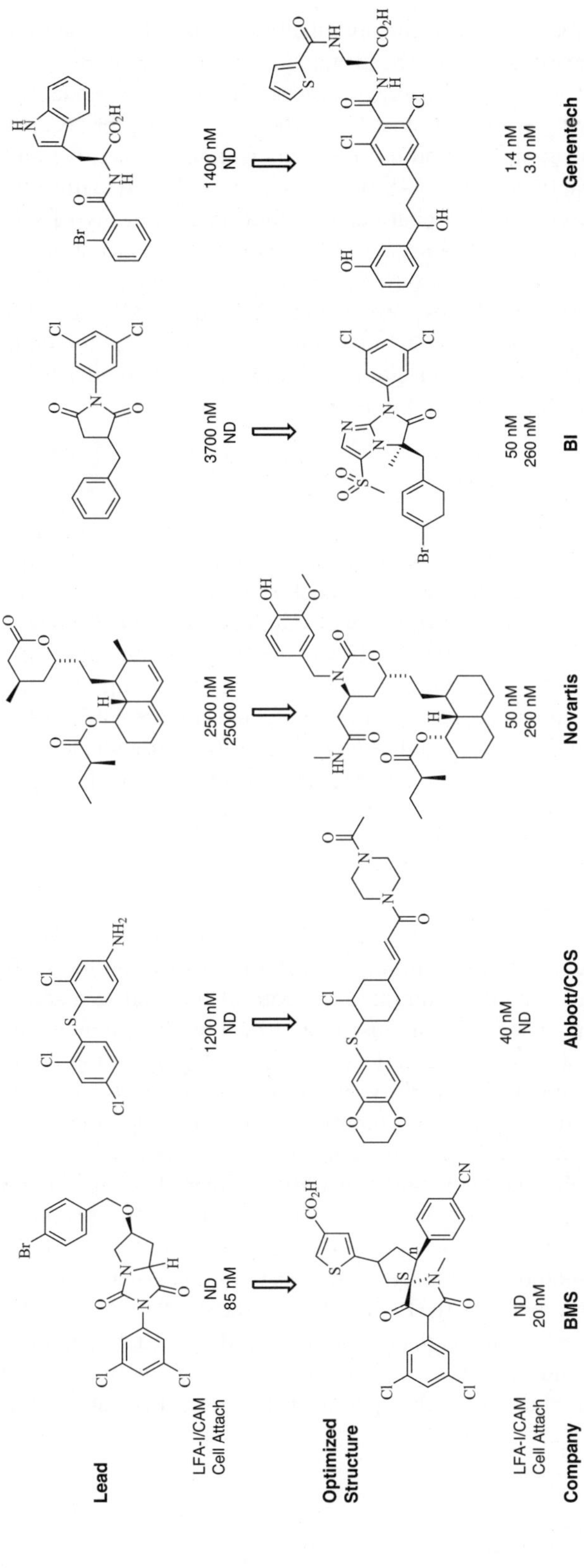

Figure 11.5 *Lead and optimized structures of LFA-1 antagonists identified by BMS, Abbott/ICOS, Novartis, BI and Genentech*

potency in cell attachment. Subsequent optimization varied the hydantoin scaffold and substituents around the ring to improve affinity for LFA-1 by more than 100 fold and identified a compound which demonstrated efficacy in vivo in a model of delayed-type hypersensitivity following an oral administration [60, 61]. A receptor occupancy assay was developed for the small molecule and used to demonstrate competitive binding between small molecule and an antibody to LFA-1 [62]. This assay was used to link pharmacokinetic circulating levels of compound to pharmacodynamic effects in vivo [63]. This PK/PD linkage between small molecule affinity for LFA-1, compound concentrations in blood and anti-inflammatory effect in vivo closes the loop from screening data with the LFA-1/ICAM Elisa to in vivo efficacy. This closure demonstrates the power of series of in vitro assays predictive of in vivo efficacy in the selection of candidate molecules for animal and human testing. The current status of the BI program is unknown, but their advanced pharmacodynamic assay efforts indicates a proximity to the clinic.

11.5.3 Abbott/ICOS

Screening in an LFA-1/ICAM-1 Elisa identified a substituted diphenyl sulfide derivative as a modest inhibitor of the protein–protein interaction (Figure 11.5) [17, 50]. NMR studies demonstrated that this molecule bound to the IDAS motif in the I-domain of LFA-1. Fragment based screening identified a number of additional functionalities which also bound to this domain and a variation of SAR by NMR was used to integrate fragment functionalities into the diphenyl sulfide lead. Additional medicinal chemistry improved the potency and optimized the pharmaceutical properties of the molecule (Figure 11.5) [49, 50]. ICOS has advanced an oral formulation of a candidate molecule, IC747, through Phase I and into Phase II clinical trials for the treatment of moderate to severe psoriasis [64]. The structure of IC747 has not been reported to date.

11.5.4 Bristol-Myers Squibb

Following on the discovery of hydantoin based LFA-1 antagonists, BMS constructed and screened a targeted combinatorial library based on the hydantoin scaffold [41–43]. Considerable medicinal chemistry migrated the core of the lead to a spiro-bicyclic lead (Figure 11.5). Like the earlier hydantoins, biophysical studies demonstrated that these molecules bind to the I-domain allosteric site (IDAS). Optimization of pharmaceutical properties introduced oral bioavailability and other pharmaceutical properties into an optimized molecule with a redefined spirocyclic sore. The optimized small molecule LFA-1 antagonist compared favorably to an anti-LFA-1 monoclonal antibody in a murine cardiac allograft transplant model. Co-administration with CTLA-4Ig along with either the small molecule LFA-1 antagonist or M17, a murine anti-LFA-1 precursor to Raptiva, resulted in similar graft protection in excess of 50 days post transplantation. This study of small molecule IDAS antagonists of LFA-1 establishes their pharmacological ‘equivalence’ to a monoclonal antibody and links the PK/PD data for the small molecule IDAS antagonists to the original Raptiva™ preclinical data with M17 validating LFA-1 as a target of interest in the pharmaceutical industry. The current status of this program within BMS has not been disclosed publicly, but has been reported in private conversations to have reached Phase II trials.

11.5.5 Genentech

A two-pronged approach which screened compound libraries in collaboration with Roche [65] in an LFA-1/ICAM Elisa and explored rational design and transfer of the ICAM-1 epitope through peptides to an optimized small molecule (Figure 11.5) [40]. The initial small molecule screening lead exhibited SAR parallels to a peptide SAR developed from the ICAM SAR. Grafting of the full ICAM/peptide SAR onto the small molecule lead enhanced the potency of the small molecules 100 fold in the Elisa assay. Further medicinal chemistry resulted in an optimized compound (Figure 11.5) which was shown to be >20 more potent than cyclosporine and 7 fold more potent than the Fab fragment of the anti-murine monoclonal anti-LFA-1 antibody MHM24 in a mixed lymphocyte reaction assay. It should be noted that MHM24 has been humanized to afford Raptiva™ [66]. In a murine delayed-type contact hypersensitivity model, the optimized lead exhibited dose dependent efficacy and proved to be pharmacologically 'equivalent' to a maximal dose of the M17 antibody, a murine precursor to Raptiva™ [40]. Traditional pharmacological binding and photochemical crosslinking studies have revealed that the compounds are direct competitive antagonists of ICAM-1 binding to LFA-1 and bind to the ICAM-1 binding site in the I-domain of the alpha subunit of LFA-1 [24]. By contrast, a recent publication suggests that the optimized Genentech compound (Figure 11.5) binds to the I like domain of the beta subunit of LFA-1 and as such functions as an antagonist of LFA-1 in the presence of manganese ions and an agonist in the presence of calcium or magnesion ions [67]. Overall, the dose dependent linkage between: the affinity of LFA-1 for Genentech's optimized small molecule in the presence of calcium, magnesium or manganese; the inhibition of LFA-1/ICAM-1 binding in the Elisa assay in the presence or absence of manganese; the functional inhibition of lymphocytes in a mixed lymphosyte reaction (MLR) in the presence of calcium and the absence of manganese; and the dose dependent inhibition of delayed-type hypersensitivity in mice links the pharmacokinetics of this compound to pharmacodynamic antagonism of LFA-1 in vitro and in vivo [24, 40].

Genentech's research has developed highly potent antagonists of the LFA-1/ICAM-1 protein–protein interaction which are unique in targeting the ICAM binding site in the I-domain of LFA-1. The company's development and commercialization of Raptiva™ for the treatment of psoriasis indicates an organizational interest in this disease. The current status of their small molecule LFA-1 antagonist program is unknown at this time.

11.6 Protein and Small Molecule Structure Activity Relationships (PSAR) in the LFA-1/ICAM-1 Interaction

Early molecular biology studies of LFA-1 and ICAM-1 by domain deletion and domain swapping were used to map the binding epitopes of both proteins [68]. These studies pointed out the importance of the I-domain of the alpha subunit of LFA-1 and the first immunoglobulin domain of ICAM-1 as critical to their interaction. Alanine point mutagenesis studies guided by NMR, X-ray crystal and homology modeling structures of the I-domain of LFA-1 and the first domain of ICAM-1 have identified the binding interface between LFA-1 and ICAM-1 [20, 69]. In ICAM-1 residues spanning 40 residues in the protein sequence (i.e. Glu34, Lys39, Met64, Tyr66, Asn68 and Gln73) were found to diminish binding to

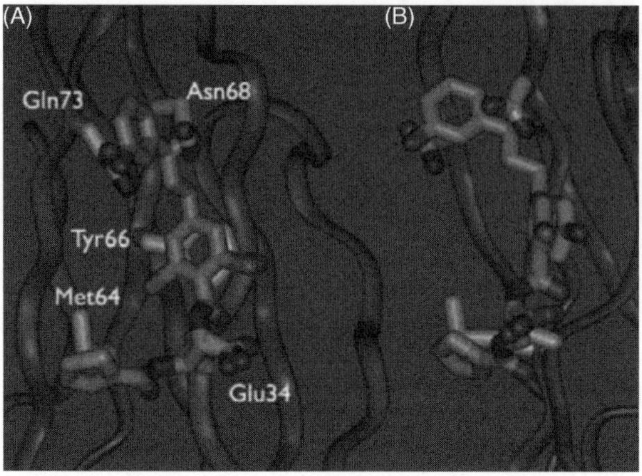

Figure 11.6 *Topdown (A) and Sideon (B) view of the structure of Genentech's optimized small molecule LFA-1 antagonist (Figure 11.5) overlaid on the epitope of the first domain of ICAM-1 (modeled from PDB1MQ8, adapted from Gadek et al. [40]). (See Plate 28.)*

LFA-1 when individually mutated to alanines. This indicates a likelihood of a diffuse or ill-defined binding epitope and does not bode well for efforts to identify a small molecule antagonist which binds to LFA-1 and inhibits the binding of ICAM-1. Interestingly, mapping these residues to a modeled structure of the first domain of ICAM-1 suggested that ICAM's tertiary fold brings the alpha-carbons of these key residues within a range easily accommodated by the footprint of a small molecule skeleton (e.g. a steroid nucleus).

Modeling of the optimized Genentech LFA-1 direct competitive antagonist targeting LFA-1's MIDAS site onto the structure of the ICAM-1 epitope (Figure 11.6) demonstrates the striking similarity in both the protein and small molecule SAR's. The H-bond donor/acceptor functionality of the Asn68/Gln73 couple is mimicked by the phenolic and benzylic hydroxy groups of the small molecule. The thioether of Met64 is seen in the compound's thiophene sidechain while the carboxylic acids of both the small molecule and Glu34 anchor the overlap. The efficacy of this compound relative to the M17 antibody demonstrates the pharmacological 'equivalence' of a small molecule ICAM mimetic to an antibody in the extent of antagonism of LFA-1 in vivo. This pharmacological 'equivalence' linking PK and PD of small molecule and antibody was the focus of research interest in the search for small molecule antagonists of LFA-1/ICAM-1 noted in the introduction to this chapter.

Interestingly, comparison of the X-ray structures for IDAS antagonists from BMS [41, 43] and Novartis [37, 46, 53] in complex with the I-domain of LFA-1 show that the overall fold of the I-domain is very similar despite the chemical and structural diversity of this group of statin, diazepane, and hydantoin/optimized spirocyclic inhibitors. Closer examination of each complex of inhibitor with I-domain in the IDAS region indicates some plasticity in the protean I-domain fold to accommodate the diverse antagonist structures (Figure 11.7).

It is interesting to speculate about the relative merits of allosteric and direct competitive antagonists of the LFA-1/ICAM-1 protein–protein interaction. Given that allosteric antagonist modify the affinity of LFA-1 for ICAM-1, it is possible to form a ternary complex of

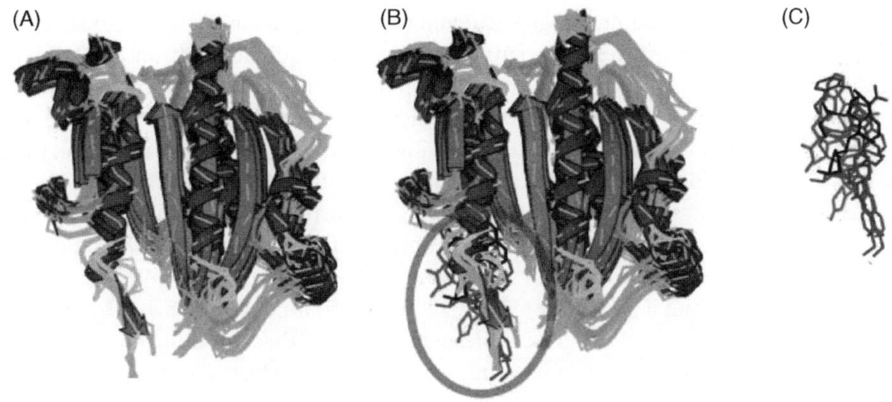

Figure 11.7 *Overlap of 5 crystal structures of I-domain and 5 different IDAS LFA-1 antagonists from BMS and Novartis. Note gross similarities in I domain folds (A) and diversity in antagonist bound structures (B/C). See text for references to structures. (See Plate 29.)*

IDAS small molecule antagonist/LFA-1/ICAM-1 which could still be biologically competent in bridging between two cells in the immune inflammatory process (Figure 11.4). With a direct competitive antagonist, 100 % occupancy of the ICAM binding site blocks 100 % of ICAM binding and block inflammation. However, even at 100 % occupancy of LFA-1's IDAS site with an allosteric antagonist, a few % of residual ICAM binding may be sufficient to sustain inflammation. The high degree of inhibition of LFA-1/ICAM association required to improve patient PASI scores in clinical trials of Raptiva™ suggest that some degree of extra engineering of may be necessary with the allosteric antagonists to assure that low levels of ternary complexes of small molecule/LFA-1/ICAM-1 are achieved which are not capable of supporting inflammation. This may require selection of compounds which form high affinity complexes with LFA-1's IDAS site and where this IDAS antagonist/LFA-1 complex has low or no residual affinity for ICAM-1. This may be nothing more than a theoretical concern and there is no indication of any level of biologically competent ternary complex in any of the IDAS antagonists identified to date.

11.7 Summary

Two classes of high affinity, high potency, drug like, small molecule antagonists of the protein–protein interaction between LFA-1 and its cognate ligand, ICAM-1, have been identified by a number of means by five independent pharmaceutical research organizations. Starting with assays which measured LFA-1 binding to ICAM-1 in either monovalent protein–protein binding assays or polyvalent cell attachment assays, each organization assembled a panel of in vitro assays to measure the binding and antagonist activities of these compounds. These lead discovery techniques were modifications of industry standard methods and the modifications to these standard methods were made to embrace and mimic the biology of the LFA-1 target. Optimization of antagonist binding affinities and the subsequent refinement of pharmaceutical properties of these molecules required a panel of assays with fidelity and dynamic range sufficient to faithfully measure the various activities of molecules which may differ by a single atom. This elucidation of an SAR for these antagonists and the 100 fold improvement in their potency from lead to optimized compound implies specific binding to LFA-1 and clearly demonstrates the class of molecules are indeed inhibitors of the protein–protein interaction and not an artifact of assay architecture. Each of these classes of antagonists required the existence and identification of a relatively small high energy binding site capable of supporting the high affinity binding that was ultimately built into the examples shown in Figure 11.5. High affinity binding to the surface of LFA-1 alters that surface by allosteric and/or electronic means to alter the shape and or surface potential of LFA-1 so that it no longer recognizes and is no longer recognized by its cognate ligands in the ICAM family [70]. The biological activities discussed for the molecules in Figure 11.5 flow from their high affinity binding to the LFA-1 target. This was first seen in the pharmacodynamic linkage of the antibody occupancy of LFA-1 correlating to pharmacokinetics in the circulating levels of antibody and the clinical improvement of patients in Phase II clinical trials. This linkage carried through the discovery of the small molecules and their efficacy in animal models of human inflammatory diseases.

There is a popular mythology that has grown up around the search for inhibitors of protein–protein interactions [71]. The quest for the unattainable small molecule grail... Even in the face of a large number of published successes described and reviewed in the literature, the 'searchers' declare that it can't be done, hasn't been done, integrins are unique in the pantheon of proteins and a number of other conditional denials. The work described and detailed in this chapter demonstrates that researchers and research organizations with a faith in technology and a commitment to gather the information necessary to solve the problem through iterative rounds of assay development, compound synthesis, data analysis, organizational analysis, program analysis, self analysis, organizational self doubt and a wary eye for the ever present artifact can succeed in identifying small molecule antagonists of large protean protein–protein interactions.

Note: The reader should be aware that cases of PML have been associated with the clinical use of the anti-LFA-1 monoclonal antibody, Raptiva, and this product has been voluntarily withdrawn from the market by Genentech (see http://www.gene.com/gene/news/press-releases/display.do?method=detail&id=12047&categoryid=4).

References

1. S.D. Marlin and T.A. Springer, Purified intercellular adhesion molecule-1 (ICAM-1) is a ligand for lymphocyte function-associated antigen 1 (LFA-1), *Cell*, 51, 813–19 (1987).
2. R.O. Hynes, Integrins: versatility, modulation, and signaling in cell adhesion, *Cell*, 69, 11–25 (1992).
3. K. Yonekawa and J.M. Harlan, Targeting leukocyte integrins in human diseases, *J. Leukoc. Biol.*, 77, 129–140 (2005).
4. M.R. Nicolls and R.G. Gill, LFA-1 (CD11a) as a therapeutic target, *Am. J. Transplant*, 6, 27–36 (2006).
5. D.M. Pariser, K.B. Gordon, K.A. Papp, *et al.*, Clinical efficacy of efalizumab in patients with chronic plaque psoriasis: results from three randomized placebo-controlled Phase III trials: part I. *J Cutan Med Surg*, 9(6), 303–12 (2005).

6. D.L. Mortensen, P.A. Walicke, X. Wang, *et al.*, Pharmacokinetics and pharmacodynamics of multiple weekly subcutaneous efalizumab doses in patients with plaque psoriasis, *J. Clin. Pharmacol.*, 45, 286–98 (2005).
7. A.B. Gottlieb, J.G. Krueger, K. Wittkowski, R. Dedrick, P.A. Walicke, M. Garovoy, Psoriasis as a model for T-cell-mediated disease: immunobiologic and clinical effects of treatment with multiple doses of efalizumab, an anti-CD11a antibody, *Arch. Dermatol.* 138, 591–600 (2002).
8. A. Joshi, R. Bauer, P. Kuebler, *et al.*, An overview of the pharmacokinetics and pharmacodynamics of efalizumab: a monoclonal antibody approved for use in psoriasis, *J. Clin. Pharmacol.*, 46, 10–20 (2006).
9. R.L. Dedrick, P. Walicke and M. Garovoy, Anti-adhesion antibodies efalizumab, a humanized anti-CD11a monoclonal antibody, *Transpl. Immunol.*, 9, 181–6 (2002).
10. D.R. Kuypers and Y.F. Vanrenterghem, Monoclonal antibodies in renal transplantation: old and new, *Nephrol. Dial. Transplant.*, 19, 297–300 (2004).
11. G.M. Gauvreau, A.B. Becker, L.P. Boulet, *et al.*, The effects of an anti-CD11a mAb, efalizumab, on allergen-induced airway responses and airway inflammation in subjects with atopic asthma, *J. Allergy Clin. Immunol.*, 12, 331–8 (2003).
12. R. Takiguchi, S. Tofte, B. Simpson, *et al.*, Efalizumab for severe atopic dermatitis: a pilot study in adults, *J. Am. Acad. Dermatol.*, 56, 222–7 (2007).
13. A. Kuek, B.L. Hazleman and A.J. Ostör, Immune-mediated inflammatory diseases (IMIDs) and biologic therapy: a medical revolution, *Postgrad. Med. J.*, 83, 251–60 (2007).
14. C. Huang and T.A. Springer, Folding of the beta-propeller domain of the integrin alphaL subunit is independent of the I domain and dependent on the beta2 subunit, *Proc. Natl. Acad. Sci. U.S.A.*, 94, 3162–7 (1997).
15. N.M. Dahms and G.W. Hart, Influence of quaternary structure on glycosylation. Differential subunit association affects the site-specific glycosylation of the common beta-chain from Mac-1 and LFA-1, *J. Biol. Chem.*, 261, 13186–96 (1986).
16. R.S. Larson, A.L. Corbi, L. Berman and T. Springer, Primary structure of the leukocyte function-associated molecule-1 alpha subunit: an integrin with an embedded domain defining a protein superfamily, *J. Cell Biol.*, 108, 703–12 (1989).
17. J.R. Huth, E.T. Olejniczak, R. Mendoza, *et al.*, NMR and mutagenesis evidence for an I domain allosteric site that regulates lymphocyte function-associated antigen 1 ligand binding, *Proc. Natl. Acad. Sci. U.S.A.*, 97, 5231–6 (2000).
18. J.P. Xiong, T. Stehle, R. Zhang, *et al.*, Crystal structure of the extracellular segment of integrin alpha Vbeta3 in complex with an Arg-Gly-Asp ligand, *Science*, 296, 151–5 (2002).
19. Q. Zang, C. Lu, C. Huang, J. Takagi and T.A. Springer, The top of the inserted-like domain of the integrin lymphocyte function-associated antigen-1 beta subunit contacts the alpha subunit beta -propeller domain near beta-sheet 3, *J. Biol. Chem.*, 275, 22202–12 (2000).
20. M. Shimaoka, T. Xiao, J.H. Liu, *et al.*, Structures of the alpha L I domain and its complex with ICAM-1 reveal a shape-shifting pathway for integrin regulation, *Cell*, 112, 99–111 (2003).
21. G. Song, Y. Yang, J.H. Liu, *et al.*, An atomic resolution view of ICAM recognition in a complex between the binding domains of ICAM-3 and integrin alphaLbeta2, *Proc. Natl. Acad. Sci. U.S.A.*, 102, 3366–71 (2005).
22. M.E. Labadia, D.D. Jeanfavre, G.O. Caviness and M.M. Morelock, Molecular regulation of the interaction between leukocyte function-associated antigen-1 and soluble ICAM-1 by divalent metal cations, *J. Immunol.* 161, 836–42 (1998).
23. E. San Sebastian, J.M. Mercero, R.H. Stote, A. Dejaegere, F.P. Cossío and X. Lopez, On the affinity regulation of the metal-ion-dependent adhesion sites in integrins, *J. Am. Chem. Soc.*, 128, 3554–63 (2006).
24. S.M. Keating, K.R. Clark, L.D. Stefanich, *et al.*, Competition between intercellular adhesion molecule-1 and a small-molecule antagonist for a common binding site on the alphal subunit of lymphocyte function-associated antigen-1, *Protein Sci.*, 15, 290–303 (2006).
25. J.C. Loftus and R.C. Liddington, Cell adhesion in vascular biology. New insights into integrin-ligand interaction, *J. Clin. Invest.*, 99, 2302–6 (1997).

26. K. Nam, V. Maiorov, B. Feuston and S. Kearsley, Dynamic control of allosteric antagonism of leukocyte function antigen-1 and intercellular adhesion molecule-1 interaction, *Proteins*, 64, 376–84 (2006).
27. M. Krummel, C. Wülfing, C. Sumen and M.M. Davis, Thirty-six views of T-cell recognition, *Philos. Trans. R. Soc. Lond. B. Biol. Sci.*, 355, 1071–6 (2000).
28. R.C. Liddington and M.H. Ginsberg, Integrin activation takes shape, *J. Cell Biol.*, 158, 833–9 (2002).
29. C.V. Carman and T.A. Springer, Integrin avidity regulation: are changes in affinity and conformation underemphasized? *Curr. Opin. Cell Biol.* 15, 547–56 (2003).
30. A. Cambi, B. Joosten, M. Koopman, *et al.*, Organization of the integrin LFA-1 in nanoclusters regulates its activity, *Mol. Biol. Cell.*, 17, 4270–81 (2006).
31. Y. van Kooyk and C.G. Figdor, Avidity regulation of integrins: the driving force in leukocyte adhesion, *Curr. Opin. Cell Biol.* 12, 542–7 (2000).
32. M.R. Tardif and M.J. Tremblay, Regulation of LFA-1 activity through cytoskeleton remodeling and signaling components modulates the efficiency of HIV type-1 entry in activated CD4+ T lymphocytes, *J. Immunol.*, 175, 926–35 (2005).
33. W.L. Connors, J. Jokinen, D.J. White, *et al.*, Two synergistic activation mechanisms of alpha2beta1 integrin-mediated collagen binding, *J. Biol. Chem.*, 282, 14675–83 (2007).
34. C.J. McCleverty, D.C. Lin and R.C. Liddington, Structure of the PTB domain of tensin1 and a model for its recruitment to fibrillar adhesions, *Protein Sci.* 16, 1223–9 (2007).
35. C. Wülfing, C. Sumen, M.D. Sjaastad, L.C. Wu, M.L. Dustin and M.M. Davis, Costimulation and endogenous MHC ligands contribute to T cell recognition, *Nat. Immunol.*, 3, 42–7 (2002).
36. D.B. McGavern, U. Christen, M.B. Oldstone, Molecular anatomy of antigen-specific CD8(+) T cell engagement and synapse formation in vivo, *Nat. Immunol.*, 3, 918–25 (2002).
37. J. Kallen, K. Welzenbach, P. Ramage, *et al.*, Structural basis for LFA-1 inhibition upon lovastatin binding to the CD11a I-domain, *J. Mol. Biol.*, 292, 1–9 (1999).
38. C.P. Edwards, K.L. Fisher, L.G. Presta and S.C. Bodary, Mapping the intercellular adhesion molecule-1 and -2 binding site on the inserted domain of leukocyte function-associated antigen-1, *J. Biol. Chem.*, 273, 28937–44 (1998).
39. G. Weitz-Schmidt, Lymphocyte function-associated antigen-1 blockade by statins: molecular basis and biological relevance, *Endothelium.* 10, 43–7 (2003).
40. T.R. Gadek, D.J. Burdick, R.S. McDowell, *et al.*, Generation of an LFA-1 antagonist by the transfer of the ICAM-1 immunoregulatory epitope to a small molecule, *Science*, 295, 1086–9 (2002).
41. D.S. Dodd, S. Sheriff, C.J. Chang, *et al.*, Design of LFA-1 antagonists based on a 2,3-dihydro-1H-pyrrolizin-5(7aH)-one scaffold, *Bioorg. Med. Chem. Lett.*, 17, 1908–11 (2007).
42. D. Potin, M. Launay, E. Nicolai, *et al.*, De novo design, synthesis, and in vitro activity of LFA-1 antagonists based on a bicyclic[5.5]hydantoin scaffold, *Bioorg. Med. Chem. Lett.*, 15, 1161–4 (2005).
43. D. Potin, M. Launay, F. Monatlik, *et al.*, Discovery and Development of 5-[(5S,9R)-9- (4-Cyanophenyl)-3-(3,5-dichlorophenyl)-1-methyl-2,4-dioxo-1,3,7-triazaspiro[4.4]non-7-yl-methyl]-3-thiophenecarboxylic acid (BMS-587101) – A Small Molecule Antagonist Leukocyte Function Associated Antigen-1, *J. Med. Chem.*, 49, 6946–9 (2006).
44. M. Okuyama, Y. Ohta, J. Kambayashi and M. Monden, Fluid shear stress induces actin polymerization in human neutrophil, *J. Cell Biochem.*, 63, 432–41 (1996).
45. A. Salas, M. Shimaoka, S. Chen, C.V. Carman and T. Springer, Transition from rolling to firm adhesion is regulated by the conformation of the I domain of the integrin lymphocyte function-associated antigen-1, *J. Biol. Chem,.* 277, 50255–62 (2002).
46. G. Weitz-Schmidt, K. Welzenbach, V. Brinkmann, *et al.*, Statins selectively inhibit leukocyte function antigen-1 by binding to a novel regulatory integrin site, *Nat. Med.* 7, 687–92 (2001).
47. G. Weitz-Schmidt, K. Welzenbach, J. Dawson and J. Kallen, Improved lymphocyte function-associated antigen-1 (LFA-1) inhibition by statin derivatives: molecular basis determined by x-ray analysis and monitoring of LFA-1 conformational changes in vitro and ex vivo, *J. Biol. Chem.* 279, 46764–71 (2004).

48. W. Yang, C.V. Carman, M. Kim, A. Salas, M. Shimaoka and T.A. Springer, A small molecule agonist of an integrin, alphaLbeta2, *J. Biol. Chem.*, 281, 37904–12 (2006).
49. M. Winn, E.B. Reilly, G. Liu, *et al.*, Discovery of novel p-arylthio cinnamides as antagonists of leukocyte function-associated antigen-1/intercellular adhesion molecule-1 interaction. 4. Structure-activity relationship of substituents on the benzene ring of the cinnamide, *J. Med. Chem.*, 44, 4393–4403 (2001).
50. G. Liu, J.R. Huth, E.T. Olejniczak, *et al.*, Novel p-arylthio cinnamides as antagonists of leukocyte function-associated antigen-1/intracellular adhesion molecule-1 interaction. 2. Mechanism of inhibition and structure-based improvement of pharmaceutical properties, *J. Med. Chem.*, 44, 1202–10 (2001).
51. K.L. Fisher, J. Lu, L. Riddle, K.J. Kim, L.G. Presta and S.C. Bodary, Identification of the binding site in intercellular adhesion molecule 1 for its receptor, leukocyte function-associated antigen 1, *Mol. Biol. Cell.*, 8, 501–15 (1997).
52. T. Ullrich, K. Baumann, K. Welzenbach, *et al.*, Statin-derived 1,3-oxazinan-2-ones as submicromolar inhibitors of LFA-1/ICAM-1 interaction: stabilization of the metabolically labile vanillyl side chain, *Bioorg. Med. Chem. Lett.*, 14, 2483–7 (2004).
53. S. Wattanasin, J. Kallen, S. Myers, *et al.*, 1,4-Diazepane-2,5-diones as novel inhibitors of LFA-1, *Bioorg. Med. Chem. Lett.*, 15, 1217–20 (2005).
54. R. Stanislaus, A.K. Singh and I. Singh. Lovastatin treatment decreases mononuclear cell infiltration into the CNS of Lewis rats with experimental allergic encephalomyelitis, *J. Neurosci. Res.*, 66, 155–62 (2001).
55. M.R. Namazi, Statins: novel additions to the dermatologic arsenal?, Exp. Dermatol., 13, 337–9 (2004).
56. J.F. Giguere and M.J. Tremblay, Statin compounds reduce human immunodeficiency virus type 1 replication by preventing the interaction between virion-associated host intercellular adhesion molecule 1 and its natural cell surface ligand LFA-1, J. Virol. 78, 12062–5 (2004).
57. L.J. Raggatt and N.C. Partridge, HMG-CoA reductase inhibitors as immunomodulators: potential use in transplant rejection, *Drugs*. 62, 2185–91 (2002).
58. T.A. Kelly, D.D. Jeanfavre, D.W. McNeil, *et al.*, Cutting edge: a small molecule antagonist of LFA-1-mediated cell adhesion, *J. Immunol.*, 163, 5173–7 (1999).
59. K. Last-Barney, W. Davidson, M. Cardozo, *et al.*, Binding site elucidation of hydantoin-based antagonists of LFA-1 using multidisciplinary technologies: evidence for the allosteric inhibition of a protein--protein interaction, *J. Am. Chem. Soc.* 123, 5643–50 (2001).
60. J.P. Wu, J. Emeigh, D.A. Gao, *et al.*, Second-generation lymphocyte function-associated antigen-1 inhibitors: 1H-imidazo[1,2-alpha]imidazol-2-one derivatives, *J. Med. Chem.*, 47, 5356–66 (2004).
61. M.J. Panzenbeck, D.D. Jeanfavre, T.A. Kelly, *et al.*, An orally active, primate selective antagonist of LFA-1 inhibits delayed-type hypersensitivity in a humanized-mouse model, *Eur. J. Pharmacol.*, 534, 233–40 (2006).
62. J.R. Woska, K. Last-Barney, R. Rothlein, *et al.*, Small molecule LFA-1 antagonists compete with an anti-LFA-1 monoclonal antibody for binding to the CD11a I domain: development of a flow-cytometry-based receptor occupancy assay, *J. Immunol. Method*, 277, 101–15 (2003).
63. G.O. Caviness, M.E. Labadia, P.A. Giblin, *et al.*, The determination and correlation of molecular and cellular equilibrium Kd and kinetic k(off) values for small molecule allosteric antagonists of LFA-1, *Biochem. Pharmacol.*, 74, 98–106 (2007).
64. www.icos.com. Third Quarter 2002 Report: http://phx.corporate-ir.net/phoenix.zhtml?c=109092&p=irol-newsArticle_print&ID=353375&highlight=
65. www.gene.com. See proxy statement: http://sec.edgar-online.com/1997/03/10/00/0000912057-97-008340/Section19.asp
66. W.A. Werther, T.N. Gonzalez, S.J. O'Connor, *et al.*, Humanization of an anti-lymphocyte function-associated antigen (LFA)-1 monoclonal antibody and reengineering of the humanized antibody for binding to rhesus LFA-1, *J. Immunol.* 157, 4986–95 (1996).
67. W. Yang, C.V. Carman, M. Kim, A. Salas, M. Shimaoka, T.A. Springer, A small molecule agonist of an integrin, alphaLbeta2, *J. Biol. Chem.* 281, 37904–12 (2006).

68. C.P. Edwards, M. Champe, T. Gonzalez, *et al.*, Identification of amino acids in the CD11a I-domain important for binding of the leukocyte function-associated antigen-1 (LFA-1) to intercellular adhesion molecule-1 (ICAM-1), *J. Biol. Chem.* 270, 12635–40 (1995).
69. C.P. Edwards, K.L. Fisher, L.G. Presta and S.C. Bodary, Mapping the intercellular adhesion molecule-1 and -2 binding site on the inserted domain of leukocyte function-associated antigen-1, *J. Biol. Chem.*, 273, 28937–28944 (1998).
70. T.R. Gadek and J.B. Nicholas, Small molecule antagonists of proteins, *Biochem. Pharmacol.* 65, 1–8 (2003).
71. B. Laudet, R. Prudent, O. Filhol and C. Cochet, Therapeutic agents targetting protein–protein interactions: myth or reality?, *Med. Sci. (Paris).* 23, 273–8 (2007).

Index

References to figures are given in italic type. References to tables are given in bold type.

Protein Surface Recognition: Approaches for Drug Discovery Ernest Giralt, Mark W. Peczuh and Xavier Salvatella. © 2011 John Wiley & Sons, Ltd